Membrane Transporters as Drug Targets

Pharmaceutical Biotechnology

Series Editor: Ronald T. Borchardt
　　　　　　　　The University of Kansas
　　　　　　　　Lawrence, Kansas

Recent volumes in this series:

Volume 4	BIOLOGICAL BARRIERS TO PROTEIN DELIVERY Edited by Kenneth L. Audus and Thomas J. Raub
Volume 5	STABILITY AND CHARACTERIZATION OF PROTEIN AND PEPTIDE DRUGS: Case Histories Edited by Y. John Wang and Rodney Pearlman
Volume 6	VACCINE DESIGN: The Subunit and Adjuvant Approach Edited by Michael F. Powell and Mark J. Newman
Volume 7	PHYSICAL METHODS TO CHARACTERIZE PHARMACEUTICAL PROTEINS Edited by James N. Herron, Win Jiskoot, and Daan J. A. Crommelin
Volume 8	MODELS FOR ASSESSING DRUG ABSORPTION AND METABOLISM Edited by Ronald T. Borchardt, Philip L. Smith, and Glynn Wilson
Volume 9	FORMULATION, CHARACTERIZATION, AND STABILITY OF PROTEIN DRUGS: Case Histories Edited by Rodney Pearlman and Y. John Wang
Volume 10	PROTEIN DELIVERY: Physical Systems Edited by Lynda M. Sanders and R. Wayne Hendren
Volume 11	INTEGRATION OF PHARMACEUTICAL DISCOVERY AND DEVELOPMENT: Case Histories Edited by Ronald T. Borchardt, Roger M. Freidinger, Tomi K. Sawyer, and Philip L. Smith
Volume 12	MEMBRANE TRANSPORTERS AS DRUG TARGETS Edited by Gordon L. Amidon and Wolfgang Sadée

A Chronological Listing of Volumes in this series appears at the back of this volume

A Continuation Order Plan is available for this series. A continuation order will bring delivery of each new volume immediately upon publication. Volumes are billed only upon actual shipment. For further information please contact the publisher.

Membrane Transporters as Drug Targets

Edited by

Gordon L. Amidon
University of Michigan
Ann Arbor, Michigan

and

Wolfgang Sadée
University of California, San Francisco
San Francisco, California

Kluwer Academic / Plenum Publishers
New York, Boston, Dordrecht, London, Moscow

Library of Congress Cataloging-in-Publication Data

Membrane transporters as drug targets / edited by Gordon L. Amidon and Wolfgang Sadée.
    p. cm. -- (Pharmaceutical biotechnology ; v. 12)
  Includes bibliographical references and index.
  ISBN 0-306-46094-7
  1. Drug targeting. 2. Drugs--Physiological transport. 3. Carrier proteins. 4. Drug carriers (Pharmacy) I. Amidon, Gordon L. II. Series.
    [DNLM: 1. Membrane Proteins--physiology. 2. Biological Transport--physiology. 3. Carrier Proteins--physiology. 4. Drug Carriers. 5. Receptors, Drug. W1 PH151N v.12 1999]
RM301.65.M45 1999
615'.7--dc21
DNLM/DLC
for Library of Congress   99-37278
                                                                                                    CIP

ISBN: 0-306-46094-7

© 1999 Kluwer Academic / Plenum Publishers, New York
233 Spring Street, New York, N.Y. 10013

10 9 8 7 6 5 4 3 2 1

A C.I.P. record for this book is available from the Library of Congress

All rights reserved

No part of this book may be reproduced, stored in a retrieval system, or transmitted in any form or by any means, electronic, mechanical, photocopying, microfilming, recording, or otherwise, without written permission from the Publisher

Printed in the United States of America

Contributors

David H. Alpers • Gastroenterology Division, Washington University School of Medicine, St. Louis, Missouri 63110

Gordon L. Amidon • College of Pharmacy, University of Michigan, Ann Arbor, Michigan 48109-1065

Matthew J. Androlewicz • Immunology Program, H. Lee Moffitt Cancer Center and Research Institute, and Department of Biochemistry and Molecular Biology, University of South Florida College of Medicine, Tampa, Florida 33612

Kenneth L. Audus • Department of Pharmaceutical Chemistry, School of Pharmacy, University of Kansas, Lawrence, Kansas 66047

Stephen A. Baldwin • School of Biochemistry and Molecular Biology, University of Leeds, Leeds, United Kingdom

Jürg Biber • Physiological Institute, University of Zurich, CH-8057, Zurich, Switzerland

Miguel A. Cabrita • Molecular Biology of Membranes Group, Membrane Transport Research Group, and Department of Biochemistry, University of Alberta, and Department of Experimental Oncology, Cross Cancer Institute, Edmonton, Alberta, Canada

Carol E. Cass • Molecular Biology of Membranes Group, Membrane Transport Research Group, and Departments of Biochemistry and Oncology, University of Alberta, and Department of Experimental Oncology, Cross Cancer Institute, Edmonton, Alberta, Canada

Ellen I. Closs • Department of Pharmacology, Johannes Gutenberg University, D-55101 Mainz, Germany

Mark J. Dresser • Department of Biopharmaceutical Sciences, University of California San Francisco, San Francisco, California 94143

Ian Forster • Physiological Institute, University of Zurich, CH-8057, Zurich, Switzerland

Kathleen M. Giacomini • Department of Biopharmaceutical Sciences, University of California San Francisco, San Francisco, California 94143

Gwyn W. Gould • Division of Biochemistry and Molecular Biology, Institute of Biomedical and Life Sciences, University of Glasgow, Glasgow G12 8QQ, Scotland

Petra Gräf • Department of Pharmacology, Johannes Gutenberg University, D-55101 Mainz, Germany

Kathryn A. Graham • Molecular Biology of Membranes Group, Membrane Transport Research Group, and Department of Oncology, University of Alberta, and Department of Experimental Oncology, Cross Cancer Institute, Edmonton, Alberta, Canada

Richard C. Graul • Departments of Biopharmaceutical Sciences and Pharmaceutical Chemistry, School of Pharmacy, University of California San Francisco, San Francisco, California 94143-0446

Mark Griffiths • School of Biochemistry and Molecular Biology, University of Leeds, Leeds, United Kingdom

Hyo-kyung Han • College of Pharmacy, University of Michigan, Ann Arbor, Michigan 48109-1065

Guido J. E. J. Hooiveld • Department of Pharmacokinetics and Drug Delivery, Groningen University Institute for Drug Exploration (GUIDE), Groningen, The Netherlands, and Department of Gastroenterology and Hepatology, University Hospital, Groningen, The Netherlands

Contributors

Ken-ichi Inui • Department of Pharmacy, Kyoto University Hospital, Sakyo-ku, Kyoto 606-8507, Japan

Peter L. M. Jansen • Department of Gastroenterology and Hepatology, University Hospital, Groningen, The Netherlands

Lori L. Jennings • Molecular Biology of Membranes Group, Membrane Transport Research Group, and Department of Oncology, University of Alberta, and Department of Experimental Oncology, Cross Cancer Institute, Edmonton, Alberta, Canada

Vashti G. Lacaille • Immunology Program, H. Lee Moffitt Cancer Center and Research Institute, and Department of Biochemistry and Molecular Biology, University of South Florida College of Medicine, Tampa, Florida 33612

Alan Y. Lee • Departments of Biopharmaceutical Sciences and Pharmaceutical Chemistry, School of Pharmacy, University of California San Francisco, San Francisco, California 94143-0446

John R. Mackey • Department of Oncology, University of Alberta, and Departments of Experimental Oncology and Medicine, Cross Cancer Institute, Edmonton, Alberta, Canada

Dirk K. F. Meijer • Department of Pharmacokinetics and Drug Delivery, Groningen University Institute for Drug Exploration (GUIDE), Groningen, The Netherlands

Michael Müller • Department of Gastroenterology and Hepatology, University Hospital, Groningen, The Netherlands

Heini Murer • Physiological Institute, University of Zurich, CH-8057 Zurich, Switzerland

Amy M. L. Ng • Membrane Transport Research Group and Department of Physiology, University of Alberta, Edmonton, Alberta, Canada

Doo-Man Oh • Department of Pharmacokinetics, Dynamics, and Metabolism, Parke-Davis Pharmaceutical Research Division, Warner-Lambert Company, Ann Arbor, Michigan 48105

Mabel W. L. Ritzel • Membrane Transport Research Group and Department of Physiology, University of Alberta, Edmonton, Alberta, Canada

Bertrand Rochat • Department of Pharmaceutical Chemistry, School of Pharmacy, University of Kansas, Lawrence, Kansas 66047

Gregory J. Russell-Jones • C/- Biotech Australia Pty Ltd, Roseville, NSW, 2069 Australia

Wolfgang Sadée • Departments of Biopharmaceutical Sciences and Pharmaceutical Chemistry, School of Pharmacy, University of California San Francisco, San Francisco, California 94143-0446

Michael J. Seatter • Division of Biochemistry and Molecular Biology, Institute of Biomedical and Life Sciences, University of Glasgow, Glasgow G12 8QQ, Scotland

Jeffrey A. Silverman • AvMax, Inc., South San Francisco, CA 94080

Johan W. Smit • Department of Pharmacokinetics and Drug Delivery, Groningen University Institute for Drug Exploration (GUIDE), Groningen, The Netherlands; present address: Division of Experimental Therapy, The Netherlands Cancer Institute, Amsterdam, The Netherlands

Yuichi Sugiyama • Graduate School of Pharmaceutical Sciences, University of Tokyo, Hongo, Bunkyo-ku, Tokyo 113-0033, Japan

Hiroshi Suzuki • Graduate School of Pharmaceutical Sciences, University of Tokyo, Hongo, Bunkyo-ku, Tokyo 113-0033, Japan

Ikumi Tamai • Department of Pharmacobio-Dynamics, Faculty of Pharmaceutical Sciences, Kanazawa University, 13-1 Takara-machi, Kanazawa 920-0934, Japan

Tomohiro Terada • Department of Pharmacy, Kyoto University Hospital, Sakyo-ku, Kyoto 606-8507, Japan

Akira Tsuji • Department of Pharmacobio-Dynamics, Faculty of Pharmaceutical Sciences, Kanazawa University, 13-1 Takara-michi, Kanazawa 920-0934, Japan

Karl Julius Ullrich • Max Planck Institute for Biophysics, 60596 Frankfurt am Main, Germany

Jessica E. van Montfoort • Department of Pharmacokinetics and Drug Delivery, Groningen University Institute for Drug Exploration (GUIDE), Groningen, The Netherlands, and Department of Gastroenterology and Hepatology, University Hospital, Groningen, The Netherlands

Mark F. Vickers • Molecular Biology of Membranes Group, Membrane Transport Research Group, and Department of Biochemistry, University of Alberta, and Department of Experimental Oncology, Cross Cancer Institute, Edmonton, Alberta, Canada

Sylvia Y. M. Yao • Membrane Transport Research Group and Department of Physiology, University of Alberta, Edmonton, Alberta, Canada

James D. Young • Membrane Transport Research Group and Department of Physiology, University of Alberta, Edmonton, Alberta, Canada

Lei Zhang • Department of Biopharmaceutical Sciences, University of California San Francisco, San Francisco, California 94143

Preface to the Series

A major challenge confronting pharmaceutical scientists in the future will be to design successful dosage forms for the next generation of drugs. Many of these drugs will be complex polymers of amino acids (e.g., peptides, proteins), nucleosides (e.g., antisense molecules), carbohydrates (e.g., polysaccharides), or complex lipids.

Through rational drug design, synthetic medicinal chemists are preparing very potent and very specific peptides and antisense drug candidates. These molecules are being developed with molecular characteristics that permit optimal interaction with the specific macromolecules (e.g., receptors, enzymes, RNA, DNA) that mediate their therapeutic effects. Rational drug design does not necessarily mean rational drug delivery, however, which strives to incorporate into a molecule the molecular properties necessary for optimal transfer between the point of administration and the pharmacological target site in the body.

Like rational drug design, molecular biology is having a significant impact on the pharmaceutical industry. For the first time, it is possible to produce large quantities of highly pure proteins, polysaccharides, and lipids for possible pharmaceutical applications. Like peptides and antisense molecules, the design of successful dosage forms for these complex biotechnology products represents a major challenge to pharmaceutical scientists.

Development of an acceptable drug dosage form is a complex process requiring strong interactions between scientists from many different divisions in a pharmaceutical company, including discovery, development, and manufacturing. The series editor, the editors of the individual volumes, and the publisher hope that this new series will be particularly helpful to scientists in the development areas of a pharmaceutical company, (e.g., drug metabolism, toxicology, pharmacokinetics and pharmacodynamics, drug delivery, preformulation, formulation, and physical and analytical chemistry). In addition, we hope this series will help to build bridges between the development scientists and scientists in discovery (e.g., medicinal chemistry, pharmacology, immunology, cell biology, molecular biology) and in manufacturing (e.g., process chemistry, engineering). The design of successful dosage forms for the next generation of drugs will require not only a high level of expertise by individual scientists, but also a high degree of interaction between scientists in these different divisions of a pharmaceutical company.

Finally, everyone involved with this series hopes that these volumes will also be useful to the educators who are training the next generation of pharmaceutical scientists. In addition to having a high level of expertise in their respective disciplines, these young scientists will need to have the scientific skills necessary to communicate with their peers in other scientific disciplines.

RONALD T. BORCHARDT
Series Editor

Preface

This monograph appears at a propitious time. Researchers from molecular biology, human genetics, bioinformatics, and drug development are converging on an integrated approach to studying membrane transporters. With the recent cloning of numerous transporter genes, substrate translocation across membranes is coming into focus at the molecular level. Moreover, with the complete sequencing of entire genomes, we have a first glimpse at not just a few transporter families, but all transporters in a living organism. This wealth of information is bound to change research directions by broadening our general understanding of the physiological role of transporters, their diversity and redundancy within an organism, and differences among various species. Soon, the entire human genome will be sequenced, and we will then deal with a large, but finite number of transporter genes. One might estimate that there are 2000–4000 human transporter/ion exchanger genes. Methodology to deal with such large gene numbers is now available, and a genomics-driven approach to transporter mechanisms and functions will likely transform this field in the near future. The chapters in this book already presage this transition, by including not just single transporters, but entire transporter gene families. Yet, only a fraction of all human transporters has been cloned, and much more work needs to be done.

While this book focuses on the interaction of drugs with transporters, scientists from various disciplines have contributed chapters providing in-depth analyses of the physiology, molecular biology, and regulation of select transporter families. Moreover, scientists traditionally involved with drug development have increasingly incorporated molecular biology into their work, a crucial step in understanding the complexities of drug absorption, distribution, and targeting in the body. This book demonstrates that these diverse disciplines are no longer separate entities, and that the boundaries among them have blurred. A physiologist needs to be concerned with the relevance of a transport function to drug targeting, whereas a scientist in drug development must delve into the molecular biology of the many transporters that could determine efficacy and, hence, clinical success of a new drug entity. Juxtaposing these various views of transporters in this book provides insight into the current state of development of this field, providing a reliable beacon for future developments. We therefore think that this monograph will remain topical for some time to come.

The book is divided into three main portions, the first dealing with background information, including drug transport methodologies and transporter classification, the second providing an integral pharmaceutical view of drug transport in several organs of the body, and the third focusing on individual transporter families. This last portion is not meant to be all-inclusive, but rather to provide examples of well-studied transporter families, presented by prominent scientists in different disciplines. In the future we intend to revisit this topic, and include additional large transporter families, some of which double as drug receptors *per se*, rather than drug targeting devices, e.g., the neurotransmitter transporters. Moreover, the ion exchangers, structurally related to transporters, interact with many drugs, and these are also not included here.

Given the number and complexity of human transporter families, and the importance they have in drug action, we can anticipate rapid expansion of this field and of the number of investigators involved in it. This will also lead to increasing focus on pharmacogenetics, i.e., the study of interindividual differences at the genetic level. Already a central theme in drug metabolism, particularly with focus on genetic polymorphism of the cytochrome P450 enzymes, virtually nothing is known about genetic polymorphism of drug transporters. For example, the intestinal H^+/dipeptide transporter appears to play a role in the oral absorption of cephalosporins, but we have not yet even tested whether some patients carry mutations that abrogate cephalosporin transport. This is certainly likely, the main question being what fraction of the population would have such genetic defects. We think that the information provided in this book will form the basis of addressing the pharmacogenetics of transporters in general. Seen from a genomics perspective, with an all-inclusive approach for all possible transporters, this might rather be called the pharmacogenomics of drug transporters. Much can be gained by understanding the entirety of human transported genes, their relevance to any chosen drug, and the genetic variations distributed throughout the patient population. This book advances the current state of the transporter field toward these goals.

GORDON L. AMIDON
WOLFGANG SADÉE

Contents

Chapter 1

Overview of Membrane Transport 1
Doo-Man Oh and Gordon L. Amidon

1. Introduction 1
2. Modes of Membrane Transport 3
   2.1. Passive Diffusion 5
   2.2. Ion Channels 6
   2.3. Facilitated Diffusion 9
   2.4. Active Transporters 10
   2.5. Secondary Active Transporters 12
   2.6. Macromolecular and Bulk Transport 12
3. Energetics of Membrane Transport 12
4. Kinetics of Membrane Transport 13
   4.1. Kinetic Equation 13
   4.2. Transport Experiments 16
5. Analysis of Membrane Transport 18
6. Methods in Membrane Transport Research 18
   6.1. Patch Clamping 19
   6.2. Fluorescence Digital Imaging Microscopy 20
   6.3. Confocal Microscopy 21
   6.4. Reconstitution 21
   6.5. Expression in *Xenopus laevis* Oocytes 21
   6.6. Cultured Cells 22
   6.7. Electrophysiological Study of Epithelial Transport 23
   6.8. Other Commonly Used Experimental Methods 24
7. Summary 24
   References 24

Chapter 2

Classification of Membrane Transporters
Wolfgang Sadée, Richard C. Graul, and Alan Y. Lee

1. Introduction	29
2. Classification of Transporters by Function and Substrate Specificity	30
3. A Genomics View of Transporters	32
4. Pharmacogenomics of Transporters	35
5. Evolution of Membrane Transporters; Sequence Analysis	36
6. Selected Examples of Membrane Transporters	39
6.1. H^+/Dipeptide Symporters	40
6.2. Facilitative Glucose Transporters and Related Sequences	44
6.3. Sodium/Glucose and Sodium/Nucleoside Cotransporters	45
6.4. Amino Acid Transporters	48
6.5. Sodium/Neurotransmitter Symporters	49
6.6. ATP-Binding Transport Protein Family	50
7. Conclusion	52
References	53

Chapter 3

Drug Transport and Targeting: Intestinal Transport
Doo-Man Oh, Hyo-kyung Han, and Gordon L. Amidon

1. Introduction	59
2. Intestinal Transporters	60
2.1. General Description of Intestinal Transporters	60
2.2. Peptide Transporter	64
2.3. Nucleoside Transporter	65
2.4. Sugar Transporter	66
2.5. Bile Acid Transporter	66
2.6. Amino Acid Transporters	67
2.7. Organic Anion Transporters	67
2.8. Vitamin Transporters	68
2.9. Phosphate Transporter	68
2.10. Bicarbonate Transporter	69
2.11. Organic Cation Transporter	69
2.12. Fatty Acid Transporter	69
2.13. P-Glycoprotein	69

3. Prodrugs: Targeting Intestinal Transporters	70
4. Disorders Related to Intestinal Transporters	75
4.1. Water and Electrolyte Transporters	75
4.2. Sugar Transporters	76
4.3. Amino Acid Transporters	76
4.4. Bile Acid Transporter	76
4.5. Vitamin Transporters	77
5. Summary	77
References	78

Chapter 4

The Molecular Basis for Hepatobiliary Transport of Organic Cations and Organic Anions
Dirk K. F. Meijer, Johan W. Smit, Guido J. E. J. Hooiveld, Jessica E. van Montfoort, Peter L. M. Jansen, and Michael Müller

1. Introduction	89
2. Candidate Proteins for Carrier-Mediated Hepatic Uptake of Cationic Drugs	93
3. Candidate Proteins Involved in Bile Canalicular Transport of Cationic Drugs	97
3.1. The Cation/Proton Antiport Secretory/Reabsorptive System	97
3.2. The mdr1-Type P-Glycoprotein Secretory System	99
3.3. Other ABC Transport Proteins in the Canaliculus that May Accommodate Cationic Drugs	103
4. Organic Cation Transport in the Intact Liver and Its Short-Term Regulation	104
5. Substrate Specificity, Driving Forces, and Multiplicity of Canalicular Organic Cation Carriers	105
6. Relation between Chemical Structure and Clearance via Bile, Urine, and Intestine	107
7. Molecular Aspects of Organic Cation Transport and Further Perspectives	111
8. Hepatic Uptake of Organic Anions: Role of Specific and Polyspecific Transport Proteins	112
8.1. Na^+/Taurocholate Cotransporter (NTCP)	113
8.2. Polyspecific Organic Anion Transport Proteins (OATPs)	115
9. Transport across the Canalicular Membrane: Role of ABC Transport Proteins	116

9.1.	MDR3 P-Glycoprotein	116
9.2.	Homologues of the Multidrug-Resistance Protein MRP1	118
9.3.	Canalicular Bile Salt Transporter (cBST/sPgp)	120
10.	Recent Results from Cholestatic Animal Models	121
	Appendix: Molecular Features of Currently Identified Carrier Proteins	
	Involved in Sinusoidal Uptake and Canalicular Transport	
	of Organic Cations and Anions	122
	References	140

Chapter 5

Affinity of Drugs to the Different Renal Transporters for Organic Anions and Organic Cations: *In Situ* K_i Values
Karl Julius Ullrich

1. Introduction	159
2. Location of the Transport Processes and Transporters	
in the Proximal Renal Tubule	160
3. Transport Processes for Net Reabsorption and Net Secretion	162
4. General Information about the K_i Values Given Here	169
5. Interaction with Contraluminal PAH Transporter	170
6. Interaction with the Contraluminal Sulfate	
and Dicarboxylate Transporters	171
7. Interaction with the Three Organic Cation Transporters	171
8. Bi- and Polysubstrates: Drugs which Interact Both with Transporters	
for Organic Anions and Transporters for Organic Cations	172
9. Conclusions and Perspectives	173
References	173

Chapter 6

Drug Disposition and Targeting: Transport across the Blood–Brain Barrier
Bertrand Rochat and Kenneth L. Audus

1. Introduction	181
2. Permeability of the Blood–Brain Barrier	183
2.1. Carrier-Mediated Transport	183
2.2. Transcytosis	190
3. Summary and Future Perspectives	193
References	193

Chapter 7

The Mammalian Facilitative Glucose Transporter (GLUT) Family
Michael J. Seatter and Gwyn W. Gould

1. Introduction 201
2. Background 202
3. The Structure of the GLUTs 203
   3.1. Predicted Secondary Structure of GLUT1 203
   3.2. Evidence in Favor of the Secondary Structure Model 204
   3.3. Biophysical Investigation of GLUT1 204
4. The Dynamics of Glucose Transport 205
   4.1. The Alternating Conformation Model 206
   4.2. Oligomerization of GLUT1 208
5. Transporter Photoaffinity Labeling 209
6. The Tissue-Specific Distribution of Glucose Transporters 211
   6.1. GLUT1 211
   6.2. GLUT2 212
   6.3. GLUT3 216
   6.4. GLUT4 218
   6.5. GLUT5 219
7. Recent Work and Future Directions 220
   References 221

Chapter 8

Cationic Amino Acid Transporters (CATs): Targets for the Manipulation of NO-Synthase Activity?
Ellen I. Closs and Petra Gräf

1. Introduction 229
2. Identification of Four Related Carrier Proteins for Cationic
   Amino Acids in Mammalian Cells 231
   2.1. Murine CAT-Proteins (mCATs) 231
   2.2. Human CAT-Proteins (hCATs) 233
3. Transport Properties of the CAT Proteins 233
   3.1. Substrate Specificity of the mCAT Proteins 234
   3.2. Kinetics of mCAT-Mediated Transport 235
   3.3. hCAT-Mediated Transport 236
4. CATs and Known Transport Systems for Cationic Amino Acids 237
5. Expression of the CAT Proteins in NO-Producing Cells 239
6. L-Arginine Transport and NO Synthesis 241

Chapter 9

Electrophysiological Analysis of Renal Na$^+$-Coupled Divalent Anion Transporters
Ian Forster, Jürg Biber, and Heini Murer

1. Introduction	251
2. Steady-State Electrophysiological Characteristics	253
2.1. Recording P_i-Induced Currents from the *Xenopus* Oocyte	253
2.2. Dose and Voltage Dependence of Steady-State Kinetics	254
2.3. Sodium Slippage in the Absence of P_i	256
2.4. The Order of Substrate Binding and Voltage Dependence Based on Steady-State Kinetics	256
3. Pre-Steady-State Electrophysiological Characteristics	256
3.1. General Properties of Pre-Steady-State Relaxations	256
3.2. Voltage Dependence of Pre-Steady-State Relaxations	258
3.3. Estimation of Transporter Number and Turnover	260
3.4. The Effects of Substrates on Pre-Steady-State Kinetics	260
4. A Kinetic Scheme for Na$^+$-Coupled P_i Cotransport	262
4.1. The Sequential, Alternating Access Model for Type II Na$^+$/P_i Cotransport	262
4.2. Site of Interaction of Protons and Foscarnet with Type II Na$^+$/P_i Cotransporters	263
5. Conclusions	264
References	265

Chapter 10

Dipeptide Transporters
Ken-ichi Inui and Tomohiro Terada

1. Introduction	269
2. Brush-Border Membrane Transport	270
2.1. Small Peptides	270
2.2. Peptide-like Drugs	271
3. Transcellular Transport	272
3.1. Caco-2 Cell Monolayers	272
3.2. Peptide Transporter in Apical Membranes	273
3.3. Peptide Transporter in Basolateral Membranes	273
4. Cloning of Peptide Transporters	274

	4.1. Structure	274
	4.2. Tissue Distribution	276
	4.3. Substrate Specificity and Recognition	276
	4.4. Structural Features	278
5.	Application to Drug Delivery	281
6.	Summary and Perspective	283
	References	283

Chapter 11

Antigenic Peptide Transporter

Vashti G. Lacaille and Matthew J. Androlewicz

1.	Introduction	289
2.	TAP and the MHC Class I Antigen Processing Pathway	290
	2.1. Introduction	290
	2.2. Current Model of MHC Class I Antigen Processing	291
3.	TAP Structure and Function	293
	3.1. TAP Substrate Specificity	294
	3.2. Characterization of the TAP Peptide-Binding Site	297
4.	Viral Inhibition of TAP	297
	4.1. Herpes Simplex Virus ICP47 Protein	298
	4.2. Human Cytomegalovirus US6 Protein	300
5.	TAP as a Drug Target	301
	5.1. Peptidomimetics	301
	5.2. Synthesis of a TAP Inhibitor	302
	5.3. Steric Requirements for Peptide Binding to TAP	303
	5.4. Incorporation of Reduced Peptide Bonds and D-Amino Acids into TAP Substrates	303
	5.5. Allosteric Modulation of TAP Activity	305
6.	Concluding Remarks	305
	References	306

Chapter 12

Nucleoside Transporters of Mammalian Cells

Carol E. Cass, James D. Young, Stephen A. Baldwin, Miguel A. Cabrita, Kathryn A. Graham, Mark Griffiths, Lori L. Jennings, John R. Mackey, Amy M. L. Ng, Mabel W. L. Ritzel, Mark F. Vickers, and Sylvia Y. M. Yao

1.	Introduction	314
	1.1. Overview	314

1.2. The Heterogeneity of Nucleoside Transport Processes	316
1.3. Recent Advances in the Molecular Biology of Nucleoside Transporters	318
2. The ENT Family of Nucleoside Transporters	321
2.1. Molecular and Functional Characteristics of the ENT Family	321
3. The CNT Family of Nucleoside Transporters	327
3.1. Molecular and Functional Characteristics of the CNT Family	327
4. "Orphan" Nucleoside Transporters	332
4.1. Na^+-Dependent Transporters	333
4.2. Organellar Transporters	335
5. Nucleoside Transporters as Drug Targets	336
5.1. Purinergic Receptors and Nucleoside Transporters	336
5.2. Anticancer Drugs and Nucleoside Transporters	338
6. Summary	343
References	344

Chapter 13

Multidrug-Resistance Transporters
Jeffrey A. Silverman

1. Introduction	353
2. MDR Gene Family	354
3. P-Glycoprotein Is an ABC Family Transporter	355
4. Structure of P-Glycoprotein	356
5. *MDR1* P-gp Is a Pump of Cytotoxic Drugs	356
6. P-Glycoprotein Is an ATP-Dependent Pump	358
7. *MDR2* P-gp Is a Phospholipid Transporter	359
8. *MDR1* Tissue Distribution	359
9. Expression of P-gp in Human Cancer	360
10. MDR2 Tissue Distribution	361
11. Modulation of P-Glycoprotein	361
12. P-Glycoprotein in Drug Absorption and Disposition	364
13. Role of P-gp in Normal Tissues and Drug Disposition	365
13.1. Intestine	365
13.2. Liver	367
13.3. Kidney	368
13.4. Blood–Brain Barrier	368
14. The Multidrug Resistance-Associated Protein	370
15. Tissue Distribution of MRP Expression	370
16. MRP Is an ATP-Dependent Drug Transporter	371

17. Modulation of MRP-Mediated Resistance	372
18. Additional Multidrug-Resistance Proteins	373
18.1. Additional MRP Homologues	373
18.2. Lung Resistance Protein	373
19. Summary	374
References	374

Chapter 14

Transporters for Bile Acids and Organic Anions
Hiroshi Suzuki and Yuichi Sugiyama

1. Introduction	387
2. Transport across the Sinusoidal Membrane	388
2.1. Na^+-Dependent Transport	388
2.2. Na^+-Independent Transport	396
3. Transport across the Canalicular Membrane	403
3.1. Transport of Bile Acids across the Bile Canalicular Membrane	404
3.2. Transport of Organic Anions across the Bile Canalicular Membrane	407
4. Concluding Remarks	424
References	424

Chapter 15

Molecular and Functional Characteristics of Cloned Human Organic Cation Transporters
Mark J. Dresser, Lei Zhang, and Kathleen M. Giacomini

I. Introduction	441
2. Background	442
3. Molecular Characteristics	447
4. Functional Characteristics of Cloned Transporters	451
4.1. Expression Systems	451
4.2. Functional Methods	455
4.3. Functional Characteristics of the Cloned Human Organic Cation Transporters	458
5. Tissue Distribution	461
6. Future Studies	463
7. Conclusions	466
References	466

Chapter 16

Organic Anion Transporters
Akira Tsuji and Ikumi Tamai

1. Introduction	471
2. Monocarboxylate Transporters in Intestine and Brain	473
2.1. Molecular Characterization of MCT1 and MCT2	473
2.2. Intestinal Transport of Lactate and Short-Chain Fatty Acids via MCT1	473
2.3. Transport of Weak Organic Acids by MCT1	474
2.4. Monocarboxylate Transport via MCT1 at the Blood–Brain Barrier	475
2.5. Anion Exchange Transport of Monocarboxylates	476
3. Organic Anion Transporters in Liver and Kidney	478
3.1. Organic Anion Transporters in Liver	479
3.2. Organic Anion Transporters in Kidney	482
3.3. Organic Anion Transporters in Other Tissues	483
4. Conclusion	485
References	485

Chapter 17

Vitamin B_{12} Transporters
Gregory J. Russell-Jones and David H. Alpers

1. Introduction	493
2. General Mechanism of Dietary Absorption of Vitamin B_{12}	495
3. Structure and Biology of the VB_{12} Transport Proteins	495
3.1. Haptocorrin (Hc Cobalophilin, R-Binders, R Protein, Nonintrinsic Factor)	495
3.2. Intrinsic Factor	499
3.3. Intrinsic Factor Receptor	502
3.4. Transcobalamin II	504
3.5. Transcobalamin II Receptor	505
3.6. Cell Models of Vitamin B_{12} Transport	507
4. The Use of Vitamin B_{12} for Transport of Pharmaceuticals	508
4.1. Conjugation of Vitamin B_{12} to Peptides and Proteins	508
4.2. *In Vitro* Transport of Peptides and Proteins Linked to VB_{12}	509
4.3. *In Vivo* Transport of Peptides and Proteins Linked to VB_{12}	509

4.4. VB_{12}-Mediated Oral Delivery of Nanoparticles	511
5. Summary	512
References	513

Index 521

1

Overview of Membrane Transport

Doo-Man Oh and Gordon L. Amidon

1. INTRODUCTION

All cells are bound by membranes, and the lipid bilayer of cell membranes is highly impermeable to most water-soluble molecules. Membrane transport processes, which have been identified in most biological events, occur during formation of electrochemical potentials, uptake of nutrients such as sugars and amino acids, removal of wastes, endocytotic internalization of macromolecules, and oxygen transport in respiration (Lehninger, 1993; Lodish *et al.,* 1995). The movement of many ions, nutrients, and metabolites across cellular membranes is catalyzed by specific transport proteins, i.e., transporters that show saturation and substrate specificity. Table I lists some transporters in the plasma membrane, mitochondria, and other organelles.

The translocation of many solute molecules and ions across biological membranes is mediated by transporters or by channels. Transporters, also known as carriers or permeases, bind a solute at one side of a membrane and deliver it to the other side. Transporters require a conformational change during the process of solute translocation across the membrane. The solute binding site of a transporter is accessible to only one side of the membrane at any time. Channels, also known as pores, are describable as tunnels across the membrane in which binding sites for the solutes to be transported are accessible from both sides of the membrane at the same time. While channels may undergo conformational changes, these

---

Doo-Man Oh • Department of Pharmacokinetics, Dynamics, and Metabolism, Parke-Davis Pharmaceutical Research Division, Warner-Lambert Company, Ann Arbor, Michigan 48105.   *Gordon L. Amidon* • College of Pharmacy, University of Michigan, Ann Arbor, Michigan 48109-1065.
Membrane Transporters as Drug Targets, edited by Amidon and Sadée. Kluwer Academic/Plenum Publishers, New York, 1999.

Table I.
Partial List of Transporters in Plasma, Mitochondria, and Organelle Membranes[a]

Site of transporter	Examples of transporters
Plasma membranes	Glucose transporter of human erythrocytes and yeast, Na^+/glucose symporter, inorganic anion transporter of erythrocytes, nucleoside transporter, lactate/H^+ transporter, Na^+/amino acid transporters, Na^+H^+ antiporter, Ca^{2+}/Na^+ antiporter, Ca^{2+}/H^+ antiporter, H^+/dipeptide transporter, Na^+/bile acid transporter, organic cation transporter, monocarboxylic acid transporter, organic anion transporter, choline transporter, neurotransmitter (serotonin) transporter, vitamin B12 transporter, multidrug-resistance transporter (P-glycoprotein), antigenic transporter, Na^+/phosphate, lactose transporter of *E. coli*, phosphoenolpyruvate/sugar phosphotransferase system
Mitochondria	Ca^{2+}, Na^+, and K^+ transport systems of the inner mitochondrial membrane, voltage-dependent anion channel, ADP/ATP translocase, citrate carrier, α-ketoglutarate carrier, phosphate carrier, uncoupling protein, ATP-Mg/phosphate carrier, dicarboxylate carrier, pyruvate carrier, carnitine, aspartate/glutamate, glutamate, ornithine/citrulline, glutamine, branched chain α-keto acid carriers
Other organelles	Phosphate transporter and K^+ channel in endoplasmic reticulum, clathrin-coated vesicles and secretory granules, vacuolar ATPase

[a]Adapted from Kaplan (1996) and Racker (1985)

changes regulate whether the channels are open or closed to solute traffic, and play little or no role in the mechanism of translocation. Channels can facilitate the transport of ions across the membranes at extremely high rate; solute flux through transporters is generally several orders of magnitude slower then through channels. Finally, cells can selectively transfer macromolecules across their membranes by very specialized membrane transport receptors, e.g., vitamin B_{12} and low-density lipoproteins (LDL).

Although primary amino acid sequences of most membrane transporters are known, their detailed structural information is not yet available because these proteins are generally difficult to purify and crystallize (Griffith and Sansom, 1998). Transporters, like enzymes, bind their substrates through many weak, noncovalent interactions with stereochemical specificity. Transporter proteins span the lipid bilayer at least once and usually several times, forming a transmembrane channel lined with hydrophilic amino acid side chains. The channel provides an alternative path for its specific substrate to move across the lipid bilayer without having to dissolve in it, resulting in an increase in the rate of transmembrane passage of the substrates.

It is not surprising that clinical disorders are possible when critical transport processes are either defective at the molecular level or not regulated properly in the physiological situation. For example, imbalance of gastric acid secretion causes gastric ulcers, and diarrhea by cholera toxin is the result of solute loss followed by water loss.

Mutations can produce defective transporters. Hereditary chloride diarrhea is the result of the loss of the Cl^-/HCO_3^- anion exchanger in the large intestine preventing normal water absorption, and Hartnup disease is due to a loss of the neutral amino acid transporter of the small intestine. Cystinuria is caused by loss of the transporter for Lys, Arg, and Cys from both intestinal and renal brush borders, resulting in kidney stones formed from the sparingly soluble cystine. Cystic fibrosis, an inherited disorder causing pancreatic, pulmonary, and sinus disease in children and young adults, is characterized by abnormal viscosity of mucous secretions caused by altered electrolyte transport across epithelial cell membranes. The protein encoded by the gene defective in cystic fibrosis, the cystic fibrosis transmembrane conductance regulator (CFTR), is a chloride channel that is regulated by cyclic AMP-dependent protein kinase phosphorylation requiring binding of ATP for channel opening (Schultz et al., 1996). Several disorders related to transport processes are listed in Table II.

P-Glycoprotein belongs to the ATP-binding cassette (ABC) family of transporters that serve a great variety of biological functions, including uptake of nutrients, extrusion of toxic compounds, secretion of toxins, transport of ions and peptides, and cell signaling. Over 50 ABC transporters have been identified, and the human multidrug-resistance transporter MDR1 has led to intense interest in studies of cancer drug resistance; since tumor cells are defective in intracellular communication, one might anticipate that plasma membranes participate in their abnormal behavior (Petty, 1993).

An understanding of the mechanism of one transporter can be aided by studies of other transport processes, because transport processes in various organisms, tissues, and cells share many common features. This chapter outlines the membrane transport processes as currently understood, and later chapters will discuss the molecular biology, structure, and regulation of transport systems.

2. MODES OF MEMBRANE TRANSPORT

Transport of ions and nutrients takes place through a variety of mechanisms. A particular nutrient, e.g., glucose of amino acid, may be transported by facilitated diffusion, Na^+ symport, or H^+ symport, depending on the cell type. Further multiple transporters may be used for one substrate in a single cell; for example, glycine is transported by four different transporters in the brain (Malandro and Kil-

Table II.
Some Disorders Related to Membrane Transport Processes[a]

Disorder	Transport defect	Molecular defect	Clinical features
Cystinuria	Renal and intestinal Cys and dibasic amino acid transport	Mutant dibasic amino acid transporter (D2H)	Renal calculi
Hartnup disorder	Renal and intestinal neutral amino acid transport	?	Usually benign; pellagra-like syndrome
Iminoglycinuria	Renal imino acid transport	?	Benign
Renal glycosuria	Renal glucose transport	Mutant SGLT2 ?	Glucosuria, benign
Nephrogenic diabetes insipidus	Water reabsorption	Mutant V_2 receptor, mutant AQP2 protein	Vasopressin-resistant polyuria, hypernatremia
Renal tubular acidosis-distal	Proton secretion	Absent H^+-ATPase, absent H^+K^+-ATPase?	Metabolic acidosis, nephrocalcinosis
Renal tubular acidosis-proximal	Bicarbonate reabsorption	?	Metabolic acidosis, Fanconi syndrome
Hereditary hypo-phosphatemia	Renal phosphate reabsorption	?	Hypophosphatemia, osteomalacia
Renal hypouricemia	Renal urate transport	?	Benign
Hyperekplexia	Gly, GABA, and Glu-gated channels	Mutant α1-subunit	Neurological and psychiatric illness
Dubin–Johnson syndrome	Bilirubin glucuronide transport	? Defect in organic anion transporter	Chronic conjugated hyperbilirubinemia, benign
Cystic fibrosis	Chloride transport	Single-gene mutation (ΔF508)in CFTR	Mucus obstruction of airways, exocrine pancreatic insufficiency, meconium ileus
Hyperkalemic periodic paralysis; normo-kalemic periodic paralysis; para-myotonia congenita	Sodium channel disease	Mutation in chromosome 17q locus	High serum potassium level, myotonia
Hypokalemic periodic paralysis	Calcium channel disease	Mutation in chromosome 1q locus	Low serum potassium level, weakness
Thomson's or Becker's myotonia congenita	Chloride channel disease	Mutation in chromosome 7q locus	Autosomal recessive disorder with myotonia and transient weakness

[a] Adapted from Schultz et al. (1996).

Table III.
Mechanisms of Membrane Transport[a]

Type	Transport protein	Saturation	Concentration gradient	Energy dependence	Remarks and examples
Simple diffusion	No	No	No	No	Oxygen, water
Ion channels	Yes	No	No	No	Na^+ channel
Facilitated diffusion	Yes	Yes	No	No	Glucose transporter
Primary active transport	Yes	Yes	Yes	Yes	H^+-ATPase, Ca^{2+}-ATPase, Na^+K^+-ATPase; energy source: ATP, light, substrate oxidation
Secondary active transport	Yes	Yes	Yes	Yes	Na^+/Ca^{2+} antiporter Na^+/amino acid symporters, Na^+/glucose symporter H^+/peptide transporter; energy source: ion gradient

[a]Adapted from Lehninger (1993).

berg, 1996). Membrane transporters found in biological membranes can be classified based on mechanisms and energetics as shown in Table III. These transport systems are categorized into two broad classes: those that require energy and those that do not. Passive transporters (or facilitated diffusion) simply facilitate diffusion of the solute across the membrane, whereas active transporters use free energy to drive solute transport against a concentration gradient. The energy input for active transport may come from light, oxidation reactions, ATP (or phosphoenolpyruvate) hydrolysis, or cotransport of some other solute. Some transporters carry out symport (cotransport), the simultaneous passage of two species in the same direction, and others mediate antiport, in which two species simultaneously move in opposite directions (Lehninger, 1993). There are superfamilies of channels and transporters comprised of functionally distinct proteins that have similar structures as revealed by their amino acid sequences. These sequence similarities suggest that there may be similar functional mechanisms within each superfamily despite differences in solute specificity.

2.1. Passive Diffusion

Passive diffusion, or simple diffusion, is the movement of a solute across a membrane down the electrochemical gradient without the assistance of a transport protein. It does not require any biological energy, but follows Fick's law:

$$V = PA \cdot \Delta C = \left(\frac{DK}{\delta}\right) A \cdot \Delta C \qquad (1)$$

where V is the transport rate, A is the surface area, ΔC is the concentration difference across the membrane, D is the diffusivity of the solute, K is the partition coefficient between membrane and water, and δ is the membrane thickness. The general plot for a simple diffusion is shown in Fig. 1. As expected from Equation (1), the transport rate is proportional to the substrate concentration, showing a straight line (Fig. 1A).

Many small lipid-soluble molecules such as oxygen, N_2, CO_2, and NH_3 are transported by simple diffusion through biological membranes.

2.2. Ion Channels

Ion transport across cell membranes plays an important role in many cell processes, and ion fluxes are closely linked to cell growth and proliferation. Important inorganic ions include Na^+, K^+, Ca^{2+}, and Cl^-; their concentrations inside and outside of a typical mammalian cell are listed in Table IV. As indicated, the high concentration of Na^+ ion outside the cell is balanced mainly by extracellular chloride ions. On the other hand, the high concentration of K^+ ion inside the cell is balanced by a variety of negatively charged intracellular ions such as Cl^-, HCO_3^-, or PO_4^{3-}, or by negatively charged organic molecules.

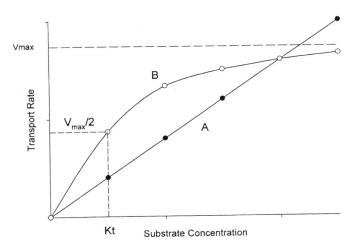

Figure 1. Concentration dependence of membrane transport: (A) Linear and (B) saturable transport processes. Simple diffusion and ion channels follow the linear dependence on the substrate concentration. Facilitated transporter and active transporters follow a saturable pattern.

Table IV.
Comparison of Concentrations of Some Important Free Ions Inside and Outside in a Typical Mammalian Cell[a]

Ion	Intracellular concentration (mM)	Extracellular concentration (mM)
Na^+	12	145
K^+	140	4
Mg^{2+}	0.5	1–2
Ca^{2+}	10^{-7}	1–2
H^+	$10^{-7.2}$	$10^{-7.4}$
Cl^-	5–15	110

[a] Membrane potential = 50–70 mV.

Various enzymes and transport mechanisms regulate the pH and the concentrations of Na^+, K^+, Ca^{2+}, and Cl^-, and other anions in cell, mitochondria, or organelle compartments. Such gradients are established at the expense of metabolic energy, e.g., ATP hydrolysis. Ionic gradients are used to transfer solutes across membranes (e.g., amino acids, sugars) into the cell by symporters, and protons out of the cell by antiporters. The outward K^+ gradient generated across the plasma membrane is a major determinant of the inside negative transmembrane potential of cells. In epithelial tissues, the polarized distribution of enzymes and ion carriers provides the driving force for movement of ions and molecules across the cell interior. Ion movements, especially proton transport, are also involved in the synthesis of ATP in mitochondria, in photosynthesis in plants, and in substrate uptake by bacteria (Evans and Graham, 1989). Ion channels in epithelial cells have been reviewed recently (Caplan, 1997; Muth et al., 1997).

Ion channels have basically two conformations: open or closed. When they are open, ions flow through the channel and produce an electric current. Gating and selectivity in various types of channels are shown in Fig. 2. Figure 2A shows a neurotransmitter-gated (or ligand-gated) channel that opens as a result of the binding of neurotransmitter molecules to sites on their extracellular surface. It is often selective to anions, as seen in the glycine receptor channel in the spinal cord. Figure 2B illustrates a channel that is selective to potassium ions and closed by the binding of an internal ligand such as ATP. The ATP-sensitive potassium channel of pancreatic β cells is an example of this type. Figure 2C shows a calcium-selective voltage-gated channel; part of the internal structure of the channel is charged and moves when the membrane potential becomes more positive inside, and this acts as the trigger for opening the channel (Aidley and Stanfield, 1996). For example, sodium channels in most nerve and muscle cell membranes control the transient sodium current during an action potential. These voltage-gated channels open as a result of a change in the potential gradient across the membrane.

Ion-selective channels, such as the acetylcholine receptors of the vertebrate synapse, play an essential role in signal transductions from one neuron to the next.

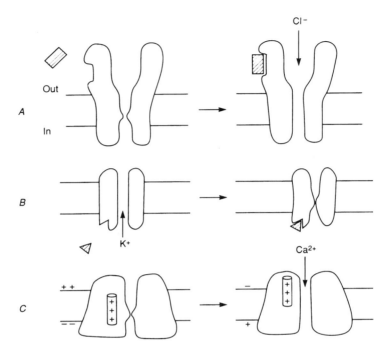

Figure 2. Gating and selectivity in various types of channels. (A) A neurotransmitter-gated (or ligand-gated) channel that opens as a result of the binding of neurotransmitter molecules to sites on their extracellular aspect. (B) A channel selective for potassium ions that is closed by the binding of an internal ligand such as ATP. (C) A calcium-selective voltage-gated channel, part of the internal structure of the channel is charged and moves when the membrane potential becomes more positive inside, and this acts as the trigger for opening the channel. Reprinted with permission from Aidley and Stanfield (1996). Copyright 1996 Cambridge University Press.

When the electrical signal carried by the presynaptic neuron reaches the synaptic end of the cell, the neurotransmitter acetylcholine is released into the synaptic cleft. Acetylcholine rapidly diffuses across the cleft to the postsynaptic neuron, where it binds to high-affinity sites on the acetylcholine receptor. This binding induces a change in receptor structure, opening a transmembrane channel in the receptor protein. The acetylcholine receptor allows Na^+ and K^+ to pass with equal ease, but other cations and all anions are unable to pass through it. The rate of Na^+ movement through the acetylcholine receptor ion channel is linear with respect to extracellular Na^+ concentration; like simple diffusion, the process is not saturable (Fig. 1A). The ion channel of the acetylcholine receptor behaves as though it provides a hydrophilic pore through the lipid bilayer through which an ion of the right size, charge, and geometry can diffuse very rapidly down its electrochemical gra-

Overview of Membrane Transport

dient. It has a gate that opens in response to stimulation by acetylcholine, and an intrinsic timing mechanism that quickly closes the gate. Thus the transient feature of the acetylcholine signal is essential to electrical signal conduction.

2.3. Facilitated Diffusion

When an energy source is not required, molecules enter and leave cells by following their electrochemical potential gradient. Facilitated diffusion, unlike simple diffusion through the phospholipid bilayer, is specifically accelerated by a specific binding between the solute and a membrane protein, and does not require energy. As shown in Fig. 3, the binding site alternates between two states—exposure to one side first, and then to the other side of the membrane. The transition between these two states is a discrete kinetic event to which a rate constant can be assigned. An example of the kinetics of a transport process is given in the following section.

Some essential features can be summarized as follows (West, 1983): (a) Facilitated diffusion by a mobile transporter system operates only so that solute flows from a higher to a lower electrochemical potential; (b) the solute flows at a rate greater than that predicted from its size or hydrophilicity; (c) the rate of penetration does not follow Fick's law, except at very low concentrations; at higher concentrations, saturation kinetics are observed, as seen in Fig. 1B; (d) competitive inhibition occurs between chemically and sterically similar substrates; (e) inhibition may be caused by other compounds, especially those reacting with or ligating reactive groups in proteins; and (f) it is often possible for the flow of substrate in one direction temporarily to drive the accumulation of an analogue (or isotope) in the opposite direction against the electrochemical potential gradient of analogues (an overshoot phenomenon).

A well-known example of facilitated diffusion is erythrocyte glucose transport. Five transport proteins capable of mediating the facilitated diffusion of glucose have been identified in various cells (Table V), all of which consist of a polypeptide chain composed of about 500 amino acids and possess a high homology, including 12 putative transmembrane-spanning segments. Glucose enters the erythrocyte by facilitated diffusion via a specific glucose transporter that allows glucose entry into the cell at a rate about 50,000 times greater than its simple diffusion through a lipid bilayer.

The erythrocyte contains another facilitated diffusion system, an anion exchanger, which is essential to CO_2 transport from tissue (such as muscle and liver) to the lungs. In this system, waste CO_2 released from respiring tissues to the blood plasma enters the erythrocyte, where it is converted into bicarbonate (HCO_3^-) by the enzyme carbonic anhydrase. The chloride–bicarbonate exchang-

Table V.
Facilitated Transporters of Glucose in Various Types of Cells[a]

Name	Distinguishing characteristics
GluT 1	Widespread distribution, but particularly high levels in plasma membranes of endothelial cells lining blood vessels and the blood–brain barrier
GluT 2	Basolateral membranes of small intestinal and renal proximal tubule epithelial cells
GluT 3	Has very high affinity for glucose and is present in plasma membranes of neurons
GluT 4	Present in plasma membranes of muscle cells and adipocytes and its expression is upregulated by insulin
GluT 5	Small intestinal and renal epithelial cells

[a]Reprinted with permission from Byrne and Schultz (1994). Copyright 1994 Lippincott-Raven Publishers.

er, also called the anion exchange protein or band 3, increases the permeability of the erythrocyte membrane to HCO_3^- by a factor of more than one million. It is an integral membrane protein that probably spans the membrane 12 times. Unlike glucose transporters, this protein mediates a bidirectional exchange; for each HCO_3^- ion that moves in one direction, one Cl^- ion must move in the opposite direction. The result of this paired movement of two monovalent anions is no net change in the charge or electrical potential across the erythrocyte membrane; thus, the process is not electrogenic.

2.4. Active Transporters

Active transport results in the accumulation of a solute on one side of a membrane and often against its electrochemical gradient. It occurs only when coupled to an exergonic process directly or indirectly. During this transport process, an energy source such as ATP, electron transport, or an electrochemical gradient of another ion is used to move ions or molecules against their electrochemical potential gradients. Active transport is necessary for maintenance of membrane potential and ion gradients, storage of energy for secondary transporters, and regulation of pH inside the cell.

In primary active transport, solute accumulation is coupled directly to an exergonic reaction (conversion of ATP to ADP + Pi). Secondary active transport occurs when uphill transport of one solute is coupled to the downhill flow of a different solute that was originally pumped uphill by primary active transport.

An example of the primary active transporter is the Na^+K^+-ATPase in the mammalian cells that is energized by ATP. Animal cells maintain a lower concentration of Na^+ and a higher concentration of K^+ intracellularly than is found in extracellular fluid. This concentration difference is established and maintained by a primary active transport system in the plasma membrane through the involvement of the enzyme Na^+K^+-ATPase, which couples the breakdown of ATP to the simultaneous movements of both Na^+ and K^+ against their concentration gradients. This transporter moves two K^+ ions inward and three Na^+ ions outward across the plasma membrane; this process is electrogenic, because three Na^+ ions move outward for every two K^+ ions that move inward. The process creates a net separation of charge across the membrane, making the inside of the cell negative relative to the outside. The resulting transmembrane potential of -50 to -70 mV is essential to the conduction of action potentials in neurons, and is also characteristic of most nonneuronal animal cells. About 25% of the energy-yielding metabolism of a human at rest goes to support the Na^+K^+-ATPase.

There are three general types of ion-pumping ATPases, as listed in Table VI. The P-type ATPases undergo reversible phosphorylation during their catalytic cycle, and are inhibited by the phosphate analogue vanadate. The Na^+K^+-ATPase is the prototype of P-type ATPases. The V-type ATPases produce gradients of protons across the membranes of a variety of intracellular organelles, including plant vacuoles. The F-type proton pumps (ATP synthases) are central to energy-conserving mechanisms in mitochondria and chloroplasts.

In animal cells, the differences in cytosolic and extracellular concentrations of Na^+ and K^+ are established and maintained by active transport via Na^+K^+-ATPase, and the resulting Na^+ gradient is used as an energy source by a variety of symport and antiport systems. The Na^+K^+-ATPase shows specific distribution patterns on the animal cell surface. In nonepithelial cells such as fibroblasts, the enzyme is evenly distributed on the cell surface, whereas in epithelial cells the location of the enzyme on the basolateral pole underlies the vectoral transport of salts, water, and organic solutes (e.g., bile salts) across the tissue.

Table VI.
Three Classes of Ion-motive ATPases[a]

	F-ATPase	V-ATPase	P-ATPase
Total size (kDa)	500	500	200–300
Subunits	8–12	3–8	1
Substrate	H^+, Na^+	H^+	H^+, Na^+, K^+, Ca^{2+}
Function	ATP synthesis	ATP hydrolysis	ATP hydrolysis
Selected inhibitor	Oligomycin azide	N-Ethylmaleimide	Vanadate
Phosphorylated intermediate	No	No	Yes
Type of membrane	Inner mitochondria	Lysosomal, endosomal, secretary vesicles	Plasma, endoplasmic reticulum

[a]Adapted from Schultz et al. (1996).

2.5. Secondary Active Transporters

There are many secondary active transport systems where the free energy for translocation is not provided directly from metabolic changes, but from the energy stored in ionic gradients. In secondary active transport, a single cotransporter couples the flow of one solute (such as H^+ or Na^+) down its concentration gradient to the pumping of a second solute (such as a sugar or an amino acid) against its concentration gradient. In intestinal epithelial cells, glucose and certain amino acids are accumulated by symport with Na^+ (Ganapathy et al., 1994). Peptide transporters in the intestine and kidney mediate small peptide transport with an inward-directed electrochemical H^+ gradient (Ganapathy and Leibach, 1996). In most animals Na^+ is the driving ion, but H^+ is common in bacteria and plants.

2.6. Macromolecular and Bulk Transport

Macromolecules and inert particles are simply too big to diffuse across a lipid bilayer. Preformed proteins are generally transported through membranes during fusion (secretion) or fission (e.g., pinocytosis) events. For example, during pinocytosis macromolecules in the extracellular fluid phase are transferred into the cytoplasm via pinocytotic vesicles budding from the plasma membrane. During exocytosis, however, storage vesicles fuse with the plasma membrane and thereby release their contents into the extracellular environment (Petty, 1993). Adsorptive pinocytosis may involve a transport receptor such as the LDL receptor during the transport process. The LDL receptor is a cell-surface integral membrane glycoprotein which recognizes LDL and mediate its endocytosis (Yamamoto et al., 1984). More specialized processes will not be further discussed in this summary.

3. ENERGETICS OF MEMBRANE TRANSPORT

Based on thermodynamics, the activation energy for translocation of a polar solute across the membrane is so large that pure lipid bilayers are virtually impermeable to polar and charged species. Transmembrane passage of polar compounds and ions, therefore, is made possible by membrane proteins that lower the activation energy for transport by providing an alternative path for specific solutes through the lipid bilayer. Facilitated diffusion (or passive transport) translocates the substrates from one side of the membrane to another without chemical alterations. The amount of energy needed for the transport of an ionized solute can be calculated from the general equation for free-energy change:

Overview of Membrane Transport

$$\Delta G_t = \Delta G° + RT \ln\left(\frac{C_{in}}{C_{out}}\right) + ZF \cdot \Delta\psi \qquad (2)$$

where R is the gas constant, T is the absolute temperature, C_{in} and C_{out} are concentrations of transported solute inside or outside the membrane, respectively, Z is the charge of the solute (ion), F is the Faraday constant, and $\Delta\psi$ is the transmembrane electrical potential. The first term in Equation (2), $\Delta G°$, equals zero because no bonds are made or broken during the transport process. For ionized solutes, the energetic cost of translocation depends on the electric potential as well as the chemical potential.

4. KINETICS OF MEMBRANE TRANSPORT

4.1. Kinetic Equation

Unlike soluble enzymes, membrane transporters are vectoral catalysts that operate between two compartments. The kinetic study of transport therefore requires specialized theoretical and practical approaches. Several reviews have described the testing and characterization of transport systems using kinetic approaches (Deves, 1991; Stein, 1986, 1990). The transport process can be modeled for a carrier-mediated transport as illustrated in Fig. 3. There are two kinetically indistinguishable models: the mobile carrier that floats across the membrane like a ferry boat (Fig. 3A) and the gated-pore model (Fig. 3B). The transporter alternates between two conformational states that differ with respect to the orientation of its binding site, which can become exposed either to the inside or outside of the cell. A transport scheme for either model can be developed with the assignment of kinetic rate constants for each step (Fig. 3C). Transport depends minimally on four consecutive events: substrate binding, translocation of the transporter–substrate complex, substrate dissociation, and return of the free transporter to the initial state. It is assumed that the rate-limiting step in the process is translocation and not association or dissociation of substrate from the transporter.

For the initial steady-state rate of influx, no substrate on the inside is assumed. The full rate equation for a carrier-mediated transport is

$$V_0 = \frac{T_{total}[S]_{out} k_1 f_2 k_{-2} f_{-1} / \{k_1(f_2 k_{-2} + f_2 k_{-1} + f_{-2} f_{-1} + k_{-2} f_{-1})\}}{(f_1 + f_{-1})(k_{-1} f_{-2} + k_{-1} k_{-2} + f_2 k_{-2}) / \{k_1(f_2 k_{-2} + f_2 k_{-1} + f_{-2} f_{-1} + k_{-2} f_{-1})\} + [S]_{out}} \qquad (3)$$

V_0 is the initial velocity of accumulation of glucose inside a cell when its concentration in the surrounding medium is $[S]_{out}$. The rate constants f are assigned to the

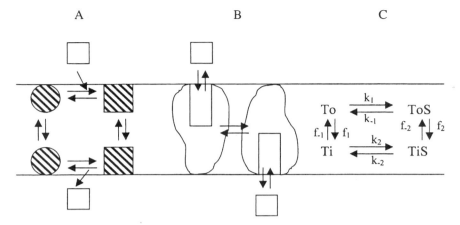

Figure 3. Models for carrier-mediated facilitated diffusion: Two extreme types of transporter are illustrated: (A) the ferryboat and (B) the gated pore. (C) A transport scheme for either model with the assumption of two discrete conformations (Ti or To) of the transporter. The rate constants f are assigned to the translocation steps and k to the dissociation steps of the transporter–substrate complexes. The substrate dissociation constants on the outer and inner faces of the membrane are $K_{S,out} = k_{-1}/k_1$ and $K_{s,in} = k_{-2}/k_2$, respectively. S represents the substrate. Based on Deves (1991) and Stein (1986).

translocation steps and k to dissociation steps of the transporter–substrate complexes. T_{total} represents total transporter concentration.

The rate equation can be simplified to an expression analogous to the Michaelis–Menten equation:

$$V_0 = \frac{V_{max}[S]}{K_t + [S]} \qquad (4)$$

where K_t (t for transport) is a constant analogous to the Michaelis–Menten constant, a combination of rate constants characterisitic of each transport system. Equation 4 can be rewritten as

$$\frac{[S]}{V_0} = \frac{K_t}{V_{max}} + \frac{1}{V_{max}}[S] \qquad (5)$$

The velocity constant V_{max} and the concentration constant K_t can be written in terms of the individual rate constants as follows:

$$V_{max} = \frac{T_{total} f_2 k_{-2} f_{-1}}{f_2 k_{-2} + f_2 k_{-1} + f_{-2} f_{-1} + k_{-2} f_{-1}} \qquad (6)$$

$$K_t = \frac{(f_1 + f_{-1})(k_{-1}f_{-2} + k_{-1}k_{-2} + f_2k_{-2})}{k_1(f_2k_{-2} + f_2k_{-1} + f_{-2}f_{-1} + k_{-2}f_{-1})} \quad (7)$$

Equation (4) describes the initial velocity, the rate observed when $[S]_{in} = 0$. Figure 1B represents a general membrane transport process. The transport rate is linear at low substrate concentrations, with a saturable pattern at higher concentrations. Because no chemical bonds are made or broken in the conversion of S_{out} into S_{in}, the process of entry is fully reversible. Although the Michaelis–Menten constant is clearly a complex kinetic parameter, it can be loosely interpreted in terms of the affinity with which a transporter binds to the substrate for translocation.

Some kinetic parameters for the sugar transporter of erythrocytes are listed in Table VII. Very different values of K_t are found for different sugars. The glucose transporter is specific for D-glucose ($K_s = 4-10$ mM). However, for the close analogues D-galactose and D-mannose, which differ only in the position of one hydroxyl group, the values of K_t are 40–60 and 14 mM, respectively. The V_{max} val-

Table VII.
Kinetic Parameters of Sugar Transport in Human Erythrocytes[a]

Sugars[b]	Temperature (°C)	K_t (mM)	V_{max} (mM/min)
D-Glucose	37	4–10	600
	25	3	150
	20	1.6	45–200
D-Galactose	37	40–60	700
	20	12	150
D-Mannose	37	14	300–700
	20	7	120
Xylose	37	60	650
Ribose	37	2000	600
L-Arabinose	37	250	700
L-Sorbose	25	3100	125
Fructose	25	9300	125

Transport mode for D-glucose[c]	K_t (mM)	V_{max} (mM/min)	V_{max}/K_t (min^{-1})
Zero-trans entry	0.5	0.38	0.74
Zero-trans efflux	2.7	3.8	1.42
Equilibrium exchange	42	50	1.18
Infinite-trans entry	0.8	23	—
Infinite trans efflux	19	37	—

[a]Reprinted with permission from Stein (1990). Copyright 1990 Academic Press.
[b]Mostly zero-trans-entry experiments.
[c]All at 0° C. From Wheeler and Whelan (1988).

ues, nevertheless, tend to be rather similar (Stein, 1990). Membrane transporters have relatively low affinities (in the range of 10 mM), in contrast to the high affinities generally displayed by membrane receptors.

4.2. Transport Experiments

By convention, the terms cis and trans refer the relative locations of substrates or inhibitors across the membrane; cis designates the compartment occupied by the reference substrate (the one whose transport is followed), while trans designates the compartment on the opposite side of the membrane (Deves, 1991). Three different types of transport experiments can be performed for carrier-mediated transport as in Fig. 3: (a) zero-trans experiments, where the substrate is initially present in only one compartment and its flux across the membrane is determined; (b) equilibrium exchange experiments, where the substrate is present in both compartments at the same concentration, but only one is labeled, and where unidirectional flux of labeled substrate is measured; and (c) infinite-trans experiments, where the unidirectional flux of substrate is measured in the presence of a saturating concentration of unlabeled substrate in the trans compartment. The transport capacity of a carrier is reflected in its maximum transport rate (the rate at saturating substrate concentration), because under this condition the system is working at its full potential.

Table VIII lists the expressions for the maximum rate and Michaelis–Menten constant for different experimental arrangements in terms of the microscopic rate constants in Fig. 3C. The maximum transport rate of zero-trans entry (or exit) depends on the rate constant for the inward translocation of the complex (f_2) and the rate constant for the outward movement of the free carrier (f_1). In the exchange mode, the maximum rate depends exclusively on the rate of translocation of the carrier–substrate complex; clearly a zero-trans arrangement cannot be used to compare relative translocation rates. An alternative is to measure exchange rates instead of zero-trans rates.

The apparent affinity of a substrate is obtained from the concentration dependence of the transport rate. A Michaelis–Menten constant K_t can be defined as the substrate concentration that can sustain a rate equal to one-half of the maximum transport rate. While the apparent affinity is dependent on the true dissociation constant of the substrate–transporter complex ($K_{s,out}$ or $K_{s,in}$), it is also influenced by the microscopic rate constants. Relative affinities, therefore, cannot be directly estimated from the K_t in zero-trans entry experiments. The relative translocation rates of the substrate–transporter complex and the free transporter may be determined by comparing the unidirectional flux of labeled substrate in the presence or absence of unlabeled analogue in the trans compartment.

Table VIII.
Maximum Rate (V_{max}) and Michaelis–Menten Constant (K_t) of Carrier Transport[a,b]

Experiment	V_{max}	K_t
Zero-trans entry	$\dfrac{f_2 f_{-1} T_{total}}{(f_2 + f_{-1})}$	$\dfrac{(f_1 + f_{-1}) K_{s,\,out}}{(f_{-1} + f_2)}$
Zero-trans exit	$\dfrac{f_1 f_{-2} T_{total}}{(f_1 + f_{-2})}$	$\dfrac{(f_1 + f_{-1}) K_{s,\,in}}{(f_{-2} + f_1)}$
Equilibrium exchange	$\dfrac{f_2 f_{-2} T_{total}}{(f_2 + f_{-2})}$	$\dfrac{(1 + f_1/f_{-1}) K_{s,\,out}}{(1 + f_2/f_{-2})}$
Infinite-trans entry	$\dfrac{f_2 f_{-2} T_{total}}{(f_2 + f_{-2})}$	$\dfrac{(f_1 + f_{-2}) K_{s,\,out}}{(f_{-2} + f_2)}$

[a] Adapted from Deves (1991).
[b] The expressions are derived under the assumption of rapid equilibrium. T_{total} is the total transporter concentration. The substrate dissociation constants on the outer and inner faces of the membrane are $K_{s,\,out} = k_{-1}/k_1$ and $K_{s,\,in} = k_{-2}/k_2$ respectively.

Some values of the K_t and V_{max} for the sugar transporter of erythrocytes are shown in Table VII. Different settings of transport experiments result in different kinetic values. The most reliable method of obtaining the kinetic parameters K_t and V_{max} of a transporter is to perform an equilibrium exchange experiment (Stein, 1990).

The observed inhibition behavior in transport experiments depends on how the inhibitor binds to the carrier: free, or complexed with the substrate (noncompetitive). It may bind to the free carrier only (competitive) or it may bind to the carrier–substrate complex only (noncompetitive). In the first case, one would therefore expect the substrate to have no effect on the potency of the inhibitor, whereas in the second case the substrate could either protect by displacing the inhibitor or facilitate the inhibition by promoting the binding of the inhibitor. The results may not be clear because the ability of the substrate to displace the inhibitor in the experiment depends on the inhibition mechanism, the relative location of the inhibitor and substrate, the relative affinity of the inhibitor for different carrier forms, and the transport mechanism.

Cotransport is analogous to an enzyme reaction involving two substrates and two products. The rate equation can be obtained, but analysis is relatively complicated because 24 rate constants are involved. An example of cotransport is the lactose permease of *Escherichia coli* (West, 1983). Detailed discussions of the kinetic treatment of cotransporters (Stein, 1986) and P-glycoproteins (Stein, 1997) have been published.

5. ANALYSIS OF MEMBRANE TRANSPORT

Membrane permeability can be determined by various experimental methods, and the models for assessing drug absorption have been recently summarized (Borchardt *et al.,* 1996). When evaluating the passage of substrates across membranes, the simplest system is composed of two well-stirred compartments separated by a semipermeable membrane (Stein, 1986). This model is easy to describe mathematically, but of course very simple. First, the assumption of complete mixing may be approximate. When convection is involved, transport is a distributed process. That is, a concentration gradient exists along the axis of convection (e.g., from the proximal to the distal end of intestine, from the arterial to the venous end of a capillary). Second, substrates traverse multiple membranes when passing from donor to receiver compartments. Finally, in many cases, there are specific transport mechanisms that move substrates across the membrane. Carrier-mediated transport across a membrane adds additional complexity to the system. For even the simplest transport, the concentration of the transporter and its affinity for the substrate must be known before it can be modeled. Other mathematical models for membrane transport in humans have been introduced for the fraction dose absorbed and the rate of drug absorption (King, 1996; Yu *et al.,* 1996).

6. METHODS IN MEMBRANE TRANSPORT RESEARCH

Recently new experimental techniques have been applied to the study of the molecular aspects of membrane transporters. Electrophysiological techniques include the patch-clamp technique, electrical noise and impedance analysis, and ion-specific microelectrodes. These methods allow one to use fluorescent dyes to examine the electrical properties of a single membrane transporter such as an ion-selective channel, and to observe real-time changes in intracellular ion activities and pH. The patch-clamp technique, for example, identifies the channels in native membranes and provides a biophysical fingerprint. Another powerful technique for the molecular and electrophysiological characterization of membrane transporters is to inject the mRNA into the *Xenopus* oocyte expression system. The development of image technologies such as fluorescence computer-assisted digital image analysis and confocal microscopy has permitted the study of subcellular localization of membrane transporters and functional interactions. The combination of these technologies with immunology has also made possible the development of probes highly specific to membrane transporters. Biochemical techniques to purify and reconstitute biological transporters in artificial membrane systems can be applied to characterize transport activities. Various methodologies in membrane transporter research (Schafer *et al.,* 1994; Schultz *et al.,* 1996), as well as biopharmaceutical studies including intestinal permeability and epithelial cell models (Borchardt *et al.,* 1996; Wills *et al.,* 1996) have been reviewed in the recent literature. A brief summary is provided below.

6.1. Patch Clamping

The patch-clamp technique (Hamill *et al.,* 1981; Sakmann and Neher, 1984; Sakmann and Neher, 1995) has made it possible to record ionic currents through single channels in the cell membrane under conditions of complete control over transmembrane voltage and ionic gradients (Yudilevich *et al.,* 1991). A diagram of the patch-clamp technique is shown in Fig. 4. The two major compoentns are ion-

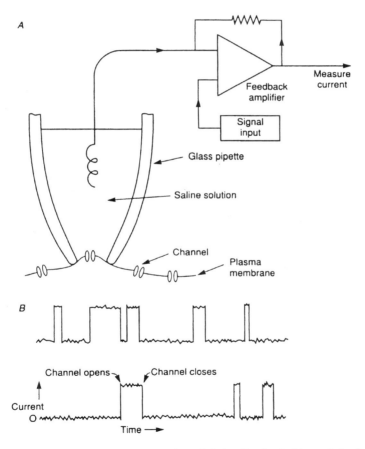

Figure 4. Patch-clamp records of channel opening and closing. Reprinted with permission from Aidley and Stanfield (1996). Copyright 1996 Cambridge University Press.

ic and capacitive currents. The capacitive current results from the charging of the membrane capacitance; part of that capacitance, however, is the result of molecular rearrangement within the channels and other polarizable membrane dielectrics. A glass pipette, usually fire polished and extremely clean, is brought to the surface of a clean membrane. After contact, usually a small amount of suction provides a high-resistance seal (several gigaohms) between the membrane and the glass, effectively isolating the patch of membrane attached to the pipette from the rest of the cell. The pipette is connected to the summing junction of the current-to-voltage transducer connected to amplifiers and signal conditioning circuits. The membrane voltage of the isolated patch can be modified by changing the voltage of the amplifier summing junction or the voltage of the bath (Yudilevich *et al.,* 1991). The patch-clamp technique has allowed the recording of a very small area of membrane, making it possible to observe the activity of one channel opening and closing as the membrane potential is controlled.

6.2. Fluorescence Digital Imaging Microscopy

Fluorescent probes are usually much more fluorescent than the constituents of most biological specimens, and the fluorescence signals can be measured continuously and nondestructively with a very good spatial and temporal resolution (Dissing and Gasbjerg, 1994). Fluorescent probes have been used for many purposes in the characterization of biological mechanisms, e.g., as markers for organelles and proteins, and for measurements of distances between sites by fluorescence energy transfer. The common feature of such applications is the main role of the dye, which is to signal its presence and location rather than to sense its environment. More recently, fluorescent probes that can sense the physical environment have been synthesized, and typically include probes for measurements of membrane potentials, electrolytes, and pH. The fluorescent cyanine dye diS-C_3(5) is the most commonly used fluorescent probe to monitor membrane potential (Schafer *et al.,* 1994). Other fluorescent dyes, such as fura-2, indo-1, and fluo-3, monitor intracellular electrolyte concentrations such as H^+ (Rink *et al.,* 1982), Ca^{2+} (Minta *et al.,* 1989), Na^+ (Minta and Tsien, 1989), and Cl^- and Mg^{2+} (Rink *et al.,* 1982; Tsien, 1989).

Fluorescence digital imaging microscopy has been introduced to describe the combined use of low-light-level image detectors, computer-assisted digital image processing, and epifluorescence microscopy (Arndt-Jovin *et al.,* 1985). Advantages of fluorescence digital imaging microscopy as a method to assess the structure and function of cells are sensitivity, selectivity, and the noninvasive nature of the method, i.e., the ability to assess quantitatively the spatial and temporal processes of living cells (Jilling and Kirk, 1994).

6.3. Confocal Microscopy

Thin optical sections parallel or perpendicular to the axis of the confocal microscope allow insight into three-dimensional tissue distribution. Further, thin optical sections with high spatial resolution enable location of transport processes, and can be used for quantitative studies of these processes (Paddock, 1991, 1996). The availability of video rate confocal microscopes introduced a high time resolution for transport kinetics (Williams, 1990). Confocal microscopy with fluorescence techniques have better advantages such as compartmental analysis, direct identification, characterization of cells, and multiparameter analysis over conventional microscopy. At the cellular and subcellular level, one can visualize that these aspects of technology, together with the current progress in molecular biology and immunology of transport proteins, can lead to significant insights into the dynamics of intracellular trafficking of membrane transporters and their regulation by intracellular parameters. The tandem scanning microscope (Boyde, 1985) and the laser scanning microscope (Williams, 1990) are frequently used confocal microscopes in the study of membrane transport in cells.

6.4. Reconstitution

A general strategy for the purification of a membrane transporter begins with identification of tissues that possess the activity of interest and are available in large quantity (Schultz *et al.,* 1996). Next, crude homogenate of the tissue is prepared and differential or gradient centrifugation or other techniques separate the major membrane fractions. The membrane of interest is identified on the basis of specific marker enzymes such as the Na^+K^+-ATPase for the basolateral membrane and alkaline phosphatase for the apical membrane of epithelial cells. Purification of the membrane is estimated by the increase in the specific activity of marker enzyme activity in the fraction as compared to the crude homogenate. Alternatively, the ion channel activity can be used to assay for purification of the membrane population. After detergent solubilization, reconstitution of a membrane transporter into planar phospholipid bilayers is possible for further characterization of transport properties (Racker, 1985; Schultz *et al.,* 1996). Some reconstituted ion channels are listed in Table IX (Schultz *et al.,* 1996).

6.5. Expression in *Xenopus laevis* Oocytes

Expression of a membrane transporter in *Xenopus laevis* oocytes by microinjection of mRNA was first used for the functional cloning of the rabbit intes-

Table IX.
Purified and Reconstituted Ion Channels[a]

Channel	Examples
Voltage-gated	Na^+ channel from rat brain, muscle, rabbit T-tubular membranes
	Ca^{2+} channel from skeletal muscle transverse tubules
	K^+ channel from squid axon membranes
Ligand-gated	Acetylcholine receptor from *Torpedo california*
	cGMP-dependent cation channel from retinal and bovine rod photoreceptor outer segment
K^+	K^+-ATP from rat liver and beef heart mitochondria and basolateral membranes of *Necturus* enterocytes
	Ca^{2+}-dependent channels from outer renal medulla, rabbit colonocytes, and tracheal smooth muscle
Other	Cl^- channels from kidney and trachea and *Torpedo california* electroplax
	CFTR from bovine tracheal epithelia
	Gap junctions from bovine lens

[a]Adapted from Dubinsky and Otilia (1996).

tine Na^+/glucose cotransporter (Hediger *et al.*, 1987), and thereafter for the functional cloning of numerous other transporters, channels, and receptors. Thus the *Xenopus* oocyte expression system is of great value for the expression cloning of the cDNA that encodes a transporter in the absence of oligonucleotides based on protein sequence data or antibodies to the transporter. The oocyte expression system offers numerous advantages, particularly availability, ease of handling and injecting, and amenability to electrophysiological measurements. One of the most powerful aspects of this technology is the ability to identify ion channels that may be expressed in the cells at very low abundance.

6.6. Cultured Cells

The application of tissue culture techniques to epithelial cells has been reviewed by Wills *et al.* (1996). Tissue culture methods provide a powerful tool for the study of morphologically complex or inaccessible tissues, such as endocrine and exocrine glands, renal tubules, airway epithelia, and ocular epithelial cells. Cell differentiation and other regulatory processes can be examined under precisely controlled conditions. This approach can produce morphologically simple monolayer epithelia that are amenable to a variety of electrophysiological, molecular, and optical methods for studies of epithelial cell function. Primary cell culture is a culture started from cells, tissues, or organs taken directly from an organism (Wills *et al.*, 1996). A continuous cell line is a transferable cell that has an essentially infinite life span. Some of the commonly used epithelial cell lines are

Caco-2 and HT-29 from human colon carcinoma. An established method using a cell monolayer on a Transwell is shown in Fig. 5. Trasient or stable expression systems of transports in HeLa (Kanai *et al.*, 1996; Shi *et al.*, 1995), COS-1 (Fang *et al.*, 1996), and CHO cells (Covitz *et al.*, 1996) can be used for characterization of membrane transporters.

6.7. Electrophysiological Study of Epithelial Transport

Transepithelial ion movements can pass along the cellular pathway (which is composed of at least two barriers, the apical and basolateral membrane) or the paracellular (between the cells) pathway. The purpose of electrophysiological studies of transepithelial transport, such as impedance analysis, is the measurement of the conductance of the different pathways. Impedance analysis is the only noninvasive electrical technique available for determining the role of epithelial structure in ion transport. It is possible to use this method to derive meaningful morphological parameters, such as the resistance and width of lateral intercellular spaces and crypt dimensions. Limitations of impedance analysis include the following: (a) It requires the use of an equivalent circuit model of resistors and capacitors which should be physiologically meaningful; (b) electric circuits often have nonunique solutions, requiring an estimate of at least one additional epithelial parameter using an alternate method; and (c) it is less sensitive with leaky epithelia. Numerous electrical methods have been used to study epithelial transport properties, including transepithelial dc measurement methods and microelectrode techniques. These approaches have generally ignored tissue architecture (Schafer

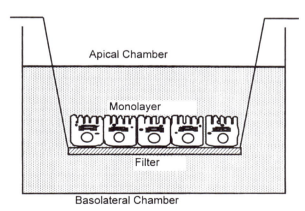

Figure 5. Cell monolayer cultured on Transwell polycarbonate filters. For example, Caco-2 cells are seeded on polycarbonate filters at a density of about 60,000 cells/cm^2. From Hidalgo (1996).

et al., 1994; Wills *et al.*, 1996). Morphometric analyses have been used with success, but this method requires fixation of the tissue. Video analysis is another recent development, but this technique can be cumbersome and, like morphometric analysis, it does not allow direct monitoring of both structural features (such as membrane areas) and membrane conductance.

6.8. Other Commonly Used Experimental Methods

Ion-selective microelectrodes, measurement of intracellular pH, antibodies to membrane proteins, and biomedical NMR spectroscopy are also described in the literature (Schafer *et al.*, 1994; Yudilevich *et al.*, 1991). Knowledge of the primary structures of transport proteins will undoubtedly lead to the development of specific tools to study individual transporters in detail using recombinant DNA techniques (Schafer *et al.*, 1994). Other *in vitro* systems, such as brush border or basolateral membrane vesicles (Biber and Murer, 1991), perfused intestinal loops, stripped intestinal mucosa, and isolated cells (Brochardt *et al.*, 1996; Wills *et al.*, 1996) are used for mucosal drug absorption.

7. SUMMARY

Pharmaceutical scientists increasingly utilize transporters for drug delivery and targeting. The biological barriers to drug delivery can basically be divided into epithelial, endothelial, elimination, and target cell barriers. Membrane transporters play an important role in drug entrance and exit from the body. In addition, it is possible to utilize transporters for drug delivery, e.g., improving oral absorption via the peptide transporter. Identification, a better understanding of their transport characteristics, and the regulation of the membrane transporters will allow the development of better drug delivery strategies.

REFERENCES

Aidley, D. J., and Stanfield, P. R., 1996, *Ion Channels: Molecules in Action,* Cambridge University Press, Cambridge.

Arndt-Jovin, D. J., Robert-Nicoud, M., Kaufman, S. J., and Jovin, T. M., 1985, Fluorescence digital imaging microscopy in cell biology, *Science* **230**:247–256.

Biber, J., and Murer, 1991, Methodological guide for studying epithelial transport with isolated membrane vesicles, in: *Cell Membrane Transport: Experimental Approaches and Methodologies* (D. L. Yudilevich, R. Deves, S. Peran, and Z. I. Cabantchik, eds.), Plenum Press, New York, pp. 163–186.

Borchardt, R. T., Smith, P. L., and Wilson, G., 1996, Models for assessing drug absorption and metabolism, Plenum Press, New York.
Boyde, A., 1985, Stereoscopic images in confocal (tandem scanning) microscopy, *Science* **230**:1270–1272.
Byrne, J. H., and Schultz, S. G., 1994, *An Introduction to Membrane Transport and Bioelectricity,* Raven Press, New York.
Caplan, M. J., 1997, Ion pumps in epithelial cells: Sorting, stabilization, and polarity, *Am. J. Physiol.* **272**:G1304–G1313.
Covitz, K. M., Amidon, G. L., and Sadee, W., 1996, Human dipeptide transporter, hPEPT1, stably transfedted into Chinese hamster ovary cells, *Pharmaceut. Res.* **13**:1631–1634.
Deves, R., 1991, Kinetics of transport: Characterizing the interaction of substrates and inhibitors with carrier systems, in: *Cell Membrane Transport: Experimental Approaches and Methodologies* (D. L. Yudilevich, R. Deves, S. Peran, and Z. I. Cabantchik, eds.), Plenum Press, New York, pp. 3–19.
Dissing, S., and Gasbjerg, P. K., 1994, Basic principles behind the application of fluorescent dyes: The study of cellular electrolyte concentrations, membrane potential and microenvironment, in: *Methods in Membrane and Transporter Research* (J. A. Schafer, G. Giebisch, P. Kristensen, and H. H. Ussing, eds.), Landes, Austin, Texas, pp. 143–176.
Dubinsky, W. P., and Otilia, M.-W., 1996, Methods of reconstitution of ion channels, in: *Molecular Biology of Membrane Transport Disorders* (S. G. Schultz, T. E. Andreoli, A. M. Brown, D. M. Fambrough, J. F. Hoffman, and M. J. Welsh, eds.), Plenum Press, New York, pp. 74–82.
Evans, W. H., and Graham, J. M., 1989, *Membrane Structure and Function,* IRL Press, Oxford.
Fang, X., Parkinson, F. E., Mowles, D. A., Young, J. D., and Cass, C. E., 1996, Functional characterization of a recombinant sodium-dependent nucleoside transporter with selectivity for pyrimidine nucleosides (cNT1rat) by transient expression in cultured mammalian cells, *Biochem. J.* **317**:457–465.
Ganapathy, V., and Leibach, F. H., 1996, Peptide transporters, *Curr. Opin. Nephrol. Hypertens.* **5**:395–400.
Ganapathy, V., Brandsch, M., and Leibach, F. H., 1994, Intestinal transport of amino acids and peptides, in: *Physiology of the Gastrointestinal Tract* (L. R. Johnson, ed.), Raven Press, New York, pp. 1773–1794.
Griffith, J., and Sansom, C., 1998, *The Transporter Factsbook,* Academic Press, San Diego, California.
Hamill, O. P., Marry, A., Neher, E., Sakmann, B., and Sigworth, F. J., 1981, Improved patch-clamp techniques for high-resolution current recording from cells and cell-free membrane patches, *Pflugers Arch.* **391**:85–100.
Hediger, M. A., Coady, M. J., Ikeda, T. S., and Wright, E. M., 1987, Expression cloning and cDNA sequencing of the Na^+/glucose co-transporter, *Nature* **330**:379–381.
Hidalgo, I. J., 1996, Cultured intestinal epithelial cell models, in: *Models for Assessing Drug Absorption and Metabolism* (R. T. Borchardt, P. L. Smith, and G. Wilson, eds.), Plenum Press, New York, pp. 35–50.
Jilling, T., and Kirk, K. L., 1994, Fluorescence digital imaging microscopy in epithelial biology, in: *Methods in Membrane and Transporter Research* (J. A. Schafer, G. Giebisch, P. Kristensen, and H. H. Ussing, eds.), Landes, Austin, Texas, pp. 177–214.

Kanai, N., Lu, R., Bao, Y., Wolkoff, A. W., and Schuster, V. L., 1996, Transient expression of oatp organic anion transporter in mammalian cells: Identification of candidate substrates, *Am. J. Physiol.* **270**:F319–325.

Kaplan, R. S., 1996, Mitochondrial transport processes, in: *Molecular Biology of Membrane Transport Disorders* (S. G. Schultz, T. E. Andreoli, A. M. Brown, D. M. Fambrough, J. F. Hoffman, and M. J. Welsh, eds.), Plenum Press, New York, pp. 277–302.

King, R. B., 1996, Modeling membrane transport, *Adv. Food Nutr. Res.* **40**:243–262.

Lehninger, A. L., 1993, *Principles of Biochemistry,* Worth, New York.

Lodish, H., Baltimore, D., Berk, A., Zipursky, S. L., Matsudaira, P., and Darnell, J., 1995, *Molecular Cell Biology,* Scientific American Books, New York.

Malandro, M. S., and Kilberg, M. S., 1996, Molecular biology of mammalian amino acid transporters, *Ann. Rev. Biochem.* **65**:305–336.

Minta, A., and Tsien, R. Y., 1989, Fluorescent indicators for cytosolic sodium, *J. Biol. Chem.* **264**:19449–19457.

Minta, A., Kao, J. P., and Tsien, R. Y., 1989, Fluorescent indicators for cytosolic calcium based on rhodamine and flourescein chromophores, *J. Biol. Chem.* **264**:8171–8178.

Muth, T. R., Dunbar, L. A., Cortois-Coutry, N., Roush, D. L., and Caplan, M. J., 1997, Sorting and trafficking of ion transport proteins in polarized epithelial cells, *Curr. Opin. Nephrol. Hypertens.* **6**:455–459.

Paddock, S. W., 1991, The laser-scanning confocal microscope in biomedical research, *Proc. Soc. Exp. Biol. Med.* **198**:772–780.

Paddock, S. W., 1996, Further developments of the laser scanning confocal microscope in biomedical research, *Proc. Soc. Exp. Biol. Med.* **213**:24–31.

Petty, H. R., 1993, *Molecular Biology of Membranes: Structure and Fuction,* Plenum Press, New York.

Racker, E., 1985, *Reconstitutions of Transporters, Receptors, and Pathological states,* Academic Press, Orlando, Florida.

Rink, T. J., Tsien, R. Y., and Pozzan, T., 1982, Cytoplasmic pH and free Mg^{2+} in lymphocytes, *J. Cell Biol.* **95**:189–196.

Sakmann, B., and Neher, E., 1984, Patch clamp techniques for studying ionic channels in excitable membranes, *Annu. Rev. Physiol.* **46**:455–472.

Sakmann, B., and Neher, E., 1995, *Single-Channel Recording,* Plenum Press, New York.

Schafer, J. A., Giebisch, G., Kristensen, P., and Ussing, H. H., 1994, *Methods in Membrane and Transporter Research,* Landes, Austin, Texas.

Schultz, S. G., Andreoli, T. E., Brown, A. M., Fambrough, D. M., Hoffman, J. F., and Welsh, M. J., 1996, *Molecular Biology of Membrane Transport Disorders,* Plenum Press, New York.

Shi, X., Bai, S., Ford, A. C., Burk, R. D., Jacquemin, E., Hagenbuch, B., Meier, P. J., and Wolkoff, A. W., 1995, Stable inducible expression of a functional rat liver organic anion transport protein in HeLa cells, *J. Biol. Chem.* **270**:25591–25595.

Stein, W. D., 1986, *Transport and Diffusion across Cell Membranes,* Academic Press, Orlando, Florida.

Stein, W. D., 1990, *Channels, Carriers, and Pumps: An Introduction to Membrane Transport,* Academic Press, San Diego, California.

Stein, W. D., 1997, Kinetics of the multidrug transporter (P-glycoprotein) and its reversal, *Physiol. Rev.* **77**:545–590.

Tsien, R. Y., 1989, Fluorescence ratio imaging of dynamic intracellular signals, *Acta Physiol. Scand. Suppl.* **582**:6.

West, I. C., 1983, *The Biochemistry of Membrane Transporter*, Chapman and Hall, London.

Wheeler, T. J., and Whelan, J. D., 1988, Infinite-cis kinetics support the carrier model for erythrocyte glucose transport, *Biochemistry* **27**:1441–1450.

Williams, D. A., 1990, Quantitative intracellular calcium imaging with laser-scanning confocal microscopy, *Cell Calcium* **11**:589–597.

Wills, N. K., Reuss, L., and Lewis, S. A., 1996, *Epithelial Transport: A Guide to Methods and Experimental Analysis,* Chapman and Hall, London.

Yamamoto, T., Davis, C. G., Brown, M. S., Schneider, W. J., Casey, M. L., Goldstein, J. L., and Russell, D. W., 1984, The human LDL receptor: A cysteine-rich protein with multiple Alu sequences in its mRNA, *Cell* **39**:27–38.

Yu, L. X., Lipka, E., Crison, J. R., and Amidon, G. L., 1996, Transport approaches to the biopharmaceutical design of oral drug delivery systems: Prediction of intestinal absorption, *Adv. Drug Delivery Rev.* **19**:359–376.

Yudilevich, D. L., Deves, R., Peran, S., and Cabantchik, Z. I., 1991, *Cell Membrane Transport: Experimental Approaches and Methodologies,* Plenum Press, New York.

2

Classification of Membrane Transporters

Wolfgang Sadée, Richard C. Graul, and Alan Y. Lee

1. INTRODUCTION

This chapter explores the various approaches to classifying membrane transporters. Even though most transporter proteins share common features, i.e., multiple α-helical transmembrane domains connected via intra- or extracellular loops, this class of proteins consists of many seemingly unrelated families. Often the primary structure of the transporter proteins and their encoding genes are dissimilar. To classify these diverse transporter families, we must consider simultaneously their evolutionary origin, topology, functional domains, substrate specificity, and three-dimensional architecture. Unfortunately, few crystal structures are available for integral membrane proteins, and none yet for α-helical polytopic transporters. Therefore, one needs to infer structure–function relationships from a plethora of experimental and theoretical approaches, each contributing only a small aspect to a full understanding of transporter structure.

A portion of this chapter is devoted to the analysis of the primary structure, i.e., the protein sequence of the transporters, exploring how the available sequence information can serve to deduce both secondary and tertiary structure of the transporters as well as their function. Moreover, in view of recent sequencing of entire genomes, transporters will be evaluated from a genomics point of view. Lastly, this

---

Wolfgang Sadée, Richard C. Graul, and Alan Y. Lee • Departments of Biopharmaceutical Sciences and Pharmaceutical Chemistry, School of Pharmacy, University of California San Francisco, San Francisco, California 94143-0446.

Membrane Transporters as Drug Targets, edited by Amidon and Sadée. Kluwer Academic/Plenum Publishers, New York, 1999.

chapter provides examples of established membrane transporter families gathered on the basis of sequence similarities. Using a bioinformatics approach, commonalities among all transporters become apparent as members of a unique class of proteins, even though their primary structures are extremely diverse. The wealth of information inherent to this approach will prove useful in drug development and targeting, as an integral portion of novel drug discovery.

2. CLASSIFICATION OF TRANSPORTERS BY FUNCTION AND SUBSTRATE SPECIFICITY

Whether solute transporters, ion exchangers, or regulatory proteins, these integral membrane proteins play important roles in nutrient uptake, osmoregulation, electrolyte control, and other fundamental cellular processes (Bell *et al.*, 1990; Kaplan, 1993; Marger and Saier, 1993; Nikaido and Saier, 1992; Sadée *et al.*, 1995; Silverman, 1991; Wright *et al.*, 1992). Many transporters additionally display ion conductances, and therefore could also be considered ion channels. As a result, the distinction between a transporter and an ion channel may not always be clear. In particular, the cystic fibrosis conductance regulator (CFTR), although evolutionarily related to multidrug-resistance (MDR) transporters, is more an ion channel than a transporter (Hasegawa *et al.*, 1992; Riordan *et al.*, 1989), despite its strikingly similar molecular architecture to that of the MDR transporters (Hyde *et al.*, 1990). Similarly, the sulfonylurea receptors are closely related homologs of MDR, yet they represent a regulatory subunit of ATP-dependent K^+ channels (Thomas *et al.*, 1996). These examples illustrate the difficulties in classifying transporters functionally or by structural similarities alone.

Before discussing the molecular structure of transporters, we need to summarize briefly the mechanisms by which transporters move solutes across lipid membranes. Translocation of polar solutes across biological membranes requires specialized transporters or channels (pores), often with binding sites accessible to solutes from either side of the membrane. For instance, bacterial porins translocate hydrophilic molecules into the periplasmic space utilizing large, water-filled channels (Nikaido and Saier, 1992). This allows them to passively convey solutes across the membrane with moderate selectivity. Porins are abundant proteins in the outer membrane of gram-negative bacteria, consisting of multiple transmembrane β-strands in the shape of a β-barrel (Jeanteur *et al.*, 1991). In contrast, transporters are thought to consist largely of multiple α-helices serving as transmembrane domains (TMDs) and require a conformational change during the process of solute translocation across the membrane. The rather more selective solute-binding site of a transporter is usually accessible only on one side of the membrane at any time. Passage of solutes through transporters is relatively slow because of the need for conformational changes of the transporter protein during solute translocation (see Sadée et al., 1995, for a review).

Transporters can be functionally classified on the basis of their energy requirements. Passive transporters of a solute across the membrane (uniporters) facilitate net solute flux down an electrochemical concentration gradient (facilitative diffusion), whereas active transporters move solutes against an electrochemical gradient across the membrane. To provide energy for active transport, solute transport must be coupled to a process yielding free energy. A number of ion pumps are primary active transporters, coupled to a chemical or photochemical reaction. For example, bacteriorhodopsin serves as a proton pump by utilizing the energy derived from light activation of the covalently bound retinal. On the other hand, Na^+/K^+-ATPase exploits chemical energy released upon ATP hydrolysis by its ATP-binding cassette (ABC). Ion pumps are often electrogenic (a net movement of charge resulting from the primary pump operation), thereby affecting the membrane potential.

Cellular uptake of numerous nutrients and drugs depends upon secondary active transporters which utilize voltage and ion gradients generated by primary active transporters/exchangers. Collectively known as cotransporters, symporters translocate two or more different solutes in the same direction, whereas antiporters couple the transport of solutes in opposite directions, exploiting chemical gradients, mostly of ions such as Na^+ and H^+ (Tse et al., 1991; Sardet et al., 1989). Prokaryotes often express H^+ symporters, whereas in eukaryotic cells Na^+ symporters are prevalent (Dimroth, 1991; Kaplan, 1993), but each type occurs in all organisms.

One can also classify transporters by their substrates rather than by their translocation mechanism. The category of glucose transporters encompasses both secondary active Na^+-glucose cotransporters present in the intestines (Hediger et al., 1987) and facilitative glucose transporters (GLUT1–5) expressed throughout the body (James et al., 1989; Bell et al., 1990). However, these two classes of glucose transporters have rather dissimilar primary structures, and consequently they are considered to represent distinct and independent transporter families with separate evolutionary origins. Thus, limiting the approach to classification by solute transport alone may conflict with mechanistic classification or with that based on primary structure.

In contrast to the glucose transporters, where distinct primary structures mediate transport of the same substrate, closely related members of a single transporter family may also recognize distinct substrates, and single point mutations can transform a secondary active cotransporter into a facilitative transporter (Nikaido and Saier, 1992). Hence, distinct substrate selectivities and mechanisms have evolved within the same transporter family. For primary active transporters, such as the multiple drug resistance transporters, ATP-binding cassettes have been fused to the transmembrane portions of a transporter. This facilitates the transfer of energy released by their ATPase activity for substrate translocation, regardless of the nature of the substrate. These considerations illustrate that transporter classification must reflect several aspects, including substrate specificity, translocation mechanism, and primary structure. Only a comprehensive view of all these aspects together provides insights into the structure and function of transmembrane transporters.

3. A GENOMICS VIEW OF TRANSPORTERS

Advances in sequencing technology have enabled a decisive step in our biological understanding of entire organisms by expanding the availability of the complete genomic sequence among a growing number of organisms (Clayton et al., 1997). The first complete genomes available in public databases include the following: *Haemophilus influenzae* [first available sequence by Fleischmann et al. (1995)], *Mycoplasma genitalium, Methanococcus jannashii, Synechocystis* sp. PCC6803, *Mycoplasma pneumoniae, Saccharomyces cerevisiae,* and *Escherichia coli*. The impact of this enormous burst of information is apparent in all fields of the biosciences. This includes profound changes in our understanding of disease and a revolution in the way new drugs are discovered and developed. In the transporter field, molecular biology, genomics, and bioinformatics have also begun to impact current research directions. Chapters in this book clearly attest to the significant advances in our understanding of the biology of transporters. We will discuss here implications of these new disciplines for research on drug transporters.

Selection of drug targets is facilitated by the availability of complete genomic sequences. For example, *Helicobacter pylori* (Tomb et al., 1997) is a pathogenic agent associated with ulcers, and its genomic sequence can now be searched for the most sensitive and selective drug targets for chemotherapy. To survive the acid environment of the stomach, *H. pylori* expresses a Ni^{2+}-dependent urease, thereby generating NH_3, which buffers the acidic pH prevalent in the stomach. To acquire the needed Ni^{2+}, the organism expresses a specialized Ni^{2+} transporter; conceivably, such a transporter could serve as a therapeutic target for selective destruction of *H. pylori*. With more and more genomic sequences available for numerous organisms relevant to human and animal diseases, a systematic rethinking of potential therapeutic targets is in order. Transporters could play a significant role as selective targets in the treatment of infectious and parasitic diseases. To illustrate, autotrophs (e.g., *Methanococcus jannashii* and *Synechocystis*) primarily require transporters regulating ionic homeostasis, whereas the heterotrophs additionally must rely on the import of nutrient substrates they cannot synthesize independently. This distinction could possibly be exploited in the treatment of infectious and parasitic diseases if one considers that the transport capacity of a heterotroph determines its overall metabolic potential. For example, inhibition of glucose transporters, such as TH11—TRYBB [vaguely similar in primary structure to dipeptide transporters (Graul and Sadée, 1997b)] could provide a successful strategy in the treatment of parasitic trypanosomal diseases.

The recent availability of genomic sequences has assisted scientists to acquire a more accurate perspective of the diversity and the quantity of transporters which coexist in a single species. An excellent overview of transporters in five diverse organisms with known genomic sequence is provided in Fig. 1 (taken from Clayton et al., 1997). While the schematics in Fig. 1 represents only a selection of expressed transporters, all functional classes of transporters are represented, with four broad substrate categories: (1) amino acids, peptides, and amines; (2) carbohydrates, organic alcohols, and acids; (3) cations; (4) anions. By searching through entire genomes, one can begin to estimate the total number of transporters contained in a given organism. The first eukaryotic genome to be fully sequenced, the yeast genome, contains ~6000 putative genes, over half of which with unknown functions (Clayton et al., 1997). This was published in a special issue of *Nature* (Yeast Genome Directory, 1997). It is now possible to estimate the approximate number of transporter genes from the available information, relying on sequence comparisons to known transporters and hydropathy analysis to establish membrane topology of the predicted proteins. Nearly ½% of all genes are represented by ABC transporters alone (29 sequences), and a similar fraction by the related multidrug-resistance transport proteins from the major facilitator superfamily (MFS-mdr, 28 sequences) (Clayton et al., 1997). Thus, yeast could be considered a model against which other organisms can be compared (Botstein et al., 1997). Transporters could account for at least 4–5% of the proteome (the totality of all expressed proteins).

If one extrapolates this estimate of transporters in yeast to the human genome, one would conservatively estimate a total number of *at least* 2000 transporters/ion exchangers. Clearly, only a fraction of these have been cloned and fully sequenced. Hence, research on transporters as drug targets per se, or as vehicles to target drugs to their active sites, has barely scratched the surface. Several general directions of future advances in this area are already apparent. Above all, one needs to clone and characterize all human transporter/ion exchanger genes. This will be achieved at a relatively rapid pace, even before the entire human genome is fully sequenced, because of the availability of EST (expressed sequence tags) databases. By rapid and partial sequencing of mRNAs expressed in various tissues and organisms, one obtains these sequence tags in very large numbers (Boguski, 1995). Whereas the common nonredundant protein databases contain ~300,000 cloned sequences of known primary structure, the publicly accessible databases contain over 2 million ESTs from all organisms, and this number is rapidly increasing. Commercial databases exceed this number significantly. The available ESTs may already cover more than 90% of all human genes.

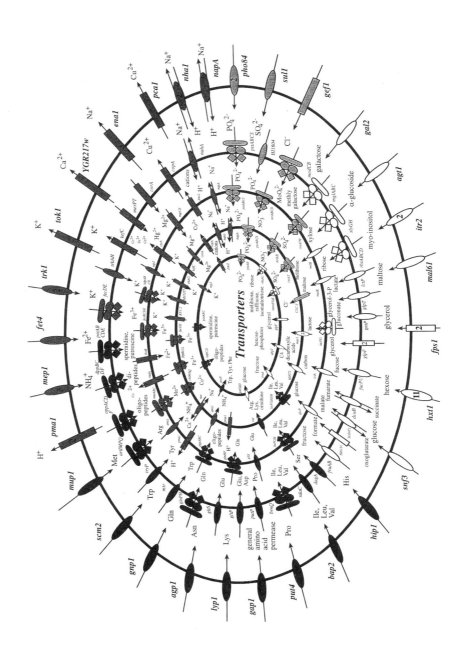

4. PHARMACOGENOMICS OF TRANSPORTERS

Pharmacogenomics is an emerging discipline which may lead to many practical applications to improve drug therapy using transporters. We can anticipate that all human transporter genes will be cloned and available in the not too distant future. To exploit this vast information while it is becoming available, we propose a broad collaborative effort among many laboratories for the genotyping and phenotyping of human transporter genes using microarray DNA analysis. Employing techniques using a small microarray chip capable of accommodating up to 400,000 DNA oligomer probes (Lockart et al., 1996) and other methods of large scale monitoring of gene expression (DeRisi et al., 1997; Schera et al., 1995; Velculesco et al., 1995; Zhang et al., 1997), it will be feasible to determine allelic distribution and transporter gene expression on an individual basis. Furthermore, the methodology for determining the proteome of a tissue (all expressed proteins) is also advancing (Ostergaard et al., 1997). This technology could be utilized to characterize transporter gene defects that could contribute to human diseases such as diabetes. In addition, it could help identify genes along with their alleles relevant to drug transport and unique expression of transporters for drug targeting to their site of action, e.g., cancer tissues.

Current knowledge on all genes responsible for drug transport, not to mention any alleles that might be defective in drug transport, is still limited. For example, the dipeptide transporter PEPT1 may contribute significantly to the oral absorption of peptoid drugs such as cephalosporins and ACE inhibitors. We anticipate that a portion of the millions of patients treated with these drugs have mutant PEPT1 alleles with defective drug transport. This could lead to failure of drug therapy against infectious diseases in this population, leading to adverse effects on disease outcome even if the microbial strain is sensitive to the antibiotic. Systematic application of microarray gene technology could provide the necessary information to identify such putative defective alleles and allow us to exclude affected patients from oral therapy with the respective drugs. Dedicated chips for genotyping of cytochrome P450 subtypes and their alleles are already available (Iarovici, 1997; Lin et al., 1996), but no efforts are currently underway to routinely measure transporter alleles. This newly emerging area, called pharmacogenetics,

---

Figure 1. Survey of membrane transporters identified from the complete sequences of five genomes: *Mycoplasma genitalium, Methanococcus jannashii, Synechocystis* PCC6803, *Hemophilus influenzae,* and *Sacharomyces cerevisiae.* Transporters are organized into the main substrate categories: amino acids, peptides, amines, carbohydrates, organic alcohols and acids, cations, and anions. Substrates and cosubstrates are indicated for each transporter. Shapes indicate functional categories. Ovals: ion-coupled permeases; circle, diamond, and oval clusters: active ABC transporters; rectangles: all other transporters. Arrows identify the transport direction, either efflux or influx. Reprinted with permission from Clayton *et al.* (1997). Copyright 1997 Macmillan Magazines Limited.

or, more broadly, pharmacogenomics, could have dramatic impact on the way we utilize existing drugs and screen for new therapeutic agents.

Lastly, we could consider transporter gene expression in specialized target tissues, including cancers. For example, we have found that the PEPT1 transporter, normally expressed in intestinal tissues, is highly expressed in several pancreatic carcinoma cell lines (Gonzales *et al.,* 1998). This could lead to a successful strategy for targeting antineoplastic agents, possibly as prodrugs that are recognized by PEPT1, to the cancer tissue. Seen more broadly, a complete analysis of all transporter genes expressed on cancer cells, including the multidrug-resistance transporters, would provide us with strong guidance in the selection of cancer chemotherapy.

Consequently, the sequencing of entire genomes coupled with the knowledge of the complete repertoire of transporters of an entire species will stimulate novel directions in drug discovery and development. Determining allelic distributions and phenotyping selected target tissues for drug transporters represent essential elements of pharmacogenomics, a new field with great potential for improving drug therapy. For the pharmaceutical industry and for the academic research laboratory, the intensive application of genomics and bioinformatics to the transporter field is compelling.

5. EVOLUTION OF MEMBRANE TRANSPORTERS; SEQUENCE ANALYSIS

Even though most transporters share similar functions and membrane topology, primary structures are extremely diverse. Thus, common sequence alignment algorithms fail to reveal significant similarities indicative of possible homology, i.e., common evolutionary ancestry. Yet, low-level similarities occur among many transporters from different gene families. Several contending hypotheses can account for these findings. One theory suggests that the various transporter families have emerged from distinct ancestral genes, and their similar topology and function drives convergence to yield similar sequences. Convergence certainly can be a strong force if one considers the physical constraints imposed on polytopic α-helical membrane proteins, including hydrophobicity and hydrophobic moment, which impart α-helical periodicity on the primary structure. However, many variations in membrane topology could lead to functional transporters, and the predominant structure with approximately 12 TMDs suggests additional commonalities not readily accounted for by convergence. Another school of thought states the opposite, believing that most transporter families derive from a common precursor, but that extensive and rapid mutational drift has led to such sequence divergence that evolutionary relationships are no longer detectable. However, primary structures are highly conserved within a transporter family, and alignments between sequences within the same transporter family, but from extremely distant organisms such as mammals and bacteria, often yield extraordinary scores consistent with a finding of certain homology. Therefore, once a gene encodes a functional transporter protein, it is likely to be conserved over a long period of time. This does not preclude the possibility that mutational drift has occurred before this stabilization into a functional structure, or that other mutational events impose drastic changes in primary structure. The possibility that a modular gene structure could have contributed to transporter diversity will be discussed below. Still another hypothesis suggests that both convergence and divergence could have played a role in generating the diversity of transporters. One might suspect that more transporter families are related to each other than is presently known, and further analysis will reveal these relationships.

In general, sequence alignments yield better statistical results with increasing sequence length, and therefore the rather large transporter proteins should readily yield to such analysis. However, there is a principal problem inherent to all polytopic membrane proteins which has not yet been explicitly discussed or resolved in the literature. This problem derives from the repetitive nature of the transmembrane topology, generated through concatenation of individual TMDs and loops. Hence, transporter proteins are highly modular, and each module is subject to similar physicochemical constraints imposed by the α-helical structure embedded in the lipid bilayer and the loops in the aqueous environment. This results in repeat sequence modules with restricted amino acid distributions, and in a low, but detectable level of sequence similarity among each TMD–loop module. Consequently, sequence similarities generated by this restricted amino acid distribution accumulate with increasing number of TMD–loop modules in the same structure. Moreover, the threshold that distinguishes between convergent and divergent evolution increases with more TMDs present because alignments scores expected on the basis of restricted amino acid sequences alone increases with increasing number of TMD modules in the proteins.

To account for the modular nature of the transporter protein structure, we have analyzed the evolution of the rather small H^+–dipeptide cotransporter family by comparing each TMD–loop module separately among several related proteins (Graul and Sadée, 1997b). The results suggested that in the process of evolution, the order of the alignable TMDs has changed either by insertions of new TMD–loop modules or by deletion or intragenic duplication. This further suggests that not only is the protein structure modular, but so might be the encoding gene. Hence, transporter genes in general might consist of modules each encoding one or more TMD–loop segments. A modular gene structure could have facilitated deletions, insertions, duplications, and unequal crossovers among different transporter genes as possible mechanisms that can account for the perplexing diversity of the many transporter families. We are now in the process of reexamining relationships among transporter families by developing approaches to sequence analysis that reflect the modular nature of the transporter proteins.

Meanwhile, with a rapidly growing database of cloned sequences, one would expect that conventional analysis can reveal relationships among distant transporter families simply because the missing links between them have been sequenced recently. Thus, relationships among many bacterial transporter families were revised by analyzing much larger databases that have permitted homologies to be established among transporters previously thought to be unrelated (Marger and Saier, 1993). As a standardized approach that can rapidly summarize all information in the sequence databases and reliably detect new structures as possible links, we routinely apply BLAST (Basic Local Alignment Search Tool) (Altschul *et al.*, 1990). To compensate for the limited information obtained with only a single starter sequence and a single BLAST run, we have devised a way of expanding the possibilities of a single BLAST sequence search. By iteratively performing BLAST beginning with a single protein starter sequence, e.g., a transporter, and then on each of its identified neighbors, we compile a list of neighbors and their respective next neighbors (Fig. 2) (Graul and Sadée, 1997a). Called INCA (Iterative Neighborhood Cluster Analysis), this Java program is accessible at http://itsa.ucsf.edu/~gram/home/inca.html. By allowing for several iterations, performing BLAST on each sequence identified with a preselected threshold score, the program converges by identifying all sequence neighbors in the database that are related to at least one other sequence in the neighborhood cluster by the minimum score. Commonly, we use a probability value $P \leq 10^{-6}$ as the cutoff score for inclusion with the cluster, meaning an alignment is likely to occur by chance (rather than because of evolutionary relationship) with a frequency of only one in

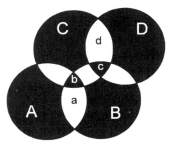

Figure 2. Schematics of INCA (Iterative Neighborhood Cluster Analysis). A single starter sequence serves to identify all related sequences in its neighborhood, A, in the first iteration. Each of the sequences in A serves as a starter of a second round of BLAST analyses, thereby identifying neighborhoods B and C, through linking sequences a and b. A third iteration of BLAST analyses on the newly identified members of B and C then finds neighborhood D, through links c and d. If no further neighborhoods are identified, INCA converges on a finite number of sequences that all belong to the overall core cluster. Selecting the appropriate stringency (we commonly set a P value of 10^{-6}) permits one to limit the search to families that are likely homologues. Sequences scoring above the set P value (representing possible homologues), are listed separately, ordered with the best matching sequences in the core cluster. Thus, one can readily identify the sequences serving as the optimal links among distant families. This is particularly useful for searching the EST databases, in a subsequent passage, after an initial run of the nonredundant protein databases, to identify possible new subtypes of a given transporter family. Redrawn from Graul and Sadée (1997a).

one million within the database analyzed. This gives us a reasonable indication that the proteins are homologous, while allowing for considerable sequence diversity. Sequences scoring with P values $>10^{-6}$ are also listed with the member of the core cluster to which they are most closely related, but they are not included in the cluster for iterative analysis.

INCA can also be applied to the EST databases. Given the diversity of sequences already available through commercial databases, identifying putative transporter genes in the EST database can be rewarding, as demonstrated by the rapid cloning of homologous genes encoding members of the multidrug-resistance-related protein (MRP) family (Kool et al., 1997). To find yet uncloned homologues one searches with several iterations the common nonredundant protein database to identify and sort all cloned and fully sequenced homologues of a starter sequence. Then, in a second passage, each putative homologue is used to search the EST database for yet uncloned homologues. The results are compiled in a list which provides the highest scoring ESTs for each cloned sequence. In our experience, this process reveals ESTs as likely homologues that would have been missed with a simple BLAST analysis. In particular, searching the EST database not only with the mammalian homologues of the starter sequence, but also with homologues from distant phyla, such as fungi, plants, and bacteria, identifies promising ESTs as possible new members of a mammalian gene family. This implies that gene duplications have occurred early in evolution, before speciation, and therefore sequences in distant phyla may be better suited for identifying members of a human gene family than a human starter sequence.

In the following, we have applied INCA to several select transporter families with a small number of member sequences and few if any known relationships to other families. Once the gene family is large, INCA can mushroom into an unwieldy document, and we are currently modifying our approach to permit INCA analysis of large families as well. Meanwhile, we have also included here a regular single BLAST analysis of one very large gene family, the multidrug-resistance transporters. The provided lists are then reduced to include only a few representative transporters, to illustrate the functional diversity and species distributions of a transporter family, and possible links to other distant families.

6. SELECTED EXAMPLES OF MEMBRANE TRANSPORTERS

This section gives a brief overview of representative membrane transporter families. This is by no means a comprehensive description of membrane transporter families, since new genes are constantly being sequenced. It merely serves to illustrate membrane transporter families as defined on the basis of sequence similarities. In particular, the alignment results juxtapose related sequences from distant taxa, such as bacteria, yeast, plants, and vertebrae. Often, detailed information is already available on some members of a transporter family, usually in bacteria, because of the ready accessibility of the protein. This information can thus be extrapolated to other family members in distant species.

6.1. H^+/Dipeptide Symporters

Oral absorption of certain drugs, such as cephalosporins and angiotensin-converting enzyme (ACE) inhibitors, appears to rely on a saturable carrier identified as a H^+/dipeptide symporter. Because these drug substrates belong to clinically important antibiotics and antihypertensives, intense efforts have focused on characterizing the responsible transporter(s). With the cloning of PEPT1 (Fei et al., 1994; Liang et al., 1995), the molecular characteristics of the intestinal transporter became available, and these will be described in the chapter by Invian and Terada. Using PEPT1 as the starter sequence, an entire family of dipeptide transporters from several species was identified (Table I). In mammals, there are at least two additional homologues, the renal PEPT2 and a rat histine/dipeptide transporter mainly expressed in the brain. The latter was identified by searching the EST databases for yet uncloned homologues of PEPT1, providing one of a growing number of examples where ESTs were used in this fashion (Yamashita et al., 1997).

The PEPT1 gene family includes further members in plants, fungi, and bac-

Table I.
The H^+/Dipeptide Symporter Family[a]

Core cluster: 44 neighbors with $P \leq 10^{-6}$	P
1 01172435 OLIGOPEPTIDE TRANSPORTER, SMALL INTEST., PET1—HUMAN	0.0
9 01172436 OLIGOPEPTIDE TRANSPORTER, KIDNEY ISOFORM, PET2—RABIT	6.4e-198
15 02506043 pH-sensing regulatory factor of peptide transporter [Homo sapiens]	2.9e-109
18 01172704 PEPTIDE TRANSPORTER PT2B—ARATH	6.9e-45
28 00544018 NITRATE/CHLORATE TRANSPORTER, CHL1—ARATH	3.9e-24
35 01172741 PEPTIDE TRANSPORTER, PTR2—CANAL	3.1e-18
42 00544192 DI-/TRIPEPTIDE TRANSPORTER, DTPT—LACLA	6.6e-06
Outside cluster	
42.40 02612908 hexuronate transporter-like protein [Bacillus subtilis]	0.0083

[a]Using PET1—HUMAN as the starter sequence, iterative BLAST (INCA, two iterations) (Graul and Sadée, 1997a) was performed on the nonredundant protein database. After the second iteration, 44 neighbors were identified within the core cluster, each sequence linked with a P value of 10^{-6}. Many more sequences were listed outside the cluster (P value above 10^{-6}). Only select examples are shown in the list. The sequences are listed in the order of their similarity to the starter sequence. In the second iteration, each additional neighbor is listed with the nearest neighbor obtained in the first iteration. Thus 42.40 (hexuronate transporter) indicates that this sequence matched best sequence 42 (DTPT—LACLA). This establishes the best links between distantly related sequences.

teria; however, the overall number of homologous sequences is limited, as determined for several other transporter gene families (Graul and Sadée, 1997b). Specifically, common-sequence search algorithms failed to yield any significant similarity to any other protein in the databases, including all other transporters. The seeming uniqueness of the dipeptide transporters is surprising because their membrane topology, deduced secondary structure, and hydropathy profiles are similar to those of other transporters, also with a proposed structure of 12 TMDs. By analyzing the transporter proteins in a modular fashion, based on analysis of the TMD–loop structures as separate sequence modules, we demonstrated that rearrangements, insertions, deletions, or duplications could have occurred in the process of evolution of the dipeptide transporter family (Graul and Sadée, 1997b). Moreover, good alignments were observed between individual TMD–loop modules of this family and those of other transporter families. However, these results failed to establish an evolutionary link to other transporter families, as the observed sequence similarities also could have occurred by convergence.

To address this issue further, we performed an INCA analysis (iterative BLAST analysis) on the nonredundant database, with PEPT1 as the starter sequence. We reasoned that if evolutionary relationships exist to other transporter families, these might be revealed by newly cloned transporter sequences that could provide the missing link, i.e., although their sequences have diverged considerably, the remaining sequence similarity would be sufficient for a finding of possible or even probable homology. Indeed, in a recent INCA run with multiple iterations to identify all neighbors, using a P value of 10^{-6} as the cutoff for inclusion with the core cluster, one transporter sequence outside the cluster was identified as a possible homolog (Table I). This was a hexuronate-like transporter protein from *Bacillus subtilis*, with a P value of 0.0083, with the nearest cluster member being the dipeptide transporter DTPT—LACLA. While this P value did not meet our criteria for inclusion with the core cluster, it was sufficiently high to warrant further investigation. The highest BLAST score for a local alignment reached 80 (BLOSSUM62 matrix), normally considered indicative of probable homology. However, care needs to be taken when comparing integral membrane proteins with restricted amino acid distribution, and it is currently not possible to answer the question of homology on the basis of such a result. Therefore, additional criteria need to be developed.

In favor of possible homology between the hexuronate transporter and DTPT is the finding that the best local alignments occur in the same regions of the primary structure; thus, one does not have to postulate rearrangements in the process of evolution because alignments between TMD/loops occur in the same order they appear in the primary sequence. To address this further, we performed a multiple alignment (ClustalW) among the hexuronate transporter, DTPT, and PEPT1. The resultant three-way alignment, shown in Fig. 3 in SeqVu format, reveals several domains that are well conserved between the hexuronate transporter and DTPT; moreover,

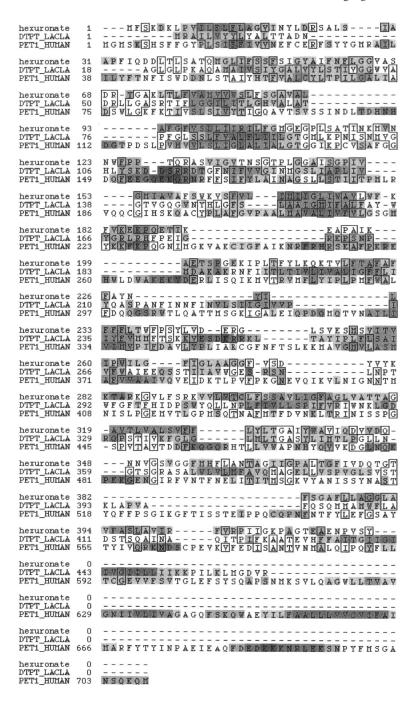

the same regions also show overlapping identities with PEPT1. This latter finding would further support a finding of possible homology among all three sequences analyzed. Some of the sequence variations observed in Fig. 3 may have resulted from an insertion/deletion of two TMD modules in the process of evolution of DTPT and PET1 and the exceedingly long extracellular loop between TMDs 9 and 10 present only in PEPT1 (Graul and Sadée, 1997b). Such rearrangements may prevent identification of the best alignments with strategies based on the entire primary structure. To illustrate the possible location of TMDs in the alignment shown in Fig. 3, we have highlighted polar and lipophilic regions of the three transporter proteins. One can see the congruence between hydropathy and sequence conservation particularly in the N-terminal portion, whereas more divergence occurs in the C-terminal portion, possibly because of the insertions/deletions.

Overall, our results suggest the possibility that the dipeptide transporters may be related to the hexuronate-like transporter of *B. subtilis*. We have also performed another iterative BLAST analysis, PSI-BLAST (position-specific iterated BLAST; go to http://www.ncbi.nlm.nih.gov/cgi-bin/BLAST/nph-psi—blast) (Altschul *et al.,* 1997), with DTPT—LACLA as the starter sequence. This approach revealed sequence similarities to the human erythroid differentiation factor and the bile acid transporter BAIG—EUBSP (*Eubacterium*). Any posssible relationships implied by such similarities require further analysis; however, it is possible that evolutionary relationships exist among dipeptide transporters and other transporter families.

What is the relevance of these findings for understanding the dipeptide transporter family? If one assumes that protein structure has been more highly conserved than primary sequence, a finding of homology would permit one to apply structural and functional results obtained with other transporter families to members of the dipeptide transporter family, or at least use such results as a basis for direct comparisons among related structures. Therefore, we can ask what are the immediate homologues of the hexuronate transporter. These turn out to include several extremely well studied transporters, including the facilitative glucose transporters which will be discussed subsequently. Yet, there are two caveats: first, our alignment does not prove homology between DTPT and the hexuronate transporter, and second, among polytopic membrane proteins we are not certain that indeed structure has been conserved more highly than sequence. Thus it is possible that the homologues DTPT and PEPT1, both clearly related and members of the dipeptide transporter family as judged by sequence similarities (Graul and Sadée, 1997b), fold differently in their tertiary structure even though their sequences are quite similar.

←

Figure 3. Multiple alignment of PET1—HUMAN, DTPT—LACLA, and the hexuronate transporter-like protein from *B. subtilis*. The alignment was performed using CLUSTALW. Identities are boxed, and hydropathy is indicated by shading, using SeqVu. From Graul and Sadée (1997b).

6.2. Facilitative Glucose Transporters and Related Sequences

Efficient glucose utilization, essential to all tissues, depends on a family of facilitative glucose transporters (Kayano et al., 1988), including one of the earliest cloned transporter genes, GTR1 (GLUT1) (Fukumoto et al., 1988; Mueckler et al., 1985). Of interest is the regulation of some of these transporters, in particular the insulin-dependent subtype GTR4 (GLUT4); upon insulin stimulation, muscle and fat cells translocate GTR4 to the plasma membrane to enable glucose uptake and utilization (Yang and Holman, 1993). Expression and regulation of GTR4 has also been studied in transgenic mice (Liu et al., 1992), a general approach increasingly used for the study of transporters. The GTR4 gene is a candidate for certain postreceptor defects in non-insulin-dependent diabetes mellitus (Buse et al., 1992; Kusari et al., 1991). This represents one of the best-studied examples of transporter regulation by cellular trafficking.

The glucose transporters (GTRs) are members of a large family of transporters with diverse functions. Using GTR1 as the starter sequence in an iterative BLAST (INCA) analysis, we found 416 neighbors (P value 10^{-6} or lower) already in the second iteration. As shown in Table II, this cluster of transporters includes members of the major facilitator family, also known as the drug-resistance translocase family of bacteria. A typical example of this family is the bicyclomycin-resistance protein BCR of *E. coli* (Bentley et al., 1993), which mediates resistance against sulfonamides, such as sulfathiazole. The NCBI Entrez database shows 200 protein neighbors for BCR (precalculated on the basis of standardized BLAST to determine significant sequence similarities). The Entrez program is extremely useful for retrieving relevant information on sequence, structure, and function of protein sequences (http://www.ncbi.nlm.nih.gov/Entrez/). Table II reveals additional relationships to pH-dependent nonspecific organic cation transporters, another family with multiple neighbors (Entrez lists 136 sequences, including renal anion transporters). The organic cation transporters affect the disposition of many clinically important drugs (Tamai et al., 1997) and are discussed in the chapter by Dresser et al.

Also included with Table II is the hexuronate transporter-like protein of *B. subtilis*, mentioned earlier as a possible evolutionary link to the dipeptide transporters. This sequence was identified as a neighbor of sequence 64 (a probable transporter of *B. subtilis*) in the second iteration with GTR1 as the starter sequence. This reveals a key feature of our INCA approach: by performing iterative BLAST analyses on each cluster member, we can find those sequences that are most closely related to each other, thereby establishing links with the highest possible similarity scores among a string of sequences. This enhances the likelihood for establishing a finding of homology among even distantly related proteins.

The diverse transporters listed in Table II illustrate evolutionary links among

Table II.
The Facilitative Glucose Transporter Family[a]

Core cluster: 416 neighbors		P
1 00121750	GLUCOSE TRANSPORT. TYPE 1, ERYTHROC., GTR1—BOVIN	0.0
34 00232207	GLUCOSE TRANSPORT. TYPE 7, HEPATIC MICROS., GTR7—RAT	2.4e-140
46 02136979	FRUCTOSE TRANSPORT PROTEIN GLUT5, small intestine, rabbit	6.3e-72
55.173 00730883	SYNAPTIC VESICLE PROTEIN 2, SYV2—RAT	1.0e-13
58 00139841	XYLOSE-PROTON SYMPORT, XYLE—ECOLI	2.8e-46
59.196 00282525	CHLORAMPHENICOL RESISTANCE PROTEIN, *Rhodococcus fascians*	3.8e-09
64 01894771	Product highly similar to metabolite transport proteins, *Bacillus subtilis*	6.6e-40
64.235 01206033	DICARBOXYLIC ACID TRANSPORTER, *Pseudomonas putida*	4.1e-09
64.249 00128511	QUINOLONE RESISTANCE NORA PROTEIN, NORA—STAAU	1.3e-08
64.257 02612908	HEXURONATE TRANSPORTER-LIKE PROTEIN, *Bacillis subtilis*	3.5e-08
108 01708578	MYO-INOSITOL TRANSPORTER 1, TR1—YEAST	3.1e-29
110 00121288	GLUCOSE FACILITATED DIFFUSION PROTEIN, GLF—ZYMMO	3.8e-29
113 01169822	GALACTOSE TRANSPORTER, GAL2—YEAST	9.7e-29
129.173 00584831	BICYCLOMYCIN RESISTANCE PROTEIN, BCR—ECOLI	2.2e-13
129.194 01170801	LINCOMYCIN RESISTANCE PROTEIN, LMRA—STRLN	1.8e-11
134.184 00116481	CITRATE-PROTON SYMPOR, CIT1—SALTY	9.3e-07
144.209 00728970	MULTI-DRUG RESISTANCE PROTEIN 2, BMR2—BACSU	2.9e-08
159.182 02828202	BILE ACID TRANSPORTER, BAIG—EUBSP	1.3e-10
176 00312958	TRYPANOSOME HEXOSE TRANSPORTER, *Trypanosoma brucei*	4.8e-14
189.4 02605501	POLYSPECIFIC ORGANIC CATION TRANSPORTER, *Homo sapiens*	7.5e-48
198.63 01346710	INORGANIC PHOSPHATE TRANSPORTE, PH84—YEAST	2e-14

[a]GTR1—BOVINE served as the starter sequence for two iterations using INCA (see footnote, Table I; same P values used); 516 neighbors with P values above 10^{-6} are not listed.

transporters with distinct translocation mechanisms and substrate specificities from extremely diverse organisms. This establishes a high degree of sequence conservation among transporter proteins belonging to the same superfamily, whereas the functions of individual members may have changed during evolution.

6.3. Sodium/Glucose and Sodium/Nucleoside Cotransporters

This family of secondary active transporters is more narrowly defined, with few if any links to other transporter families. A representative list of family members is shown in Table III, derived from exhaustive iterative BLAST analysis. In addition to nucleoside (e.g., Pajor and Wright, 1992) and glucose transporters, we find iodide and proline transporters (Hediger *et al.*, 1989), each driven by a sodium gradient. Again, these related sequences are distributed over distant species in different phyla, including the Archaea (sodium/proline symporter of *Methanobacterium thermoaquaticum*). The high-affinity sodium/glucose cotransporter NAG1

Table III.
Sodium/Glucose and Nucleoside Cotransporters[a]

Core cluster: 101 neighbors			P
1	00127827	SODIUM/NUCLEOSIDE COTRANSPORTER (NANU—RABIT)	0.0
2	00400337	SODIUM/GLUCOSE COTRANSP. 2, low affinity (NAG2—HUMAN)	0.0
3	00445299	SODIUM GLUCOSE COTRANSPORTER (*Homo sapiens*)	0.0
5	01709218	SODIUM/GLUCOSE COTRANSPORTER 1 (NAG1—RAT)	3.3e-257
22.2	00127803	SODIUM/GLUCOSE COTRANSP. 1, high affinity (NAG1—HUMAN)	9.7e-196
23	02347048	SODIUM MYO-INOSITOL TRANSPORTER (*Rattus norvegicus*)	3.9e-137
30	02136179	SODIUM IODIDE SYMPORTER (*Homo sapiens*)	1.7e-12
35	02622992	SODIUM/proline symporter (*Methanobacterium thermoautotrophicum*)	7.7e-09
35.3	02649519	PANTOTHENATE PERMEASE (*Archaeoglobus fulgidus*)	2.1e-40
38	01172009	SODIUM/PANTOTHENATE SYMPORTER (PANF—HAEIN)	1.7e-07
39	02650635	PROLINE PERMEASE (putP-1), (*Archaeoglobus fulgidus*)	4.4e-07
39.2	02649629	PROLINE PERMEASE (putP-2), (*Archaeoglobus fulgidus*)	1.6e-29
53.23	01172774	SODIUM/PROLINE SYMPORTER (PUTP—HAEIN)	7.2e-198
P value above 10^{-6}:			
23.84	00121751	GLUCOSE TRANSPORTER TYPE 1, ERYTHROC. (GTR1—HUMAN)	0.073
35.50.12.30	00400622	SODIUM- AND CHLORIDE-DEP. GABA TR.1 (NTG1—MOUSE)	0.29
39.2.59	01170643	GLUTATHIONE-REGUL. POTASSIUM-EFFLUX (KEFX—HAEIN)	0.0011
39.2.69	01706439	DICARBOXYLIC AMINO ACID PERMEASE (DIP5—YEAST)	0.51
41.25.93	02498121	PUTATIVE ALLANTOIN PERMEASE (ALLP—ECOLI)	0.29

[a] An INCA run was performed (as described in Table I) with ten iterations. The program converged with five iteration after which no new neighbors were identified with P value 10^{-6}. The numbers in front of the sequences indicate the string of best matches among aligned sequences. Thus, the best string of links between 39.2.59 (KEFX—HAEIN) and the starter sequence NANU—RABIT is through sequence 39 (proline permease putP-1), and 39.2 (proline permease putP-2). The P value of 0.0011 for the alignment between KEFX—HAEIN and putP-2 indicates that this represents the weakest link in the string.

actively drives glucose, together with sodium in a ratio of 1:2, into the cell, and it plays a major role in intestinal glucose absorption. Thus, genetic defects of NAG1 are associated with congenital glucose–galactose malabsorption syndrome (Turk *et al.,* 1991).

Also listed in Table III are possibly related sequences outside the core cluster, with intermediate alignments scores. Of principal interest among these is the facilitative glucose transporter GTR1—HUMAN described in the preceding subsection. Even though the sodium/glucose cotransporter NAG1 and the facilitative GTR1 share the same substrate, they are considered to belong to distinct families with different evolutionary origin. Our finding here, based on the INCA analysis summarized in Table III, reveals that GTR1 shows sequence similarity with a rat myo-inositol transporter, a member of the sodium/glucose cotransporter family, which suggests a possible relationship. However, in contrast to our finding of a possible relationship among the hexuronate transporter and dipeptide transporters (Fig. 3), BLAST alignments between GTR1 and the myo-inositol transporter have lower scores (maximally 62, BLOSSUM62 matrix). Even though this could still

Classification of Membrane Transporters

support a finding of possible or even probable homology, the alignments are in different locations of the respective primary structures. This introduces further doubt as to the significance of such alignments, since any modular rearrangements of the primary structure in the process of evolution have yet to be documented for this class of transporter. On the basis of these results alone, a possible evolutionary relationship is doubtful. However, if there indeed is a relationship, we anticipate that further cloned sequences will become available that could provide the missing link with convincing alignment scores. In the future, we will periodically perform INCA runs with the same family of transporters to search for such missing links. Resolving this question would enhance our understanding of transporter evolution, including possible modular rearrangements of the gene structure.

Another interesting link to a distinct transporter family is suggested by the similarity of a glutathione-regulated potassium efflux—system, KEFX—HAEIN (Fleischmann et al., 1995). This sequence has been obtained by random sequencing of *H. influenzae*, and its function inferred by sequence comparisons as a K^+/H^+ exchanger. KEFX—HAEIN is hence a member of the large superfamily of ion exchangers. Many of these are direct drug receptor targets in human patients, rather than functional drug transporters, involved in electrolyte regulation and maintenance of membrane potential (Sadée et al., 1995). These include Na^+/K^+-ATPases and hormone-regulated Na^+/H^+-antiporters (Tse et al., 1991; Sardet et al., 1989). Na^+/H^+ exchangers play a significant role in hypertension and are viable drug target receptors (Siffert and Dusing, 1995; Wakabayashi et al., 1997). A possible relationship between Na^+/glucose or nucleoside cotransporters and ion exchangers would be of considerable interest, but again, alignment scores are insufficient at present to imply a finding of homology.

Further nucleoside transporters appear to belong to yet another gene family, summarized in Table IV. Lack of sequence similarity of this small family, with only 21 sequences identified in the neighborhood cluster, to other transporters, in particular to the Na^+/nucleoside family listed in Table III, is surprising. This could represent a case of functional convergence, or extreme divergence due to mutational drift. The nucleoside transporters are discussed in the chapter by Cass *et al.*

Table IV.
Sodium nucleoside cotransporters related to the nucleoside permease, NUPC—ECOLI[a]

Core cluster: 21 neighbors			P
1	02507087	NUCLEOSIDE PERMEASE NUPC (NUPC—ECOLI)	9.3e-244
13	02072786	Na^+/nucleoside cotransporter (*Homo sapiens*)	1.5e-23
19	02731439	Na^+-dependent purine specific transporter (*Homo sapiens*)	5.7e-18
20	02665908	Na^+/nucleoside cotransporter hcnt2 (*Homo sapiens*)	5.7e-18

[a]INCA was allowed to run ten iterations, but after iteration 2 no new neighbors were found in the core cluster ($P \leq 10^{-6}$).

6.4. Amino Acid Transporters

Not surprisingly, amino acid transporters play an important role in nutrient uptake, particularly in heterotrophs, but also in autotrophs. Table V lists members of a family, related to the yeast starter sequence ALP1 (basic amino acid permease), of many amino acid permeases. One of its mammalian homologues, CTR1—RAT (Table V), is a high-affinity, low-capacity permease for cationic amino acids in nonhepatic tissues, and it may also serve as an ecotropic retroviral leukemia receptor (Kim et al., 1991). This is an example that transporters can also serve as entry vectors facilitating viral infections. CTR1 has been identified as a cation amino acid transporter (system y+) of the rat blood–brain barrier (Stoll et al., 1993). This provides an example of the biological and physiological relevance of transporters at the blood–brain barrier.

The sodium/dicarboxylate symporter family contains a number of pharmaceutically relevant transporters (Table VI). Mediating uptake of L-serine, L-alanine, and L-glutamate in adipose and other tissues, AAAT—MOUSE represents another transporter type that is activatable by insulin (Liao and Lane, 1995), in addition to GTR4. Hence, hormonal regulation of cellular trafficking represents a general mechanism for regulating transport functions.

Also related to this family are the neuronal excitatory amino acid transporters (EAT) (Kanai and Hediger, 1992; Arriza et al., 1994). Acting as sodium symporters, the EATs are essential for terminating the postsynaptic action of glutamate by rapid removal of glutamate from the synaptic cleft (Arriza et al., 1994). As glutamate plays a pivotal role in excitatory neurotoxicity and memory functions,

Table V.
Basic amino acid permeases related to ALP1—YEAST as the starter sequence[a]

Core cluster: 163 neighbors			P
1	01703255	BASIC AMINO-ACID PERMEASE, ALP1—YEAST	0.0
3	00729014	ARGININE PERMEASE, CAN1—YEAST	7.5e-253
5	01708894	LYSINE-SPECIFIC PERMEASE, LYP1—YEAST	4.2e-224
10	01706439	DICARBOXYLIC AMINO ACID PERM., DIP5—YEAST	3.4e-102
11	00544369	GENERAL AMINO-ACID PERMEASE, GAP1—YEAST	2.4e-95
18	01709941	PROLINE-SPECIFIC PERMEASE, PUT4—YEAST	3.6e-87
24	00586506	LEU/VAL/ILE AMINO-ACID PERMEASE BAP	21.2e-84
44	02828523	GABA PERMEASE , GABP—BACSU	1.7e-63
48	00130068	PHENYLALANINE-SPECIFIC PERM., PHEP—ECOLI	2.2e-62
52.91	01706187	HIGH-AFF. CATION. AMINO ACID TR., CTR1—RAT	8e-19
64.95	01172565	PUTRESCINE-ORN. ANTIPORTER, POTE—HAEIN	2.7e-10
74	00732281	D-SER/D-ALA/GLY TRANSPORT, CYCA—ECOLI	1.9e-37
91	00403170	ETHANOLAMINE PERMEASE, Rhodococcus erythropolis	3.8e-07

[a]INCA was performed with two iterations.

Table VI.
Dicarboxylate transporters, including excitatory amino acid transporters[a]

Core cluster: 97 neighbors			P
1	00585038	C4-DICARBOXYLATE TRANSPORT PROTEIN DCTA—ECOLI	3.9e-277
11	00121466	PROTON/SODIUM-GLUTAMATE SYMPORT PROT., DCTA—ECOLI	2.1e-98
20	01169458	EXCITATORY AMINO ACID TRANSPORTER 1, glial EAT1—HUMAN	2.1e-30
28	01706561	EXCITATORY AMINO ACID TRANSPORTER 3, EAT3—RAT	1.9e-27
35	02465213	EAAT4—RAT	3.1e-26
51	02315861	NEUTRAL AMINO ACID TRANSPORTER, *Oryctolagus cuniculus*	8.8e-25
71	01173365	NEUTRAL AMINO ACID TRANSPORTER, SATT—HUMAN	2.8e-21
73	01703032	INSULIN-ACTIVATED AMINO ACID TRANSPORT., AAAT—MOUSE	8.2e-20

[a]INCA was run in two iterations.

cloning the genes encoding this neuronal transporter subfamily has generated much interest in providing potential drug targets.

6.5. Sodium/Neurotransmitter Symporters

Apparently unrelated to the EATs, the large family of neurotransmitter transporters terminates trans-synaptic signal transmission, commonly by reuptake of the neurotransmitter into the presysnaptic terminal (Amara, 1992; Worrall and Williams, 1994) (Table VII). Main members of this family include the dopamine (Giros *et al.*, 1992; Kilty *et al.*, 1991; Shimada *et al.*, 1991), norepinephrine (Pacholczyk *et al.*, 1991), GABA (Guastella *et al.*, 1990), and glycine (Kim *et al.*, 1994; Smith *et al.*, 1992) transporters, many of which represent important drug receptors. These include antidepressant drugs and stimulants, including several drugs of abuse (Erickson, 1996). For example, the dopamine transporter has been identified as the main target receptor for cocaine, which prevents dopamine reuptake and thereby enhances dopaminergic neurotransmission. Polymorphism of the dopamine transporter gene has been suggested, but any relationship to addictive liability remains obscure (Vandenbergh *et al.*, 1992). Further, the dopamine transporter appears to play a role in MPP^+-induced neurotoxicity by concentrating this toxin into the target cell (Kitayama *et al.*, 1992a). There appears to be a precipitous decrease of dopamine transporter mRNA content in human substantia nigra with age (Bannon *et al.*, 1992), which may affect dopaminergic neurotransmission in the elderly. Because of their therapeutic importance, the neurotransmitter transporter family has been studied in great detail, but a validated structural model of the mature protein has yet to be published. Numerous studies using site-directed mutagenesis have provided indirect information on the structure–function relationships of neurotransmitter transporters (e.g., Kitayama *et al.*, 1992b). Future ad-

Table VII.
Sodium/neurotransmitter symporters[a]

Core cluster: 196 neighbors			P
1	00266666	Na$^+$/Cl$^-$-DEPENDENT GABA TRANSPORTER 1, NTG1—HUMAN	0.0
13	00400621	Na$^+$/Cl$^-$-DEPENDENT CREATINE TRANSPORTER 1, NTCR—RABIT	5.9e-201
15	01352527	Na$^+$/Cl$^-$-DEPENDENT BETAINE TRANSPORTER, NTBE—RAT	2.9e-200
19	01352535	Na$^+$/Cl$^-$-DEPENDENT TAURINE TRANSPORTER, NTTA—HUMAN	1.9e-197
32	00128617	Na$^+$/Cl$^-$-DEPENDENT PROLINE TRANSPORTER, NTPR—RAT	3.7e-172
34	00266667	Na$^+$-DEPENDENT DOPAMINE TRANSPORTER, NTDO—HUMAN	4.6e-172
40	01709356	Na$^+$-DEPENDENT NORADRENALINE TRANSP., NTNO—BOVIN	9.9e-170
44	01709361	Na$^+$-DEPENDENT SEROTONIN TRANSPORTER, NTSE—DROME	1.1e-168
49	00128615	Na$^+$/Cl$^-$-DEPENDENT GLYCINE TRANSPORTER, NTGL—RAT	9.1e-167
60	00400630	Na$^+$-DEPENDENT SEROTONIN TRANSPORTER, NTSE—HUMAN	4.1e-154
67	1575565	INEBRIATED, *Drosophila melanogaster*	2.9e-143
71	00128611	Na$^+$-DEPENDENT CHOLINE TRANSPORTER, NTCH—RAT	1.9e-143

[a]INCA was run in two iterations. Note that the serotonin transporter of *Drosophila* was more closely related to NTG1 (starter sequence) than the human serotonin receptor. This result could suggest that the gene duplication leading to these two transporters occurred before species divergence. Therefore, finding human homologues through INCA could be enhanced by comparing sequences from distant phyla.

vances will require crystallization or other structural analysis that could reveal the folding of these multi-TMD proteins into functional structures within the membrane.

6.6. ATP-Binding Transport Protein Family

Many active transporters contain an ATP-binding cassette (ABC) (Hyde *et al.*, 1990) covalently linked to membrane spanning segments (Table VIII). While the ATPase activity inherent to ABC is thought to provide the energy required for solute translocation against a gradient, the mechanism of energy coupling remains to be clarified in molecular detail. Typically, two six-TMD membrane segments and two ABCs combine to yield the functional primary active transporter. Gene organization for encoding the requisite four protein modules is quite variable. In bacteria one often finds four distinct genes, one for each of the two membrane modules and the two ABCs, that need to be expressed simultaneously for functional transport. The human antigenic peptide transporters, e.g., TAP1 and TAP2, are heterodimers consisting of one membrane module and one ABC each (Neefjes *et al.*, 1993; Spies *et al.*, 1990; Spies *et al.*, 1992), whereas the multidrug-resistance proteins are encoded by a single gene (Cole *et al.*, 1992; Chen *et al.*, 1986). These have been linked to tumor resistance against chemotherapeutic agents. Conversely, normal bone marrow cells have been made resistant to anticancer drugs

Table VIII.
ATP-Binding Transport Protein Family[a]

		P
gi|2506118|MDR1—HUMAN	MULTI-DRUG RESISTANCE PROTEIN 1	0.0
sp|P21439|MDR3—HUMAN	MULTI-DRUG RESISTANCE PROTEIN 3	0.0
sp|P34712|MDR1—CAEEL	MULTI-DRUG RESISTANCE PROTEIN 1	0.0
pir|A42150	P-glycoprotein atpgp1, *Arabidopsis thaliana*	2.6e-287
sp|P36619|PMD1—SCHPO	LEPTOMYCIN B RESISTANCE PROTEIN	4.2e-237
sp|P12866|STE6—YEAST	MATING FACTOR A SECRETION PROTEIN	7.1e-117
sp)P33310|MDL1—YEAST	ATP-DEPENDENT PERMEASE MDL1	2.7e-78
sp|Q03518|TAP1—HUMAN	ANTIGEN PEPTIDE TRANSPORTER	15.0e-71
sp|P23702|HLYB—ACTAC	LEUKOTOXIN SECRETION ATP-BIND. PROT.	7.1e-70
sp|P10089|HLY2—ECOLI	HEMOLYSIN SECRETION ATP-BINDING P	.9.7e-70
sp|Q02592|HMT1—SCHPO	HEAVY METAL TOLERANCE PROT. PREC.	1.4e-66
sp|Q03519|TAP2—HUMAN	ANTIGEN PEPTIDE TRANSPORTER 2	1.7e-65
sp|P18768|CHVA—AGRTU	BETA-(1—/2)GLUCAN EXPORT ATP-BIND.P.	2.7e-64
sp|P40416|ATM1—YEAST	MITOCHONDR. TRANSPORTER ATM1 PREC.	3.8e-60
sp|P18770|CYAB—BORPE	CYCLOLYSIN SECRETION ATP-BIND. PROT.	1.6e-58
sp|P33527|MRP1—HUMAN	MULTI-DRUG RESISTANCE-ASSOC. PROT.	7.3e-56
pir||S71839	Canalicular multidrug resistance protein, rat	7.5e-56
sp|Q09428|SUR—HUMAN	SULFONYLUREA RECEPTOR	9.3e-49
gi|2340166	Glutathione S-conjugate transporter	2.9e-47
sp|P53049|YOR1—YEAST	OLIGOMYCIN RESIST. ATP-DEP. TR.	3.3e-46
sp|Q00564|LCNC—LACLA	LACTOCOCCIN A TRANSPORT ATP-BIND. P.	3.9e-44
sp|P39109|YCFI—YEAST	METAL RESISTANCE PROTEIN	2.4e-43
sp|P22520|CVAB—ECOLI	COLICIN V SECRETION ATP-BINDING PROT.	9.3e-39
sp|P13569|CFTR—HUMAN	CYSTIC FIBROSIS TRANSMEMBRANE COND.	6.9e-34
sp|Q10418|MESD—LEUME	MESENTERICIN Y105 TRANSP. ATP-BIND. P	7.0e-34
gi|184141	TAP2, *Homo sapiens*	4.1e-31
sp|Q03024|APRD—PSEAE	ALKALINE PROTEASE SECRET. ATP-B.P.	3.6e-30
sp|P33116|SPAT—BACSU	SUBTILIN TRANSPORT ATP-BINDING PROT.	3.2e-29

[a] A single BLAST was performed with MDR1—HUMAN as the starter sequence. Performing INCA on this family would yield an extremely large file unless severe restrictions are imposed. We are currently modifying the INCA Web page (http://itsa.ucsf.edu/~gram/home/inca.html) to permit its application to such large families. BLAST identifies 2000 sequence neighbors. The score is a probability value.

by retroviral transfer of the human MDR1 gene (Sorrentino *et al.,* 1992) with the goal to prevent lethal toxicity caused by very high drug dosages. This strategy could improve anticancer efficacy. Related to the MDRs are the multidrug-resistance proteins (e.g., Buchler *et al.,* 1996). These important transporters are discussed in the chapter by Silverman.

Because the ABC gene module is shared by numerous proteins, a single BLAST analysis reveals ~2000 neighbors listed in Entrez. A selection of these is provided in Table VIII. One finds the mammalian multidrug-resistance proteins (MDRs), multidrug-resistance-associated proteins (MRPs), and antigenic peptide transporters (TAPs). High similarity of MDRs also exists to the antidiabetic sulfonylurea receptors (SUR—HUMAN), which are not considered transporters per se, but regulate the activity of K^+ channels, specifically in Langerhans β cells. Thus, the SURs are considered a subunit of ATP-sensitive K^+ channels which regulate insulin release. Mutations of the SUR gene that inactivate the ABC module are the cause of familial hyperinsulinemic hypocalcemia of infancy (Thomas et al., 1996). Similarly, the related cystic fibrosis membrane regulator is not a transporter in the strict sense, but rather a channel involved in the transport of chloride ions. Many mutations of CFTR have been identified that cause cystic fibrosis, the most common genetic disease in the White population with a prevalence of 1 in 2000 live births (see Sadée et al., 1995, and Taylor, 1992, for review). These mutations associated with functional changes have revealed much insight into the structure of the CFTR gene product. A majority of cases involve a single amino acid deletion, dF508, which does not inactivate CFTR function, but rather prevents its transfer to the plasma membrane. Possibly, in vivo gene therapy could overcome defects associated with CFTR gene mutations (Hyde et al., 1993).

The ABC transport family also contains peptide and protein transporters, providing an alternative path for the import and export of polypeptides. For example, CVAB—ECOLI is a secretion ATP-binding protein responsible for the export of colicin V (Gilson et al., 1990; Havarstein et al., 1994), a polypeptide toxin (103 amino acids long) disrupting membrane potential. Further, STE6—YEAST secretes the farnesylated polypeptide mating factor A (McGrath and Varshavsky, 1989; Kuchler et al., 1989), and ARPD—PSEAE secretes a protein, namely alkaline protease (Duong et al., 1992). These results document the functional diversity of a common sequence motif which has been conserved over extended evolutionary time periods.

7. CONCLUSION

In summation, classification of membrane transporters is a challenging issue. One must consider not only their primary structure, but one must also ponder their function, topology, evolutionary origins, substrate specificity, and structural architecture. It is not known to what extent convergence and divergence played a role in the evolutionary progression of membrane transporters. A major question remaining to be resolved is whether common ancestry indeed implies similar folding into a common molecular architecture, as commonly assumed for soluble proteins. Most transporter gene families share a similar membrane topology, and the primary structure is highly conserved among members of the same gene family. Transporter families often include diverse functions in terms of substrate specificity and translocation mechanism. However, sequence similarities among different transporter families are often limited. It is difficult to establish evolutionary relationships between distant transporter families because of the recurrent TMD–loop structure, with restricted amino acid content. For a better understanding of these polytopic membrane proteins, we must determine how their structural diversity has evolved, and how the primary structures fold into the mature transport protein. Much remains to be learned about the biology and structure of membrane transporters.

ACKNOWLEDGMENT

Part of this work was supported by a National Institute of Health research grant, GM37188.

REFERENCES

Altschul, S. F., Gish, W., Miller, W., Myers, E. W., and Lipman, D. J., 1990, Basic local alignment search tool, *J. Mol. Biol.* **215**:403–410.
Amara, S. G., 1992, Neurotransmitter transporters. A tale of two families, *Nature* **360**:420–421.
Arriza, J. L., Fairman, W. A., Wadiche, J. I., Murdoch, G. H., Kavanaugh, M. P., and Amara, S. G., 1994, Functional comparisons of three glutamate transporter subtypes cloned from human motor cortex, *J. Neurosci.* **14**:5559–5569.
Bannon, M. J., Poosch, M. S., Xia, Y., Goebel, D. J., Cassin, B., and Kapatos, G., 1992, Dopamine transporter mRNA content in human substantia nigra decreases precipitously with age, *Proc. Natl. Acad. Sci. USA* **89**:7095–7099.
Bell, G. I., Kayano, T., Buse, J. B., Burant, C. F., Takeda, J., Lin, D., Fukumoto, H., and Seino, S., 1990, Molecular biology of mammalian glucose transporters, *Diabetes Care* **13**:198–208.
Bentley, J., Hyatt, L. S., Ainley, K., Parish, J. H., Herbert, R. B., and White, G. R., 1993, Cloning and sequence analysis of an *Escherichia coli* gene conferring bicyclomycin resistance, *Gene* **127**:117–120.
Boguski, M. S., 1995, The turning point in genome research, *Trends Biochem. Sci.* **20**:295–296.
Botstein, D., Chervitz, S. A., and Cherry, J. M., 1997, Yeast as a model organism, *Science* **277**:1259–1260.
Buchler, M., Konig, J., Brom, M., Kartenbeck, J., Spring, H., Horie, T., and Keppler, D., 1996, cDNA cloning of the hepatocyte canalicular isoform of the multi-drug resistance protein, cMRP, reveals a novel conjugate export pump deficient in hyperbilirubinemic mutant rats, *J. Biol. Chem.* **271**:15091–15098.
Buse, J. B., Yasuda, K., Lay, T. P., Seo, T. S., Olson, A. L., Pessin, J. E., Karam, J. H., Seino, S., and Bell, G. I., 1992, Human GLUT4/muscle-fat glucose-transporter gene. Characterization and genetic variation, *Diabetes* **41**:1436–1445.
Chen, C. J., Chin, J. E., Ueda, K., Clark, D. P., Pastan, I., Gottesman, M. M., and Ronin-

son, I. B., 1986, Internal duplication and homology with bacterial transport proteins in the mdr1 (P-glycoprotein) gene from multi-drug-resistant human cells, *Cell* **47**:381–389.

Clayton, R. A., White, O., Ketchum, K. A., and Venter, J. C., 1997, The first genome from the third domain of life, *Nature* **387**:459–462.

Cole, S. P., Bhardwaj, G., Gerlach, J. H., Mackie, J. E., Grant, C. E., Almquist, K. C., Stewart, A. J., Kurz, E. U., Duncan, A. M., and Deeley, R. G., 1992, Overexpression of a transporter gene in a multi-drug-resistant human lung cancer cell line, *Science* **258**:1650–1654.

DeRisi, J. L., Iyer, V. R., and Brown, P. O., 1997, Exploring the metabolic and genetic control of gene expression on a genomic scale, *Science* **278**:680–686.

Dimroth, P., 1991, Na^+-coupled alternative to H^+-coupled primary transport systems in bacteria, *BioEssays* **13**:463–468.

Duong, F., Lazdunski, A., Cami, B., and Murgier, M., 1992, Sequence of a cluster of genes controlling synthesis and secretion of alkaline protease in *Pseudomonas aeruginosa*: Relationships to other secretory pathways, *Gene* **121**:47–54.

Erickson, C. K., 1996, Review of neurotransmitters and their role in alcoholism treatment, *Alcohol Alcoholism* **31 (Suppl. 1)**:5–11.

Fei, Y. J., Kanai, Y., Nussberger, S., Ganapathy, V., Leibach, F. H., Romero, M. F., Singh, S. K., Boron, W. F., and Hediger, M. A., 1994, Expression cloning of a mammalian proton-coupled oligopeptide transporter, *Nature* **368**:563–566.

Fleischmann, R. D., Adams, M. D., White, O., Clayton, R. A., Kirkness, E. F., Kerlavage, A. R., Bult, C. J., Tomb, J. F., Dougherty, B. A., Merrick, J. M., *et al.*, 1995, Whole-genome random sequencing and assembly of *Haemophilus influenzae*, *Science* **269**:496–512.

Fukumoto, H., Seino, S., Imura, H., Seino, Y., and Bell, G. I., 1988, Characterization and expression of human HepG2/erythrocyte glucose-transporter gene, *Diabetes* **37**:657–661.

Gilson, L., Mahanty, H. K., and Kolter, R., 1990, Genetic analysis of an MDR-like export system: The secretion of colicin V, *EMBO J.* **9**:3875–3894.

Giros, B., el Mestikawy, S., Godinot, N., Zheng, K., Han, H., Yang-Feng, T., and Caron, M. G., 1992, Cloning, pharmacological characterization, and chromosome assignment of the human dopamine transporter, *Mol. Pharmacol.* **42**:383–390.

Gonzales, D. E., Covitz, K.-M. Y., Sadée, W., and Mrsny, R. J., 1998, An oligopeptide transporter is expressed at high levels in pancreatic carcinoma cell lines AsPc-1 and Capan-2, *Cancer Res.* **58**:519–525.

Graul, R. C., and Sadée, W., 1997a, Evolutionary relationships among proteins probed by an iterative neighborhood cluster analysis (INCA). Alignment of bacteriorhodopsins with the yeast sequence YRO2, *Pharmaceut. Res.* **14**:1533–1541.

Graul, R. C., and Sadée, W., 1997b, Sequence alignments of the H^+-dependent oligopeptide transporter family PTR: Inferences on structure and function of the intestinal PET1 transporter, *Pharmaceut. Res.* **14**:388–400.

Guastella, J., Nelson, N., Nelson, H., Czyzyk, L., Keynan, S., Miedel, M. C., Davidson, N., Lester, H. A., and Kanner, B. I., 1990, Cloning and expression of a rat brain GABA transporter, *Science* **249**:1303–1306.

Hasegawa, H., Skach, W., Baker, O., Calayag, M. C., Lingappa, V., and Verkman, A. S., 1992, A multifunctional aqueous channel formed by CFTR, *Science* **258**:1477–1479.

Havarstein, L. S., Holo, H., and Nes, I. F., 1994, The leader peptide of colicin V shares consensus sequences with leader peptides that are common among peptide bacteriocins produced by gram-positive bacteria, *Microbiology* **140**:2383–2389.

Hediger, M. A., Coady, M. J., Ikeda, T. S., and Wright, E. M., 1987, Expression cloning and cDNA sequencing of the Na^+/glucose cotransporter, *Nature* **330**:379–381.

Hediger, M. A., Turk, E., and Wright, E. M., 1989, Homology of the human intestinal Na^+/glucose and Escherichia coli Na^+/proline cotransporters, *Proc. Natl. Acad. Sci. USA* **86**:5748–5752.

Hyde, S. C., Emsley, P., Hartshorn, M. J., Mimmack, M. M., Gileadi, U., Pearce, S. R., Gallagher, M. P., Gill, D. R., Hubbard, R. E., and Higgins, C. F., 1990, Structural model of ATP-binding proteins associated with cystic fibrosis, multi-drug resistance and bacterial transport, *Nature* **346**:362–365.

Hyde, S. C., Gill, D. R., Higgins, C. F., Trezise, A. E., MacVinish, L. J., Cuthbert, A. W., Ratcliff, R., Evans, M. J., and Colledge, W. H., 1993, Correction of the ion transport defect in cystic fibrosis transgenic mice b gene therapy, *Nature* **362**:250–255.

Iarovici, D., 1997, Single blood tests might predict drug's effect on patients, *J. NIH Res.* **9**:34–35.

James, D. E., Strube, M., and Mueckler, M., 1989, Molecular cloning and characterization of an insulin-regulatable glucose transporter, *Nature* **338**:83–87.

Jeanteur, D., Lakey, J. H., and Pattus, F., 1991, The bacterial porin superfamily: Sequence alignment and structure prediction, *Mol. Microbiol.* **5**:2153–2164.

Kanai, Y., and Hediger, M. A., 1992, Primary structure and functional characterization of a high-affinity glutamate transporter, *Nature* **360**:467–471.

Kaplan, J. H., 1993, Molecular biology of carrier proteins, *Cell* **72**:13–18.

Kayano, T., Fukumoto, H., Eddy, R. L., Fan, Y. S., Byers, M. G., Shows, T. B., and Bell, G. I., 1988, Evidence for a family of human glucose transporter-like proteins. Sequence and gene localization of a protein expressed in fetal skeletal muscle and other tissues, *J. Biol. Chem.* **263**:15245–15248.

Kilty, J. E., Lorang, D., and Amara, S. G., 1991, Cloning and expression of a cocaine-sensitive rat dopamine transporter, *Science* **254**:578–579.

Kim, J. W., Closs, E. I., Albritton, L. M., and Cunningham, J. M., 1991, Transport of cationic amino acids by the mouse ecotropic retrovirus receptor, *Nature* **352**:725–728.

Kim, K. M., Kingsmore, S. F., Han, H., Yang-Feng, T. L., Godinot, N., Seldin, M. F., Caron, M. G., and Giros, B., 1994, Cloning of the human glycine transporter type 1: Molecular and pharmacological characterization of novel isoform variants and chromosomal localization of the gene in the human and mouse genomes, *Mol. Pharmacol.* **45**:608–617.

Kitayama, S., Shimada, S., and Uhl, G. R., 1992a, Parkinsonism-inducing neurotoxin MPP^+; uptake and toxicity in noneuronal COS cells expressing dopamine transporter cDNA, *Ann. Neurobiol.* **32**:109–111.

Kitayama, S., Shimada, S., Xu, H., Markham, L., Donovan, D. M., and Uhl, G. R., 1992b, Dopamine transporter site-directed mutations differentially alter substrate transport and cocaine binding, *Proc. Natl. Acad. Sci. USA* **89**:7782–7785.

Kool, M., de Haas, M., Scheffer, G. L., Scheper, R. J., van Eijk, M. J., Juijn, J. A., Baas, F., and Borst, P., 1997, Analysis of expression of cMOAT (MRP2), MRP3, MRP4, and MRP5, homologues of the multi-drug resistance-associated protein gene (MRP1), in human cancer cell lines, *Cancer Res.* **57:**3537–3547.

Kuchler, K., Sterne, R. E., and Thorner, J., 1989, *Saccharomyces cerevisiae* STE6 gene product: A novel pathway for protein export in eukaryotic cells, *EMBO J.* **8:**3973–3984.

Kusari, J., Verma, U. S., Buse, J. B., Henry, R. R., and Olefsky, J. M., 1991, Analysis of the gene sequences of the insulin receptor and the insulin-sensitive glucose transporter (GLUT-4) in patients with common-type non-insulin-dependent diabetes mellitus, *J. Clin. Invest.* **88:**1323–1330.

Liang, R., Fei, Y. J., Prasad, P. D., Ramamoorthy, S., Han, H., Yang-Feng, T. L., Hediger, M. A., Ganapathy, V., and Leibach, F. H., 1995, Human intestinal H^+/peptide cotransporter. Cloning, functional expression, and chromosomal localization, *J. Biol. Chem.* **270:**6456–6463.

Liao, K., and Lane, M. D., 1995, Expression of a novel insulin-activated amino acid transporter gene during differentiation of 3T3-L1 preadipocytes into adipocytes, *Biochem. Biophys. Res. Commun.* **208:**1008–1015.

Lin, C. Y., Hahnenberger, K. M., Cronin, M. T., Lee, D., Sampas, N. M., and Kanemoto, R., 1996, A method for genotyping CYP2D6 and CYP2C19 using GeneChip^R probe array hybridization, Poster reprint, ISSX Meeting.

Liu, M. L., Olson, A. L., Moye-Rowley, W. S., Buse, J. B., Bell, G. I., and Pessin, J. E., 1992, Expression and regulation of the human GLUT4/muscle-fat facilitative glucose transporter gene in transgenic mice, *J. Biol. Chem.* **267:**11673–11676.

Lockart, D. J., *et al.*, 1996, Expression monitoring by hybridization to high-density oligonucleotide arrays, *Nature Biotech.* **14:**1675–1680.

Marger, M. D., and Saier, M. H., Jr., 1993, A major superfamily of transmembrane facilitators that catalyse uniport, symport and antiport, *Trends Biochem. Sci.* **18:**13–20.

McGrath, J. P., and Varshavsky, A., 1989, The yeast STE6 gene encodes a homologue of the mammalian multi-drug resistance P-glycoprotein, *Nature* **340:**400–404.

Mueckler, M., Caruso, C., Baldwin, S. A., Panico, M., Blench, I., Morris, H. R., Allard, W. J., Lienhard, G. E., and Lodish, H. F., 1985, Sequence and structure of a human glucose transporter, *Science* **229:**941–945.

Neefjes, J. J., Momburg, F., and Hammerling, G. J., 1993, Selective and ATP-dependent translocation of peptides by the MHC-encoded transporter, *Science* **261:**769–771.

Nikaido, H., and Saier, M. H., Jr., 1992, Transport proteins in bacteria: Common themes in their design, *Science* **258:**936–942.

Ostergaard, M., Rasmussen, H. H., Nielsen, H. V., Vorum, H., Orntoft, T. F., Wolf, H., and Celis, J. E., 1997, Proteome profiling of bladder squamous cell carcinomas: Identification of markers that define their degree of differentiation, *Cancer Res.* **57:**4111–4117.

Pacholczyk, T., Blakely, R. D., and Amara, S. G., 1991, Expression cloning of a cocaine- and antidepressant-sensitive human noradrenaline transporter, *Nature* **350:**350–354.

Pajor, A. M., and Wright, E. M., 1992, Cloning and functional expression of a mammalian Na^+/nucleoside cotransporter. A member of the SGLT family, *J. Biol. Chem.* **267:**3557–3560.

Riordan, J. R., Rommens, J. M., Kerem, B., Alon, N., Rozmahel, R., Grzelczak, Z., Zielenski, J., Lok, S., Plavsic, N., Chou, J. L., et al., 1989, Identification of the cystic fibrosis gene: Cloning and characterization of complementary DNA, *Science* **245:** 1066–1073.

Sadée, W., Drubbisch, V., and Amidon, G. L., 1995, Biology of membrane transport proteins, *Pharmaceut. Res.* **12:**1823–1837.

Sardet, C., Franchi, A., and Pouyssegur, J., 1989, Molecular cloning, primary structure, and expression of the human growth factor-activatable Na^+/H^+ antiporter, *Cell* **56:**271–280.

Schena, M., Shalon, D., Davis, R. W., and Brown, P. O., 1995, Quantitative monitoring of gene expression patterns with a complimentary DNA microarray, *Science* **270:**467–470.

Shimada, S., Kitayama, S., Lin, C. L., Patel, A., Nanthakumar, E., Gregor, P., Kuhar, M., and Uhl, G., 1991, Cloning and expression of a cocaine-sensitive dopamine transporter complementary DNA, *Science* **254:**576–578.

Siffert, W., and Dusing, R., 1995, Sodium-proton exchange and primary hypertension. An update, *Hypertension* **26:**649–655.

Silverman, M., 1991, Structure and function of hexose transporters, *Annu. Rev. Biochem.* **60:**757–794.

Smith, K. E., Borden, L. A., Hartig, P. R., Branchek, T., and Weinshank, R. L., 1992, Cloning and expression of a glycine transporter reveal colocalization with NMDA receptors, *Neuron* **8:**927–935.

Sorrentino, B. P., Brandt, S. J., Bodine, D., Gottesman, M., Pastan, I., Cline, A., and Nienhuis, A. W., 1992, Selection of drug-resistant bone marrow cells *in vivo* after retroviral transfer of human MDR1, *Science* **257:**99–103.

Spies, T., Bresnahan, M., Bahram, S., Arnold, D., Blanck, G., Mellins, E., Pious, D., and DeMars, R., 1990, A gene in the human major histocompatibility complex class II region controlling the class I antigen presentation pathway, *Nature* **348:**744–747.

Spies, T., Cerundolo, V., Colonna, M., Cresswell, P., Townsend, A., and DeMars, R., 1992, Presentation of viral antigen by MHC class I molecules is dependent on a putative peptide transporter heterodimer, *Nature* **355:**644–646.

Stoll, J., Wadhwani, K. C., and Smith, Q. R., 1993, Identification of the cationic amino acid transporter (system y+) of the rat blood–brain barrier, *J. Neurochem.* **60:**1956–1959.

Tamai, I., Yabuuchi, H., Nezu, J., Sai, Y., Oku, A., Shimane, M., and Tsuji, A., 1997, Cloning and characterization of a novel human pH-dependent organic cation transporter, OCTN1, *FEBS Lett.* **419:**107–111.

Taylor, R., 1992, All dressed up: Cystic fibrosis research steps out, *J. NIH Res.* **4(11):**55–59.

Thomas, P. M., Wohllk, N., Huang, E., Kuhnle, U., Rabl, W., Gagel, R. F., and Cote, G. J., 1996, Inactivation of the first nucleotide-binding fold of the sulfonylurea receptor, and familial persistent hyperinsulinemic hypoglycemia of infancy, *Am. J. Hum. Genet.* **59:**510–518.

Tomb, J. F., White, O., Kerlavage, A. R., Clayton, R. A., Sutton, G. G., Fleischmann, R. D., Ketchum, K. A., Klenk, H. P., Gill, S., Dougherty, B. A., Nelson, K., Quackenbush, J., Zhou, L., Kirkness, E. F., Peterson, S., Loftus, B., Richardson, D., Dodson, R., Khalak, H. G., Glodek, A., McKenney, K., Fitzegerald, L. M., Lee, N., Adams, M. D., Venter, J. C., *et al.,* 1997, The complete genome sequence of the gastric pathogen *Helicobacter pylori, Nature* **388**:539–547.

Tse, C. M., Ma, A. I., Yang, V. W., Watson, A. J., Levine, S., Montrose, M. H., Potter, J., Sardet, C., Pouyssegur, J., and Donowitz, M., 1991, Molecular cloning and expression of a cDNA encoding the rabbit ileal villus cell basolateral membrane Na^+/H^+ exchanger, *EMBO J.* **10**:1957–1967.

Turk, E., Zabel, B., Mundlos, S., Dyer, J., and Wright, E. M., 1991, Glucose/galactose malabsorption caused by a defect in the Na^+/glucose cotransporter, *Nature* **350**:354–356.

Vandenbergh, D. J., Persico, A. M., and Uhl, G. R., 1992, A human dopamine transporter cDNA predicts reduced glycosylation, displays a novel repetitive element and provides racially-dimorphic TaqI RFLPs, *Brain Res. Mol. Brain Res.* **15**:161–166.

Velculescu, V. E., Zhang, L., Vogelstein, B., and Kinzler, K. W., 1995, Serial analysis of gene expression, *Science* **270**:484–487.

Wakabayashi, S., Shigekawa, M., and Pouyssegur, J., 1997, Molecular physiology of vertebrate Na^+/H^+ exchangers, *Physiol. Rev.* **77**:51–74.

Worrall, D. M., and Williams, D. C., 1994, Sodium ion-dependent transporters for neurotransmitters: A review of recent developments, *Biochem. J.* **297**:425–436.

Wright, E. M., Hager, K. M., and Turk, E., 1992, Sodium cotransport proteins, *Curr. Opin. Cell. Biol.* **4**:696–702.

Yamashita, T., Shimada, S., Guo, W. Sato, K. Kohmura, E., Hayakawa, T., Takagi, T., and Tohyama, M., 1997, Cloning and functional expression of a brain peptide/histidine transporter, *J. Biol. Chem.* **272**:10205–10211.

Yang, J., and Holman, G. D., 1993, Comparison of GLUT4 and GLUT1 subcellular trafficking in basal and insulin-stimulated 3T3-L1 cells, *J. Biol. Chem.* **268**:4600–4603.

Yeast Genome Directory, 1997, *Nature* **387**(663275).

Zhang, L., Zhou, W., Velculescu, V. E., Kern, S. E., Hruban, R. H., Hamilton, S. R., Vogelstein, B., and Kinzler, K. W., 1997, Gene expression profiles in normal and cancer cells, *Science* **276**:1268–1272.

3

Drug Transport and Targeting

Intestinal Transport

*Doo-Man Oh, Hyo-kyung Han,
and Gordon L. Amidon*

1. INTRODUCTION

The intestine is the major absorption site for nutrients such as amino acids and sugars. Water and electrolytes, such as Na^+, K^+, HCO_3^-, Cl^-, Ca^{2+}, and Fe^{2+}, are also transported via the intestine to body fluids. One or more mechanisms transport these solutes and ions across the intestinal epithelia, including passive diffusion, facilitated diffusion, and active transport. In addition, pinocytosis occurs at the base of microvilli and may also contribute to the uptake of protein (Johnson, 1997).

Different regions of the intestine exhibit distinct transporters and significant variability in permeability. For example, the bile acid transporters are only found in the lower part of the small intestine, the ileum. Membrane transporters in the intestine also show axial heterogeneity. If a pathological process destroys primarily brush border membrane cells, the absorptive capacity of the epithelium will be impacted to a much greater extent than the secretory capacity. There are several clinical disorders that involve membrane transporters. Hartnup disease, for example, is a hereditary disorder in which the active transport of dipolar amino acids is de-

---

Doo-Man Oh • Department of Pharmacokinetics, Dynamics, and Metabolism, Parke-Davis Pharmaceutical Research Division, Warner-Lambert Company, Ann Arbor, Michigan 48105. *Hyo-kyung Han and Gordon L. Amidon* • College of Pharmacy, University of Michigan, Ann Arbor, Michigan 48109-1065.
Membrane Transporters as Drug Targets, edited by Amidon and Sadée. Kluwer Academic/Plenum Publishers, New York, 1999.

ficient in both renal tubules and the small intestine. The disease is usually benign, because protein digestive products are also absorbed in the form of small peptides.

While a variety of drugs are absorbed from the intestine by passive diffusion, the absorption of hydrophilic drugs is enhanced by specific transporters in the intestine. Many peptide-like drugs, such as cephalosporins and angiotensin-converting enzyme (ACE) inhibitors, depend on the intestinal dipeptide transporter for efficient absorption. Nucleosides and their analogues for antiviral and anticancer drugs also depend on the nucleoside transporters to be taken up. Therefore, utilization of the intestinal epithelial transporters to facilitate the absorption of drugs or prodrugs appears to be an attractive strategy for improving the bioavailability of poorly absorbed drugs (Tsuji and Tamai, 1996).

The purpose of this chapter is to generally describe the pharmaceutical and pharmacological relevance of intestinal transporters. Detailed information on individual intestinal transporters are described elsewhere in this book. In this chapter, selected transporters will be discussed as drug delivery and targeting sites for better bioavailability.

2. INTESTINAL TRANSPORTERS

2.1. General Description of Intestinal Transporters

There are a variety of membrane transporters in the intestinal membrane as shown in Table I. Various solutes, such as amino acids, sugars, peptides, nucleosides, bile acids, inorganic phosphate, organic anions and cations, as well as several vitamins, are absorbed by their own specific transporters. Many transporters, including the following, are located in the brush border (or apical) membrane: Na^+/B amino acid transporter, β-amino acid transporter, Na^+/glucose cotransporter, fructose transporter, and Na^+/nucleoside transporter. Some transporters are only located in the basolateral membrane: Na^+/A amino acid transporter, Na^+/ASC amino acid transporters, GLUT2 hexose transporters, and Na^+-independent folic acid transporter. Among others, the following exist in both membranes: y^+ amino acid transporters, Na^+/H^+ antiporters, and inorganic phosphate transporter.

Many transporters have isoforms. For example, the facilitated glucose transporters (GLUT) have five different isoforms (Wright *et al.*, 1994). Among them, GLUT5 is found in the brush border membrane, and GLUT2 is located in the basolateral membrane of the human intestine. In the brush border membrane one can find the Na^+/glucose cotransporter (SGLT1), which is responsible for sugar absorption. Glucose and galactose are transported into the enterocyte across the brush border membrane by SGLT1, and then they are transported out across the basolateral membrane by a facilitated sugar transporter (GLUT2). Fructose is tak-

Table I.
Membrane Transporters Expressed in the Intestines

Transporters	Substrates/Specificity	Site[a]	Expressed tissues
1. H^+/Peptide Transporter			
PEPT1	Di-, tripeptides, β-lactam	A	Human intestine, PEPT1 (Liang et al., 1995)
PEPT2	Antibiotics, enalapril	BL?	Human kidney, PEPT2 (Liu et al., 1995), rabbit intestine (Fei et al., 1994); rat intestine (Miyamoto et al., 1996), kidney, PepT1 (Saito et al., 1995)
2. Nucleoside transporters			Ubiquitous, rabbit brain, rabbit cortical synaptosomes, human erythrocytes, hepatocytes
N1	Na^+-dependent, purine-selective	A	Rat liver, cNT1 (Che et al., 1995); rat intestine and liver, SPNTint (Yao et al., 1996), human intestine N1(cif) (Patil and Unadkat, 1997)
N2	Na^+-dependent, pyrimidine nucleosides and adenosine	A	Human kidney, hCNT1 (Ritzel et al., 1997), rat jejunum, rCNT1 (Huang et al., 1994), human intestine, N2(cit)
N3/N4	Broad-selective, Na^+-dependent, purine and pyrimidine nucleosides	A	Intestine (Huang et al., 1994)
es, ei	Broad-selective, Na^+-independent, purine and pyrimidine nucleosides, a low affinity and a high capacity, equilibrative	BL	Human intestine (Chandrasena et al., 1997)
3. Sugar transporters			
Na/D-Glucose	Cotransporter	A	Rat, pig, human intestine, SGLT1 (Hediger et al., 1987)
Hexose	Facilitated diffusion	BL	Human liver, GLUT2 Fukumoto et al., 1988), intestine
D-Fructose	Facilitated diffusion	A	Human jejunum, GLUT5 (Kayano et al., 1990); rat (Rand et al., 1993), rabbit (Miyamoto et al., 1994)
4. Bile acid transporter	Na^+-dependent, taurocholate, glucocorticoid sensitive	A	Hamster ileum, IBAT (Wong et al., 1994); rat ileum (Schneider et al., 1995) and kidney (Christie et al., 1996), ASBT; human liver, NTCP (Hagenbuch and Meier, 1994); rat liver, Nctp (Hagenbuch et al., 1991), LBAT
5. Amino acid transporter			Ubiquitous
Na^+/ B	dipolar α-AA, neutral AA, Arg, Ser	A	Rabbit intestine, riATBO (Kekuda et al., 1997)
Na^+/ $B^{0,+}$	Cl^--dependent, dipolar α-AA, basic AA, Cys	A	—

(continued)

Table I. (*Continued*)
Membrane Transporters Expressed in the Intestines

Transporters	Substrates/Specificity	Site[a]	Expressed tissues
$b^{0,+}$	Na^+-independent, dipolar α-AA, basic AA, Cys	A	Rat, rabbit kidney (NBAT)
y^+	Na^+-independent, basic AA	A, BL	Rat intestine (rBAT), rat kidney
$Na^+/Cl^-/$ Imino	proline, hydroxyproline	A	—
$Na^+/Cl^-/\beta$	β-AA	A	—
$Na^+/K^+/X^-_{AG}$	glutamate, aspartate	A	Rabbit intestine, EAAC1 (Hediger et al., 1995), rat brain and kidney
Na^+/A	dipolar α-AA, imino acids	BL	—
Na^+/ASC	3- and 4- carbon dipolar AA	BL	—
asc	3- and 4- carbon dipolar AA	BL	—
L	Neutral AA with hydrophobic side chains	A, BL	—
6. Organic Anion Transporters			Human liver, OATP Kullak-Ublick et al., 1996), NTCP (Hagenbuch and Meier, 1994); rat liver, oatp Jacquemin et al., 1994), kidney, OAT-K1(Saito et al., 1996); brain
H^+/Monocarboxylic acid	lactic acid, pyruvic acid	A, BL?	Rat intestine, MCT1 Takanaga et al., 1995); human (Garcia et al., 1994); erythrocytes muscle, hamster liver, MCT2 (Garcia et al., 1995)
H^+/SCFA	HCO_3^--dependent, acetic acid, propionic acid, butyric acid	A	—
Na^+/dicarboxylic acid	—		Human kidney (Pajor, 1996), human intestine, rabbit (Pajor, 1995); X. laevis intestine, (L. Bai and Pajor, 1997)
7. Vitamin transporters			
H^+/Nicotinic acid	—	A	—
HCO_3^-/Nicotinic acid	—	A A	—
OH^-/Folic acid	$H^+/$	A A	—
Folic acid	—	BL	Mouse intestine, IFC1(RFC1) (Said et al., 1996)
8. Phosphate transporters	—	A	Flounder intestine, flounder kidney, Na-Pi-I
Na^+/phosphate	—	BL	Na-Pi-II (Werner et al., 1994)
9. Organic cation transporters	small organic cation		Human liver, hOCT1 (Zhang et al., 1997), rat kidney, rOCT (Grundemann et al., 1994), OCT2 (Okuda et al., 1996), NKT (Lopez-Nieto et al., 1997), intestine

(*continued*)

Table I. (*Continued*)
Membrane Transporters Expressed in the Intestines

Transporters	Substrates/Specificity	Site[a]	Expressed tissues
Choline	tetraethylammonium, acetylcholine, N-methylnicotinamide	A	—
10. Fatty acid transporter	oleic acid	A	Small intestine, FAT; liver and intestine, FABP
11. P-glyco-protein	ATP-dependent	A	Ubiquitous, MDR1(Chen *et al.*, 1986)
12. Na^+/H^+/antiporter			Ubiquitous (Orlowski *et al.*, 1992; Sardet *et al.*, 1989; Tse *et al.*, 1991, 1992) (Shallat *et al.*, 1995)
NHE-1		A, BL	Ubiquitous
NHE-2	specific to intestine	BL	Rat intestine (Bookstein *et al.*, 1997)
NHE-3	electroneutral	A	Intestine and kidney (Bookstein *et al.*, 1994)
13. prostaglandin transporter	Na^+/Cl^--dependent, PGE1, PGE2, PGD2, PGF2α		Small intestine, colon, human kidney kidney (hPGT) (Lu *et al.*, 1996)
14. Bicarbonate $Na^+/HCO32^-$	—		Salamander kidney (Romero *et al.*, 1997)

[a]Location of intestinal transporters in the epithelia: A, apical (or brush border) membrane, BL, basolateral membrane.

en up by GLUT5 at the brush border membrane and transported into blood by GLUT2 in the basolateral membrane.

Similarly, four isoforms of Na^+/H^+ antiporters (NHE) have been cloned (Orlowski *et al.*, 1992; Sardet *et al.*, 1989; Shallat *et al.*, 1995; Tse *et al.*, 1991, 1992). NHE-1 is found in both epithelial and nonepithelial cells, and most likely represents the ubiquitous housekeeper involved in regulation of intracellular pH, cell volume, and growth. In polarized epithelia, NHE-1 is localized to the basolateral membrane. NHE-2 is relatively specific for the intestine (Bookstein *et al.*, 1997). NHE-3 is an apical Na^+/H^+ antiporter that mediates electroneutral sodium absorption in the intestine (Bookstein *et all*, 1994).

In the small intestine, Na^+-nutrient-coupled transporters (Hopfer, 1987) and Na^+/H^+ exchangers are found primarily in mature villus cells and absent from crypt cells. This cross-sectional difference may have important clinical implications; if a pathological process destroys primarily surface cells, the absorptive capacity of the epithelium will be impacted to a much larger extent than the secretory function.

It is apparent that the jejunum, cecum, proximal colon, and distal colon exhibit distinctly different transporters (Sellin, 1996). Small intestinal transport is dominated by Na^+-nutrient-coupled absorption and electroneutral sodium absorption mediated by Na^+/H^+ exchanger. The distal colon demonstrates the classical amiloride-inhibitable electrogenic sodium transport pathway. The cecum absorbs sodium by an electrogenic, apical nonselective cation channel. The variation in transport characteristics at different regions is indicative of diverse physiologic function and perhaps pathologic response. Clearly the epithelial response to a particular agonist will be constrained by the presence or absence of specific sets of transporters. The absorption of vitamin B_{12} and bile salts is confined to the terminal ileum. The major site of iron absorption is in the duodenum, with the jejunum and ileum contributing little, if any, to iron transport. In addition to site dependence of membrane transporters, there may be variations in expression of transporters between individuals.

Epithelial transport, such as intestinal salt and water transport, is regulated by several peptide hormones. The intestinal peptide guanylin is ideally suited to play a pivotal role in this regulation. Guanylin is produced by the epithelium and appears to be secreted mucosally to act locally on the apical receptor. The guanylin receptor is a member of the guanylate cyclase (GC-C) family of proteins. Elevation of intracellular cyclic GMP by guanylin mediates the stimulation of Cl^- secretion, which results in increased intestinal fluid secretion. Uroguanylin, a member of the guanylin peptide family, is also an endogenous activator of intestinal guanylate cyclase (GC-C). A cDNA encoding a precursor for human uroguanylin was cloned from a human colon cDNA library and sequenced (Miyazato *et al.*, 1996). Human uroguanylin mRNA is expressed in the gastric fundus and pylorus as well as in the intestine (Miyazato *et al.*, 1997). Overproduction of guanylin is expected to elicit secretory diarrhea similar to that caused by the bacteria that produce peptide analogues of these endogenous peptide hormones (Forte and Currie, 1995).

2.2. Peptide Transporter

It is well known that H^+-dependent peptide transport is located in the brush border membrane of the intestine. A H^+/peptide transporter (PEPT1) has been cloned from human (Liang *et al.*, 1995), rabbit (Fei *et al.*, 1994), and rat (Saito *et al.*, 1995) intestines. Also a Na^+-independent glutathione transport system has been confirmed in human small intestine epithelial cells (Iantomasi *et al.*, 1997).

Peptide transporters have a broad substrate specificity, and mediate the membrane transport of various peptidomimetic drugs as well as small intact peptides. Many β-lactam antibiotics can be transported by a peptide transporter in intestinal tissue preparations, isolated intestinal brush border membrane vesicles the *Xenopus* oocyte expression system (Fei *et al.*, 1994; Liang *et al.*, 1995; Tamai *et al.*, 1997a,b), and Caco-2 cells (Tsuji, 1995). Aminocephalosporins with a free α-amino group, including the following, are good substrates for the peptide transporter: cephradine (Inui *et al.*, 1992; Okano *et al.*, 1986), cefadroxil (Tamai *et al.*, 1994) cephalexin (Tamai *et al.*, 1997a), cephradine (Tomita *et al.*, 1995), cefadroxil (Ganapathy *et al.*, 1995), cephaloglycine (Tamai *et al.*, 1997a), and cefaclor (Tamai *et al.*, 1997a). Cephalosporins without a free α-amino group at the N-terminus have been taken up via the peptide transporter (Ganapathy *et al.*, 1997; Tamai *et al.*, 1997a): cefixime (Oh *et al.*, 1993), ceftibuten (Tamai *et al.*, 1995b; Terada *et al.*, 1997). Transport activity for cephalosporins and dipeptides was completely eliminated by prehybridization of the mRNA with antisense oligonucleotide against the 5′-coding region of rabbit PetT1 cDNA, clearly revealing the involvement of PepT1 (Sai *et al.*, 1996). Aminopenicillins (Boll *et al.*, 1994), such as ampicillin (Oh *et al.*, 1992), amoxicillin, and cyclacillin (Ganapathy *et al.*, 1995), are also absorbed by the peptide transporter. It has been demonstrated that the peptide transporter takes up angiotensin-converting enzyme inhibitors (Amidon and Lee, 1994; Boll *et al.*, 1994; Tamai *et al.*, 1997a) such as captopril (Hu and Amidon, 1988), enalapril (Friedman and Amidon, 1989b), lisinopril (Friedman and Amidon, 1989a), quinapril (Yee and Amidon, 1995), benazepril (Yee and Amidon, 1995), and fosinopril (Friedman and Amidon, 1989b). The peptide transporter is also involved in the absorption of renin inhibitors (Hashimoto *et al.*, 1994; Kramer *et al.*, 1990), anticancer drugs (Takano *et al.*, 1994), and peptidomimetic thrombin inhibitors (Walter *et al.*, 1995).

Recently, arphamenine A, an Arg-Phe analogue without a peptide bond, was shown to be a substrate for the peptide transporter (Enjoh *et al.*, 1996). An amino acid prodrug of acyclovir with no peptide bond is also transported by the peptide transporter (Han, 1997), suggesting a much wider substrate specificity of the peptide transporter than previously suspected.

Substrate specificity for the peptide transporter can be diminished or abolished in one of three ways: esterification of the free carboxylic acid moiety, introduction of a second negative group, and intramolecular steric hindrance of the free carboxylic acid by either side chain with a positively charged nitrogen function or groups capable of hydrogen bond formation (Swaan and Tukker, 1997). Therefore, PEPT1 seems to play important roles in nutritional and pharmacologic therapies through its broad substrate specificity. Design of prodrugs targeting the peptide transporter can be a useful strategy for improving the absorption of small polar drugs that exhibit very poor bioavailability.

2.3. Nucleoside Transporter

There are two broad types of adenosine transporters involved in facilitated-diffusion and active processes driven by the transmembrane sodium gradient (Thorn and Jarvis, 1996). Facilitated-diffusion adenosine carriers may be sensitive (*es*) or insensitive (*ei*) to nanomolar concentrations of the transport inhibitor nitrobenzylthioinosine (NBMPR). The es transporter has a broad substrate specificity, and there is increasing evidence to suggest the presence of isoforms of the es transporter in different cells and species, based on kinetic and molecular properties. The ei transporter also has a broad substrate specificity with a lower affinity for some nucleoside permeants than the es carrier. Sodium-dependent adenosine transport is catalyzed by four distinct systems, N1, N2, N3, and N4. Among them, the rat intestinal N2 transporter (rCNT1) has been cloned (Huang *et al.*, 1994). Recently it has been demonstrated that when *Xenopus laevis* oocytes are microinjected with human jejunal mRNA, four nucleoside transporters are expressed simultaneously, namely the N1 and N2 Na^+-dependent nucleoside transporters and the *es* and the NBMPR-insensitive (*ei*) Na^+-independent transporters (Chandrasena *et al.*, 1997). It was further hypothesized that the concentrative transporters in the brush border membrane and equilibrative transporters in the basolateral membrane are arranged in series in the human jejunal epithelium to allow efficient vectorial transport of nucleosides from the lumen to the blood.

In mammals, the nucleoside transport is an important determinant of the pharmacokinetics (plasma and tissue concentration, disposition) and *in vivo* biological activity of adenosine, as well as nucleoside analogues used in antiviral and anticancer therapies.

2.4. Sugar Transporter

Intestinal sugar transport involves two different types of transporters; one is the Na^+/glucose cotransporter (SGLT1) and the other type is represented by two distinct facilitated transporters in the intestine: the basolateral facilitated glucose transporter (GLUT2) and the brush border fructose transporter (GLUT5). The sugar transporter is specific for D-glucose. D-Galactose and D-mannose, which differ only in the position of one hydroxyl group, have much higher Kt values (lower affinity). The L-forms of sugars have about a 1000 times lower affinity than D-forms for the transporter.

Modification of parent compounds to sugar analogues like *p*-nitrophenyl-β-D-glucopyranoside (Mizuma *et al.*, 1992) or β-D-galactopyranoside (Mizuma *et al.*, 1994a) has been employed to facilitate intestinal absorption and tissue distribution of less permeable compounds by utilizing sugar transporters.

2.5. Bile Acid Transporter

Bile acids are efficiently absorbed by an active, Na^+-dependent transporter in the ileum of mammals to preserve their bile salts via enterohepatic recirculation. The ileal Na$^+$/bile acid cotransporter (IBAT) plays a critical role in the reabsorption of bile acids from the small intestine.

Peptide–bile acid conjugates have been shown to inhibit Na$^+$-dependent ^{3}H-taurocholate uptake into brush border membrane vesicles isolated from rabbit ileum in a concentration-dependent manner. The oligopeptides may be made enterally absorbable by coupling to modified bile acid molecules making use of the specific Na$^+$/ileal acid cotransporter. This finding may be of importance for the design and development of orally active peptide drugs (Kramer et al., 1994).

The use of an inhibitor such as S-8921 to ileal Na$^+$/bile acid cotransporter could decrease serum cholesterol in the normal condition (Hara et al., 1997). For example, crilvastatin, a new drug of the pyrrolidone family, did not alone change the activity of cholesterol 7-α-hydroxylase in the liver despite the marked reduction in both hepatic cholesterogenesis and intestinal absorption of dietary cholesterol (Hajri et al., 1997).

2.6. Amino Acid Transporters

There are many different amino acid transporters in the intestine. Among them, systems B, IMINO, α, A, and ASC are Na$^+$-dependent, and systems y$^+$, b, asc, and L are Na$^+$-independent transporters. Amino acid transporters have highly restrictive substrate specificities.

Gabapentin is absorbed by the large neutral amino acid transporter (Stewart et al., 1993). Several amino acid analogues such as L-α-methyl-dopa (Amidon et al., 1986; Hu and Borchardt, 1990), L-dopa (Shindo et al., 1973), and baclofen (Cercos-Fortea et al., 1995) are also absorbed by the large neutral amino acid transporter. D-Cyclosporin has been reported to be transported by a proton-coupled amino acid transporter (Thwaites et al., 1995).

2.7. Organic Anion Transporters

A proton-coupled monocarboxylic acid specific transport mechanism for pravastatin is likely to be present in the intestine (Tamai et al., 1995a). It was suggested that monocarboxylic acids are transported by at least two independent transporters, namely a proton-coupled transporter for most monocarboxylic acids, including mevalonic acid, pravastatin, and acetic acid, and an anion antiporter for acetic acid, but not for mevalonic acid or pravastatin. Activation of anion antiporter can induce HCO$_3^-$ secretion in intact intestine (Tamai et al., 1997b). Human kidney Na$^+$/dicarboxylate cotransporter (hNaDC-1) has been cloned and also expressed in the intestine (Pajor, 1996). cDNA encoding a novel rat organic anion transporter, OAT-K1, has been cloned. OAT-K1 is expressed exclusively in the renal proximal tubules, and mediates the transport of methorexate. Monocarboxylic acid-type drugs such as benzoic acid (Tsuji et al., 1994) and salicylic acid (Takanaga et al., 1994) are transported by a H^+/monocarboxylic acid transporter in Caco-2 cells.

2.8. Vitamin Transporters

Some water-soluble vitamins, such as thiamine (Na^+-dependent), vitamin C (Na^+-dependent), folic acid, and vitamin B_{12}, can be transported by specific transporters.

A recent study suggests that a specific mechanism is present for the absorption of retinol bound to retinol-binding protein (RBP) (Dew and Ong, 1997). Nicotinic acid is absorbed by two independent active transport mechanisms from small intestine, i.e., a proton cotransporter and an anion antiporter. It was suggested that the pH dependence observed in the intestinal absorption of nicotinic acid is ascribed partly to pH-sensitive and partly to carrier-mediated transport mechanisms in the brush border membrane (Takanaga et al., 1996). Various monocarboxylic acid-like drugs, such as valproic acid, salicylic acid, and penicillins, are absorbed via H^+/nicotinic acid cotransporter and/or HCO_3^-/nicotinic acid exchanger as well as via lactic acid and/or short-chain fatty acid (SCFA) transporters (Tsuji and Tamai, 1996).

Carrier-mediated folate transport in isolated luminal epithelial cells was characterized in terms of pH-dependent and non-pH-dependent components on the basis of their differential sensitivity to the inhibitor stilbene. Methotrexate is an analogue of folic acid, and its intestinal absorption mechanism is similar to the pH-dependent, carrier-mediated absorption of folic acid (Zimmerman, 1992a, 1992b).

2.9. Phosphate Transporter

The intestinal phosphate transporter is Na^+-dependent and located in the brush border membrane. Foscarnet, an antiviral drug, was absorbed in the small intestine via a carrier-mediated mechanism shared with phosphate (Swaan and Tukker, 1995; Tsuji and Tamai, 1989). Fosfomycin, a water-soluble antibiotic, is taken up by Na^+/phosphate transporter (Ishizawa et al., 1990, 1991, 1992). Considering the structure of phosphate-mimic drugs such as phosphonoacetic acid and phosphonopropionic acid, relatively small molecules containing a phosphate moiety may be utilized as substrates for the intestinal sodium-dependent phosphate transporter, resulting in enhanced intestinal absorption.

Drug Transport and Targeting

2.10. Bicarbonate Transporter

Bicarbonate transporters are the principal regulators of pH in animal cells, and play a vital role in acid–base movement in the stomach, pancreas, intestine, kidney, reproductive system, and central nervous system. The functional family of HCO_3^- transporters includes $Cl^-–HCO_3^-$ exchangers (cloned), three Na^+/HCO_3^- cotransporters, a K^+/HCO_3^- cotransporter, and a Na-driven $Cl^-–HCO_3^-$ exchanger. The renal electrogenic sodium bicarbonate cotransporter (NBC) was electrogenic, Na^+- and HCO_3^--dependent, and blocked by the anion-transport inhibitor DIDS (Romero et al., 1997).

2.11. Organic Cation Transporter

Facilitated diffusion, a carrier-mediated transport system for choline, has been suggested in the isolated brush border membrane vesicles from rat small intestine (Saitoh et al., 1992).

2.12. Fatty Acid Transporter

Human intestinal epithelial cells have been shown to have a saturable component in the transport of long-chain fatty acids such as palmitate (16:0) and oleate (18:1), but not short-chain fatty acids (Trotter et al., 1996).

Intestinal fatty acid-binding protein (I-FABP) may be involved in the uptake and/or specific targeting of fatty acid to subcellular membrane sites (Hsu and Storch, 1996).

2.13. P-Glycoprotein

P-Glycoprotein (P-gp) actively pumps a number of antineoplastic drugs, such as etoposide, out of cancer cells, and causes multidrug resistance. P-gp is also expressed at the brush border membrane of the small intestine under normal physiological conditions. Many investigators have studied the potential role of P-glycoprotein in the intestine as a limiting factor in drug absorption and bioavailability of certain drugs (Aungst and Saitoh, 1996; Hunter et al., 1993; Karlsson et al., 1993; Leu and Huang, 1995). Both the broad substrate specificity of P-gp and the strategic localization of this transport protein at the brush border in the intestinal mucosa may be crucial factors providing an active barrier against efficient absorption of many xenobiotics and various clinically important drugs. Recent findings using mdr1a knockout mice clearly showed that mdr1a P-gp is involved in drug counterflux phenomena at the level of the intestinal mucosa (Mayer et al., 1996; Schinkel et al., 1997; Sparreboom et al., 1997).

The drugs secreted by P-glycoprotein into intestinal lumen include anthracyclines (vincristine) (Hunter et al., 1991; Meyers et al., 1991), paclitaxel (Taxol) (Sparreboom et al., 1997), digoxin (Mayer et al., 1996), PSC833 (Mayer et al., 1997), ofloxacin (chiral fluoroquinolone) (Rabbaa et al., 1996, 1997), ciprofloxacin (Cavet et al., 1997), quinidine (Su and Huang, 1996), etoposide (Hunter et al., 1993; Leu and Huang, 1995; Wils et al., 1994), pristinamycin (Phung-Ba et al., 1995), cyclosporin A (Augustijns et al., 1993), peptides (Burton et al., 1993; Saitoh and Aungst, 1997), celoprolol (Karlsson et al., 1993), and other organic cations (Hsing et al., 1992; Saitoh and Aungst, 1995).

Some inhibitors to P-glycoprotein can increase bioavailability by blocking secretion. Inhibition of P-glycoprotein by a variety of modulators (verapamil, 1,9-dideoxyforskolin, nifedipine, and taxotere) has been associated with an increased vinblastine absorptive permeability (Hunter et al., 1993). A ten-fold increased oral bioavailability of paclitaxel was reported in mice treated with the P-glycoprotein blocker SDZ PSC 833, as shown in Fig. 1 (van Asperen et al., 1997). The use of P-glycoprotein-inhibiting agents such as quinidine increased the bioavailability of etoposide (Leu and Huang, 1995).

There is some evidence to suggest the existence of other efflux systems in the intestine. The intestinal absorption of some β-lactam antibiotics such as cephaloridine and cefoperazone was decreased by an energy-dependent efflux system in rat intestine. This energy-demanding efflux system was distinct from P-glycoprotein-mediated transport. A free α-amino group in the molecule is an important factor for reducing an affinity with the efflux system (Saitoh et al., 1997).

3. PRODRUGS: TARGETING INTESTINAL TRANSPORTERS

Gastrointestinal transport of drugs can be improved by prodrug strategies. Traditionally, the chemical modification of a drug aims at the enhancement of solubility and permeability. For water-insoluble hydrophobic drugs, amino acid derivatives have been synthesized to increase solubility, and targeted to be converted back to the parent drugs by intestinal brush border membrane enzymes (Amidon et al., 1985; Stewart et al., 1986). For hydrophilic drugs, a common prodrug approach is to use simple esters of a polar parent drug to increase the hydrophobicity of the drug. Parent drugs containing —OH or —COOH groups can often be converted to ester-type produrgs from which the active drugs are regenerated by esterases within the body. Another strategy for hydrophilic drugs is to design suit-

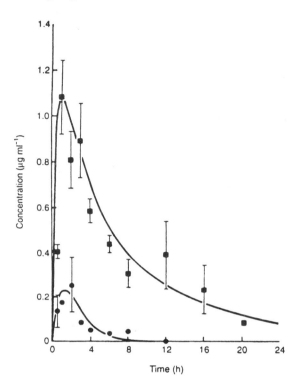

Figure 1. Plasma concentration–time curve after oral administration of paclitaxel to mice. Per kilogram of body weight, the mice received either 10 mg of paclitaxel alone (●, control group) or in combination with 50 mg of SDZ PSC 833 as an inhibitor (■). Symbols and error bars represent mean ± standard error. Reprinted with permission from van Asperen *et al.* (1997). Copyright 1997 Churchill Livingstone, Inc.

able substrates utilizing a specific intestinal transporter. It would be better if the designed prodrug is not a substrate for any secretory transporters such as P-glycoprotein.

Choosing a proper intestinal transporter is an important step toward a successful prodrug strategy. The target transporter should have a high capacity for intestinal absorption, and targeting more than one transporter, if possible, may be more efficient to increase the absorption from the intestine. However, the transporter needs to have a broad specificity, tolerating chemical modifications to prodrugs. In this regard, peptide or nucleoside transporters are more suitable targets for structural modification than glucose and amino acid transporters. It would be beneficial if the target transporter is located in both apical and basolateral membranes, enhancing systemic bioavailability. For example, glucose can be taken up by SGLT and GLUT5 in the apical membrane, and further transported to blood by GLUT2 at a fast rate. The intestinal peptide transporter PEPT1 can be a target transporter because most digested oligopeptides are primarily absorbed by H^+/peptide cotransporter and because its specificity is known to be broad. Chemical modification strategies may include the modification of amide bonds or the design of compounds bearing nonpeptide templates such as SB 208651, a potent orally active GPIIb/IIIa antagonist (Samanen et al., 1996). There are many reviews of the peptide prodrug approach to increasing oral bioavailability (J. P. F. Bai et al., 1994; Stewart and Taylor, 1995; Yee and Amidon, 1995).

Some examples of peptidyl prodrugs that utilize the H^+/peptide transporter for improving bioavailability are briefly introduced below. A schematic representation of a general prodrug strategy to increase intestinal drug absorption is given in Fig. 2.

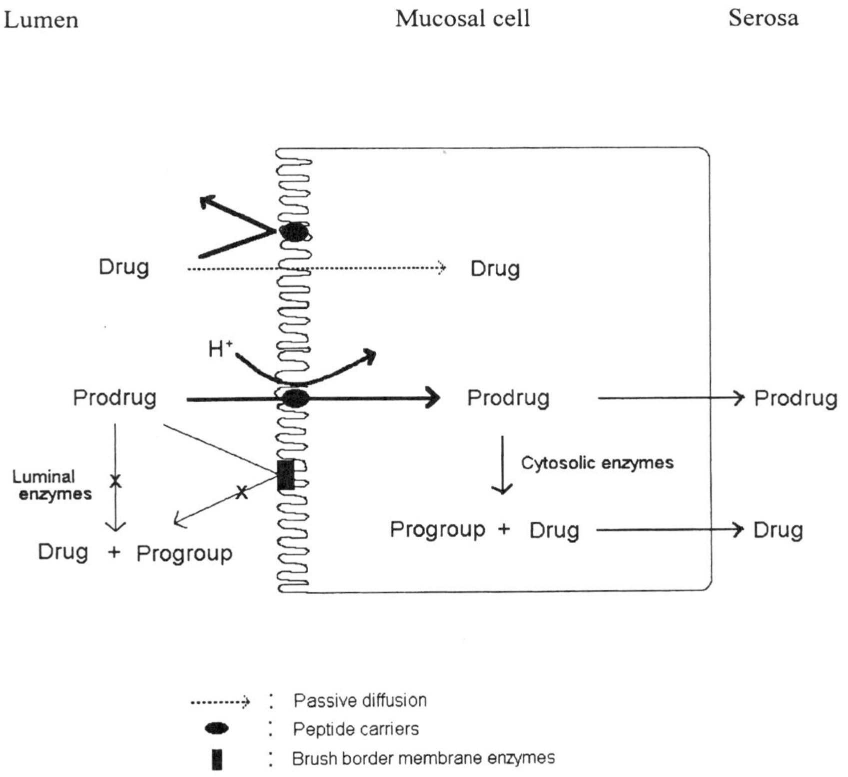

Figure 2. Schematic representation of the prodrug strategy to increase intestinal drug absorption (Han, 1997).

Table II.
Dimensionless Wall Permeability (P_W^*) of L-a-Methyldopa and Its Dipeptidyl Derivatives[a]

Compound	Permeability at given concentation[b]		
	At 1 mM	At 0.1 mM	At 0.01 mM
L-α-Methyldopa (I)	0.41 (0.11)	0.4 (0.22)	0.43 (0.14)
Gly-I		4.34 (0.27)	
Pro-I		1.68 (0.23)	
I-Pro		5.41 (0.55)	
Phe-I		5.29 (1.57)	
I-Phe	4.30 (0.30)	10.22 (0.45)	10.9 (1.8)

[a]Taken from Hu et al. (1989).
[b]Mean (SE).

The poorly absorbed active ACE inhibitors have been designed as ester prodrugs to potentiate their biological *in vivo* activity. Enalapril, the ethyl ester prodrug of enalaprilat, is absorbed much better than its parent drug. Higher bioavailability of enalapril is partly due to the higher lipophilicity of the prodrug, resulting in a greater degree of membrane partitioning. The intestinal absorption of the ACE inhibitors indicates that esterification of one of the carboxylic acid groups results in a more peptide-like compound that is transported by the peptide carrier (Friedman and Amidon, 1989b). This conclusion is consistent with other results with β-lactam antibiotics that suggest that a free N-terminal α-amino group is not an absolute requirement for peptide transporter (P. F. Bai et al., 1991; Oh et al., 1993). The peptide transporter thus appears to have a broad specificity, and use of prodrugs designed to exploit this broad specificity may be applicable to other drugs in the di- and tripeptide size range. An example of this strategy for poorly absorbed polar drugs targeted to the peptide transporter is L-α-methyldopa (Hu et al., 1989). Various dipeptides of L-α-methyldopa have been synthesized and demonstrated to have a significantly higher membrane permeability than the parent drug and to be metabolized to the active drug in the mucosal tissue (Table II). The use of its peptidyl prodrug has increased the *in vivo* systemic availability of L-α-methyldopa to nearly 100% as shown in Fig. 3 (Yee and Amidon, 1995). These results show that a transporter-targeted prodrug strategy is effective in improving the bioavailability through increased membrane permeability of the polar amino acid drug, L-α-methyldopa (Amidon and Lee, 1994; Tsuji et al., 1990). D-Phenylglycine prodrug of L-α-methyldopa has also been shown to be a feasible delivery tool in carrying the parent drug through the intestine (Wang et al., 1996). Once amino acid prodrugs are transported across the brush border membrane, reconversion to the amino acid can be accomplished by peptidases. For example, the structure–activity relationships suggest that prolidase may be a useful target for the reconversion of proline-containing prodrugs after transport into the intestinal mucosal cell by the peptide transporter (J. P. Bai et al., 1992).

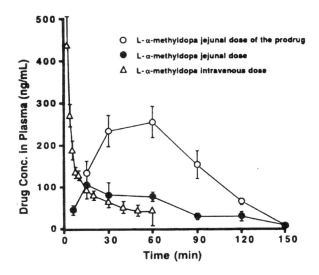

Figure 3. Plasma profiles of L-α-methyldopa following intravenous dose of L-α-methyldopa and jejunal dose of L-α-methyldopa-phenylalanine and L-α-methyldopa. Reprinted with permission from Yee and Amidon (1995). Copyright 1995 American Chemical Society.

Amino acid ester prodrugs have been investigated for improving oral absorption (Amidon et al., 1998). Permeability studies in the rat jejunum have indicated that the wall permeability of the L-valyl ester of acyclovir (valcyclovir) was approximately tenfold higher than that of acyclovir, while the glycyl ester showed a fourfold enhancement, and the D-valyl ester exhibited much smaller permeability (Fig. 4). Correspondingly, the L-valyl ester of azidothymidine (AZT) enhanced the permeability of AZT threefold (Han, 1997).

Other approaches have also appeared in the literature. Chemically modified thyrotropin-releasing hormone derivative with lauric acid (Lau-TRH) was rapidly bound to the brush border membrane in the small intestine, where Lau-TRH is converted to TRH, and it is efficiently transported by an oligopeptide transporter which exists in the upper small intestine (Tanaka et al., 1996). To overcome the low bioavailability of oral L-dopa due to decarboxylation in the gut wall, a tripeptide prodrug of L-dopa, p-glu-L-dopa, was designed to be absorbed via the intestinal peptide transporter (Fig. 5) so as to minimize the decarboxylation in the gut wall, and to be converted to L-dopa by peptidases, with cleavage by pyroglutamyl aminopeptidase I to L-dopa-pro as the rate-limiting step (J. P. Bai, 1995). A strategy for the enhancement of intestinal absorption by derivatization to sugar analogues has also been applied to peptides. Mono- or disaccharide derivatives of Tyr-Gly-Gly appeared on the serosal side, and no metabolites were detected following the addition of these compounds to intestinal mucosa. The coupling of unstable

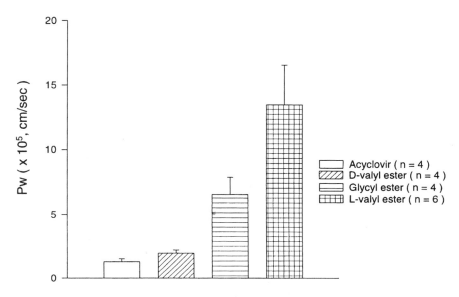

Figure 4. Intestinal permeability of acyclovir prodrugs in the rat. Reprinted with permission from Amidon *et al.* (1998). Copyright 1998 Munksgard International Publishers.

peptides with sugars has been shown to improve both hydrolytic stability and membrane permeation (Mizuma *et al.*, 1994b).

In summary, the transporters are feasible targets when the passive transmembrane transport is negligible because of its charge or hydrophilicity. Designing a prodrug as an efficient substrate to a specific transporter is a potentially powerful approach to improving the absorption of the small to moderately sized peptide and peptidomimetic drugs.

4. DISORDERS RELATED TO INTESTINAL TRANSPORTERS

4.1. Water and Electrolyte Transporters

Sophisticated models of epithelial ion secretion have clearly facilitated our understanding of the underlying pathophysiology of diarrhea diseases. Cholera-induced diarrhea involves complex steps, including electrogenic chloride secretion. The etiology of inflammatory bowel disease is usually deemed multifactorial, and presumed to result in part from the loss of absorptive function in an injured epithelium by inflammation and ulceration.

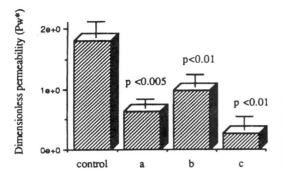

Figure 5. Effects of inhibitors on the *in situ* intestinal permeability of *p*-Glu-L-dopa-Pro: (a) 20 mM captopril, (b) 80 mM Gly-Gly and 5 mM Gly-Pro, and (c) 2 mM cephradine. From J. B. Bai (1995).

4.2. Sugar Transporters

Carbohydrates remaining in the intestinal lumen increase the osmotic pressure of the lumen, and cause bacterial infections due to defects in digestion and absorption. The osmotic retention of water in the lumen leads to diarrhea. Intolerance to glucose and galactose has been documented in rare instances, and in these patients the transporter for glucose has not been found. These patients show no symptoms when fed fructose. This can be easily explained by the fact that glucose and chemically related sugars are absorbed by a Na^+-dependent secondary active transport process (SGLT), whereas fructose is absorbed by Na^+-independent facilitated diffusion (GLUT).

4.3. Amino Acid Transporters

Intestinal malabsorption of amino acids occurs in various hereditary diseases. Cystinuria is a disease characterized partly by defective transport of cystine in the proximal renal tubule and the small bowel (Johnson, 1997). Cystinuria patients lack the basic amino acid transporter. As shown in Fig. 6, the dipeptide Arg–Leu is well absorbed in these patients, but amino acid arginine is not absorbed. Hartnup disease is a hereditary condition in which the active transport of several neutral amino acids is deficient in both renal tubules and the small intestine. It exhibits a pellagra-like syndrome, but is benign, and patients fare quite well by absorbing oligopeptides.

4.4. Bile Acid Transporter

Primary bile acid malabsorption is an idiopathic intestinal disorder associated with congenial diarrhea, steatorrhea, interruption of the enterohepatic circula-

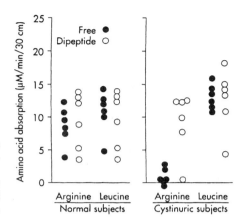

Figure 6. Jejunal absorption of free arginine and leucine during perfusion of solutions containing L-arginine (1 mM) and L-leucine (1 mM) or L-arginyl-L-leucine (1 mM). Results are from studies carried out in six normal and six cystinuric patients. From Johnson (1997).

tion of bile acids, and reduced plasma cholesterol levels. Primary bile acid malabsorption is caused by mutations in the ileal sodium-dependent bile acid transporter gene (Oelkers *et al.*, 1997). A dysfunctional isoform of human ileal Na^+/bile acid cotransporter has been identified in a patient diagnosed with Crohn disease (Wong *et al.*, 1995).

4.5. Vitamin Transporters

Congenital errors of folate metabolism can be related either to defective transport of folate through various cells or to defective intracellular utilization of folate due to enzyme deficiencies. Defective transport of folate across the intestine and the blood–brain barrier has been reported in the condition called "congenital malabsorption of folate." This disease is characterized by a severe megaloblastic anemia of early appearance associated with mental retardation. Anemia is folate-responsive, but neurologic symptoms are only poorly improved because of the inability to maintain adequate levels of folate in the CSF. A familial defect of cellular uptake has been described in a family with a high frequency of aplastic anemia or leukemia. An isolated defect in folate transport into CSF has been identified in a patient suffering from a cerebellar syndrome and pyramidal tract dysfunction (Zittoun, 1995).

5. SUMMARY

A wide variety of transporters are found in the intestine, and are involved in the membrane transport of daily nutrients as well as drugs. These intestinal transporters are located in the brush border membrane as well as basolateral membrane. Each transporter exhibits its own substrate specificity, and some have broader specificities than others. In addition, the distribution and characteristics of the intestinal transporters exhibit regional differences along the intestine, implying diverse physiologic functions and in some cases pathologic responses. Indeed several genetic disorders have been shown to result from deficient intestinal transporters.

The development of prodrugs that target to intestinal transporters has been successful in improving oral absorption. For example, the intestinal peptide transporter is utilized in order to increase the bioavailability of several classes of peptidomimetic drugs, especially ACE inhibitors and β-lactam antibiotics. The bioavailability of poorly absorbed drugs can be improved by utilization of the transporters responsible for the intestinal absorption of various solutes and/or by inhibiting the transporter involved in the efflux system.

Recent advances in gene cloning and molecular biology techniques make it possible to study the characteristics and distribution of transporters at the molecular level. Based on molecular characterizations of membrane transporters and accumulated biochemical data on their specificities and kinetics, structural modification and targeting of a specific transporter is a promising strategy for the design of drugs that improve bioavailability and tissue distribution.

REFERENCES

Amidon, G. L., and Lee, H. J., 1994, Absorption of peptide and peptidomimetic drugs, *Annu. Rev. Pharmacol. Toxicol.* **34**:321–341.

Amidon, G. L., Stewart, B. H., and Pogany, S., 1985, Improving the intestinal mucosal cell uptake of water insoluble compounds, *J. Controlled Release* **2**:13–26.

Amidon, G. L., Merfeld, A. E., and Dressman, J. B., 1986, Concentration and pH dependency of alpha-methyldopa absorption in rat intestine, *J. Pharm. Pharmacol.* **38**:363–368.

Amidon, G. L., Han, H., Oh, D.-M., Walter, E., and Hilfinger, J. M., 1998, Oral administration of peptide and protein drugs, in: *Peptide and Protein Drug Delivery, Alfred Benzon Symposium 43* (S. Frokjar, L. Christrup, and P. Krogsgaard-Larsen, eds.), Munksgaard, Copenhagen, pp. 146–152.

Augustijns, P. F., Bradshaw, T. P., Gan, L. S., Hendren, R. W., and Thakker, D. R., 1983, Evidence for a polarized efflux system in Caco-2 cells capable of modulating cyclosporin A transport, *Biochem. Biophys. Res. Commun.* **197**:360–365.

Aungst, B. J., and Saitoh, H., 1996, Intestinal absorption barriers and transport mechanisms, including secretory transport, for a cyclic peptide, fibrinogen antagonist, *Pharmaceut. Res.* **13**:114–119.

Bai, J. P., 1995, pGlu-L-Dopa-Pro: A tripeptide prodrug targeting the intestinal peptide transporter for absorption and tissue enzymes for conversion, *Pharmaceut. Res.* **12**:1101–1104.

Bai, J. P., Hu, M., Subramanina, P., Mosberg, H. I., and Amidon, G. L., 1992, Utilization of peptide carrier system to improve intestinal absorption: Targeting prolidase as a pro-drug-converting enzyme, *J. Pharmaceut. Sci.* **81**:113–116.

Bai, J. P. F., Stewart, B. H., and Amidon, G. L., 1994, Gastrointestinal transport of peptide and protein drugs and prodrugs, in: *Pharmacokinetics of Drugs* (P. G. Welling and L. P. Balant, eds.), Springer-Verlag, Berlin, pp. 189–206.

Bai, L., and Pajor, A. M., 1997, Expression cloning of NaDC-2, an intestinal Na(+)- or Li(+)-dependent dicarboxylate transporter, *Am. J. Physiol.* **273**:G267–G274.

Bai, P. F., Subramanian, P., Mosberg, H. I., and Amidon, G. L., 1991, Structural requirements for the intestinal mucosal-cell peptide transporter: The need for N-terminal alpha-amino group, *Pharmaceut. Res.* **8**:593–599.

Boll, M., Markovich, D., Weber, W. M., Korte, H., Daniel, H., and Murer, H., 1994, Expression cloning of a cDNA from rabbit small intestine related to proton-coupled transport of peptides, beta-lactam antibiotics and ACE-inhibitors, *Pflugers Arch. Eur. J. Physiol.* **429**:146–149.

Bookstein, C., DePaoli, A. M., Xie, Y., Niu, P., Musch, M. W., Rao, M. C., and Chang, E. B., 1994, Na^+/H^+ exchangers, NHE-1 and NHE-3, of rat intestine. Expression and localization, *J. Clin. Invest.* **93**:106–113.

Bookstein, C., Xie, Y., Rabenau, K., Musch, M. W., McSwine, R. L., Rao, M. C., and Chang, E. B., 1997, Tissue distribution of Na^+/H^+ exchanger isoforms NHE2 and NHE4 in rat intestine and kidney, *Am. J. Physiol.* **273**:C1496–C1505.

Burton, P. S., Conradi, R. A., Hilgers, A. R., and Ho, N. F., 1993, Evidence for a polarized efflux system for peptides in the apical membrane of Caco-2 cells, *Biochem. Biophys. Res. Commun.* **190**:760–766.

Cavert, M. E., West, M., and Simmons, N. L., 1997, Fluoroquinolone (ciprofloxacin) secretion by human intestinal epithelial (Caco-2) cells, *Br. J. Pharmacol.* **121**:1567–1578.

Cercos-Fortea, T., Polache, A., Nacher, A., Cejudo-Ferragud, E., Casabo, V. G., and Merino, M., 1995, Influence of leucine on intestinal baclofen absorption as a model compound of neutral alpha-aminoacids, *Biopharmaceut. Drug Disposition* **16**:563–577.

Chandrasena, G., Giltay, R., Patil, S. D., Bakken, A., and Unadkat, J. D., 1997, Functional expression of human intestinal Na^+-dependent and Na^+-independent nucleoside transporters in *Xenopus laevis* oocytes, *Biochem. Pharmacol.* **53**:1909–1918.

Che, M., Ortiz, D. F., and Arias, I. M., 1995, Primary structure and functional expression of a cDNA encoding the bile canalicular, purine-specific Na(+)-nucleoside cotransporter, *J. Biol. Chem.* **270**:13596–13599.

Chen, C. J., Chin, J. E., Ueda, K., Clark, D. P., Pastan, I., Gottesman, M. M., and Roninson, I. B., 1986, Internal duplication and homology with bacterial transport proteins in the mdr1 (P-glycoprotein) gene from multidrug-resistant human cells, *Cell* **47**:381–389.

Christie, D. M., Dawson, P. A., Thevananther, S., and Shneider, B. L., 1996, Comparative analysis of the ontogeny of a sodium-dependent bile acid transporter in rat kidney and ileum, *Am. J. Physiol.* **271**:G377–G385.

Dew, S. E., and Ong, D. E., 1997, Absorption of retinol from the retinol:retinol-binding protein complex by small intestinal gut sheets from the rat, *Arch. Biochem. Biophys.* **338**:233–236.

Enjoh, M., Hashimoto, K., Arai, S., and Shimizu, M., 1996, Inhibitory effect of arphamenine A on intestinal dipeptide transport, *Biosch. Biotechnol. Biochem.* **60**:1893–1895.

Fei, Y. J., Kanai, Y., Nussberger, S., Ganapathy, V., Leibach, F. H., Romero, M. F., Singh, S. K., Boron, W. F., and Hediger, M. A., 1994, Expression cloning of a mammalian proton-coupled oligopeptide transporter, *Nature* **368**:563–566.

Forte, L. R., and Currie, M. G., 1995, Guanylin: A peptide regulator of epithelial transport, *FASEB J.* **9**:643–650.

Friedman, D. I., and Amidon, G. L., 1989a, Intestinal absorption mechanism of dipeptide angiotensin converting enzyme inhibitors of the lysyl-proline type: Lisinopril and SQ 29,852, *J. Pharmaceut. Sci.* **78**:995–998.

Friedman, D. I., and Amidon, G. L., 1989b, Passive and carrier-mediated intestinal absorption components of two angiotensin converting enzyme (ACE) inhibitor prodrugs in rats: Enalapril and fosinopril, *Pharmaceut. Res.* **6**:1043–1047.

Fukumoto, H., Seino, S., Imura, H., Seino, Y., Eddy, R. L., Fukushima, Y., Byers, M. G., Shows, T. B., and Bell, G. I., 1988, Sequence, tissue distribution, and chromosomal localization of mRNA encoding a human glucose transporter-like protein, *Proc. Natl. Acad. Sci. USA* **85**:5434–5438.

Ganapathy, M. E., Brandsch, M., Prasad, P. D., Ganapathy, V., and Leibach, F. H., 1995, Differential recognition of beta-lactam antibiotics by intestinal and renal peptide transporters, PEPT 1 and PEPT 2, *J. Biol. Chem.* **270**:25672–25677.

Ganapathy, M. E., Prasad, P. D., Mackenzie, B., Ganapathy, V., and Leibach, F. H., 1997, Interaction of anionic cephalosporins with the intestinal and renal peptide transporters PEPT 1 and PEPT 2, *Biochim. Biophys. Acta* **1324**:296–308.

Garcia, C. K., Goldstein, J. L., Pathak, R. K., Anderson, R. G., and Brown, M. S., 1994, Molecular characterization of a membrane transporter for lactate, pyruvate, and other monocarboxylates: Implications for the Cori cycle, *Cell* **76**:865–873.

Garcia, C. K., Brown, M. S., Pathak, R. K., and Goldstein, J. L., 1995, cDna cloning of Mct2, a second monocarboxylate transporter expressed in different cells that Mct1, *J. Biol. Chem.* **270**:1843–1849.

Grundemann, D., Gorboulev, V., Gambaryan, S., Veyhl, M., and Koepsell, H., 1994, Drug excretion mediated by a new prototype of polyspecific transporter, *Nature* **372**:549–552.

Hagenbuch, B., and Meier, P. J., 1994, Molecular cloning, chromosomal localization, and functional characterization of a human liver Na^+/bile acid cotransporter, *J. Clin. Invest.* **93**:1326–1331.

Hagenbuch, B., Stieger, B., Foguet, M., Lubbert, H., and Meier, P. J., 1991, Functional expression cloning and characterization of the hepatocyte Na^+/bile acid cotransport system, *Proc. Natl. Acad. Sci. USA* **88**:10629–10633.

Hajri, T., Chanussot, F., Ferezou, J., Riottot, M., Lafont, H., Laruelle, C., and Lutton, C., 1997, Reduced cholesterol absorption in hamsters by crilvastatin, a new 3-hydroxy-3-methylglutaryl coenzyme A reductase inhibitor, *Eur. J. Pharmacol.* **320**:65–71.

Han, H.-k., 1997, A new discovery of improved oral drug absorption targeting the hPEPT1 transporter, Ph.D. dissertation, University of Michigan, Ann Arbor, Michigan.

Hara, S., Higaki, J., Higashino, K., Iwai, M., Takasu, N., Miyata, K., Tonda, K., Nagata, K., Goh, Y., and Mizui, T., 1997, S-8921, an ileal Na^+/bile acid cotransporter inhibitor decreases serum cholesterol in hamsters, *Life Sci.* **60**:365–370.

Hashimoto, N., Fujioka, T., Toyoda, T., Muranushi, N., and Hirano, K., 1994, Renin inhibitor: Transport mechanism in rat small intestinal brush-border membrane vesicles, *Pharmaceut. Res.* **11**:1448–1451.
Hediger, M. A., Coady, M. J., Ikeda, T. S., and Wright, E. M., 1987, Expression cloning and cDNA sequencing of the Na$^+$/glucose co-transporter, *Nature* **330**:379–381.
Hediger, M. A., Kanai, Y., You, G., and Nussberger, S., 1995, Mammalian ion-coupled solute transporters, *J. Physiol.* **482**:7S–17S.
Hopfer, J., 1987, Membrane transport mechanisms for hexose and amino acids in the small intestine, in: *Physiology of the Gastrointestinal Tract* (L. R. Johnson, ed.), Raven Press, New York, pp. 1499–1526.
Hsing, S., Gatmaitan, Z., and Arias, I. M., 1992, The function of Gp170, the multidrug-resistance gene product, in the brush border of rat intestinal mucosa, *Gastroenterology* **102**:879–885.
Hsu, K. T., and Storch, J., 1996, Fatty acid transfer from liver and intestinal fatty acid-binding proteins to membranes occurs by different mechanisms, *J. Biol. Chem.* **271**:13317–13323.
Hu, M., and Amidon, G. L., 1988, Passive and carrier-mediated intestinal absorption components of captopril, *J. Pharmaceut. Sci.* **77**:1007–1011.
Hu, M., and Borchardt, R. T., 1990, Mechanism of L-alpha-methyldopa transport through a monolayer of polarized human intestinal epithelial cells (Caco-2), *Pharmaceut. Res.* **7**:1313–1319.
Hu, M., Subramanian, P., Mosberg, H. I., and Amidon, G. L., 1989, Use of the peptide carrier system to improve the intestinal absorption of L-alpha-methyldopa: Carrier kinetics, intestinal permeabilities, and *in vitro* hydrolysis of dipeptidyl derivatives of L-alpha-methyldopa, *Pharmaceut. Res.* **6**:66–70.
Huang, Q. Q., Yao, S. Y., Ritzel, M. W., Paterson, A. R., Cass, C. E., and Young, J. D., 1994, Cloning and functional expression of a complementary DNA encoding a mammalian nucleoside transport protein, *J. Biol. Chem.* **269**:17757–17760.
Hunter, J., Hirst, B. H., and Simmons, N. L., 1991, Epithelial secretion of vinblastine by human intestinal adenocarcinoma cell (HCT-8 and T84) layers expressing P-glycoprotein, *Br. J. Cancer* **64**:437–444.
Hunter, J., Hirst, B. H., and Simmons, N. L., 1993, Drug absorption limited by P-glycoprotein-mediated secretory drug transport in human intestinal epithelial Caco-2 cell layers, *Pharmaceut. Res.* **10**:743–749.
Iantomasi, T., Favilli, F., Marraccini, P., Magaldi, T., Bruni, P., and Vincenzini, M. T., 1997, Glutathione transport system in human small intestine epithelial cells, *Biochim. Biophys. Acta* **1330**:274–283.
Inui, K., Yamamoto, M., and Saito, H., 1992, Transepithelial transport of oral cephalosporins by monolayers of intestinal epithelial cell line Caco-2: Specific transport systems in apical and basolateral membranes, *J. Pharmacol. Exp. Ther.* **261**:195–201.
Ishizawa, T., Tsuji, A., Tamai, I., Terasaki, T., Hosoi, K., and Fukatsu, S., 1990, Sodium and pH dependent carrier-mediated transport of antibiotic, fosfomycin, in the rat intestinal brush-border membrane, *J. Pharmacobio-Dynamics* **13**:292–300.
Ishizawa, T., Hayashi, M., and Awazu, S., 1991, Effect of carrier-mediated transport system on intestinal fosfomycin absorption *in situ* and *in vivo*, *J. Pharmacobio-Dynamics* **14**:82–86.

Ishizawa, T., Sadahiro, S., Hosoi, K., Tamai, I., Terasaki, T., and Tsuji, A., 1992, Mechanisms of intestinal absorption of the antibiotic, fosfomycin, in brush-border membrane vesicles in rabbits and humans, *J. Pharmacobio-Dynamics* **15**:481–489.

Jacquemin, E., Hagenbuch, B., Stieger, B., Wolkoff, A. W., and Meier, P. J., 1994, Expression cloning of a rat liver Na(+)-independent organic anion transporter, *Proc. Natl. Acad. Sci. USA* **91**:133–137.

Johnson, L. R., 1997, Digestion and absorption, in: *Gastrointestinal Physiology* (L. R. Johnson, ed.), Mosby-Year Book, St. Louis, Missouri, pp. 113–134.

Karlsson, J., Kuo, S. M., Ziemniak, J., and Artursson, P., 1993, Transport of celiprolol across human intestinal epithelial (Caco-2) cells: Mediation of secretion by multiple transporters including P-glycoprotein, *Br. J. Pharmacol.* **110**:1009–1016.

Kayano, T., Burant, C. F., Fukumoto, H., Gould, G. W., Fan, Y. S., Eddy, R. L., Byers, M. G., Shows, T. B., Seino, S., and Bell, G. I., 1990, Human facilitative glucose transporters. Isolation, functional characterization, and gene localization of cDNAs encoding an isoform (GLUT5) expressed in small intestine, kidney, muscle, and adipose tissue and an unusual glucose transporter pseudogene-like sequence (GLUT6), *J. Biol. Chem.* **265**:13276–13282.

Kekuda, R., Torres-Zamorano, V., Fei, Y. J., Prasad, P. D., Li, H. W., Mader, L. D., Leibach, F. H., and Ganapathy, V., 1997, Molecular and functional characterization of intestinal Na(+)-dependent neutral amino acid transporter B0, *Am. J. Physiol.* **272**:G1463–1472.

Kramer, W., Girbig, F., Futjahr, U., Kleemann, H. W., Leipe, I., Urbach, H., and Wagner, A., 1990, Interaction of renin inhibitors with the intestinal uptake system for oligopeptides and beta-lactam antibiotics, *Biiochim. Biophys. Acta* **1027**:25–30.

Kramer, W., Wess, G., Neckermann, G., Schubert, G., Fink, J., Girbig, F., Gutjahr, U., Kowalewski, S., Baringhaus, K. H., Boger, G., *et al.*, 1994, Intestinal absorption of peptides by coupling to bile acids, *J. Biol. Chem.* **269**:10621–10627.

Kullak-Ublick, G. A., Beuers, U., Meier, P. J., Domdey, H., and Paumgartner, G., 1996, Assignment of the human organic anion transporting polypeptide (OATP) gene to chromosome 12p12 by fluorescence *in situ* hybridization, *J. Hepatol.* **25**:985–987.

Leu, B. L., and Huang, J. D., 1995, Inhibition of intestinal P-glycoprotein and effects on etoposide absorption, *Cancer Chemother. Pharmacol.* **35**:432–436.

Liang, R., Fei, Y. J., Prasad, P. D., Ramamoorthy, S., Han, H., Yang-Feng, T. L., Hediger, M. A., Ganapathy, V., and Leibach, F. H., 1995, Human intestinal H^+/peptide cotransporter. Cloning, functional expression, and chromosomal localization. *J. Biol. Chem.* **270**:6456–6463.

Liu, W., Liang, R., Ramamoorthy, S., Fei, Y. J., Ganapathy, M. E., Hediger, M. A., Ganapathy, V., and Leibach, F. H., 1995, Molecular cloning of PEPT 2, a new member of the H^+/peptide cotransporter family, from human kidney, *Biochim. Biophys. Acta* **1235**:461–466.

Lopez-Nieto, C. E., You, G., Bush, K. T., Barros, E. J., Beier, D. R., and Nigam, S. K., 1997, Molecular cloning and characterization of NKT, a gene product related to the organic cation transporter family that is almost exclusively expressed in the kidney, *J. Biol. Chem.* **272**:6471–6478.

Lu, R., Kanai, N., Bao, Y., and Schuster, V. L., 1996, Cloning, *in vitro* expression, and tissue distribution of a human prostaglandin transporter cDNA(hPGT), *J. Clin. Invest.* **98**:1142–1149.

Mayer, U., Wagenaar, E., Beijnen, J. H., Smit, J. W., Meijer, D. K., van Asperen, J., Borst, P., and Schinkel, A. H., 1996, Substantial excretion of digoxin via the intestinal mucosa and prevention of long-term digoxin accumulation in the brain by the mdr 1 a P-glycoprotein, *Br. J. Pharmacol.* **119**:1038–1044.

Mayer, U., Wagenaar, E., Dorobek, B., Beijnen, J. H., Borst, P., and Schinkel, A. H., 1997, Full blockade of intestinal P-glycoprotein and extensive inhibition of blood–brain barrier P-glycoprotein by oral treatment of mice with PSC833, *J. Clin. Invest.* **100**:2430–2436.

Meyers, M. B., Scotto, K. W., and Sirotnak, F. M., 1991, P-glycoprotein content and mediation of vincristine efflux: Correlation with the level of differentiation in luminal epithelium of mouse small intestine, *Cancer Commun.* **3**:159–165.

Miyamoto, K., Tatsumi, S., Morimoto, A., Minami, H., Yamamoto, H., Stone, K., Taketani, Y., Nakabou, Y., Oka, T., and Takeda, E., 1994, Characterization of the rabbit intestinal fructose transporter (GLUT5), *Biochem. J.* **303**:877–883.

Miyamoto, K., Shiraga, T., Morita, K., Yamamoto, H., Haga, H., Taketani, Y., Tamai, I., Sai, Y., Tsuji, A., and Takeda, E., 1996, Sequence, tissue distribution and developmental changes in rat intestinal oligopeptide transporter, *Biochim. Biophys. Acta* **1305**:34–38.

Miyazato, M., Nakazato, M., Yamaguchi, H., Date, Y., Kojima, M., Kangawa, K., Matsuo, H., and Matsukura, S., 1996, Cloning and characterization of a cDNA encoding a precursor for human uroguanylin, *Biochem. Biophys. Res. Commun.* **219**:644–648.

Miyazato, M., Nakazato, M., Matsukura, S., Kangawa, K., and Matsuo, H., 1997k, Genomic structure and chromosomal localization of human uroguanylin, *Genomics* **43**:359–365.

Mizuma, T., Ohta, K., Hayashi, M., and Awazu, S., 1992, Intestinal active absorption of sugar-conjugated compounds by glucose transport system: Implication of improvement of poorly absorbable drugs, *Biochem. Pharmacol.* **43**:2037–2039.

Mizuma, T., Ohta, K., and Awazu, S., 1994a, The beta-anomeric and glucose preferences of glucost transport carrier for intestinal active absorption of monosaccharide conjugates, *Biochim. Biophys. Acta* **1200**:117–122.

Mizuma, T., Sakai, N., and Awazu, S., 1994b, Na(+)-dependent transport of aminopeptidase-resistant sugar-coupled tripeptides in rat intestine, *Biochem. Biophys. Res. Commun.* **203**:1412–1416.

Oelkers, P., Kirby, L. C., Heubi, J. E., and Dawson, P. A., 1997, Primary bile acid malabsorption caused by mutations in the ileal sodium-dependent bile acid transporter gene (SLC10A2), *J. Clin. Invest.* **99**:1880–1887.

Oh, D.-M., Sinko, P. J., and Amidon, G. L., 1992, Characterization of the oral absorption of some penicillins: Determination of intrinsic membrane absorption parameters in the rat intestine in situ, *Int. J. Pharmaceut.* **85**:181–187.

Oh, D. M., Sinko, P. J., and Amidon, G. L., 1993, Characterization of the oral absorption of some beta-lactams: Effect of the alpha-amino side chain group, *J. Pharmaceut. Sci.* **82**:897–900.

Okano, T., Inui, K., Maegawa, H., Takano, M., and Hori, R., 1986, H^+ coupled uphill transport of aminocephalosporins via the dipeptide transport system in rabbit intestinal brush-border membranes, *J. Biol. Chem.* **261**:14130–14134.

Okuda, M., Saito, H., Urakami, Y., Takano, M., and Inuui, K., 1996, cDNA cloning and functional expression of a novel rat kidney organic cation transporter, Oct2, *Biochem. Biophys. Res. Commun.* **224**:500–507.

Orlowski, J., Kandasamy, R. A., and Shull, G. E., 1992, Molecular cloning of putative members of the Na/H exchanger gene family. cDNA cloning, deduced amino acid sequence, and mRNA tissue expression of the rat Na/H exchanger NHE-1 and two structurally related proteins, *J. Biol. Chem.* **267**:9331–9339.

Pajor, A. M., 1995, Sequence and functional characterization of a renal sodium/dicarboxylate cotransporter, *J. Biol. Chem.* **270**:5779–5785.

Pajor, A. M., 1996, Molecular cloning and functional expression of a sodium-dicarboxylate cotransporter from human kidney, *Am. J. Physiol.* **270**:F642–F648.

Patil, S. D., and Unadkat, J. D., 1997, Sodium-dependent nucleoside transport in the human intestinal brush-border membrane, *Am. J. PHysiol.* **272**:G1314–G1320.

Phung-Ba, V., Warnery, A., Scherman, D., and Wils, P., 1995, Interaction of pristinamycin IA with P-glycoprotein in human intestinal epithelial cells, *Eur. J. Pharmacol.* **288**: 187–192 [Erratum, *Eur. J. Pharmacol.* **289**: 479 (1995)].

Rabbaa, L., Dautrey, S., Colas-Linhart, N., Carbon, C., and Farinotti, R., 1996, Intestinal elimination of floxacin enantiomers in the rat: Evidence of a carrier-mediated process, *Antimicrobial Agents Chemother.* **40**:2126–2130.

Rabbaa, L., Dautrey, S., Colas-Linhart, N., Carbon, C., and Farinotti, R., 1997, Absorption of ofloxacin isomers in the rat small intestine, *Antimicrobial Agents Chemother.* **41**: 2274–2277.

Rand, E. B., Depaoli, A. M., Davidson, N. O., Bell, G. I., and Burant, C. F., 1993, Sequence, tissue distribution, and functional characterization of the rat fructose transporter GLUT5, *Am. J. PHysiol.* **264**:G1169–G1176.

Ritzel, M. W., Yao, S. Y., Huang, M. Y., Elliott, J. F., Cass, C. E., and Young, J. D., 1997, Molecular cloning and functional expression of cDNAs encoding a human Na^+-nucleoside cotransporter (hCNT1), *Am. J. Physiol.* **272**:C707–C714.

Romero, M. F., Hediger, M. A., Boulpaep, E. L., and Boron, W. F., 1997, Expression cloning and characterization of a renal electrogenic $Na^+/HCO\{NE\}_3^-$ cotransporter, *Nature* **387**:409–413.

Sai, Y., Tamai, I., Sumikawa, H., Hayashi, K., Nakanishi, T., Amano, O., Numata, M., Iseki, S., and Tsuji, A., 1996, Immunolocalization and pharmacological relevance of oligopeptide transporter PepT1 in intestinal absorption of beta-lactam antibiotics, *FEBS Lett.* **392**:25–29.

Said, H. M., Nguyen, T. T., Dyer, D. L., Cowan, K. H., and Rubin, S. A., 1996, Intestinal folate transport: Identification of a cDNA involved in folate transport and the functional expression and distribution of its mRNA, *Biochim. Biophys. Acta* **1281**:164–172.

Saito, H., Okuda, M., Terada, T., Sasaki, S., and Inui, K., 1995, Cloning and characterization of a rat H^+/peptide cotransporter mediating absorption of beta-lactam antibiotics in the intestine and kidney, *J. Pharmacol. Exp. Ther.* **275**:1631–1637.

Saito, H., Masuda, S., and Inui, K., 1996, Cloning and functional characterization of a novel rat organic anion tansporter mediating basolateral uptake of methotrexate in the kidney, *J. Biol. Chem.* **271**:20719–20725.

Saitoh, H., and Aungst, B. J., 1995, Possible involvement of multiple P-glycoprotein-mediated efflux systems in the transport of verapamil and other organic cations across rat intestine, *Pharmaceut. Res.* **12**:1304–1310.

Saitoh, H., and Aungst, B. J., 1997, Prodrug and analog approaches to improving the intestinal absorption of a cyclic peptide, GPIIb/IIIa receptor antagonist, *Pharmaceut. Res.* **14**: 1026–1029.

Saitoh, H., Kobayashi, M., Sugawara, M., Iseki, K., and Miyazaki, K., 1992, Carrier-mediated transport system for choline and its related quaternary ammonium compounds on rat intestinal brush-border membrane, *Biochim. Biophys. Acta* **1112**:153–160.

Saitoh, H., Fujisaki, H., Aungst, B. J., and Miyazaki, K., 1997, Restricted intestinal absorption of some beta-lactam antibiotics by an energy-dependent efflux system in rat intestine, *Pharmaceut. Res.* **14**:645–649.

Sumanen, J., Wilson, G., Smith, P. L., Lee, C. P., Bondinell, W., Ku, T., Rhodes, G., and Nichols, A., 1996, Chemical approaches to improve the oral bioavailability of peptidergic molecules, *J. Pharm. Pharmacol.* **48**:119–135.

Sardet, C., Franchi, A., and Pouyssegur, J., 1989, Molecular cloning, primary structure, and expression of the human growth factor-activatable Na^+/H^+ antiporter, *Cell* **56**:271–280.

Schinkel, A. H., Mayer, U., Wagenaar, E., Mol, C. A., van Deemter, L., Smit, J. J., van der Valk, M. A., Voordouw, A. C., Spits, H., van Tellingen, O., Zijlmans, J. M., Fibbe, W. E., and Borst, P., 1997, Normal viability and altered pharmacokinetics in mice lacking mdr1-type (drug-transporting) P-glycoproteins, *Proc. Natl. Acad. Sci. USA* **94**:4028–4033.

Sellin, J. H., 1996, The pathophysiology of diarrhea, in: *Molecular Biology of Membrane Transport Disorders* (S. G. Schultz, T. E. Andreoli, A. M. Brown, D. M. Fambrough, J. F. Hoffman, and M. J. Welsh, eds.), Plenum Press, New York, pp. 541–563.

Shallat, S., Schmidt, L., Reaka, A., Rao, D., Chang, E. B., Rao, M. C., Ramaswamy, K., and Layden, T. J., 1995, NHE-1 isoform of the Na^+/H^+ antiport is expressed in the rat and rabbit esophagus, *Gastroenterology* **109**:1421–1428.

Shindo, H., Komai, T., and Kawai, K., 1973, Studies on the metabolism of D- and L-isomers of 3,4-dihydroxyphenylalanine (DOPA). V. Mechanism of intestinal absorption of D- and L-DOPA-14C in rats, *Chem. Pharmaceut. Bull.* **21**:2031–2038.

Shneider, B. L., Dawson, P. A., Christie, D. M., Hardikar, W., Wong, M. H., and Suchy, F. J., 1995, Cloning and molecular characterization of the ontogeny of a rat ileal sodium-dependent bile acid transporter, *J. Clin. Invest.* **95**:745–754.

Sparreboom, A., van Asperen, J., Mayer, U., Schinkel, A. H., Smit, J. W., Meijer, D. K., Borst, P., Nooijen, W. J., Beijnen, J. H., and van Tellingen, O., 1997, Limited oral bioavailability and active epithelial excretion of paclitaxel (Taxol) caused by P-glycoprotein in the intestine, *Proc. Natl. Acad. Sci. USA* **94**:2031–2035.

Stewart, B. H., and Taylor, M. D., 1995, Prodrug approaches for improving peptidomimetic drug absorption, in: *Peptide-Based Drug Design* (M. D. Taylor and G. L. Amidon, eds.), American Chemistry Society, Washington, D.C., pp. 199–217.

Stewart, B. H., Amidon, G. L., and Brabec, R. K., 1986, Uptake of prodrugs by rat intestinal mucosal cells: Mechanism and pharmaceutical implications, *J. Pharmaceut. Sci.* **75**:940–945.

Stewart, B. H., Kugler, A. R., Thompson, P. R., and Bockbrader, H. N., 1993, A saturable transport mechanism in the intestinal absorption of gabapentin is the underlying cause of the lack of proportionality between increasing dose and drug levels in plasma, *Pharmaceut. Res.* **10**:276–281.

Su, S. F., and Huang, J. D., 1996, Inhibition of the intestinal digoxin absorption and exsorption by quinidine, *Drug Metab. Disposition* **24**:142–147.

Swaan, P. W., and Tukker, J. J., 1995, Carrier-mediated transport mechanism of foscarnet (trisodium phosphonoformate hexahydrate) in rat intestinal tissue, *J. Pharmacol. Exp. Ther.* **272**:242–247.

Swann, P. W., and Tukkar, J. J., 1997, Molecular determinants of recognition for the intestinal peptide carrier, *J. Pharmaceut. Sci.* **86**:596–602.

Takanaga, H., Tamai, I., and Tsuji, A., 1994, pH-dependent and carrier-mediated transport of salicylic acid across Caco-2 cells, *J. Pharm. Pharmacol.* **46**:567–570.

Takanaga, H., Tamai, I., Inaba, S., Sai, Y., Higashida, H., Yamamoto, H., and Tsuji, A., 1995, cDNA cloning and functional characterization of rat intestinal monocarboxylate transporter, *Biochem. Biophys. Res. Commun.* **217**:370–377.

Takanaga, H., Maeda, H., Yabuuchi, H., Tamai, I., Higashida, H., and Tsuji, A., 1996, Nicotinic acid transport mediated by pH-dependent anion antiporter and proton cotransporter in rabbit intestinal brush-border membrane, *J. Pharm. Pharmacol.* **48**:1073–1077.

Takano, M., Tomita, Y., Katsura, T., Yasuhara, M., Inui, K., and Hori, R., 1994, Bestatin transport in rabbit intestinal brush-border membrane vesicles, *Biochem. Pharmacol.* **47**:1089–1909.

Tamai, I., Tomizawa, N., Kadowaki, A., Terasaki, T., Nakayama, K., Higashida, H., and Tsuji, A., 1994, Functional expression of intestinal dipeptide/beta-lactam antibiotic transporter in *Xenopus laevis* oocytes, *Biochem. Pharmacol.* **48**:881–888.

Tamai, I., Takanaga, H., Maeda, H., Ogihara, T., Yoneda, M., and Tsuji, A., 1995a, Protoncotransport of pravastatin across intestinal brush-border membrane, *Pharmaceut. Res.* **12**:1727–1732.

Tamai, I., Tomizawa, N., Takeuchi, T., Nakayama, K., Higashida, H., and Tsuji, A., 1995b, Functional expression of transporter for beta-lactam antibiotics and dipeptides in *Xenopus laevis* oocytes injected with messenger RNA from human, rat and rabbit small intestines, *J. Pharmacol. Exp. Ther.* **273**:26–31.

Tamai, I., Nakanishi, T., Hayashi, K., Terao, T., Sai, Y., Shiraga, T., Miyamoto, K., Takeda, E., Higashida, H., and Tsuji, A., 1997a, The predominant contribution of oligopeptide transporter PepT1 to intestinal absorption of beta-lactam antibiotics in the rat small intestine, *J. Pharm. Pharmacol.* **49**:796–801.

Tamai, I., Takanaga, H., Maeda, H., Yabuuchi, H., Sai, Y., Suzuki, Y., and Tsuji, A., 1997b, Intestinal brush-border membrane transport of monocarboxylic acids mediated by proton-coupled transport and anion antiport mechanisms, *J. Pharm. Pharmacol.* **49**:108–112.

Tanaka, K., Fujita, T., Yamamoto, Y., Murakami, M., Yamamoto, A., and Muranishi, S., 1996, Enhancement of intestinal transport of thyrotropin-releasing hormone via a carrier-mediated transport system by chemical modification woth lauric acid, *Biochim. Biophys. Acta* **1283**:119–126.

Terada, T., Saito, H., Mukai, M., and Inui, K., 1997, REcognition of beta-lactam antibiotics by rat peptide transporters, PEPT1 and PEPT2, in LLC-PK1 cells, *Am. J. Physiol.* **273**:F706–F711.

Thorn, J. A., and Jarvis, S. M., 1996, Adenosine transporters, *Gen. Pharmacol.* **27**:613–620.

Thwaites, D. T., Armstrong, G., Hirst, B. H., and Simmons, N. L., 1995, D-Cycloserine transport in human intestinal epithelial (Caco-2) cells: Mediation by a H(+)-coupled amino acid transporter, *Br. J. Pharmacol.* **115**:761–766.

Tomita, Y., Takano, M., Yasuhara, M., Hori, R., and Inui, K., 1995, Transport of oral cephalosporins by the H^+/dipeptide cotransporter and distribution of the transport activity in isolated rabbit intestinal epithelial cells, *J. Pharmacol. Exp. Ther.* **272**:63–69.

Trotter, P. J., Ho, S. Y, and Storch, J., 1996, Fatty acid uptake by Caco-2 human intestinal cells, *J. Lipid Res.* **37**:336–346.

Tse, C. M., Ma, A. I., Yang, V. W., Watson, A. J., Levine, S., Montrose, M. H., Potter, J., Sardet, C., Pouyssegur, J., and Donowitz, M., 1991, Molecular cloning and expression of a cDNA encoding the rabbit ileal villus cell basolateral membrane Na^+/H^+ exchanger, *EMBO J.* **10**:1957–1967.

Tse, C. M., Brant, S. R., Walker, M. S., Pouyssegur, J., and Donowitz, M., 1992, Cloning and sequencing of a rabbit cDNA encoding an intestinal and kidney-specific Na^+/H^+ exchanger isoform (NHE-3), *J. Biol. Chem.* **267**:9340–9346.

Tsuji, A., 1995, Intestinal absorption of b-lactam antibiotics, in: *Peptide-Based Drug Design* (M. D. Taylor and G. L. Amidon, eds.), American Chemistry Society, Washington, D. C., pp. 299–316.

Tsuji, A., and Tamai, I., 1989, Na^+ and pH dependent transport of forsarnet via the phosphate carrier system across intestinal brush-border membrane, *Biochem. Pharmacol.* **38**:1019–1022.

Tsuji, A., and Tamai, I., 1996, Carrier-mediated intestinal transport of drugs, *Pharmaceut. Res.* **13**:963–977.

Tsuji, A., Tamai, I., Nakanishi, M., and Amidon, G. L., 1990, Mechanism of absorption of the dipeptide alpha-methyldopa-phe in intestinal brush-border membrane vesicles, *Pharmaceut. Res.* **7**:308–309.

Tsuji, A., Takanaga, H., Tamai, I., and Terasaki, T., 1994, Transcellular transport of benzoic acid across Caco-2 cells by a pH-dependent and carrier-mediated transport mechanism, *Pharmaceut. Res.* **11**:30–37.

van Asperen, J., van Tellingen, O., Sparreboom, A., Schinkel, A. H., Borst, P., Nooijen, W. J., and Beijnen, J. H., 1997, Enhanced oral bioavailability of paclitaxel in mice treated with the P-glycoprotein blocker SDZ PSC 833, *Br. J. Cancer* **76**:1181–1183.

Walter, E., Kissel, T., Reers, M., Dickneite, G., Hoffmann, D., and Stuber, W., 1995, Transepithelial transport properties of peptidomimetic thrombin inhibitors in monolayers of a human intestinal cell line (Caco-2) and their correlation to *in vivo* data, *Pharmaceut. Res.* **12**:360–365.

Wang, H. P., Lu, H. H., Lee, J. S., Cheng, C. Y., Mah, J. R., Ku, C. Y., Hsu, W., Yen, C. F., Lin, C. J., and Kuo, H. S., 1996, Intestinal absorption studies on peptide mimetic alpha-methyldopa prodrugs, *J. Pharm. Pharmacol.* **48**:270–276.

Werner, A., Murer, H., and Kinne, R. K., 1994, Cloning and expression of a renal $Na-P_i$ cotransport system from flounder, *Am. J. Physiol.* **267**:F311–F317.

Wils, P., Phung-Ba, V., Warnery, A., Lechardeur, D., Raeissi, S., Hidalgo, I. J., and Scherman, D., 1994, Polarized transport of docetaxel and vinblastine mediated by P-glycoprotein in human intestinal epithelial cell monolayers, *Biochem. Pharmacol.* **48**:1528–1530.

Wong, M. H., Oelkers, P., Craddock, A. L., and Dawson, P. A., 1994, Expression cloning and characterization of the hamster ileal sodium-dependent bile acid transporter, *J. Biol. Chem.* **269:**1340–1347.

Wong, M. H., Oelkers, P., and Dawson, P. A., 1995, Identification of a mutation in the ileal sodium-dependent bile acid transporter gene that abolishes transport activity, *J. Biol. Chem.* **270:**27228–27234.

Wright, E. M., Hirayama, B. A., Loo, D. D., Turk, E., and Hager, K., 1994, Intestinal sugar transport, in: *Physiology of the Gastrointestinal Tract* (L. R. Johnson, ed.), Raven Press, New York, pp. 1751–1772.

Yao, S. Y., Ng, A. M., Ritzel, M. W., Gati, W. P., Cass, C. E., and Young, J. D., 1996, Transport of adenosine by recombinant purine- and pyrimidine-selective sodium/nucleoside cotransporters from rat jejunum expressed in *Xenopus laevis* oocytes, *Mol. Pharmacol.* **50:**1529–1535.

Yee, S., and Amidon, G. L., 1995, Oral absorption of angiotensin-converting enzyme inhibitors and peptide prodrugs, in: *Peptide-Based Drug Design* (M. D. Taylor and G. L. Amidon, eds.), American Chemistry Society, Washington, D.C., pp. 137–147.

Zhang, L., Dresser, M. J., Gray, A. T., Yost, S. C., Terashita, S., and Giacomini, K. M., 1997, Cloning and functional expression of a human liver organic cation transporter, *Mol. Pharmacol.* **51:**913–921.

Zimmerman, J., 1992a, Drug interactions in intestinal transport of folic acid and methotrexate. Further evidence for the heterogeneity of folate transport in the human small intestine, *Biochem. Pharmacol.* **44:**1839–1842.

Zimmerman, J., 1992b, Methotrexate transport in the human intestine. Evidence for heterogeneity, *Biochem. Pharmacol.* **43:**2377–2383.

Zittoun, J., 1995, Congenital errors of folate metabolism, *Baillieres Clin. Haematol.* **8:**603–616.

4

The Molecular Basis for Hepatobiliary Transport of Organic Cations and Organic Anions

*Dirk K. F. Meijer, Johan W. Smit,
Guido J. E. J. Hooiveld,
Jessica E. van Montfoort,
Peter L. M. Jansen, and Michael Müller*

1. INTRODUCTION

Of the presently prescribed drugs, many contain one or more tertiary or quaternary amine groups or anionic functional groups such as carboxylic (COOH) and sulfonic acid (SO_3^-) groups. Under physiological conditions, quaternary amines bear a constant positive charge, whereas tertiary amines can be partly protonated depending on the pK_a of the drug (see Fig. 1). Sulfonic acid groups have a perma-

---

Dirk K. F. Meijer • Department of Pharmacokinetics and Drug Delivery, Groningen University Institute for Drug Exploration (GUIDE), Groningen, The Netherlands. *Johan W. Smit* • Department of Pharmacokinetics and Drug Delivery, Groningen University Institute for Drug Exploration (GUIDE), Groningen, The Netherlands; present address: Division of Experimental Therapy, The Netherlands Cancer Institute, Amsterdam, The Netherlands. *Guido J. E. J. Hooiveld and Jessica E. van Montfoort* • Department of Pharmacokinetics and Drug Delivery, Groningen University Institute for Drug Exploration (GUIDE), Groningen, The Netherlands, and Department of Gastroenterology and Hepatology, University Hospital, Groningen, The Netherlands. *Peter L. M. Jansen and Michael Müller* • Department of Gastroenterology and Hepatology, University Hospital, Groningen, The Netherlands.

Membrane Transporters as Drug Targets, edited by Amidon and Sadée. Kluwer Academic/Plenum Publishers, New York, 1999.

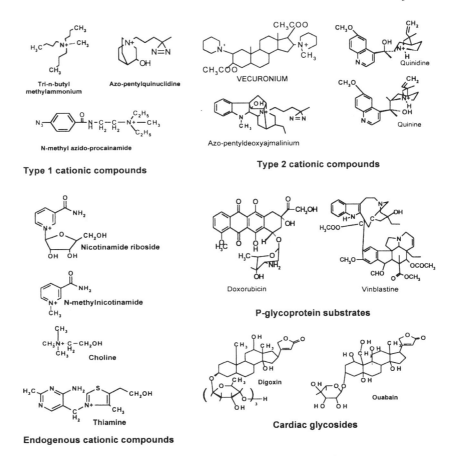

Figure 1. Chemical structures of various organic cations, P-glycoprotein substrates, and cardiac glycosides. The upper panels show relatively small organic cations (type 1), some of which are aliphatic, and the bulkier (type 2) organic cations. Endogenous substrates for carrier-mediated uptake systems are depicted in the lower left panel. Two P-gp substrates contain secondary or tertiary amine functions and are mainly protonized at physiological pH to cationic molecules. Cardiac glycosides (lower right panel) are uncharged compounds that interact with type 2 cationic drugs with regard to hepatic uptake and possibly are substrates for P-gp-mediated transport at the canalicular domain of the hepatocyte.

nently anionic charge at physiological pH, while COOH groups can be partly dissociated to the anionic COO^- moiety. Elimination of organic cations and organic anions from the body is a concerted action of the liver, kidneys, and intestines. Transcellular transport of drugs in these organs involves membrane-embedded transport proteins both at the uptake and the excretion level.

The translocation of solutes across plasma membranes can either occur via channel or pore proteins, or transport proteins. Pore-forming proteins differ from transport proteins in that they merely permit passive transmembrane permeation of ions and small molecules and allow extremely fast passage of these solutes (Chaudhary et al., 1992). Transport proteins, unlike channels, require conformational change during the catalysis of solute translocation across the plasma membrane. This translocational process is orders of magnitudes slower than transport via channel proteins. Furthermore, channels have different secondary structures, consisting mainly of β-barrels in the transmembrane domains (TMDs) (Nikaido and Saier, 1992). The TMDs of transport proteins consist of multiple (helices (Henderson, 1993).

To catalyze the transmembrane movement of substrates, energy is necessary to drive the process of translocation. Depending on the energy sources used by a specific transporter (ATP, the membrane potential, or chemical gradients consisting of (in)organic ions), transporters can be classified into different groups or families.

Passive transporters allow the membrane passage of a substrate in the sense that transport is facilitated and driven by the substance's own gradient across the membrane. Primary or secondary active transporters, on the other hand, can drive solute movement across the biomembrane against an electrochemical gradient. These systems therefore use a free energy source to drive this translocation process, for example, a sodium gradient, which can drive glucose transport via the Na^+-glucose symport system. Another class of transport proteins are the primary active transporters like bacteriorhodopsin. These transporters couple energy derived from (photo)chemical energy to ion translocation across the membrane. The so-called ATP binding cassette (ABC) family of transporters also belong to this class of transport proteins (Higgins, 1992).

Transport proteins have also been classified according to sequence similarities between certain parts of the proteins. For example, ATP-binding domains turn out to be highly conserved between different members of protein families (Hyde et al., 1990). Alignment of the primary amino acid sequence may be used to group proteins with regard to certain primary structures and/or functional similarities. Also, hydropathic profile comparison may show that proteins from diverse species have a secondary structure from which similarities in topological features and functions can be deduced.

Many transport proteins exist that have diverse primary structure, topology, and function. Some of these polypeptides exhibit a dual function in which more than one functional transport mode can be distinguished. The ABC superfamily of transporters comprise a large superfamily of proteins with very diverse functions. Important examples of ABC transporters include the *ste6* gene product, which is responsible for peptide secretion for the yeast mating a-factor, the TAP1,2 proteins, which are involved in MHC-class antigenic peptide transport, the cystic fibrosis transmembrane regulator (CFTR), as well as multidrug-resistance proteins from eukaryotes and prokaryotes (P-glycoprotein and pfMDR) (Higgins, 1992; Gottesman and Pastan, 1993).

An interesting field with regard to organic cation transport is that of neurotransmitter transport systems. These proteins usually are involved in the reuptake of neurotransmitters and their precursors. Several neurotransmitter transport proteins have been discovered for choline. It has become clear that these types of systems are very similar to, among others, the dopamine transporter (Giros and Caron, 1996). The dopamine transporter may also be involved in transport of the neurotoxin $MPTP^+$ (Kitayama et al., 1992).

Recently, a cDNA sequence was cloned that encodes a biogenic amine transporter (Liu et al., 1992). The primary structure of this protein shares some sequence similarity with tetracycline transporters of prokaryotes (Liu et al., 1992).

Both uptake mechanisms and secretory mechanisms have been shown to function in gut, liver, and kidney. Since living organisms have been exposed to many cationic food components, such as plant alkaloids, which can be quite harmful to the organism, these excretory organs have become adequately equipped with a detoxifying apparatus. This include metabolic conversion processes and secretory systems to remove potentially life-threatening agents from the body (Meijer and Groothuis, 1991; Meijer, 1989). The liver is the first organ to be exposed to such agents after gastrointestinal absorption. Against this background, it is not surprising that multiple carrier-mediated transport systems for cationic compounds have evolved at the hepatic uptake and excretion levels during evolution.

Uptake into hepatocytes of small (type 1) cationic drugs is likely to be mediated by the organic cation transporter 1 (oct1) (Gründemann et al., 1994), a transporter that probably has a similar function in the small intestine and kidneys. Bulkier (type 2) cationic drugs can bind to this carrier and thereby inhibit type 1 cation uptake, but are not transported by it (Koepsell, 1998).

Hepatic uptake of bulkier (type 2) organic cations into hepatocytes was proposed to be mediated by so-called "organic anion transporting polypeptides" (oatp's) (Bossuyt et al., 1996a,b; Kullak-Ublick et al., 1995). The oatp family is also responsible for the hepatic uptake of many anionic drugs as well as certain (but not all) bile acids.

Less detailed information is available about transport proteins involved in the secretion of drugs into bile. Hepatobiliary transport of such compounds seems a highly concentrative "uphill" process indicating that energy-consuming carrier systems should be involved in the secretion of cationic drugs. Bile secretion of cationic drugs occurs not only against its own concentration gradient, but also against the electrochemical membrane gradient (inside negative) (Meijer et al., 1990). To overcome this thermodynamic barrier, energy-consuming carrier systems are a necessity. In contrast, secretion of organic anions is down the electrochemical gradient.

ABC transporters that have been detected in liver (Lomri *et al.,* 1996) are likely candidates for such concentrative secretion processes of drugs into bile. In several studies, evidence was found for a role of P-glycoprotein (P-gp; indicating permeability) in drug secretion into bile (Smit *et al.,* 1998a,b). Other members of the ABC family of transporters, such as the so-called sister of P-gp (sPgp) (Childs *et al.,* 1995) and members of the multidrug-resistance-related protein (MRP) family MRP2 (and MRP3?) (Almquist *et al.,* 1995; Müller *et al.,* 1996a), may also contribute to hepatobiliary trafficking of amphiphilic drugs. Besides these primary active transporters, secondary active transport may be involved in the total biliary secretion of cationic drugs (Moseley *et al.,* 1996a).

This review describes the elimination of cationic and anionic drugs from the body that takes place in the liver as mediated by various transport systems.

The literature on the mechanisms of organic cation and organic anion transport, i.e., uptake, sequestration, and secretion, in the liver has been recently reviewed (Müller and Jansen, 1997; Meijer *et al.,* 1997). Here we discuss the current state of the art of carrier-mediated uptake processes of organic compounds in the hepatocyte as well as recent identification of various carrier systems involved in the biliary secretion of drugs.

2. CANDIDATE PROTEINS FOR CARRIER-MEDIATED HEPATIC UPTAKE OF CATIONIC DRUGS

As mentioned above, Gründemann *et al.* (1994) succeeded in cloning a polyspecific transmembrane protein from a rat kidney cDNA library which transports, among others, tetraethylammonium (TEA) (see Fig. 2). This so-called rat organic cation transporter 1 (roct1) is expressed in renal proximal tubule cells and in hepatocytes as well as in the small intestinal enterocytes as detected by Northern blot analysis. roct1 most likely functions in hepatic uptake of organic cationic compounds, of both hydrophilic and hydrophobic nature. The protein is comprised of 556 amino acid residues and has a molecular mass of approximately 61 kDa (Gründemann *et al.,* 1994). Very recently it was found that roct1 transports not only TEA, but also NMN, 1-methyl-4-phenylpyridinium (MPP), and choline, as determined by transport studies utilizing the *Xenopsis laevis* oocyte system. The roct1 also catalyzes efflux of MPP. This could be trans-stimulated by MPP itself and by TEA. Earlier described inhibitors of organic cation transport like cyanine863, quinine, and d-tubocurarine also seem to be transported by the roct1. This transport system is not dependent on Na^+ and H^+ gradients and is electrogenic. Interestingly, at low substrate concentration, the translocation process is voltage-dependent, whereas at high substrate concentration, no voltage dependence could be seen (Busch *et al.,* 1996a). This resolves the apparent discrepancy of previous studies

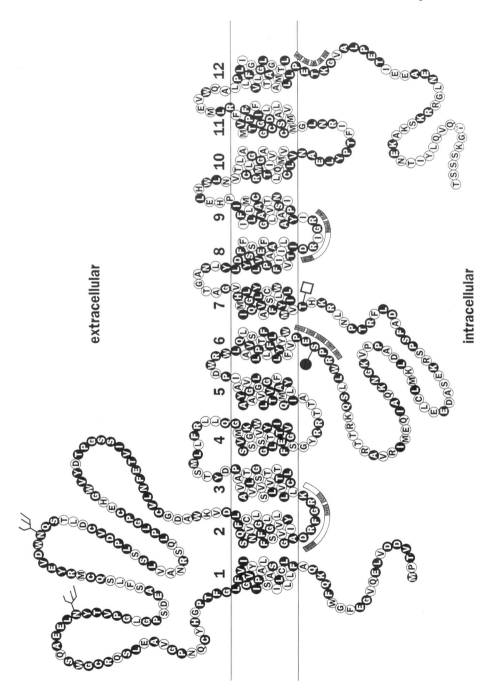

Hepatobiliary Transport of Organic Cations and Anions   95

in which a voltage dependence of transport was reported for tributylmethylammonium (TBuMA) in sinusoidal plasma membrane vesicles (performed with tracer concentrations), whereas in uptake studies with this model compound in isolated hepatocytes performed at much higher concentrations it was shown that the uptake was independent of the membrane potential (Steen and Meijer, 1990).

Collectively, it seems that the mutual inhibition patterns for organic cations during hepatic transport as described in previous studies can at least be partly explained by competitive interaction at the level of hepatic uptake via roct1 (Fig. 3).

MPP is a well-known substrate of the monoamine transport system in neuronal vesicles. It is interesting that Streich *et al.* (1996) recently found expression of an extraneuronal monoamine transporter designated as uptake 2 in human glioma cells (Streich *et al.*, 1996). This transporter has features similar to roct1 and is electrogenic. It is proposed that this oct1-like protein could contribute to inactivation of neuronal released norepinephrine via removal of norepinephrine from the synaptic cleft (Streich *et al.*, 1996).

In contrast, Moseley *et al.* (1997) reported that MPP transport in canalicular liver plasma membrane (cLPM) vesicles is dependent on proton gradient and found an electroneutral cation:proton exchange mechanism similar to transport characteristics established for TEA and TBuMA (Moseley *et al.*, 1992, 1996a). The characteristics of TEA transport in basolateral membrane vesicles of hepatocytes are in line with hepatocyte uptake studies for the cationic drugs azidoprocainamide ethobromide (aPAEB) (Mol *et al.*, 1992), TBuMA (Steen *et al.*, 1991), vecuronium (Mol *et al.*, 1988), and azopentyldeoxyajmalinium (APDA) (Müller, 1988). For all of these compounds at least two uptake systems could be kinetically demonstrated.

Recently Bossuyt *et al.* (1996a) found, using the *X. laevis* oocyte expression cloning system, that APDA as well as uncharged steroidal compounds like aldosterone and cortisol can also be transported via the so-called roatp. This carrier was previously assumed to be involved in uptake of organic anionic drugs and therefore was called "organic anion transporting polypeptide" from rat liver. The transport of the above-mentioned cationic drug and the uncharged steroidal compounds by roatp, as well as mutual inhibitory effects, led the authors to redefine the carrier as a "multispecific transport system" that somehow accommodates hydrophobic drugs irrespective of charge. Such a multispecific system was proposed earlier on the basis of kinetic studies (Meijer and Ziegler, 1993; Steen *et al.*, 1992; Yousef *et al.*, 1990). However, oatp seems to have a higher structural specificity

←

Figure 2. The primary sequence and proposed topology of roct1. Amino acids which are conserved among roct1, moct1, hOCT1, roct2, hOCT2, and pOCT2 are shown in black. Four short intracellular consensus sequences with nutrient transporters from the MFS superfamily (■□) and two conserved potential glycosylation sites (ψ) are marked. One conserved protein kinase C-dependent phosphorylation site (●) and one conserved tyrosine kinase-dependent phosphorylation site (large box) are indicated.

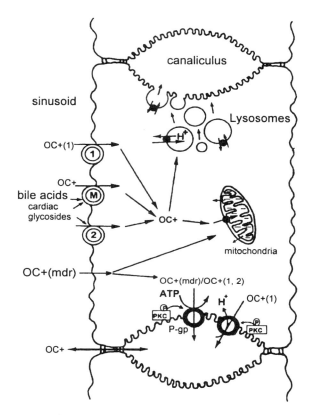

Figure 3. Scheme depicting carrier-mediated transport of organic cations at the sinusoidal and the canalicular level of the hepatocyte. OC+, Organic cation; (1), type 1 uptake system; (2), type 2 uptake system, which is inhibited by cardiac glycosides; M, multispecific uptake system, which also accommodates bile acids and cardiac glycosides; PKC, protein kinase C; P-gp, P-glycoprotein; OC+(mdr), cationic MDR substrates; OC+(1,2), type 1 and type 2 organic cations, with regard to sinusoidal uptake. Besides carrier-mediated transport, passive membrane diffusion is also indicated for cationic MDR substrates. Intracellular sequestration induces carrier-mediated electrogenic uptake in mitochondria and H^+ antiport in endosomes/lysosomes. These acidified vesicles may insert H^+ antiport carriers in the canalicular membrane (upper part) that may lead to ATP-independent H^+/cation exchange and excretion at the canalicular level that operates in concert with ATP-dependent P-gp-mediated secretion into bile.

than earlier assumed, as indicated by the largely varying apparent K_m values found for the steroids ouabain (~1.7 mM), aldosterone (15 nM), and cortisol (13μM) (Bossuyt *et al.*, 1996a). One possible explanation is that the transport protein exhibits specific but completely separate binding sites for anionic, cationic, and uncharged drugs. In view of the observed interactions between these compounds, the binding sites for the various types of drugs may overlap with regard to a putative hydrophobic binding pocket.

Anyway, roatp could be a second candidate for the hepatic uptake of certain organic cations next to roct1. It may be involved in the uptake of relatively bulky and hydrophobic cationic drugs (Bossuyt et al., 1996a). The driving forces for roct1 and roatp mediating organic cation transport remain to be identified. Very likely, a direct ATP dependence is not involved.

The relatively large renal clearance of type 1 organic cations in (wild-type) mice represents an interesting point. As mentioned earlier, it has been proposed from rat studies that the relative lipophilicity of cationic drugs is an important factor determining the relative contribution of liver, kidney, or intestine to the total clearance (Neef, 1983; Neef and Meijer, 1984; Neef et al., 1984). Relatively hydrophobic (lipophilic) cationic drugs seem to be preferably cleared via the liver, and small hydrophilic cationic drugs are predominantly cleared via the kidney. The intestine takes an intermediate position between hepatic and renal clearance with regard to the influence of the hydrophobicity/hydrophilicity balance. In the present studies we observed in (wild-type) mice that the cationic drugs follow a similar excretion pattern. The murine renal clearance of the small (type 1) organic cations was 60- to 90-fold higher than the renal clearance of type 2 cation vecuronium, while in the wild-type mice the intestinal clearance of the type 1 compounds was 5- to 10-fold higher than that of vecuronium. The bulky (type 2) organic cation vecuronium, however, was efficiently eliminated by the liver, resulting in a high hepatic clearance which was at least 3- to 6-fold higher than that of the type 1 organic cations (see Fig. 3).

3. CANDIDATE PROTEINS INVOLVED IN BILE CANALICULAR TRANSPORT OF CATIONIC DRUGS

To identify potential carriers that are involved in solute secretion into bile, extensive studies have been performed to functionally characterize transport in canalicular liver plasma membrane (cLPM) vesicles (e.g. Kamimoto et al., 1989; Moseley et al., 1996a). Evidence is available for the involvement of primary active transporters like P-gp (Kamimoto et al., 1989) or other ATP-dependent transporters as well as ATP-independent transporter proteins (Kwon et al., 1996; Moseley et al., 1996a, 1997) in organic cation transport across the canalicular membrane.

3.1. The Cation/Proton Antiport Secretory/Reabsorptive System

The molecular mechanism of cationic drug transport across the canalicular membrane was studied by us in a canalicular membrane vesicle-enriched liver plasma membrane preparation (cLPM) (Moseley et al., 1996a). A potential role of an organic cation:proton exchanger that can mediate the transport of TBuMA in cLPM vesicles was proposed. The observed transport was dependent on an outwardly directed pH gradient and could be cis-inhibited by several cationic model drugs such as decynium22 and vecuronium. Addition of ATP did not stimulate TBuMA uptake into cLPM vesicles. [^{3}H]-TBuMA uptake into cLPM vesicles was trans-stimulated upon preloading vesicles with nonradioactive labeled TBuMA. This indicated that this organic cation antiporter differs from the primary active P-glycoprotein and can function bidirectionally. Transport of quaternary amines such as TBuMA in cLPM vesicles (Moseley et al., 1996a) was indeed shown to be mediated by a potential secondary active transporter that catalyzes cation:proton exchange. Consequently, it appears that this process is fundamentally different from primary active ATP-dependent transport and that the organic cation:proton exchanger can function as a bidirectional transporter for organic cations such as TBuMA.

The potential homology to bacterial secondary transporters (see also below) could imply that, besides exchange with protons, organic cation:sodium-ion exchange is distinctly possible. In view of the Na$^+$ concentration gradient between bile and the cytoplasm, an inwardly directed sodium gradient may play a role apart from the supposed proton gradient. Although the composition of primary (canalicular) bile is not exactly known, an inwardly directed proton gradient from bile to cytoplasm is unlikely. Only fusion of acidified vesicles with the canalicular membrane could afford ideal local conditions for efficient organic cation:H$^+$ antiport in the biliary direction (Moseley and Van Dyke, 1995). Another possibility is that this transporter is rather involved in canalicular reabsorption in order to retain valuable endogenous cationic compounds in the liver.

Interestingly, the canalicular organic cation:proton exchanger exhibits transport characteristics similar to monoamine transport proteins that are localized in synaptic vesicles of neural cells (Schuldiner et al., 1995). These so-called vesicular monoamine transporters (VMATs) are involved in the intraneuronal reuptake of neurotransmitters into synaptic vesicles (Liu et al., 1992; Schuldiner et al., 1995), a process that is coupled to the efflux of at least two H$^+$ ions (Moriyama et al., 1993). It has been shown that VMAT1 catalyzes transport of MPP (Moriyama et al., 1993; Yelin and Schuldiner, 1996). Recently, Moseley et al. (1997) showed that the neurotoxin MPP is also a substrate for the canalicular organic cation:proton exchanger, a feature that was earlier established for the renal TEA:proton exchanger (Ayer Lazaruk and Wright, 1990). This suggests that the organic cation: proton exchanger may have some homology to the VMATs.

Interestingly, the predicted sequences of VMATs show significant homology to bacterial secondary drug-extruding transporters (Schuldiner et al., 1995). These proteins couple the exchange of protons to the extrusion of potential toxic compounds and have been termed toxin-extruding antiporters (TEXANS) (Schuldiner et al., 1995). The VMATs have been the only mammalian TEXANs identified thus far. However, it is certainly not excluded that in the near future other mammalian TEXANs will be identified and that the canalicular organic cation proton exchanger will turn out to be a member of a larger family of toxin extruders.

3.2. The mdr1-Type P-Glycoprotein Secretory System

Schinkel et al. (1994) developed a mouse deficient in mdr1a P-gp [*mdr1a* ($-/-$) mice] to be able to investigate the pharmacological role of mdr1a P-gp. It was found that these mice lacked efficient protection at the level of the blood–brain barrier against the neurotoxin ivermectine. Furthermore, it was shown that at the intestinal mucosa the mdr1a P-gp can effectively reduce the oral absorption of uncharged lipophilic drugs (Sparreboom et al., 1996). It also contributes largely to the disposition of iv-administered (toxic) compounds such as digoxin (Mayer et al., 1996). The importance of mdr1-type P-gp in limiting the oral availability of the anticancer drug paclitaxel (Taxol®) was substantiated by van Asperen et al. (1997). In the latter study it was shown that efficient inhibition of mdr1a P-gp in the intestinal mucosa of mice by the cyclosporine analogue PSC833 resulted in a dramatic increase in the bioavailability of orally administered paclitaxel. Such useful drug interactions may be of value in oral anticancer therapy if bioavailability of the particular drug is low and variable.

The mdr1a P-gp-deficient mouse model provided an elegant model to further substantiate the contribution of drug-transporting P-gp to elimination of iv-administered cationic drugs from the body (Smit et al., 1998b). Absence of the mdr1a P-gp resulted in a significant reduction of the biliary and intestinal clearance of organic cations compared to the wild type. This implies that P-gp is involved both in the hepatobiliary as well as in the intestinal secretion of cationic compounds like TBuMA, azidoprocainamide methoiodide (APM), and vecuronium in mice. We hypothesized that the residual biliary output of cationic compounds in mdr1a P-gp-deficient mice could be mediated by mdr1b P-gp and/or the functionally described organic cation:proton antiporter that was mentioned earlier.

The marked reduction of the biliary and intestinal clearance of the type 1 cationic drugs compared to the wild type was rather unexpected. Physicochemically these compounds seem to differ considerably from the presently known hydrophobic bulky P-gp substrates. Yet our findings clearly indicate that murine mdr1a P-gp plays a role in the hepatobiliary as well as in the intestinal secretion of type 1 organic cations like TBuMA and APM. As mentioned above, hepatic clearance of the bulkier type 2 organic cation vecuronium was also reduced in the absence of mdr1a P-gp. This, however, is less surprising since it was shown earlier that bulky type 2 organic cations such as azopentyldeoxyajmalinium (APDA) and pentylquinidine are substrates for the rat mdr1b P-gp (Müller et al., 1994a). Urinary clearance of small type 1 cationic compounds was also affected by the disruption of the *mdr1a* gene, although for TBuMA the difference between wild-type and *mdr1a* ($-/-$) mice did not reach statistical significance (see Fig. 4). At least the effect on APM renal clearance data does fit in with the present hypothesis that mdr1-type P-gp is involved in organic cation secretion in the renal proximal tubular cells. For example, it has been recently shown that transcellular transport of the organic cation cimetidine across the epithelial apical membrane is mediated by mdr1-type P-gp's (Dudley and Brown, 1996; B.-F. Pan *et al.*, 1994). Yet, as discussed later, *mdr1a/1b* gene "disruption" in mice did not result in lower renal clearance of cationic drugs.

In contrast to the findings in *mdr1a* ($-/-$) mice, our data obtained from rat cLPM vesicle studies indicated an absence of ATP-dependent transport, suggesting that P-gp is not involved in TBuMA transport at the canalicular membrane level. However, since the absence of mdr1a P-gp in mice clearly decreased biliary output and total hepatic clearance of TBuMA, as well as of APM and vecuronium, it is likely that P-gp can mediate the canalicular transport of cationic drugs into bile in the intact organ. It is distinctly possible that in the vesicle studies, ATP-driven TBuMA transport was not observed due to the activity of the canalicular organic cation:proton exchanger that can transport TBuMA bidirectionally (Moseley *et al.*, 1996a). Such coexistence of the vectorial and ATP-dependent transporter and the bidirectionally operating organic cation:proton exchanger in the canalicular plasma membrane vesicles may lead to an efficient shunting of vectorial TBuMA transport across the apical hepatocyte membrane. In the intact organ such canalicular backflux may be limited due to insertion of TBuMA into biliary micelles. Indeed, we showed earlier that the biliary excretion rate of TBuMA can be stimulated by micelle formation, and bile flow stimulates bile acid taurocholate (Steen *et al.*, 1993). The effect of bile salts may be due to "dilution" of canalicular bile and/or binding of the cationic drug to biliary micelles, which would both reduce canalicular reapsorption. Alternatively, bile acids may stimulate protein kinase C and thereby stimulate the cell to transport bile (Rao *et al.*, 1997).

In their initial study with mdr1a P-gp-deficient mice, Schinkel *et al.* (1994) provided evidence that in the absence of mdr1a P-gp, mdr1b P-gp may be (over)expressed in the liver and in the kidney, but not in the small intestine. To further analyze the pharmacological role of the so-called mdr1-type (drug-transporting) P-gp's in mice, mice with simultaneously disrupted *mdr1a* and *mdr1b* genes were generated in the Netherlands Cancer Institute. It was shown unequivocally that these *mdr* double-knockout mice completely lack mdr1a/mdr1b P-gp's (Schinkel *et al.*, 1997). Surprisingly, these deficient mice turned out to be healthy and did not show an aberrant phenotype under laboratory conditions. Also, bile flow and bile constituents were similar in *mdr1a/1b* ($-/-$) mice compared to the wild type. If the mdr1-type P-gp's serve to protect the body against potential hazardous compounds by (a) counteracting the passive influx of orally ingested agents, (b) increasing their rate of elimination, and (c) at the same time restricting

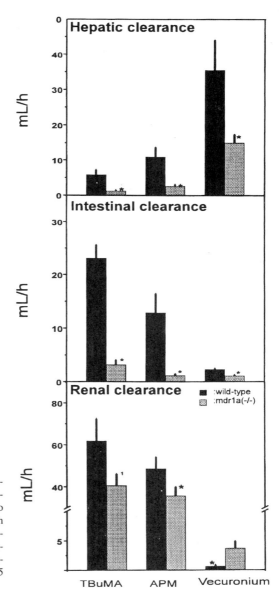

Figure 4. Hepatic, intestinal, and renal drug clearances after the iv administration of the tested cationic drugs to wild-type and $mdr1a$ $(-/-)$ mice with a cannulated gall bladder. The parameters are calculated from three to six independent experiments. Values are expressed as the mean $\pm$SE. $*p < 0.05$ (two-tailed Student's t test).

their distribution to the brain, their function may only be clearly exhibited if the body is challenged with P-gp substrates. These may include not only various amphiphilic drugs, but also normal food constituents (e.g., plant alkaloids). Consequently, pharmacological and physiological functions of P-gp are not clearly separated.

Using *mdr1a/1b* (−/−) mice (Smit *et al.*, 1998c), we provided direct evidence for an important role of drug-transporting P-gp in hepatic as well as intestinal secretion of cationic drugs. Hepatic clearance values were up to 70% reduced compared to wild type, whereas intestinal clearance was reduced even further (>90%). Unexpectedly, the renal clearance of the type 1 cations was about 150% increased in the absence of both mdr1a and mdr1b P-gp. Vecuronium renal clearance was even increased ~5-fold compared to controls. In the *mdr1a/1b* (−/−) mice, the site of elimination of TBuMA as well as that of APM shifted largely to renal clearance. In fact, in the total body clearance, the reduced hepatic and intestinal clearance (Fig. 4) was entirely compensated for by the increased renal clearance. Since plasma protein binding of some of the tested cationic drugs is rather low, and consequently the glomerular filtration rate of the investigated compounds is likely to be unchanged, this implies that either the net renal secretion is increased or apical reabsorption of cationic drugs is reduced at the level of the proximal tubular cells. It was therefore considered whether the disruption of the *mdr1a* and *mdr1b* genes and the total absence of mdr1-type P-gp may result in secondary alterations that influence organic cation elimination from the body. Such alterations may include (over)expression of other carriers. This aspect was recently studied by Schinkel *et al.* (1997). mRNA analysis of *mdr1a/1b* (−/−) mouse tissues revealed that the expression of *oct1, cftr, mrp*, and *sPgp* was not altered. The normal viability and fertility of the *mdr1a/1b* (−/−) mice furthermore suggested that mdr1-type P-gp is not essential under standard laboratory conditions. Yet, no phenotypic alterations have been discovered in the *mdr1a/1b* (−/−) mice (Schinkel *et al.*, 1997). However, it is certainly not yet excluded that the observed increase in organic cation secretion via the kidneys is related to secondary effects resulting from the complete absence of mdr1-type P-gp.

We further studied mdr1-type P-gp-mediated transport of cationic drugs in more detail in *mdr1a-, mdr1b-*, or *MDR1*-transfected cell lines in order to demonstrate that mdr1-type P-gp is directly involved in organic cation transport across the membrane. Transcellular transport of cationic drugs was studied in epithelial LLC-PK$_1$ cells transfected with various cDNAs encoding mdr1-type P-gps (Smit *et al.*, 1998d). This system can be elegantly used to investigate vectorial transport across polarized epithelial cells and also to study substrate specificity of transporters as well as the influence of inhibitors (Fauth *et al.*, 1988; Fouda *et al.*, 1990; Pfaller *et al.*, 1990; Takano *et al.*, 1992).

Apical directed transport of all the tested cationic compounds was significantly increased when P-gp was expressed at the apical domain in polarized grown epithelial LLC-PK$_1$ cells. This indicates that P-gp can mediate the transport of aliphatic as well as of bulkier cationic compounds in such a cell system.

Interestingly, van Helvoort *et al.* (1996) recently showed that MDR1 and mdr1a P-gp's can mediate the transmembrane translocation of short-chain NBD-phospholipids and NBD-glycolipids. Thus, mdr1-type P-gp's may be involved in the translocation of certain phopholipids similar to the earlier defined activity mdr2 P-gp (Smit *et al.*, 1993). However, organic cation excretion in the *mdr2* ($-/-$) is not affected. At first sight it is difficult to envision why mdr1 P-gp would play a facilitator role in the excretion of phospholipid from the liver and at the same time would counteract reabsorption of phospholipid at the intestinal level. Perhaps mdr1-type P-gp's in the liver and intestine are both under regulatory control in order to balance the input/output of lipids in the body.

The recent observation of an increased apical directed secretion of the tested cationic drugs in the presence of mdr1-type P-gp's (Smit *et al.*, 1998d) support our *in vivo* data (Smit *et al.*, 1998b,c) and indicate that various P-gp isoforms are involved in the elimination of a wide variety of amphiphilic cations from the body. It remains to be studied whether mdr2 P-gp and/or sister of P-gp (sPgp) are also involved in biliary drug secretion of cationic drugs such as TBuMA, APM, and vecuronium.

3.3. Other ABC Transport Proteins in the Canaliculus that May Accommodate Cationic Drugs

Other ABC transporters may be involved in organic cation secretion into bile (for review see Lomri *et al.*, 1996). Briefly, the so-called sister of P-gp (sPgp) has been shown to be highly expressed at the hepatocyte canalicular membrane (Childs *et al.*, 1995) and recent findings indicate that it represents the canalicular bile acid transporter (cBAT) (Gerloff *et al.*, 1997), which is indeed mutated in patients with a genetic disorder in bile salt secretion (Strautnieks *et al.*, 1997). It is not excluded, however, that sPgp also accommodates certain hydrophobic organic cations. Furthermore, the multidrug resistance-related protein 1 (MRP1), which can confer multidrug resistance if overexpressed in tumor cells (Cole *et al.*, 1992), has been localized in epithelial tissue. MRP1 is exclusively present in the lateral domain of the hepatocyte plasma membrane (Roelofsen *et al.*, 1997), whereas the MRP homologue cMOAT is highly expressed at the apical domain of the hepatocyte plasma membrane (Paulusma *et al.*, 1996). Interestingly, several authors provided evidence that MRP1 may also transport cationic substances such as vincristine (Slapak *et al.*, 1996) and Taxol. cMOAT (MRP2) could function similarly to MRP1, taking into account its structural and functional homology to MRP1, and therefore may contribute to the overall biliary secretion of cationic drugs such as doxorubicin. Recent findings show that cationic agents can be (co)transported together with glutathione, but that the binding sites for glutathione and the cationic drugs in mrp2 are different (Loe et al., 1996a).

4. ORGANIC CATION TRANSPORT IN THE INTACT LIVER AND ITS SHORT-TERM REGULATION

Posttranscriptional regulation of protein function by protein kinase C (PKC)-mediated phosphorylation is a well-established process that regulates many cellular protein functions (for review see Newton, 1995; Nishizuka, 1988). Roelofsen et al. (1991) showed that canalicular multispecific organic anion transporter (cMOAT)-mediated transport of the glutathione S conjugate of chlorodinitrobenzene (GS-DNP) can be increased by stimulating PKC activity using phorbolmyristate-acetate (PMA) and can be reduced upon addition of the potent PKC inhibitor staurosporin (Roelofsen et al., 1991). A potential influence of PKC-mediated phosphorylation on canalicular organic cation transport was described by us earlier (Steen et al., 1993). The biliary secretion of the cationic drug TBuMA was stimulated dose dependently by PMA and vasopressin, whereas staurosporin decreased the TBuMA biliary excretion rate. cAMP-mediated phosphorylation did not seem to play a significant role in the biliary excretion of organic cations. It was speculated that an ATP-dependent transport protein localized at the hepatocyte canalicular plasma membrane domain is involved in the cationic drug secretion into bile. However, as mentioned above, besides from P-gp, various other ATP-dependent transporters are potential candidates.

These findings were strikingly similar to the data on organic anion excretion presented by Roelofsen et al. (1991). These marked effects of the PKC-influencing agents on the biliary excretion rate of TBuMA may imply that phosphorylation of the transport protein(s) involved in the excretion of TBuMA can occur. This process may serve the purpose of a shorter term regulation of carrier activity.

Extensive data are available concerning the involvement of PKC in regulating cytoplasmic protein function (see for review Fukami and Nishizuka, 1992; Nishizuka, 1995). PKC-mediated phosphorylation has also been shown to modulate the function of many membrane-embedded proteins (Liu, 1996; Oike et al., 1993; Oishi et al., 1990; M. Pan and Stevens, 1995; Yang et al., 1996), including that of the P-gp transport function in multidrug-resistant (MDR) cells (Blobe et al., 1993; Chambers et al., 1990a,b, 1993). The stimulation of P-gp function may therefore lead to an increased resistance of tumor cells that express P-gp, but the evidence is contradictory (Oude Elferink et al., 1995; Meijer et al., 1997).

Since P-gp is also a canalicular membrane protein that very likely mediates, at least partly, the canalicular transport of amphiphilic cationic drugs, the potential regulation of its activity by PKC-mediated phosphorylation is an attractive hypothesis. Yet, the role of PKC-mediated phosphorylation of P-gp and the potential role of phosphorylation in the modulation of P-gp activity are still a matter of debate. For instance, Goodfellow *et al.* (1996) showed that the replacement of potential PKC phosphorylation sites (serines) with alanine or glutamate residues did not alter P-gp transport function. Insertion of a permanent negative charge by glutamate, which mimics the introduction of a net negative charge normally introduced by phosphorylation of the serine residues, did not alter P-gp function. This, however, leaves open the possibility that true PKC-mediated P-gp phosphorylation modulates the transport activity of this transporter (Ahmad *et al.*, 1994). More recently, Yang *et al.* (1996) showed in drug-sensitive MCF-7 and in P-gp-overexpressing multidrug-resistant MCF-7/AdrR cells that P-gp was highly phosphorylated upon PKC stimulation by PMA (see Fine *et al.*, 1996, and Yang *et al.*, 1996, for recent review). Therefore, P-gp is an attractive candidate for cationic drug transport both with regard to localization (Croop *et al.*, 1989) and its potential regulatory phosphorylation sites (Fine *et al.*, 1996).

5. SUBSTRATE SPECIFICITY, DRIVING FORCES, AND MULTIPLICITY OF CANALICULAR ORGANIC CATION CARRIERS

As mentioned above, the biliary secretion of organic cationic compounds is a highly concentrative process against a steep "uphill" concentration gradient. Similarly, in cLPM vesicle studies, transport of organic cationic compounds into the vesicular space is an energy-consuming process that occurs very likely against an uphill concentration gradient. This implies that these transport systems extrude cationic drugs into bile at the cost of cellular energy sources such as ATP or at the expense of the electrochemical gradient across the canalicular membrane. As mentioned above, several ABC transporters have now been identified at the apical domain of liver parenchymal cells and some of them are likely to contribute to the secretion of cationic drugs (for review see Lomri *et al.*, 1996). We collected indirect evidence that at least two systems seem to be involved at the level of biliary excretion of cationic drugs (Smit *et al.*, 1998a). In earlier studies it was shown that coadministration of cationic drugs and MDR-modulating agents in the isolated perfused rat liver model leads to a significant decrease in the biliary secretion of cationic drugs/P-gp substrates such as colchicine and doxorubicin (Speeg and Maldonado, 1994; Speeg *et al.*, 1992; Watanabe *et al.*, 1992). These authors suggested that the observed interactions are due to interaction at the level of P-gp-mediated organic cation secretion into bile.

We extended these observations by investigating interactions between a series of potential P-gp inhibitors and three organic cationic model compounds: the P-gp substrate doxorubicin, the small (type 1) organic cation TBuMA, and the bulkier (type 2) organic cation rocuronium (Fig. 5). The interactions between these agents at the level of biliary excretion (expressed as size reduction in biliary output of the tested organic cations compared to control) were investigated in rat liver perfusion studies (Smit et al., 1998a). This study clearly revealed interaction during membrane transport at the bile canalicular level during coadministration of cationic drugs. A significant correlation between the size reduction of biliary output of the investigated model compounds doxorubicin, rocuronium, and TBuMA and the lipophilicity of a series of lipophilic P-gp substrates was found. This indicates that the relative affinity of cationic drugs for the supposed transport protein is influenced by the hydrophobic interaction with the particular transporters. Alternatively, this physicochemical feature may influence the partitioning of cationic drugs into the lipid phase of the membrane, a process that may be involved in the association of substrates to drug-binding site(s) in the potential transport pro-

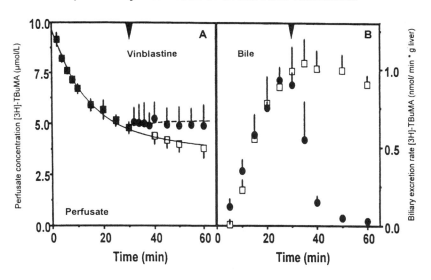

Figure 5. The perfusate concentration versus time curve (A) and the biliary excretion rate versus time curve (B) of [^{3}H]-TBuMA in the isolated perfused rat liver. TBuMA was administered as a bolus of 1.2 μmole at $t = 0$ min ($C_0 \sim 10$ μM). Data are the mean ± SEM of three to six independent experiments. At the designated time point indicated by the arrow, when 60% of TBuMA is taken up in the liver, the inhibitor vinblastine (4.8 μmole) is added to the perfusate. A profound decrease in biliary output of TBuMA is observed without any change in bile flow, indicating an interaction at the bile canalicular level. (□: control experiment, ●: vinblastine experiment)

tein.

Since an excess of the competing agents only produced a *partial* inhibition of cationic drug secretion into bile, and since transport inhibition was not mutual in some cases, we postulated that besides P-glycoprotein other carriers are probably involved in drug secretion. One example could be the organic cation:proton antiporter described above. An alternative explanation for the lack of mutual inhibition in the dose range of the agents that was chosen is a major difference in the relative carrier affinity of the tested cationic drugs compared to the various inhibitors. High-affinity compounds will successfully compete with low-affinity substrates but not vice versa. The binding of cationic drugs to the supposed carriers is likely to be influenced by hydrophobic interaction of the drug and the protein molecules. This aspect may explain the apparent correlation between the lipophilicity of the agents and their inhibitory potency.

Although the extrapolation of findings from the studies in rats to the human situation is risky, the data could imply that major interactions may also occur during the hepatobiliary elimination of cationic drugs in patients. One example relates to the advocated combined use of antineoplastic drugs and of MDR reversal agents, which not only may improve the effective concentration of the cytostatic agents in multidrug-resistant tumor cells, but also may lead to severe interactions at the level of hepatic, intestinal, and renal elimination or at the level of the blood–brain barrier. This could result in an increased residence time in the body and consequently overdosing of the drug, as well as to toxic side effects due to abnormal accumulation in healthy tissues.

6. RELATION BETWEEN CHEMICAL STRUCTURE AND CLEARANCE VIA BILE, URINE, AND INTESTINE

The organ-specific expression and cellular localization of H^+ cation antiport systems as well as the various isoforms of organic cation transporter (oct), oatp, and mdr carrier proteins in liver, kidney, and intestine provide a first clue in explaining the observed differences in excretory patterns for organic cations in hepatic, renal, and intestinal elimination routes.

Drug excretion is often a concerted action of kidney, liver, and intestines (Meijer, 1989; Meijer *et al.*, 1997). One ultimate goal of the study of membrane transport of drugs is to determine which factors determine the relative contribution of excretory organs in the total body clearance of various classes of drugs. The proper prediction of such excretory patterns should be based, on one hand, on knowledge concerning the molecular features of the transported substrates (charge, hydrophobicity/hydrophilicity balance, functional groups, spatial structure, etc.). On the other hand, a number of physiological factors that are increasingly being characterized at the molecular level are supposed to play a role in determining the respective elimination routes. Only a few groups have pioneered in very systematic studies concerning structure–transport relationship studies in drug elimination, among them teams from London, Jena, and Frankfurt as well as our laboratories (Barth *et al.*, 1996; R. L. Smith, 1966; Ullrich, 1997). The following aspects determining the relative contribution of liver, kidneys and intestines in the elimination can be mentioned:

- Relative organ blood flow (at high clearance values)
- Binding to plasma proteins (at intermediate and low clearance values)
- The involvement of passive fluxes in the organs (i.e., glomerular filtration)
- The interorgan expression of the putative carriers involved, including their membrane domain localization
- The relative affinity of the substrates for the various carriers involved (determining carrier occupation and possible saturation)
- Short-term/long-term regulation of carrier activity during physiological and pathological conditions that, in principle, can be organ-specific
- The impact of reabsorptive mechanisms in net secretion from the organs
- The binding of secreted agents in the luminal compartment (biliary micelles/vesicles/intestinal contents), processes that may limit reabsorption

Our study concerning the excretory profiles of a series of cationic drugs, using 14 organic cations with increasing molecular weight and lipophilicity, revealed that liver, kidney, and intestine differently dispose such agents as related to their relative lipophilicity.

The agents with relatively low molecular weight and lipophilicity are predominantly excreted by the kidneys. At increasing M_r of the agents, the small intestine and, in particular, the liver play a more dominant role in their overall excretion (Neef and Meijer, 1984). The organ clearance values depicted are corrected for passive filtration and differences in plasma protein binding. Therefore, they reflect net secretory processes mediated by carrier transport that occurs at two sequential steps: cellular uptake and the net excretion at the apical domains of the particular cell types (the latter process is the net result of secretion and reabsorption processes).

It stands to reason that the impact of hydrophobicity on elimination routes has something to do with the affinity of these agents for the uptake and secretion carriers involved. Indeed, for several cloned membrane carriers, the importance of hydrophobic interactions has been demonstrated, e.g., the organic cation transporters oct1 and oct2 (Koepsell, 1998), and the multispecific organic anion transporter peptide (oatp), which, as mentioned above, also accommodates certain uncharged and cationic drugs (Bossuyt *et al.*, 1996a,b). Consequently, the relative expression and cellular localization of these carriers in the three excretory organs is one factor in determining the elimination routes. The relative affinity of substrates for these carriers is a second factor. Of note, the high affinity of a substrate for a carrier does not always imply efficient transport. For instance, the so-called type 2 organic cations strongly inhibit type 1 transport but the converse is not true (Meijer *et al.*, 1991). This may be due to binding to an allosteric binding site at the type 1 carrier (Koepsell, 1998). In fact, recent oocyte studies with the oct1 carrier, which accommodates the type 1 organic cations, indicate that type 2 compounds are bound with high affinity, but are not transported (Koepsell, 1998; Nagel *et al.*, 1997). This and other recent observations in various laboratories clearly indicate that the functionally defined type 1 and type 2 organic cation uptake systems, as inferred from earlier studies in isolated perfused liver and hepatocyte studies (Meijer *et al.*, 1991; Steen *et al.*, 1992), have an apparent molecular basis: oct1 in liver and intestine as well as oct1 and oct2 in the rat kidney accommodate type 1 compounds, while bulky (type 2) organic cations as well as their inhibitors, cardiac glycosides, can be transported by oatp1 (Meier, 1996). The recently cloned oatp2 even preferentially accommodates cardiac glycosides, exhibiting a 1000-fold difference in K_m between oubain and digoxin (Noé *et al.*, 1997). A similar difference was earlier found between these cardiac glycosides with regard to inhibition of type 2 organic cation uptake into intact liver (Steen *et al.*, 1992) and isolated hepatocytes. Cardiac glycosides and basic drugs mutually influence each others' hepatic uptake. We recently detected an interesting stereospecificity in inhibition of hepatic uptake of cardiac glycosides by the (dia)-stereoisomers quinine and quinidine, quinine being a much more potent inhibitor (Hedman and Meijer, 1997). We are presently studying whether this stereospecific interaction may reflect differences in affinity to the isoforms of oatp. Interactions between cardiac glycosides and basic drugs represent one of the clinically relevant drug interactions that have been reported to occur both at the renal and hepatic level in patients (Hedman *et al.*, 1990).

What can be concluded now from knowledge on organic cation carriers with regard to the contribution of renal, hepatic and intestinal routes? If we assume that the OCT, OATP, P-gp, and perhaps the MRP families as well as proton–antiport systems play a significant role, at least a partial clarification of the above-mentioned excretory patterns of organic cations can be provided (Fig. 6).

In the liver, uptake of both type 1 and type 2 organic cations can be explained by the presence of oct1 and oatp1, respectively. The oatp isoform oatp2, which is particularly highly expressed in liver, could very well represent the type 2 carrier since it accommodates cardiac glycosides and perhaps also cationic drugs. Very small organic cations such as tetraethylmethylammonium are not excreted in bile (Moseley *et al.*, 1996a). This could be due to the presence of a (proton-dependent) carrier-mediated reabsorption process at the canalicular level as driven by the negative membrane potential and a moderate proton gradient from outside to inside (Moseley *et al.*, 1996a). Affinity for ATP-dependent systems of such cations is probably small. Hydrophobic moieties in the cationic drugs may be instrumental

HEPATOBILIARY TRANSPORT OF ORGANIC CATIONS

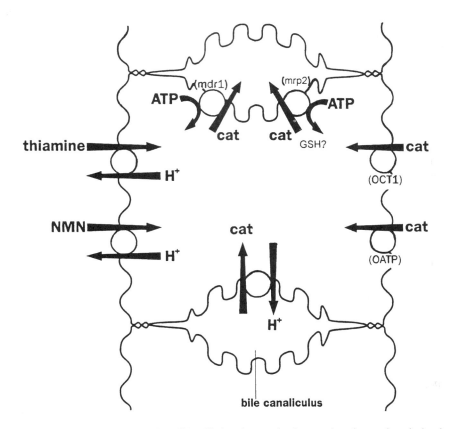

Figure 6. Schematic representation of identified carrier proteins for organic cations at the apical and basolateral plasma membrane domains of the hepatocyte. Separate transport processes are present for hepatic uptake of the endogenous cations N-methyl-nicotinamide (NMN) and thiamine. Choline may be accommodated by the oct1 carrier, which also transports type 1 cationic drugs, while type 2 agents can be transported by oatp isoforms. For secretion into bile, the ATP-dependent carrier P-glycoprotein (mdr1) and multidrug resistance-related protein (mrp2) are probably involved. Mrp2-mediated transport may involve GSH cotransport. The H^+ antiporter is a non-ATP-dependent process that is bidirectional. Bile-to-liver reabsorptive transport is likely, taking into account the inside negative membrane potential and inside-to-outside H^+ gradient.

in both for P-gp-mediated transport as well as binding to biliary micelles. Therefore, tributylmethylammonium may be efficiently excreted into bile whereas tetramethyl- or tetraethylammonium are not (Sohn *et al.,* 1982).

In contrast, small organic cations in the kidney not only can be readily taken up in tubular cells via the oct1/oct2 carriers, but can also be effectively transported out of the cells in the primary urine through the well-defined proton-antiport system. In contrast to the liver, an "inside-to-outside" proton gradient is present in the kidney (Pritchard and Miller, 1997; Ullrich, 1997). Larger cationic drugs will have problems entering tubular cells since the oatp's are likely to be located at the apical membrane (Bossuyt *et al.*, 1996a,b; Noé *et al.*, 1997) and are not recognised as substrates by oct1 and oct2 (Koepsell, 1998). If they were to some extent to enter the tubular cells, they might be transported into primary urine by the P-gp that is expressed in the particular cell type (Simmons *et al.*, 1997). There is little binding of excreted cations in the urinary space. Therefore, in some cases, organic cations such as choline could undergo significant reabsorption. Substrate specificity for this carrier for other organic cations has been recently reviewed (Ullrich, 1997).

With regard to the small intestine, only oct1 and not oatp is present at basolateral domains of the mucosa cells. These cells are important for direct secretion of organic cations from blood into the intestinal lumen (Lauterbach, 1984). The mdr1a isoform is certainly present at the brush border domain, and mediates secretion from the mucosal cells into the intestines in addition to H^+-antiport systems that may operate here due to the lumen-to-cell H^+ gradient. These factors may explain why small organic cations are not efficiently excreted in the gut whereas agents with intermediate lipophilicity that are substrates for oct1 as well as P-glycoproteins are secreted. However, for larger and bulkier organic cations, oatp uptake systems, as present in liver, are not expressed in the gut and due to this no extra intestinal excretion may occur if hydrophobicity of the organic cations reaches higher values. Although there is still a lot to be learned in order to clarify these excretion patterns, it is clear that we are closer now to creating predictive rules that in principle can describe the respective roles of liver, intestine, and kidney in drug elimination.

7. MOLECULAR ASPECTS OF ORGANIC CATION TRANSPORT AND FURTHER PERSPECTIVES

As discussed above, TBuMA transport into bile may be under the regulatory control of PKC. Taking together our findings of a highly enhanced apically directed transport of TBuMA in LLC-PK$_1$ cells that express various P-gp isoforms, it is quite well possible that the PKC-modulatory effects detected earlier (Steen *et al.*, 1993) are related to phosphorylation of P-gp. In further studies, the (transfected) LLC-PK$_1$ cells could be of use to substantiate the role of PKC modulation on P-gp-mediated transport of cationic drugs. Among others, the influence of phosphorylation on substrate specificity, intrinsic transport activity, as well as the actual mechanism of membrane translocation of various categories of cationic drugs should be further elucidated. Studies are being initiated in our laboratories to construct expression systems to produce efficiently various ABC transporters in sufficient amounts to allow protein reconstitution and peptide mapping as shown recently by Popkov *et al.* (1998). Such studies may provide a deeper insight in the translocation mechanism of cationic drugs as well as provide an excellent tool to elucidate the driving forces that energize the transport of cationic drugs.

An overview of hepatic drug transporters that are involved in the hepatobiliary secretion of amphiphilic drugs is presented in Fig. 6.

The mdr1-type P-gp-deficient mouse proved to be a very valuable model to establish the involvement of P-glycoprotein in cationic drug elimination from the body. Further efforts should be made to clarify whether other ABC transporters contribute to the overall excretion of cationic drugs. The mdr1-type P-gp-deficient mice may also provide an elegant model to investigate other non-P-gp excretory systems. For instance, an interesting aspect that warrants further investigation is whether such systems are differentially (up)regulated in the absence of mdr1-type P-gp. This may improve our understanding of potential drug interactions and adaptive responses in the body during chronic drug treatment.

Besides the involvement of primary active transport systems in drug excretion such as P-gp, secondary active (e.g., ΔpH-driven) transport of cationic drugs has been demonstrated at least in the kidney. More effort should be made to identify ultimately the supposed organic cation:H^+ antiporter in liver as well as to determine its coding sequence. With the availability of the cDNAs encoding the potential secondary active organic cation transporter from liver or kidney, progress can be made to define further the importance of proton antiport of cationic drugs in the various excretory organs as well as at the blood–brain barrier.

As depicted in Fig. 6 for the hepatocytes, multiple transport systems, both primary and secondary active transporters, have been identified in epithelial cells of the kidney and the intestine. Understanding the mechanisms that control the transcriptional regulation of the expression of these proteins that are involved in drug disposition is a major challenge. The study of short-term regulation of transport protein expression may improve our understanding of circadian variations in transcellular transport. Furthermore, studies should be performed to assess the regulation of and variations in protein expression due to differences in age, gender, species, chronic drug exposure, and pathology (for a recent review see Müller and Jansen, 1998).

8. HEPATIC UPTAKE OF ORGANIC ANIONS: ROLE OF SPECIFIC AND POLYSPECIFIC TRANSPORT PROTEINS

To secrete bile and to excrete anionic metabolites of toxic substances, hepatocytes must transport bile salts, phospholipids, and other solutes from blood to bile. In recent years various rat and human basolateral transporters for organic solutes have been cloned and characterized (Müller and Jansen, 1997). These transporters comprise the sodium-dependent transporter for the uptake of bile salts (NTCP), transporters for amphiphilic substrates such as the human organic anion transporting polypeptide (OATP), its isoforms, and the earlier mentioned organic cation transporter (OCT1) (see Fig. 7).

8.1. Na$^+$/Taurocholate Cotransporter (NTCP)

The rat liver ntcp is a 362-amino acid, seven-transmembrane-domain glycoprotein with an apparent molecular mass of 51 kDa, localized exclusively in the basolateral membrane of hepatocytes (Ananthanarayanan *et al.*, 1994; Stieger *et al.*, 1994). Ntcp preferentially mediates Na$^+$-dependent transport of conjugated bile salts such as taurocholate and this transport comprises the predominant if not

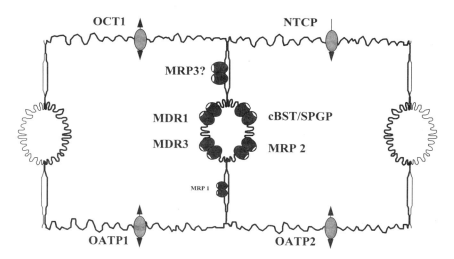

Figure 7. Current concept of carrier-mediated transport of cationic compounds into the liver. Organic anion-transporting polypeptides (OATPs) are sinusoidal uptake systems that have been cloned recently and may also accommodate amphiphilic cations and uncharged compounds. OATP homologues (OATPs) are present that may transport bulky cationic drugs into the hepatocyte. OCT1 represents the organic cation transporter, which is responsible for the uptake of a wide variety of endogenous and exogenous cations. NTCP stands for the Na–taurocholate cotransporting polypeptide, which transports most conjugated bile acids into the hepatocyte. At the canalicular level, the mdr1-type P-glycoprotein (MDR1) and, presumably, the organic cation:proton exchanger are involved in the excretion of catonic compounds. MDR3/mdr2 is involved in phospholipid secretion into bile. MRP2/mrp2 (multidrug-resistance-related protein) is similar to cMOAT (the canalicular multispecific organic anion transporter). MRP1 has only been detected in the lateral plasma membrane. cBAT is the proposed canalicular bile acid transporter and may be identical to sPgp; the function of MRP3 is unknown. Modified from Müller *et al.* (1996b).

exclusive fraction in hepatic bile salt uptake (Hagenbuch et al., 1996; Meier, 1995). The human liver NTCP consists of 349 amino acids and exhibits a 77% homology with rat ntcp. It also transports conjugated bile salts, but NTCP has a higher affinity for taurocholate than ntcp (K_M-6 μM compared with 25 μM) (Hagenbuch and Meier, 1996). Ntcp and NTCP appear to belong to the sodium/bile acid symporter family (SBF) in the superfamily of sodium/solute symporters (SSF) (Reizer et al., 1994). To this family also belong the rat, hamster, and human 348-amino acid apical sodium-dependent bile acid cotransporters (INTCP, ileal-localized NTCP) (Dawson and Oelkers, 1995; Shneider et al., 1995) located in enterocyte membranes. INTCP is also present in apical membranes of the renal proximal convoluted tubule (Christie et al., 1996) and in the apical membranes of cholangiocytes (Lazarides et al., 1996). The human INTCP shares 35% identity with the human NTCP.

Ntcp expression is strongly dependent on the differentiation state of hepatocytes: *Ntcp* mRNA is almost not detectable in primary cultured rat hepatocytes until 72 hr after plating (Liang et al., 1993). In developing rat liver it is detected after 18–21 days of gestation (Hardikar et al., 1995). It is not expressed in human hepatoma cells (Boyer et al., 1993; Kullak-Ublick et al., 1996b) and transiently decreases after partial hepatectomy in rat liver (Green et al., 1994). Recently, genomic cloning and analysis of the 5'-upstream DNA sequence of the *ntcp* gene has been performed (Karpen et al., 1996). The rat *ntcp* gene spans 16.5 kb of rat chromosome 6q24 (Cohn et al., 1995) and is organized as five exons separated by four introns (Karpen et al., 1996).

Basal *ntcp* expression resides in a relatively short promoter region that also directs liver-specific expression. Three elements within this promoter element were determined to contribute to *ntcp* expression: a TATA element, an HNF1-binding site, and a novel palindrome bound by an unknown liver-enriched factor (Karpen et al., 1996). Interestingly, the *ntcp* gene contains a sequence that is identical in composition to a bile acid-responsive element in the gene encoding for cholesterol-7αhydroxylase. This suggests a possible downregulation of its expression by bile salts, similar to the feedback inhibition of cholesterol-7αhydroxylase. Whereas variations of hepatic bile salt fluxes between 0% and 300% did not modulate the level and the activity of ntcp (Koopen et al., 1997), the expression of *ntcp* is rapidly downregulated under cholestatic conditions and in cultured hepatocytes (reviewed in Meier, 1995). The hormone prolactin is the only known inducer of *ntcp* expression. The intracellular mechanisms underlying the upregulation of *ntcp* by prolactin have recently been investigated (Ganguly et al., 1997). A time-dependent increase in nuclear translocation of phosphorylated liver Stat5 (a member of the signal transducers and activators of transcription family) correlated with suckling-induced increases in serum prolactin levels. Nuclear Stat5 exhibited specific DNA-binding ability toward interferon γ-activated sequence-like elements in the *ntcp* promoter. Thus, prolactin acts via the prolactin receptor to facilitate Stat5 binding to the *ntcp* promoter and to transcriptionally regulate *ntcp* (Ganguly et al., 1997).

8.2. Polyspecific Organic Anion Transport Proteins (OATPs)

The contribution of the cloned organic anion transporting polypeptides oatp1 (Hagenbuch et al., 1996; Jacquemin et al., 1994; Kullak-Ublick et al., 1994) and OATP (Kullak-Ublick et al., 1995) to overall hepatic bile salt uptake is not properly understood (for recent reviews see Hagenbuch and Meier, 1996; Meier, 1995; Wolkoff, 1996). The rat oatp1 is a polyspecific transport system that accepts a large variety of structurally unrelated and differently charged amphipathic compounds (Bossuyt et al., 1996a). Originally thought to mediate Na^+-independent transport of organic anions such as bile salts (taurocholate; $K_M \approx 50$ μM) and bromosulfophthalein (BSP; $K_M \approx 1.5$ μM), it also accepts uncharged compounds and even permanently charged bulky organic cations (type 2 cations) (Meijer et al., 1990) and it mediates steroid transport (Bossuyt et al., 1996a). The role of oatp1 in the transport of bilirubin, however, is unclear (Kanai et al., 1996a; Kullak-Ublick et al., 1994). Kanai et al. (1996a) used a vaccinia-based transient expression system to study oatp1 function in HeLa cells and to compare its functional characteristics with those found in the oocyte system. They found that unconjugated bilirubin and bilirubin mono- and diglucuronide appeared not to be transported by oatp1, either with or without the presence of albumin (Kanai et al., 1996a). More recently, another member of the rat oatp family was cloned (Noé et al., 1997), named oatp2. Oatp2 is a polypeptide of 661 amino acids and is homologous to other members of the oatp gene family of membrane transporters. It contains 12 predicted transmembrane domains, five potential glycosylation sites, and six potential protein kinase C phosphorylation sites. Northern blot analysis demonstrated a high *oatp2* expression in brain, liver, and kidney, but not in heart, spleen, lung, skeletal muscle, and testes. Oatp2 exhibits similar substrate specificity as oatp1 but has higher affinity for digoxin. This makes oatp2 more likely *the* polyspecific uptake transporter of the hepatocyte (see Fig. 7).

Rat *oatp1* undergoes a time-related expression in rat liver during development, and its gene transcription precedes *ntcp* (Dubuisson et al., 1996). The 2.5-kb *oatp1* transcript, barely detectable during the end of gestation, was first detected by Northern blotting on fetal day 16, whereas *ntcp* mRNA expression is detected later, on day 20. Thereafter, mRNA expression remains stable during the perinatal period and increases dramatically after weaning. This increase of expression after weaning could be attributed to a bile salt-inducible effect. It may represent an adaptive mechanism to physiological cholestasis, since at that time the serum bile salt concentration is highly increased.

The human OATP (Kullak-Ublick et al., 1995), which localizes to chromosome 12p12 (Kullak-Ublick et al., 1996a), has a similar, but not identical substrate specificity as oatp1 (Bossuyt et al., 1996b) and it is very likely that the human OATP and the rat oatp1 and even oatp2 are different gene products (Kullak-Ublick et al., 1995; Meier, 1995). In recent studies, the promoter of the human *OATP* gene has been partly analyzed. It could be demonstrated that its activity was stimulated in HepG2 cells by taurocholate, suggesting that under cholestatic conditions an increased bile salt concentration may induce *OATP* expression. Since OATP is a bidirectional transporter, this will lead to a decrease of intracellular toxic bile salt levels (Kullak-Ublick et al., 1997a). Due to their ability to facilitate bidirectional transport (Meier, 1995; Shi et al., 1995), OATPs may function as an extrusion transporter at high intracellular substrate concentrations. Thus, OATPs may work as "uptake" systems under normal conditions or as "export" or overflow systems for amphipathic compounds under pathological conditions such as cholestasis.

9. TRANSPORT ACROSS THE CANALICULAR MEMBRANE: ROLE OF ABC TRANSPORT PROTEINS

Most canalicular transport systems involved in bile formation belong to the ATP-binding cassette (ABC) transporter superfamily (Ananthanarayanan et al., 1994), one of the largest superfamilies of proteins in prokaryotes and eukaryotes (Allikmets et al., 1996; Decottignies and Goffeau, 1997). With respect to bile formation, two subclusters appear to be most important, the P-glycoprotein (P-gp) and the multidrug-resistance protein (MRP) clusters (Dean and Allikmets, 1995).

9.1. MDR3 P-Glycoprotein

The P-glycoproteins, the human MDR1 and MDR3 and their mouse and rodent homologues mdr1a, mdr1b, and mdr2 (for review see Meijer et al., 1997) were the first ABC transporters recognized in canalicular membranes of normal hepatocytes. However, in rodent liver, mdr2 is the predominant *mdr* gene product, whereas in human liver the expression of *MDR1* and *MDR3* is more equal (Silverman and Schrenk, 1997).

The function of mdr2 became apparent after producing knockout mice in which the *mdr2* gene is disrupted, $mdr2(-/-)$ (Borst and Schinkel, 1996; Smit et al., 1993), leading to a complete absence of phospholipid in bile (Smit et al., 1993). This suggests that mdr2 and its human homologue MDR3 function in hepatobiliary lipid transport as ATP-dependent phosphatidylcholine (PC) flippases (Oude Elferink et al., 1995) (for recent review see Oude Elferink et al., 1997). Further support came from *in vitro* studies using secretory vesicles from the yeast mutant sec6–4 overexpressing *mdr2* (Ruetz and Gros, 1995; Ruetz and Gros, 1994) and from studies with inside-out canalicular membrane vesicles from rat liver (Nies *et al.*, 1996). Using these model systems, it has been demonstrated that the bile salt taurocholate enhances mdr2-mediated phosphatidylcholine translocation activity (Nies *et al.*, 1996; Ruetz and Gros, 1995). The current working hypothesis for the mechanism of phospholipid secretion is that MDR3/mdr2 flips PC from the inner to the outer leaflet of the lipid bilayer. Bile salt micelles then selectively solubilize PC. Another possibility is that lipids are concentrated in microdomains in the exoplasmic hemileaflet of the canalicular membrane (Crawford *et al.*, 1995; Oude Elferink *et al.*, 1997). Bile salts then solubilize PC from these microdomains.

In the liver the *mdr2* is expressed in canalicular membranes just as for the *mdr1* P-gp's (Buschman *et al.*, 1992) (for MDR3 see Smit *et al.*, 1994). A gender and zonal difference in *mdr2* expression has been discribed by Furuya *et al.* (1994) in rat liver. Expression of *mdr2* in males (Sprague-Dawley) was severalfold lower than in females. By *in situ* hybridization *mdr2* mRNA was shown to be zonally distributed, with the perivenular region (zone 3) of the hepatic acinus having the highest concentration (Furuya *et al.*, 1994). Experiments using hepatocytes isolated from predominantly the periportal zone 1 and perivenous zone 3 hepatocytes confirmed these findings: low *mdr2* expression in zone 1 and high in zone 3 hepatocytes (Furuya *et al.*, 1994).

Recent studies have provided evidence that also various other ABC transporters interact with physiological lipids (Ruetz *et al.*, 1997; van Helvoort *et al.*, 1996). MDR1 is a lipid translocase of broad specificity, while MDR3 specifically translocates phosphatidylcholine. Evidence came from experiments using fluorescent short-chain analogues of various membrane lipids and polarized kidney cells transfected with *MDR1* or *MDR3*. More recently it has been demonstrated that a synthetic PC analog (edelfosine) is a substrate for MDR3, but also for MDR1 and other ABC transporters such as the multidrug resistance protein (MRP1) or the yeast MDR1-homologue STE6 (Ruetz *et al.*, 1997). These results suggest an analogy in the transport mechanisms of these ABC transporters. It has also led to speculation that ABC transporters such as MDR1 and MRP1 may function as an aspecific protection system to maintain cellular membrane integrity by flipping toxic hydrophobic compounds back to the exoplasmic hemileaflet.

Cholesterol secretion is reduced but can be increased in these knockout mice by enriching the bile with taurocholate (Oude Elferink *et al.*, 1996). Interestingely the canalicular GSH secretion is also defective, possibly due to secondary effects of the mutation (Smit *et al.*, 1993). These mice cannot be called cholestatic because the bile acid secretion is normal and bile flow is elevated (Smit *et al.*, 1993). However, histologically there are features as observed in cholestasis such as bile ductular proliferation and feathery degeneration of hepatocytes (Mauad *et al.*, 1994). This is explained by the secretion of bile acids unaccompanied by phospholipids. Older mice with the gene disruption develop liver tumors (Mauad et al., 1994). The secretion of bile salts unaccompanied by phospholipids and glutathione is apparently very toxic to the liver and causes the type of damage seen in cholestasis. A recent paper and a preliminary report have provided evidence for MDR3 dysfunction in a subgroup of patients with the syndrome of progressive familial intrahepatic cholestasis (PFIC-3) (Deleuze et al., 1996; Jacquemin et al., 1997; Müller and Jansen, 1998). In contrast to PFIC-1 and PFIC-2, in PFIC-3 serum levels of γ-glutamyltranspeptidase are increased. In patients with PFIC-3 the *MDR3* mRNA is mutated and MDR3 is lacking (Deleuze et al., 1996; Jacquemin et al., 1997) with, as consequence, an impaired or absent phosphatidylcholine secretion, while the canalicular bile salt secretion is unaffected.

Drugs such as clofibrate and ciprofibrate known to produce peroxisome proliferation (Lake, 1995) induce *mdr2* expression and enhance the phospholipid secretion in bile (Chianale et al., 1996). Furthermore, protoporphyria-inducing drugs, such as griseofulvin, induce *mdr2* expression and to a lesser extent *mdr1a* expression (Preisegger et al., 1996). After several (8–12) weeks of feeding mice with griseofulvin, Mallory bodies appeared in some hepatocytes and a decrease or loss of P-gp staining was seen in Mallory bodies-containing hepatocytes despite an overall increased *mdr2* mRNA. The authors speculated that protoporphyrin or other bile constituents could be involved in the *mdr2* gene regulation (Preisegger et al., 1996). A potential role of mdr2 in the secretion of protoporphyrin has also been suggested (Beukeveld et al., 1996). Finally, during cholestasis P-gp's also are upregulated, at least in rodents and nonhuman primates (Schrenk et al., 1993).

9.2. Homologues of the Multidrug-Resistance Protein MRP1

At least three MRP/mrp isoforms are present in hepatocytes: human MRP1, MRP2, and MRP3 (in rodents mrp1, mrp2, and mrp3) (Allikmets et al., 1996; Büchler et al., 1996; Kool et al., 1997; Mayer et al., 1995; Müller et al., 1996b; Paulusma et al., 1996). MRP1 has been demonstrated *in vitro* to function as an ATP-dependent transporter for amphipathic anionic conjugates such as glutathione S conjugates (leukotriene C_4, GS-DNP) (Leier et al., 1994; Loe et al., 1996a; Müller et al., 1994b), oxidized glutathione (GSH) (Leier et al., 1996), glucuronate, sulfate conjugates (Jedlitschky et al., 1996), and even GSH itself (Zaman et al., 1995). The natural substrates of MRP1 are unknown, but may be derived from the cellular GSH-dependent detoxification of toxins such as reactive oxygen species, electrophils, metals, or oxyanions (Müller et al., 1996a,b). Normal human, mouse, and rat liver have a very low *MRP1* (human) and *mrp1* (rodent) expression, in fact lower than in most other tissues (Cole et al., 1992; Müller et al., 1996b; Paulusma et al., 1996; Stride et al., 1996; Zaman et al., 1993). In contrast, MRP1 is highly increased in both HepG2 and SV40 large T-antigen immortalized human hepatocytes (Roelofsen et al., 1997). MRP1 is localized in lateral membranes of adjacent cells, whereas plasma membrane staining was absent in separate cells. These results suggest that MRP1-mediated organic anion transport is important in proliferating hepatocytes, but not in quiescent cells (Roelofsen et al., 1997).

In contrast to the very low expression of MRP1/mrp1 in normal hepatocytes, MRP2/mrp2 is present at a high levels in the canalicular membrane. The open reading frame of the *mrp2* mRNA of 4623 bp encodes a predicted protein of 1541 amino acids. Sequence analysis demonstrated that MRP2/mrp2 shares only a limited part of its structure with the MRP1 (47.8% identity; 67.9% similarity for mrp2) (Büchler et al., 1996; Keppler and König, 1997a). Northern blot analysis of rat tissues revealed high expression of two to three transcripts, ranging from 5.5 to 9.5 kb in liver (Büchler et al., 1996; Paulusma et al., 1996). Expression of the 5.5-kb transcript was also found in kidney and intestine, albeit at a lower level than in liver. Messenger RNA levels of *mrp2* were extremely low in livers from TR^- and EHBR rat livers (Büchler et al., 1996; Paulusma et al., 1996). The nature of the mutation has been analyzed for TR^- rats by sequencing the mutated *mrp2* after PCR amplification of the respective mRNA from mutant rat liver (Paulusma et al., 1996). The mutation appears to be due to a base deletion and a respective frameshift which leads to an early stop codon. No active mrp2 is formed and the untranslated mutated *mrp2* mRNA appears to be rapidly degraded. The recently cloned human MRP2 (Taniguchi et al., 1996) is the homologue of the canalicular mrp2 in rat liver. *MRP2* is mainly expressed in liver and kidney and was mapped to chromosome 10q24 (Taniguchi et al., 1996). In the human counterpart of the TR^- rat, patients with Dubin–Johnson syndrome (conjugated hyperbilirubinemia), MRP2 is lacking (Kartenbeck et al., 1996; Paulusma et al., 1997). This defect is therefore associated with a deficiency in the secretion of amphiphilic anionic conjugates into the bile.

Using the above-mentioned mutant rat strains, the substrate specificity of mrp2 has been extensively characterized (Oude Elferink et al., 1995). Different organic anions have been identified as substrates in these studies such as the glutathione S^- conjugates of chlorodinitrobenzene (GS-DNP) and of bromosulfofthalein, leukotriene C_4 (LTC_4), or sulfated and glucuronidated bile salts, bilirubin diglucuronides, and the unmetabolized antibiotic ceftriaxone (Oude Elferink et al., 1995). These are all poorly excreted by the mutant rats. Interestingly, however, the secretion of some is more impaired than the secretion of others: for example, the secretion of carboxyfluorescein is almost nil and the secretion of bilirubin ditaurate is almost normal (Jansen et al., 1993). This suggests that for some substrates alternative canalicular transporters are available. These may be members of the expanding MRP family of ABC transporters.

By screening databases of human expressed sequence tags, three new homologues of *MRP1* and *MRP2* have been identified: *MRP3, MRP4,* and *MRP5* (Kool *et al.,* 1997). Similar to *MRP2, MRP3* is mainly expressed in the liver (Allikmets *et al.,* 1996; Kool *et al.,* 1997). *MRP4* is expressed only at very low level in a few tissues, and *MRP5,* like *MRP1,* is expressed in almost every tissue tested. It has been further demonstrated that in cell lines selected for a low level of drug resistance, several MRP-related genes can be upregulated simultaneously (Kool *et al.,* 1997).

9.3. Canalicular Bile Salt Transporter (cBST/sPgp)

Two ATP-dependent transport systems, one for monovalent bile salts (Adachi *et al.,* 1991; Büchler *et al.,* 1994; Böhme *et al.,* 1994; Kast *et al.,* 1994; Müller *et al.,* 1991; Nishida *et al.,* 1991; Stieger *et al.,* 1992; Wolters *et al.,* 1992) and one for divalent sulfated or glucuronidated bile salt conjugates, mediate canalicular bile salt secretion (Müller and Jansen, 1997). The transporter for bivalent bile salts is identical with the cloned MRP2/mrp2 (Paulusma *et al.,* 1996). The ATP-dependent transporter for monovalent bile salts, cBST, has only been characterized functionally. However, recent data strongly suggest that sister of P-glycoprotein (sPgp), another member of the P-glycoprotein cluster of ABC transporter proteins, functions as cBST. The gene for sister of P-gp (*spgp*) was originally identified by low-stringency screening of a pig liver library (Childs *et al.,* 1995). Northern blot analysis revealed the expression of a 5.2-kb transcript almost exclusively in the liver. The protein was detected in canalicular membranes of hepatocytes and was shown to be developmental regulated (Childs *et al.,* 1996; Müller *et al.,* 1996c). In rat liver, the mRNA levels of *spgp* and *mdr2* (*pgp3*) are approximately equivalent and thus both gene products probably are the predominant P-gp's in normal rat liver and are much more abundant than the *mdr1a/b* gene products. Most recently it has been demonstrated by using membrane vesicles from *spgp*-overexpressing Sf9 insect cells that sPgp is able to transport taurocholate in an ATP-dependent manner (K_M = 4.3 μM) (Gerloff *et al.,* 1997).

A yeast gene (*BST1p*) has been identified to encode for an ATP-dependent transporter for taurocholate (Ortiz *et al.,* 1997). BST1p, a member of the MRP/CFTR subfamily (Decottignies and Goffeau, 1997), is localized to the yeast vacuole. Disruption of this gene completely abrogated vacuolar ATP-dependent taurocholate transport, but did not affect ATP-dependent transport of glutathione S conjugates, which is mediated by YCF1, the yeast MRP1 homologue (Ortiz *et al.,* 1997). ATP-dependent transport of bile salts by the vacuolar fraction was independent of the vacuolar proton ATPase, responded to changes in the osmotically sensitive intravesicular space, and was saturable, exhibiting a K_M of 63 μM for taurocholate. The protein, of 1661 amino acids, is much longer than ABC proteins belonging to the MDR cluster and even about 100 amino acids longer than other

MRP1 homologues. The function of this protein in yeast is unknown and further studies to identify rodent or human homologues are needed to clarify a putative role of a "BST1p" homologue as canalicular bile salt transporter.

10. RECENT RESULTS FROM CHOLESTATIC ANIMAL MODELS

Three animal models have been used to characterize changes in transporter expression and activity in cholestatic livers: (1) Ethinylestradiol-treated rats; (2) bile duct-ligated rats, and (3) endotoxin-treated rats. Bile duct ligation of rats resulted in a rapid downregulation of *ntcp* transcription as well as ntcp protein concomitantly with a reduced Na^+-dependent bile salt transport capacity (Gartung *et al.*, 1996). In endotoxemic rat livers both uptake and secretion processes of bile salts and polyanionic compounds are impaired (Bolder *et al.*, 1997; Green *et al.*, 1996; Moseley *et al.*, 1996b). High levels of endotoxin (15 mg/kg) caused 16 hr after treatment a more than 90% decrease of *ntcp* expression and a similar decrease in protein expression (Green *et al.*, 1996). In another study using lower levels of endotoxin (3 mg/kg) bile salt-independent bile flow was reduced by 51% (Bolder *et al.*, 1997) comparable with a 50% reduced secretion of bilirubin (Roelofsen *et al.*, 1994). Impairment was maximal 12–16 hr after endotoxin injection and recovered thereafter. In basolateral plasma membrane vesicles, sodium-dependent transport for bile salts was reduced by around 40% (Bolder *et al.*, 1997) due to the downregulation of ntcp (Green *et al.*, 1996; Moseley *et al.*, 1996b). In addition, ATP-dependent transport was greatly decreased in canalicular vesicles prepared from endotoxemic animals for substrates of cBST as well as of MRP2. A recent study demonstrated that all three cholestatic models resulted in a marked decrease in mrp2 protein (Trauner *et al.*, 1997). *Mrp2* mRNA levels diminished intensely after endotoxin and bile duct ligation, but did not change after ethinylestradiol. In contrast, protein levels of C-CAM105 remained unchanged in livers of endotoxin- (Moseley *et al.*, 1996b) and ethinylestradiol-treated animals (Trauner *et al.*, 1997). Using specific antibodies against sPgp, we recently found decreased levels of the putative bile salt transporter sPgp in liver membrane subfractions of endotoxin-treated rats (Vos *et al.*, 1998). However, the decrease was not as pronounced as found for mrp2 (Müller *et al.*, 1997). This is in line with a minor decrease in bile salt secretion (20%) compared to organic anion secretion (50%) in septic rat livers (Roelofsen *et al.*, 1994). The canalicular membrane localization of the sPgp as assessed by confocal scanning laser microscopy appeared to be dramatically changed ("fuzzy" staining) (Vos *et al.*, 1998), similar to that found for mrp2 (Trauner *et al.*, 1997). In conclusion, cholestasis due to endotoxemia and bile duct obstruction is the result of decreased levels of membrane transporters such as the putative bile salt transporter sPgp (cBST) or the organic polyanion transporter mrp2.

APPENDIX: MOLECULAR FEATURES OF CURRENTLY IDENTIFIED CARRIER PROTEINS INVOLVED IN SINUSOIDAL UPTAKE AND CANALICULAR TRANSPORT OF ORGANIC CATIONS AND ANIONS

Human NTCP (Data fom Hagenbuch and Meier, 1994)

M_r	50 kDa (349 amino acids [aa])
M_r deglycosylated	33–35 kDa
Protein structure	Seven transmembrane domains Five potential N-linked glycosylation sites
Genetic organization	
Size of mRNA	1.7 kb
Genomic DNA	Gene is located on chromosome 14
Related transporters	Rat ntcp, 77% aa identity
Organ distribution	High expression in liver; no other data available
Membrane domain localization	Basolateral membrane of hepatocytes
Ion dependence	Na^+-dependent
Substrates	Conjugated bile salts such as taurocholate

Rat ntcp

M_r	51 kDa (362 aa) (Hagenbuch *et al.*, 1991; Stieger *et al.*, 1994)
M_r deglycosylated	33–35 kDa
Protein structure	Seven putative transmembrane domains (Hagenbuch *et al.*, 1991) Two N-linked glycosylation sites at positions 5 and 11 (Hagenbuch *et al.*, 1991; Stieger *et al.*, 1994)

Genetic organization	
Size of mRNA	1.8 kb (Hagenbuch *et al.*, 1991; Stieger *et al.*, 1994)
Genomic DNA	Gene contains five exons, interspersed with four introns that together extend over 16.5 kb of genomic DNA, located on chromosome 6q24 (Cohn *et al.*, 1995; Karpen *et al.*, 1996)
Related transporters	Human NTCP, 77% aa identity
Organ distribution	Expression in liver, kidney, duodenum, and ileum; no expression in lung, brain, heart, and skeletal muscle (Hagenbuch *et al.*, 1991)
Membrane domain localization	Basolateral membrane of hepatocytes (Stieger *et al.*, 1994; Anantharayan *et al.*, 1994)
Ion dependence	Na^+-dependent (Hagenbuch *et al.*, 1991)
Substrates	Conjugated bile salts such as taurocholate (Hagenbuch *et al.*, 1996; Meier, 1995; Boyer *et al.*, 1994)

Human OATP

M_r	74 kDa (670 aa) (Kullak-Ublick *et al.*, 1995)
M_r deglycosylated	59–61 kDa
Protein structure	Ten to 12 putative membrane-spanning domains (Kullak-Ublick *et al.*, 1995)
	Eight potential N-linked glycosylation sites (Kullak-Ublick *et al.*, 1995)
	Three or four possible N-linked glycosylation sites (Kullak-Ublick *et al.*, 1995)
Genetic organization	
Size of mRNA	2.7–7.4 kb (Kullak-Ublick *et al.*, 1995)
Genomic DNA	Gene is located on chromosome 12p12 (Kullak-Ublick *et al.*, 1996a)
Related transporters	Rat oatp1, 67% aa identity
	Rat oatp2 73% aa identity

Organ distribution	High expression only in liver; intermediate to low expression in brain, lung, kidney, and testis (Kullak-Ublick et al., 1995)
Membrane domain localization	No data available
Ion dependence	Na^+- and Cl^--independent, transport is highly temperature dependent (Kullak-Ublick et al., 1995, 1996; Bossuyt et al., 1996b)
Substrates	Structurally unrelated and differently charged amphipathic agents including organic anions such as bile salts and bromosulfophthalein (BSP), uncharged compounds such as ouabain, and permanently charged bulky organic cations such as APDA (Kullak-Ublick et al., 1995, 1997b; Bossuyt et al., 1996b)

Rat oatp1

M_r	80 kDa (670 aa) (Jacquemin et al., 1994; Meier, 1995; Bergwerk et al., 1996)
M_r deglycosylated	65–75 kDa
Protein structure	Twelve transmembrane domains (Jacquemin et al., 1994; Meier, 1995; Hagenbuch, 1997) Four potential N-linked glycosylation sites located in extracellular loops 2 and 5 (Jacquemin et al., 1994; Hagenbuch, 1997)
Genetic organization	
Size of mRNA	1.5–4.0 kb (Jacquemin et al., 1994; Bergwerk et al., 1996; Lu et al., 1996)
Genomic DNA	No data available
Related transporters	Rat oatp2, 77% aa identity Human OATP, 67% aa identity
Organ distribution	High expression in liver and kidney; low expression in colon, brain, lung, and skeletal

	muscle (Jacquemin et al., 1994; Bergwerk et al., 1996; Lu et al., 1996)
Membrane domain localization	Basolateral membrane of hepatocytes, apical brush border membrane of proximal tubule epitheium, apical surface of the choroid plexus (Meier, 1995; Bergwerk et al., 1996; Angeletti et al., 1997)
Ion dependence	Na^+-independent, transport is not electrogenic, but is highly temperature dependent (Shi et al., 1995; Kullak-Ublick et al., 1994)
Transport direction	Bidirectional (Shi et al., 1995)
Substrates	Structurally unrelated and differently charged amphipathic agents including organic anions such as bile salts and bromosulfophthalein, uncharged compounds such as ouabain and cortisol, and possibly permanently charged bulky organic cations such as APDA (Jacquemin et al., 1994; Kullak-Ublick et al., 1994; Hagenbuch et al., 1996; Kanai et al., 1996b; Bossuyt et al., 1996a; Meier et al., 1997)

Rat oatp2 (Data from Noé et al., 1997)

M_r	73 kDa (661 aa)
M_r deglycosylated	No data available
Protein structure	Twelve predicted transmembrane domains Five possible N-linked glycosylation sites in the extracellular loops Six potential protein kinase C phosphorylation sites at residues 4, 383, 580, 596, 634, 645
Genetic organization Size of mRNA Genomic DNA	 1.4–4.4 kb No data available

Related transporters	Rat oatp1, 77% aa identity Human OATP, 73% aa identity
Organ distribution	High expression in liver, kidney, and brain; no expression in testis, lung, spleen, heart, and skeletal muscle
Membrane domain localization	No data available
Ion dependence	Na^+-independent
Substrates	Bile salts, estrogen conjugates, and cardiac glycosides

Human OCT1

M_r	61 kDa (554 aa) (Zhang *et al.*, 1997; Gorboulev *et al.*, 1997)
M_r deglycosylated	45–50 kDa
Protein structure	Eleven or 12 transmembrane domains (Zhang *et al.*, 1997; Gorboulev *et al.*, 1997) Three potential N-linked glycosylation sites at positions 71, 96, and 112 (Zhang *et al.*, 1997; Gorboulev *et al.*, 1997) Nine potential phosphorylation sites at residues 285, 291, 296, 321, 327, 333, 340, 347, 524; phophorylated by protein kinase A, protein kinase C, casein kinase A, and/or tyrosine kinase (Zhang *et al.*, 1997; Gorboulev *et al.*, 1997)
Genetic organization Size of mRNA	2.0–2.5 kb (Zhang *et al.*, 1997; Gorboulev *et al.*, 1997)
Genomic DNA Related transporters	rOCT1, 78% aa identity rOCT2, 64% aa identity hOCT2, 70% aa identity

Organ distribution	High expression only in liver; intermediate expression in kidney, heart, and skeletal muscle; low expression in brain and placenta; no expression in lung, pancreas, and intestine (compiled from Zhang et al., 1997, and Gorboulev et al., 1997)
Membrane domain localization	No data available
pH dependence	Uptake of 1-methyl-4-phenylpyridinium increases as pH increases between 6.4 and 7.4 (Zhang et al., 1997) and does not significantly change between pH 7.4 and 8.5 (Gorboulev et al., 1997)
Substrates	Small organic cations like 1-methyl-4-phenylpyridinium, tetraethylammonium, and N-1-methylnicotinamide, inhibition by other small and large organic cations (Zhang et al., 1997; Gorboulev et al., 1997)

Rat oct1

M_r	62 kDa (556 aa) (Gründemann et al., 1994)
M_r deglycosylated	No data available
Protein structure	Eleven or 12 putative transmembrane protein regions (Gründemann et al., 1994; Okuda et al., 1996; Gorboulev et al., 1997) Three potential N-linked glycosylation sites in the first large hydrophilic loop between transmembrane segments 1 and 2 (Gründemann et al., 1994)
Genetic organization	
Size of mRNA	1.9–4.8 kb (Gründemann et al., 1994)
Genomic DNA	Gene is located on chromosome 1q11–12 (Koehler et al., 1996)
Related transporters	Rat oct2, 67% aa identity

	Human OCT1, 78% aa identity
	Human OCT2, 68% aa identity
Organ distribution	Expression detected in liver, kidney, colon, and intestine (Gründemann et al., 1994)
Membrane domain localization	Basolateral membrane of hepatocytes and proximal tubule epithelium (Gründemann et al., 1994; Martel et al., 1996; Busch et al., 1996a)
Ion dependence	Sensitive to membrane potential, Na^+-, Ca^2- and Mg^{2+}-independent (Gründemann et al., 1994; Busch et al., 1996a,b)
Substrates	Small exogenous and endogenous organic cations including 1-methyl-4-phenylpyridinium, tetraethylammonium, N-1-methylnicotinamide, dopamine, norepinephrine, serotonin, histamine, acetylcholine (Gründemann et al., 1994; Busch et al., 1996a,b; Nagel et al., 1997).

Human OCT2 (Data from Gorboulev et al., 1997)

M_r	63 kDa (555 aa)
Protein structure	Twelve transmembrane domains
	Three potential N-linked glycosylation sites at positions 72, 97, and 113
	Six potential phosphorylation sites at residues 286, 327, 334, 348, 525 and 544; phosphorylated by protein kinase A, protein kinase C, casein kinase A, and/or tyrosine kinase
Genetic organization	
Size of mRNA	2.5 kb
Genomic DNA	No data available
Related transporters	Rat oct2, 81% aa identity
	Rat oct1, 68% aa identity
	Human OCT1, 70% aa identity

Organ distribution	High expression only in kidney; low expression in spleen, placenta, small intestine, and brain
Membrane domain localization	No data available
pH dependence	Identical inward currents induced by choline at pH 7.5 and 8.5
Substrates	Small organic cations like choline, 1-methyl-4-phenylpyridinium, tetraethylammonium, and N-1-methylnicotinamide, inhibition by other small and large organic cations

Rat oct2

M_r	66 kDa (593 aa) (Okuda et al., 1994)
M_r deglycosylated	No data available
Protein structure	Twelve putative membrane-spanning domains (Okuda et al., 1994)
	Two potential N-linked glycosylation sites in the first large hydrophilic loop between trans-membrane segments 1 and 2 (Okuda et al., 1994)
	Five potential phosphorylation sites at residues 286, 334, 348, 553, and 586; phophorylated by protein kinase A and/or protein kinase C (Okuda et al., 1994)
Genetic organization	
Size of mRNA	2.2 kB (1) (Okuda et al., 1994)
Genomic DNA	No data available
Related transporters	Rat oct1, 67% aa identity
	Human OCT1, 64% aa identity
	Human OCT2, 81% aa identity
Organ distribution	High expression only in kidney; no expression in liver, lung, small intestine, spleen, brain, and

	heart (Okuda et al., 1994; Gründemann et al., 1997)
Membrane domain localization	Membrane of distal tubule epithelium (Gorboulev et al., 1997)
Ion dependence	Conflicting data regarding pH dependence (Okuda et al., 1996; Gründemann et al., 1997)
Substrates	Tetraethylammonium, inhibition by cimetidine, procainamide, and quinidine (Okuda et al., 1996; Gründemann et al., 1997)

Human MDR1

M_r	170 kDa (1280 aa) (Gottesman and Pastan, 1988)
M_r deglycosylated	120–140 kDa
Protein structure	Twelve transmembrane domains; composed of two homologous halves each spanning the plasma membrane bilayer six times and containing an ATP-binding/utilization domain (Gottesman and Pastan, 1988)
	Three N-linked glycosylation sites located in the amino-terminal half in the first extracytoplasmic loop (Schinkel et al., 1993)
	Four phosphorylation sites at residues 661, 667, 671, 683; phosphorylated by cAMP-dependent protein kinase (PKA) and/or protein kinase C (PKC); at least three different novel kinases remain to be identified (reviewed in (Gottesman et al., 1995)
Genetic organization	
Size of mRNA	4.5 kb (Roninson et al., 1986; Shen et al., 1986)
Genomic DNA	Gene contains 28 exons, interspersed with introns that together extend over 120 kb of genomic DNA, located on chromosome 7q21.1 (Lincke et al., 1991; Chen et al., 1990; Chen et al., 1989; Fairchild et al., 1987; Fojo et al., 1986)
Related transporters	Human MDR3, 77% aa identity

	Mouse mdr1a, 87% aa identity
	Mouse mdr1b, 81% aa identity
Organ distribution	High expression in liver, jejunum/ileum, colon, adrenal, and kidney; intermediate expression in uterus in pregnancy, spleen, and lung; low expression in esophagus, stomach, ovary, testis, bladder, uterus, prostate, brain, and skeletal muscle (compiled from Silverman and Schrenk, 1997; Schinkel *et al.*, 1995; Thiebaut *et al.*, 1987, 1989)
Membrane domain localization	Brush border of renal proximal tubules, biliary membrane of hepatocytes, apical membrane of mucosal cells in the intestine, adrenal gland, placental trophoblasts, thus secretory epithelium
Driving force	ATP binding and hydrolysis, cofactor Mg^{2+} is required (Ambudkar, 1995; Ambudkar *et al.*, 1992; Shapiro and Ling, 1994; Sharom *et al.*, 1993; Senior *et al.*, 1995)
Substrates	Structurally unrelated, mainly amphipillic hydrophobic agents including anthracyclines, *vinca* alkaloids, epipodophyllotoxins, antibiotics; inhibitory (reversal) agents include calcium channel blockers (verapamil, azidopine), antiarrhythmics (quinidine), immunosuppressants (cyclosporine)

Rodent mdr1 Family

M_r	170 kDa [mouse mdr1a: 1276 aa (Devault and Gros, 1990; Hsu *et al.*, 1990); mouse mdr1b: 1276 aa (Gros *et al.*, 1986); rat mdr1b: 1277 aa (Silverman *et al.*, 1991); the rat mdr1a has not been cloned yet]
M_r deglycosylated	120–140 kDa
Structure	Twelve transmembrane domains; composed of two homologous halves each spanning the plasma

	membrane bilayer six times and containing an ATP-binding/utilization domain (Gottesman and Pastan, 1988)
	Two to four N-linked glycosylation sites located in the first putative extracellular loop (Gottesman et al., 1995)
	For most mdr gene products the actual sites of phosphorylation have not been reported (Gottesman et al., 1995); mouse mdr1b contains two phosphorylation sites at residues 669 and 681, phosphorylated by PKC and PKA, respectively (Takanishi et al., 1997)
Genetic organization	
Size of mRNA	4.3–4.6 kb (Silverman et al., 1991; Gros et al., 1986)
Genomic DNA	Mouse mdr gene cluster extends over 775 kb and is located on chromosome 5; the mouse mdr1b gene extends over 68 kb; the rat mdr1b gene is located on chromosome 4q11–12 (Raymond and Gros, 1989; Raymond et al., 1990; Borst et al., 1993; Popescu et al., 1993)
Related transporters (% aa identity)	Human MDR1: mouse mdr1a 87%, mouse mdr1b 81%, rat mdr1b 80%
	Human MDR3: mouse mdr1a 74%, mouse mdr1b 71%, rat mdr1b 71%
	High expression: mouse mdr1a in liver, jejunum/ileum, and colon; mouse mdr1b in adrenal and uterus in pregnancy; rat mdr1a in jejunum/ileum; rat mdr1b in liver and lung
	Intermediate expression: mouse mdr1a in testis, uterus, uterus in pregnancy, lung, and brain; mouse mdr1b in liver, stomach, ovary, kidney, and uterus; rat mdr1a in liver, kidney, lung, and brain; rat mdr1b in jejunum/ileum
	Low expression: mouse mdr1a in stomach, adrenal, ovary, kidney, spleen, and skeletal muscle; mouse mdr1b in jejunum/ileum, colon, testis, spleen, lung, brain, and skeletal muscle; rat mdr1a in spleen and skeletal muscle; rat mdr1b in kidney and spleen

	No rat mdr1b expression in adrenal, brain, and skeletal muscle Compiled from Schinkel *et al.* (1995) and Silverman and Schrenk (1997) and references therein
Localization	Secretory epithelium; the distribution data of MDR1 compared to mdr1a/1b suggest that md1a/1b together fulfil the same function as MDR1 in humans
Driving force	ATP binding and hydrolysis, cofactor Mg^{2+} is required (see human MDR1)
Substrates	Structurally unrelated, mainly amphiphilic hydrophobic agents including anthracyclines, *vinca* alkaloids, epipodophyllotoxins, antibiotics; Inhibitory (reversal) agents include calcium channel blockers (verapamil, azidopine), antiarrhythmics (quinidine), immunosupressants (cyclosporine, PSC833)

Human MDR3

M_r	170 kDa (1279 aa) (van der Bliek *et al.*, 1987, 1988)
M_r deglycosylated	140 kDa
Protein structure	Twelve transmembrane domains; composed of two homologous halves each spanning the plasma membrane bilayer six times and containing an ATP-binding/utilization domain (van der Bliek *et al.*, 1988; Chen *et al.*, 1986)
Genetic organization size of mRNA Genomic DNA	 4.5 kb (van der Bliek *et al.*, 1987) Gene contains 28 exons, interspersed with introns that together extend over 74 kb of genomic DNA, located on chromosome 7q21.1 (van der

Related transporters	Bliek et al., 1987; Lincke et al., 1991; Chin et al., 1989) Human MDR1, 77% aa identity Mouse mdr2, 91% aa identity
Organ distribution	High expression only in liver; low expression in adrenal, kidney, spleen, and skeletal muscle; no expression in esophagus, stomach, colon, ovary, testis, bladder, lung, and brain (compiled from (van der Bliek et al., 1987; Chin et al., 1989; Smit et al., 1994; Silverman and Schrenk, 1997)
Membrane domain localization	Canalicular membrane of hepatocytes, throughout the lobule (Smit et al., 1994)
Driving force	ATP binding and hydrolysis, cofactor Mg^{2+} is required (Ruetz and Gros, 1994)
Substrates	Phosphatidylcholine (Ruetz and Gros, 1994; de Vree et al., 1998; Smith et al., 1994; Smit et al., 1993), inhibition by verapamil and vanadate (Ruetz and Gros, 1994)

Rodent mdr2

M_r	170 kDa [mouse mdr2: 1276 aa (Gros et al., 1988), rat mdr2: 1278 aa (Brown et al., 1993)]
M_r deglycosylated	140 kDa
Protein structure	Twelve transmembrane domains; composed of two homologous halves each spanning the plasma membrane bilayer six times and containing an ATP-binding/utilization domain (Gros et al., 1988; Brown et al., 1993; Gottesman and Pastan, 1988)

Genetic organization
  Size of mRNA
  Genomic DNA
	
	4.5 kb (Brown et al., 1993; Croop et al., 1989)
	Mouse mdr gene cluster extends over 775 kb and

	is located on chromosome 5; rat mdr2 is located on chromosome 4q11–12 (Raymond et al., 1990; Borst et al., 1993; Zimonjic et al., 1996)
Related transporters (% aa identity)	Human MDR1: mouse mdr2 75%, rat mdr2 74% Human MDR2: mouse mdr2 91%, rat mdr2 91%
Organ distribution	High expression in liver; low expression in adrenal, spleen, lung, brain, and skeletal muscle; no expression in jejunum/ileum, kidney, and uterus (adapted from Silverman and Schrenk, 1997, and references therein)
Membrane domain localization	Canalicular membrane of hepatocytes, throughout the lobule (Buschman et al., 1992; Smit et al., 1994)
Driving force	ATP binding and hydrolysis, cofactor Mg^{2+} is required (Ruetz and Gros, 1994)
Substrates	Phosphatidylcholine (Ruetz and Gros, 1994, 1995; Smit et al., 1993; de Vree et al., 1998), inhibition by verapamil and vanadate (Ruetz and Gros, 1994)

Human MRP1

M_r	190 kDa (1531 aa) (Cole et al., 1992; Cole and Deeley, 1993; Grant et al., 1994; Zaman et al., 1994; Krishnamachary and Center, 1993)
M_r deglycosylated	150–180 kDa
Protein structure	Fifteen or 17 transmembrane domains; two membrane-spanning domains each followed by a nucleotide-binding domain (consisting of six and four, or six and six transmembrane segments), plus an additional, extremely hydrophobic NH_2-terminal membrane-spanning domain of five transmembrane helices; this NH_2-terminus of the protein is located

	extracytosolically (Hipfner *et al.*, 1997; Bakos *et al.*, 1996; Tusnady *et al.*, 1997; Loe *et al.*, 1996)
	Three N-linked glycosylation sites at residues 19, 23, and 1006 (Hipfner *et al.*, 1997)
	Protein phosphorylation has been demonstrated (on serine residues), but the specific amino acids in MRP1 that are phosphorylated and the kinases responsible have not been identified (Krishnamachary and Center, 1993; Ma *et al.*, 1995; Almquist *et al.*, 1995)
Genetic organization	
Size of mRNA	6.5 kb (Cole *et al.*, 1992)
Genomic DNA	Gene contains 31 exons, interspersed with introns that together extend over 200 kb of genomic DNA, located on chromosome 16p13.13–13.12 (Cole *et al.*, 1997; Grant *et al.*, 1997; Slovak *et al.*, 1993)
Related transporters	Human MDR1, 23% aa identity
	Human MRP2/cMOAT, 49% aa identity
	Mouse mrp1, 88% aa identity
Organ distribution	High expression in esophagus, jejunum/ilieum, colon, testis, kidney, adrenal, lung, skeletal muscle, and heart; intermediate expression in stomach, ovary, bladder, prostate, and spleen; low expression in liver, uterus, and brain (compiled from Loe *et al.*, 1996; Flens *et al.*, 1996; Kool *et al.*, 1997)
Membrane domain localization	Both plasma membrane as well as cytoplasmatic staining is observed (Loe *et al.*, 1996; Flens *et al.*, 1996) MRP1 is present in many epithelia
Driving force	ATP binding and hydrolysis, cofactor Mg^{2+} is required (Cole *et al.*, 1992; Fairchild *et al.*, 1987; Higgins, 1992)
Substrates	Structurally diverse range of organic anionic conjugates and certain antineoplastic drugs in the presence of reduced glutathione (GSH) (Müller *et al.*, 1994b; Loe *et al.*, 1996; Stride

et al., 1997; Zaman et al., 1995; Taguchi et al., 1997), inhibition by the leukotriene receptor antagonist MK 571 (Jedlitschky et al., 1996)

Mouse mrp1

M_r	190 kDa (1528 aa) (Stride et al., 1996)
M_r deglycosylated	150–180 kDa
Protein structure	Based on hydropathy alignment data: 15 or 17 transmembrane domains; two membrane-spanning domains each followed by a nucleotide-binding domain (consisting of six and four, or six and six transmembrane segments), plus an additional, extremely hydrophobic NH_2-terminal membrane-spanning domain of five transmembrane helices; this NH_2-terminus of the protein is located extracytosolically (Stride et al., 1996; Hipfner et al., 1997; Loe et al., 1996b)
Genetic organization	
Size of mRNA	6.0–6.5 kb (Stride et al., 1996)
Genomic DNA	No data available
Related transporters	Human MRP1, 88% aa identity Human MRP2/cMOAT, 48% aa identity Rat mrp2, 47% aa identity
Organ distribution	High expression in testis, kidney, muscle, and heart; intermediate expression in spleen, lung, and brain; low expression in liver and stomach; no expression in jejunum/ileum (compiled from Stride et al., 1996)
Membrane domain localization	No data available
Driving force	ATP binding and hydrolysis, cofactor Mg^{2+} is required (Stride et al., 1997)
Substrates	Structurally diverse range of organic anionic

conjugates and certain antineoplastic drugs in the presence of reduced glutathione (GSH) (Stride et al., 1997; Wijnholds et al., 1997), inhibition by the leukotriene receptor antagonist MK 571 (Jedlitschky et al., 1996)

Human MRP2

M_r	200 kDa (1545 aa) (Keppler et al., 1997; Paulusma et al., 1997; Taniguchi et al., 1996; Buchler et al., 1996; Keppler and Konig, 1997)
M_r deglycosylated	150–180 kDa
Protein structure	Fifteen to 17 transmembrane domains; two membrane-spanning domains each followed by a nucleotide-binding domain (consisting of six and four/six transmembrane segments), plus an additional NH_2-terminal membrane-spanning domain (based on sequence homology between related proteins like MRP1 and rat mrp2)
Genetic organization	
Size of mRNA	6.5 kb (Taniguchi et al., 1996)
Genomic DNA	Gene is located on chromosome 10q24 (Taniguchi et al., 1996)
Related transporters	Human MRP1, 49% aa identity Human MDR1, 25% aa identity Rat mrp2/cmoat, 78% aa identity
Organ distribution	High expresion only in liver; low expression in duodenum and kidney; no expression in colon, adrenal, testis, bladder, spleen, lung, brain, and skeletal muscle (compiled from Kool et al., 1997)
Membrane domain localization	Canalicular membrane of hepatocytes (Paulusma et al., 1997; Kartenbeck et al., 1996)
Driving force	ATP binding and hydrolysis [MRP2 is member of the ABC superfamily (Higgins, 1992)]

Substrates	Glutathione, glucuronate, and sulfate conjugates of endogenous and exogenous compounds, unconjugated amphiphilic anions (Keppler and Konig, 1997; Kartenbeck *et al.*, 1996; Jedlitschky *et al.*, 1997; Paulusma and Oude Elferink, 1997; Oude Elferink *et al.*, 1995), inhibition by the leukotriene receptor antagonist MK 571 (Buchler *et al.*, 1996)

Rat mrp2/cmoat

M_r	200 kDa (1541 aa) (Paulusma *et al.*, 1996; Büchler *et al.*, 1996; Ito *et al.*, 1997)
M_r deglycosylated	150–180 kDa
Protein structure	Fifteen to 17 transmembrane domains; two membrane-spanning domains each followed by a nucleotide-binding domain (consisting of six and four/six transmembrane segments), plus an additional NH_2-terminal membrane-spanning domain (based on sequence homology between related proteins like MRP1 and mouse mrp1)
Genetic organization	
Size of mRNA	5.5–9.0 kb (Paulusma *et al.*, 1996; Büchler *et al.*, 1996; Ito *et al.*, 1997)
Genomic DNA	No data available
Related transporters	Mouse mrp1, 47% aa identity Human MRP1, 46% aa identity Human MRP2/cMOAT, 78% aa identity
Organ distribution	High expression only in liver; low expression in kidney, duodenum, and ileum (Paulusma *et al.*, 1996; Ito *et al.*, 1997)
Membrane domain localization	Canalicular membrane of hepatocytes, brush border membrane of proximal tubule epithelium (Paulusma *et al.*, 1996; Büchler *et al.*, 1996; Mayer *et al.*, 1995; Schaub *et al.*, 1997)

Driving force	ATP binding and hydrolysis [MRP2 is member of the ABC superfamily (Higgins, 1992)]
Substrates	Glutathione, glucuronate, and sulfate conjugates of endogenous and exogenous compounds, unconjugated amphiphilic anions (Keppler and Konig, 1997; Jedlitschky *et al.*, 1997; Paulusma and Oude Elferink, 1997; Oude Elferink *et al.*, 1995); inhibition by the leukotriene receptor antagonist MK 571 (Büchler *et al.*, 1996)

REFERENCES

Adachi, Y., Kobayashi, H., Kurumi, Y., Shouji, M., Kitano, M., and Yamamoto, T., 1991, ATP-dependent taurocholate transport by rat liver canalicular membrane vesicles, *Hepatology* **14**:655–659.

Ahmad, S., Safa, A. R., and Glazer, R. I., 1994, Modulation of P-glycoprotein by protein kinase Cα in a baculovirus expression system, *Biochemistry* **33**:10313–10318.

Allikmets, R., Gerrard, B., Hutchinson, A., and Dean, M., 1996, Characterization of the human ABC superfamily: Isolation and mapping of 21 new genes using the Expressed Sequence Tags database, *Hum. Mol. Genet.* **5**:1649–1655.

Almquist, K. C., Loe, D. W., Hipfner, D. R., Mackie, J. E., Cole, S. P. C., and Deeley, R. G., 1995, Characterisation of the M_r 190,000 multidrug resistance protein (MRP) in drug-selected and transfected human tumor cells, *Cancer Res.* **55**:102–110.

Ambudkar, S. V., 1995, Purification and reconstitution of functional human P-glycoprotein, *J. Bioenerg. Biomembr.* **27**:23–29.

Ambudkar, S. V., Lelong, I. H., Zhang, J., Cardarelli, C. O., Gottesman, M. M., and Pastan, I., 1992, Partial purification and reconstitution of the human multidrug-resistance pump: Characterization of the drug-stimulatable ATP hydrolysis, *Proc. Natl. Acad. Sci. USA* **89**:8472–8476.

Ananthanarayanan, M., Ng, O. C., Boyer, J. L., and Suchy, F. J., 1994, Characterization of cloned rat liver Na^+–bile acid cotransporter using peptide and fusion protein antibodies, *Am. J. Physiol.* **267**:G637–G643.

Ayer Lazaruk, K. D., and Wright, S. H., 1990, MPP^+ is transported by the TEA^+–H^+ exchanger of renal brush-border membrane vesicles, *Am. J. Physiol.* **258**:F597–F605.

Angeletti, R. H., Novikoff, P. M., Juvvadi, S. R., Fritschy, J. M., Meier, P. J., and Wolkoff, A. W., 1997, The choroid plexus epithelium is the site of the organic anion transport protein in the brain, *Proc. Natl. Acad. Sci. USA* **94**:283–286.

Bakos, E., Hegedus, T., Hollo, Z., Welker, E., Tusnady, G. E., Zaman, G. J. R., Flens, M. J., Varadi, A., and Sarkadi, B., 1996, Membrane topology and glycosylation of the human multidrug resistance-associated protein, *J. Biol. Chem.* **271**:12322–12326.

Barth, A., Fleck, C., and Klinger, W., 1996, Development of organic anion transport in the liver, *Exp. Toxicol. Pathol.* **48**:421–432.

Bergwerk, A. J., Shi, X. Y., Ford, A. C., Kanai, N., Jacquemin, E., Burk, R. D., Bai, S., Novikoff, P. M., Stieger, B., Meier, P. J., Schuster, V. L., and Wolkoff, A. W., 1996, Immunologic distribution of an organic anion transport protein in rat liver and kidney, *Am. J. Physiol.* **271**:G231–G238.

Beukeveld, G. J. J., In 't Veld, G., Havinga, R., Groen, A. K., Wolthers, B. G., and Kuipers, F., 1996, Relationship between biliary lipid and protoporphyrin secretion; potential role of mdr2 P-glycoprotein in hepatobiliary organic anion transport, *J. Hepatol.* **24**: 343–352.

Blobe, G. C., Sachs, C. W., Khan, W. A., Fabbro, D., Stabel, S., Wetsel, W. C., Obeid, L. M., Fine, R. L., and Hannun, Y. A., 1993, Selective regulation of expression of protein kinase C (PKC) isoenzymes in multidrug-resistant MCF-7 cells. Functional significance of enhanced expression of PKC α, *J. Biol. Chem.* **268**:658–664.

Böhme, M., Müller, M., Leier, I., Jedlitschky, G., and Keppler, D., 1994, Cholestasis caused by inhibition of the adenosine triphosphate-dependent bile salt transport in rat liver, *Gastroenterology* **107**:255–265.

Bolder, U., Ton-Nu, H.-T., Schteingart, C. D., Frick, E., and Hofmann, A. F., 1997, Hepatocyte transport of bile acids and organic anions in endotoxemic rats: Impaired uptake and secretion, *Gastroenterology* **112**:214–225.

Borst, P., and Schinkel, A. H., 1996, What have we learnt thus far from mice with disrupted P-glycoprotein genes? *Eur. J. Cancer* **32A**:985–990.

Borst, P., Schinkel, A. H., Smit, J. J. M., Wagenaar, E., van Deemter, L., Smith, A. J., Eijdems, E. W. H. M., Baas, F., and Zaman, G. J. R., 1993, Classical and novel forms of multidrug resistance and the physiological functions of P-glycoproteins in mammals, *Pharmacol. Ther.* **60**:289–299.

Bossuyt, X., Müller, M., Hagenbuch, B., and Meier, P. J., 1996a, Polyspecific drug and steroid clearance by an organic anion transporter of mammalian liver, *J. Pharmacol. Exp. Ther.* **276**:891–896.

Bossuyt, X., Müller, M., and Meier, P. J., 1996b, Multispecific amphipathic substrate transport by an organic anion transporter of human liver, *J. Hepatol.* **25**:733–738.

Boyer, J. L., Hagenbuch, B., Ananthanarayanan, M., Suchy, F., Stieger, B., and Meier, P. J., 1993, Phylogenic and ontogenic expression of hepatocellular bile acid transport, *Proc. Natl. Acad. Sci. USA* **90**:435–438.

Boyer, J. L., Ng, O. C., Ananthanarayanan, M., Hofmann, A. F., Schteingart, C. D., Hagenbuch, B., Stieger, B., and Meier, P. J., 1994, Expression and characterization of a functional rat liver Na^+ bile acid cotransport system in COS-7 cells, *Am. J. Physiol.* **266**:G382–G387.

Brown, P. C., Thorgeirsson, S. S., and Silverman, J. A., 1993, Cloning and regulation of the rat mdr2 gene, *Nucleic Acids Res.* **21**:3885–3891.

Büchler, M., Böhme, M., Ortlepp, H., and Keppler, D., 1994, Functional reconstitution of ATP-dependent transporters from the solubilized hepatocyte canalicular membrane, *Eur. J. Biochem.* **224**:345–352.

Büchler, M., König, J., Brom, M., Kartenbeck, J., Spring, H., Horie, T., and Keppler, D., 1996, cDNA cloning of the hepatocyte canalicular isoform of the multidrug resistance protein, cMrp, reveals a novel conjugate export pump deficient in hyperbilirubinemic mutant rats, *J. Biol. Chem.* **271**:15091–15098.

Busch, A. E., Quester, S., Ulzheimer, J. C., Waldegger, S., Gorboulev, V., Arndt, P., Lang,

F., and Koepsell, H., 1996a, Electrogenic properties and substrate specificity of the polyspecific rat cation transporter rOCT1, *J. Biol. Chem.* **271**:32599–32604.

Busch, A. E., Quester, S., Ulzheimer, J. C., Gorboulev, V., Akhoundova, A., Waldegger, S., Lang, F., and Koepsell, H., 1996b, Monoamine neurotransmitter transport mediated by the polyspecific cation transporter rOCT1, *FEBS Lett.* **395**:153–156.

Buschman, E., Arceci, R. J., Croop, J. M., Che, M., Arias, I. M., Housman, D. E., and Gros, P., 1992, mdr2 encodes P-glycoprotein expressed in the bile canalicular membrane as determined by isoform-specific antibodies, *J. Biol. Chem.* **267**:18093–18099.

Chambers, T. C., Chalikonda, I., and Eilon, G., 1990a, Correlation of protein kinase C translocation, P-glycoprotein phosphorylation and reduced drug accumulation in multi-drug resistant human KB cells, *Biochem. Biophys. Res. Commun.* **169**:253–259.

Chambers, T. C., McAvoy, E. M., Jacobs, J. W., and Eilon, G., 1990b, Protein kinase C phosphorylates P-glycoprotein in multidrug resistant human KB carcinoma cells, *J. Biol. Chem.* **265**:7679–7686.

Chambers, T. C., Pohl, J., Raynor, R. L., and Kuo, J. F., 1993, Identification of specific sites in human P-glycoprotein phosphorylated by protein kinase C, *J. Biol. Chem.* **268**:4592–4595.

Chaudhary, P. M., Mechetner, E. B., and Roninson, I. B., 1992, Expression and activity of the multidrug resistance P-glycoprotein in human peripheral blood lymphocytes, *Blood* **80**:2735–2739.

Chen, C. J., Chin, J. E., Ueda, K., Clark, D. P., Pastan, I., Gottesman, M. M., and Roninson, I. B., 1986, Internal duplication and homology with bacterial transport proteins in the mdr1 (P-glycoprotein) gene from multidrug-resistant human cells, *Cell* **47**:381–389.

Chen, C. J., Clark, D., Ueda, K., Pastan, I., Gottesman, M. M., and Roninson, I. B., 1990, Genomic organization of the human multidrug resistance (MDR1) gene and origin of P-glycoproteins, *J. Biol. Chem.* **265**:506–514.

Chianale, J., Vollrath, V., Wielandt, A. M., Amigo, L., Rigotto, A., Nervi, F., Gonzalez, S., Andrade, L., Pizarro, M., and Accatino, L., 1996, Fibrates induce mdr2 gene expression and biliary phospholipid secretion in the mouse, *Biochem. J.* **314**:781–786.

Childs, S., Yeh, R. L., Georges, E., and Ling, V., 1995, Identification of a sister gene to P-glycoprotein, *Cancer Res.* **55**:2029–2034.

Childs, S., Yeh, R. L., and Ling, V., 1996, Characterization of *spgp*, a novel P-glycoprotein gene from liver, *Proc. AACR* **37**:326

Chin, J. E., Soffir, R., Noonan, K. E., Choi, K., and Roninson, I. B., 1989, Structure and expression of the human MDR (P-glycoprotein) gene family, *Mol. Cell. Biol.* **9**:3808–3820.

Christie, D. M., Dawson, P. A., Thevananther, S., and Schneider, B. L., 1996, Comparative analysis of the ontogeny of a sodium-dependent bile acid transporter in rat kidney and ileum, *Am. J. Physiol.* **271**:G377-G385.

Cohn, M. A., Rounds, D. J., Karpen, S. J., Ananthanarayanan, M., and Suchy, F. J., 1995, Assignment of a rat liver Na^+/bile acid cotransporter gene to chromosome 6q24, *Mamm. Genome* **6**:60.

Cole, S. P. C., and Deeley, R. G., 1993, Multidrug resistance-associated protein: Sequence correction [Letter], *Science* **260**:879.

Cole, S. P. C., Bhardwaj, G., Gerlach, J. H., Mackie, J. E., Grant, C. E., Almquist, K. C.,

Stewart, A. J., Kurz, E. U., Duncan, A. M. V., and Deeley, R. G., 1992, Overexpression of a transporter gene in a multidrug-resistant human lung cancer cell line, *Science* **258**:1650–1654.
Crawford, J. M., Möckel, G.-M., Crawford, A. R., Hagen, S. J., Hatch, V. C., Barnes, S., Godleski, J. J., and Carey, M. C., 1995, Imaging biliary lipid secretion in the rat: Ultrastructural evidence for vesiculation of the hepatocyte canalicular membrane, *J. Lipid Res.* **36**:2147–2163.
Croop, J. M., Raymond, M., Haber, D., Devault, A., Arceci, R. J., Gros, P., and Housman, D. E., 1989, The three mouse multidrug resistance (mdr) genes are expressed in a tissue-specific manner in normal mouse tissues, *Mol. Cell. Biol.* **9**:1346–1350.
Dawson, P. A., and Oelkers, P., 1995, Bile acid transporters, *Curr. Opin. Lipidol.* **6**:109–114.
Dean, M., and Allikmets, R., 1995, Evolution of ATP-binding cassette transporter genes, *Curr. Opin. Genet. Dev.* **5**:779–785.
Decottignies, A., and Goffeau, A., 1997, Complete inventory of the yeast ABC proteins, *Nature Genet.* **15**:137–145.
Deleuze, J. F., Jacquemin, E., Dubuisson, C., Cresteil, D., Dumont, M., Erlinger, S., Bernard, O., and Hadchouel, M., 1996, Defect of multidug-resistance 3 gene expression in a subtype of progressive familial intrahepatic cholestasis, *Hepatology* **23**:904–908.
de Vree, J. M. L., Jacquemin, E., Sturm, E., Cresteil, D., Bosma, P. J., Aten, J., Deleuze, J. F., Desrochers, M., Burdelski, M., Bernard, O., Oude Elferink, R. P. J., and Hadchouel, M., 1998, Mutations in the MDR3 gene cause progressive familial intrahepatic cholestasis, *Proc. Natl. Acad. Sci. USA* **95**:282–287.
Dubuisson, C., Cresteil, D., Desrochers, M., Decimo, D., Hadchouel, M., and Jacquemin, E., 1996, Ontogenic expression of the Na^+-independent organic anion transporting polypeptide (oatp) in rat liver and kidney, *J. Hepatol.* **25**:932–940.
Dudley, A. J., and Brown, C. D. A., 1996, Mediation of cimetidine secretion by P-glycoprotein and a novel H^+-coupled mechanism in cultured renal epithelial monolayers of $LLC-PK_1$ cells, *Br. J. Pharmacol.* **117**:1139–1144.
Devault, A., and Gros, P., 1990, Two members of the mouse mdr gene family confer multidrug resistance with overlapping but distinct drug specificities, *Mol. Cell. Biol.* **10**:1652–1663.
Fairchild, C. R., Ivy, S. P., Kao-Shan, C. S., Whang-Peng, J., Rosen, N., Israel, M. A., Melera, P. W., Cowan, K. H., and Goldsmith, M. E., 1987, Isolation of amplified and overexpressed DNA sequences from adriamycin-resistant human breast cancer cells, *Cancer Res.* **47**:5141–5148.
Fauth, C., Rossier, B., and Roch-Ramel, F., 1988, Transport of tetraethylammonium by a kidney epithelial cell line ($LLC-PK_1$), *Am. J. Physiol.* **254**:F351–F357.
Fine, R. L., Chambers, T. C., and Sachs, C. W., 1996, P-glycoprotein, multidrug resistance and protein kinase C, *Stem Cells* **14**:47–55.
Flens, M. J., Zaman, G. J. R., van der Valk, P., Izquierdo, M. A., Schroeijers, A. B., Scheffer, G. L., van der Groep, P., de Haas, M., Meijer, C. J. L. M., and Scheper, R. J., 1996, Tissue distribution of the multidrug resistance protein, *Am. J. Pathol.* **148**:1237–1247.
Fojo, A., Lebo, R., Shimizu, N., Chin, J. E., Roninson, I. B., Merlino, G. T., Gottesman, M. M., and Pastan, I., 1986, Localization of multidrug resistance-associated DNA se-

quences to human chromosome 7, *Somat. Cell Mol. Genet.* **12**:415–420.

Fouda, A.-K., Fauth, C., and Roch-Ramel, F., 1990, Transport of organic cations by kidney epithelial cell line LLC-PK$_1$, *J. Pharmacol. Exp. Ther.* **252**:286–292.

Fukami, Y., and Nishizuka, Y., 1992, The protein kinase C family in signal transduction and cellular regulation, in: *Adenine Nucleotides in Cellular Energy Transfer and Signal Transduction* (S. Papa, A. Azzi, and J. M. Tager, eds.), Birkhäuser, Basel, pp. 201–205.

Furuya, K. N., Gebhardt, R., Schuetz, E. G., and Schuetz, J. D., 1994, Isolation of rat pgp3 cDNA: Evidence for gender and zonal regulation of expression in the liver, *Biochim. Biophys. Acta* **1219**:636–644.

Ganguly, T. C., O'Brien, M. L., Karpen, S. J., Hyde, J. F., Suchy, F. J., and Vore, M., 1997, Regulation of the rat liver sodium-dependent bile acid cotransporter gene by prolactin. Mediation of transcriptional activation by stat5, *J. Clin. Invest.* **99**:2906–2914.

Gartung, C., Ananthanarayanan, M., Rahman, M. A., Schuele, S., Nundy, S., Soroka, C. J., Stolz, A., Suchy, F. J., and Boyer, J. L., 1996, Down-regulation of expression and function of the rat liver Na^+/bile acid cotransporter in extrahepatic cholestasis, *Gastroenterology* **110**:199–209.

Gerloff, T., Stieger, B., Hagenbuch, B., Landmann, L., and Meier, P. J., 1997, The sister P-glycoprotein of rat liver mediates ATP-dependent taurocholate (TCA) transport, *Hepatology* **26**(No. 4, Pt. 2):358A (Abstract).

Giros, B., and Caron, M. G., 1996, Molecular characterization of the dopamine transporter, *Trends Pharmacol. Sci.* **14**:43–49.

Goodfellow, H. R., Sardini, A., Ruetz, S., Callaghan, R., Gros, P., Mcnaughton, P. A., and Higgins, C. F., 1996, Protein kinase C-mediated phosphorylation does not regulate drug transport by the human multidrug resistance P-glycoprotein, *J. Biol. Chem.* **271**:13668–13674.

Gorboulev, V., Ulzheimer, J. C., Akhoundova, A., Ulzheimer-Teuber, I., Karbach, U., Quester, S., Baumann, C., Lang, F., Busch, A. E., and Koepsell, H., 1997, Cloning and characterization of two human polyspecific organic cation transporters, *DNA Cell Biol.* **16**:871–881.

Gottesman, M. M., and Pastan, I., 1988, The multidrug transporter, a double-edged sword, *J. Biol. Chem.* **263**:12163–12166.

Gottesman, M. M., and Pastan, I., 1993, Biochemistry of multidrug resistance mediated by the multidrug transporter, *Annu. Rev. Biochem.* **62**:385–427.

Gottesman, M. M., Hrycyna, C. A., Schoenlein, P. V., Germann, U. A., and Pastan, I., 1995, Genetic analysis of the multidrug transporter, *Annu. Rev. Genet.* **29**:607–649.

Grant, C. E., Valdimarsson, G., Hipfner, D. R., Almquist, K. C., Cole, S. P. C., and Deeley, R. G., 1994, Overexpression of multidrug resistance-associated protein (MRP) increases resistance to natural product drugs, *Cancer Res.* **54**:357–361.

Grant, C. E., Kurz, E. U., Cole, S. P. C., and Deeley, R. G., 1997, Analysis of the intron–exon organization of the human multidrug-resistance protein gene (MRP) and alternative splicing of its mRNA, *Genomics* **45**:368–378.

Green, R. M., Lipin, A. I., Pelletier, E. M., Hagenbuch, B., Meier, P. J., Beier, D., and Gollan, J. L., 1994, Hepatic regeneration is associated with a marked reduction in mRNA expression of the basolateral Na^+-taurocholate transporter, *Gastroenterology* **106**(No. 4, Part 2):A901 (Abstract).

Green, R. M., Beier, D., and Gollan, J. L., 1996, Regulation of hepatocyte bile salt trans-

porters by endotoxin and inflammatory cytokines in rodents, *Gastroenterology* **111:** 193–198.
Gros, P., Croop, J., and Housman, D., 1986, Mammalian multidrug resistance gene: Complete cDNA sequence indicates strong homology to bacterial transport proteins, *Cell* **47:**371–380.
Gros, P., Croop, J., Roninson, I., Varshavsky, A., and Housman, D. E., 1986, Isolation and characterization of DNA sequences amplified in multidrug-resistant hamster cells, *Proc. Natl. Acad. Sci. USA* **83:**337–341.
Gros, P., Raymond, M., Bell, J., and Housman, D., 1988, Cloning and characterization of a second member of the mouse mdr gene family, *Mol. Cell. Biol.* **8:**2770–2778.
Gründemann, D., Gorboulev, V., Gambaryan, S., Veyhl, M., and Koepsell, H., 1994, Drug excretion mediated by a new prototype of polyspecific transporter, *Nature* **372:**549–552.
Gründemann, D., Babin-Ebell, J., Martel, F., Örding, N., Schmidt, A., and Schömig, E., 1997, Primary structure and functional expression of the apical organic cation transporter from kidney epithelial LLC-PK$_1$ cells, *J. Biol. Chem.* **272:**10408–10413.
Hagenbuch, B., 1997, Molecular properties of hepatic uptake systems for bile acids and organic anions, *J. Membr. Biol.* **160:**1–8.
Hegenbuch, B., and Meier, P. J., 1994, Molecular cloning, chromosomal localization, and functional characterization of human liver Na$^+$/bile acid contransporter, *J. Clin. Invest.* **93:**326–1331.
Hagenbuch, B., and Meier, P. J., 1996, Sinusoidal (basolateral) bile salt uptake systems of hepatocytes, *Semin. Liver Dis.* **16:**129–136.
Hagenbuch, B., Stieger, B., Foguet, M., Lübbert, H., and Meier, P. J., 1991, Functional expression cloning and characterization of the hepatocyte Na$^+$/bile acid cotransport system, *Proc. Natl. Acad. Sci. USA* **88:**10629–10633
Hagenbuch, B., Scharschmidt, B. F., and Meier, P. J., 1996, Effect of antisense oligonucleotides on the expression of hepatocellular bile acid and organic anion uptake systems in Xenopus laevis oocytes, *Biochem. J.* **316:**901–904.
Hardikar, W., Ananthanarayanan, M., and Suchy, F. J., 1995, Differential ontogenic regulation of basolateral and canalicular bile acid transport proteins in rat liver, *J. Biol. Chem.* **270:**20841–20846.
Hedman, A., and Meijer, D. K. F., 1997, The stereoisomers quinine and quinidine exhibit a marked stereoselectivity in the inhibition of hepatobiliary transport of cardiac glycosides, *J. Hepatol.* **28:**240–249.
Hedman, A., Angelin, B., Arvidsson, A., Dahlqvist, R., and Nilsson, B., 1990, Interactions in the renal and biliary elimination of digoxin: Stereoselective difference between quinine and quinidine, *Clin. Pharmacol. Ther.* **47:**20–26.
Henderson, P. J. F., 1993, The 12-transmembrane helix transporters, *Curr. Opin. Cell Biol.* **5:**708–721.
Higgins, C. F., 1992, ABC transporters: From microorganisms to man, *Annu. Rev. Cell Biol.* **8:**67–113.
Hipfner, D. R., Almquist, K. C., Leslie, E. M., Gerlach, J. H., Grant, C. E., Deeley, R. G., and Cole, S. P. C., 1997, Membrane topology of the multidrug resistance protein (MRP). A study of glycosylation-site mutants reveals an extracytosolic NH$_2$ terminus, *J. Biol. Chem.* **272:**23623–23630.

Hsu, S. I., Cohen, D., Kirschner, L. S., Lothstein, L., Hartstein, M., and Horwitz, S. B., 1990, Structural analysis of the mouse mdrla (P-glycoprotein) promoter reveals the basis for differential transcript heterogeneity in multidrug-resistant J774.2 cells, *Mol. Cell. Biol.* **10**:3596–3606.

Hyde, S. C., Emsley, P., Hartshorn, M. J., Mimmack, M. M., Gileadi, U., Pearce, S. R., Gallagher, M. P., Gill, D. R., Hubbard, R. E., and Higgins, C. F., 1990, Structural model of ATP-binding proteins associated with cystic fibrosis, multidrug resistance and bacterial transport, *Nature* **346**:362–365

Ito, K., Suzuki, H., Hirohashi, T., Kume, K., Shimizu, T., and Sugiyama, Y., 1997, Molecular cloning of canalicular multispecific organic anion transporter defective in EHBR, *Am. J. Physiol.* **272**:G16–G22.

Jacquemin, E., Hagenbuch, B., Stieger, B., Wolkoff, A. W., and Meier, P. J., 1994, Expression cloning of a rat liver Na^+-independent organic anion transporter, *Proc. Natl. Acad. Sci. USA* **91**:133–137.

Jacquemin, E., de Vree, J. M. L., Sturm, E., Cresteil, D., Bosma, P. J., Aten, J., Deleuze, J.-F., Desrochers, M., Burdelski, M., Bernard, O., Hadchouel, M., and Oude Elferink, R. P. J., 1997, Mutations in the MDR3 gene are responsible for a subtype of progressive familial intrahepatic cholestasis (PFIC), *Hepatology* **26**(No. 4, Pt. 2):248A (Abstract).

Jansen, P. L. M., Van Klinken, J.-W., Van Gelder, M., Ottenhoff, R., and Oude Elferink, R. P. J., 1993, Preserved organic anion transport in mutant TR^- rats with a hepatobiliary secretion defect, *Am. J. Physiol.* **265**:G445–G452.

Jedlitschky, G., Leier, I., Buchholz, U., Barnouin, K., Kurz, G., and Keppler, D., 1996, Transport of glutathione, glucuronate, and sulfate conjugates by the MRP gene-encoded conjugate export pump, *Cancer Res.* **56**:988–994.

Jedlitschky, G., Leier, I., Buchholz, U., Hummel-Eisenbeiss, J., Burchell, B., and Keppler, D., 1997, ATP-dependent transport of bilirubin glucuronides by the multidrug resistance protein MRP1 and its hepatocyte canalicular isoform MRP2, *Biochem. J.* **327**:305–310.

Kamimoto, Y., Gatmaitan, Z., Hsu, J., and Arias, I. M., 1989, The function of Gp170, the multidrug resistance gene product, in rat liver canalicular membrane vesicles, *J. Biol. Chem.* **264**:11693–11698.

Kanai, N., Lu, R., Bao, Y., Wolkoff, A. W., and Schuster, V. L., 1996a, Transient expression of oatp organic anion transporter in mammalian cells: Identification of candidate substrates, *Am. J. Physiol.* **270**:F319–F325.

Kanai, N., Lu, R., Bao, Y., Wolkoff, A. W., Vore, M., and Schuster, V. L., 1996b, Estradiol 17-beta-D-glucuronide is a high-affinity substrate for oatp organic anion transporter, *Am. J. Physiol.* **270**:F326–F331.

Karpen, S. J., Sun, A.-Q., Kudish, B., Hagenbuch, B., Meier, P. J., Ananthanarayanan, M., and Suchy, F. J., 1996, Multiple factors regulate the rat liver basolateral sodium-dependent bile acid cotransporter gene promotor, *J. Biol. Chem.* **271**:15211–15221.

Kartenbeck, J., Leuschner, U., Mayer, R., and Keppler, D., 1996, Absence of the canalicular isoform of the MRP gene-encoded conjugate export pump from the hepatocytes in Dubin–Johnson syndrome, *Hepatology* **23**:1061–1066.

Kast, C., Stieger, B., Winterhalter, K. H., and Meier, P. J., 1994, Hepatocellular transport of bile acids. Evidence for distinct subcellular localizations of electrogenic and ATP-

dependent taurocholate transport in rat hepatocytes, *J. Biol. Chem.* **269:**5179–5186.
Keppler, D., and Konig, J., 1997, Hepatic canalicular membrane. 5. Expression and localization of the conjugate export pump encoded by the MRP2 (cMRP/cMOAT) gene in liver, *FASEB J.* **1:**509–516.
Keppler, D., Konig, J., and Buchler, M., 1997, The canalicular multidrug resistance protein, cMRP/MRP2, a novel conjugate export pump expressed in the apical membrane of hepatocytes, *Adv. Enzyme Regul.* **37:**321–333.
Koehler, M. R., Gorboulev, V., Koepsell, H., Steinlein, C., and Schmid, M., 1996, Roct1, a rat polyspecific transporter gene for the excretion of cationic drugs, maps to chromosome 1q11–12, *Mamm. Genome* **7:**247–248.
Koepsell, H., 1998, Organic cation transporters in intestine, kidney, liver and brain, *Annu. Rev. Physiol.* **60:**243–266.
Kitayama, S., Shimada, S., and Uhl, G. R., 1992, Parkisonism-inducing neurotoxin MPP^+: Uptake and toxicity in non-neuronal COS cells expressing dopamine transporter cDNA, *Ann. Neurobiol.* **32:**109–111.
Kool, M., de Haas, M., Scheffer, G. L., Scheper, R. J., van Eijk, M. J. T., Juijn, J. A., Baas, F., and Borst, P., 1997, Analysis of expression of cMOAT (MRP2), MRP3, MRP4, and MRP5, homologues of the multidrug resistance-associated protein gene (MRP1), in human cancer cell lines, *Cancer Res.* **57:**3537–3547.
Koopen, N. R., Wolters, H., Müller, M., Schippers, I. J., Havinga, R., Roelofsen, H., Vonk, R., Stieger, B., Meier, P. J., and Kuipers, F., 1997, Hepatic bile salt flux does not modulate level and activity of the sinusoidal Na^+–taurocholate cotransporter (ntcp) in rats, *J. Hepatol.* **27:**699–706.
Krishnamachary, N., and Center, M. S., 1993, The MRP gene associated with a non-P-glycoprotein multidrug resistance encodes a 190-kDa membrane bound glycoprotein, *Cancer Res.* **53:**3658–3661.
Kullak-Ublick, G.-A., Hagenbuch, B., Stieger, B., Wolkoff, A. W., and Meier, P. J., 1994, Functional characterization of the basolateral rat liver organic anion transporting polypeptide, *Hepatology* **20:**411–416.
Kullak-Ublick, G.-A., Hagenbuch, B., Stieger, B., Schteingart, C. D., Hofmann, A. F., Wolkoff, A. W., and Meier, P. J., 1995, Molecular and functional characterization of an organic anion transporting polypeptide cloned from human liver, *Gastroenterology* **109:**1274–1282.
Kullak-Ublick, G.-A., Beuers, U., Meier, P. J., Domdey, H., and Paumgartner, G., 1996a, Assignment of the human organic anion transporting polypeptide (OATP) gene to chromosome 12p12 by fluorescence *in situ* hybridization, *J. Hepatol.* **25:**985–987.
Kullak-Ublick, G.-A., Beuers, U., and Paumgartner, G., 1996b, Molecular and functional characterization of bile acid transport in human hepatoblastoma HepG2 cells, *Hepatology* **23:**1053–1060.
Kullak-Ublick, G.-A., Beuers, U., Fahney, C., Hagenbuch, B., Meier, P. J., and Paumgartner, G., 1997a, Identification and characterization of the promotor of the human organic anion transporting polypeptide gene, *J. Hepatol.* **26**(Suppl. 1)**:**71 (Abstract).
Kullak-Ublich, G.-A., Glasa, J., Boker, C., Oswald, M., Grutzner, U., Hagenbuch, B., Stieger, B., Meier, P. J., Beuers, U., Kramer, W., Wess, G., and Paumgartner, G., 1997b,

Chlorambucil-taurocholate is transported by bile acid carriers expressed in human hepatocellular carcinomas, *Gastroenterology* **113**:1295–1305.

Kwon, Y., Lee, R. D., and Morris, M. E., 1996, Hepatic uptake of choline in rat liver basolateral and canalicular membrane vesicle preparations, *J. Pharmacol. Exp. Ther.* **279**: 774–781.

Lake, B. G., 1995, Mechanisms of hepatocarcinogenicity of peroxisome-proliferating drugs and chemicals, *Annu. Rev. Pharmacol. Toxicol.* **35**:483–507.

Lauterbach, F., 1984, Intestinal permeation of organic bases and quaternary ammonium compounds, in: *Handbook of Experimental Pharmacology, Vol. 70/II. Pharmacology of Intestinal Permeation* (T. Z. Csaky, ed.), Springer-Verlag, Berlin, pp. 271–284.

Lazarides, K., Pham, L., de Groen, P., Dawson, P., and Larusso, N., 1996, Rat cholangiocytes express the ileal Na^+-dependent taurocholate co-transporter, *Hepatology* **24**(No. 4, Pt. 2):351A (Abstract).

Leier, I., Jedlitschky, G., Buchholz, U., Cole, S. P. C., Deeley, R. G., and Keppler, D., 1994, The MRP gene encodes an ATP-dependent export pump for leukotriene C_4, and structurally related conjugates, *J. Biol. Chem.* **269**:27807–27810.

Leier, I., Jedlitschky, G., Buchholz, U., Center, M., Cole, S. P. C., Deeley, R. G., and Keppler, D., 1996, ATP-dependent glutathione disulphide transport mediated by the MRP gene-encoded conjugate export pump, *Biochem. J.* **314**:433–437.

Liang, D., Hagenbuch, B., Stieger, B., and Meijer, P. J., 1993, Parallel decrease of Na^+-taurocholate cotransport and its encoding mRNA in primary cultures of rat hepatocytes, *Hepatology* **18**:1163–1166.

Lincke, C. R., Smit, J. J. M., van der Velde-Koerts, T., and Borst, P., 1991, Structure of the human MDR3 gene and physical mapping of the human MDR locus, *J. Biol. Chem.* **266**:5303–5310.

Liu, J. P., 1996, Protein kinase C and its substrates, *Mol. Cell. Endocrinol.* **116**:1–29.

Liu, Y., Peter, D., Roghani, A., Schuldiner, S., Privé, G. G., Eisenberg, D., Brecha, N., and Edwards, R. H., 1992, A cDNA that suppresses MPP^+ toxicity encodes a vesicular amine transporter, *Cell* **70**:539–551.

Loe, D. W., Almquist, K. C., Deeley, R. G., and Cole, S. P. C., 1996a, Multidrug resistance protein (MRP)-mediated transport of leukotriene C_4 and chemotherapeutic agents in membrane vesicles. Demonstration of glutathione-dependent vincristine transport, *J. Biol. Chem.* **271**:9675–9682.

Loe, D. W., Deeley, R. G., and Cole, S. P. C., 1996b, Biology of the multidrug resistance-associated protein, MRP, *Eur. J. Cancer* **32A**:945–957.

Lomri, N., Fitz, J. G., and Scharschmidt, B. F., 1996, Hepatocellular transport: Role of ATP-binding cassette proteins, *Semin. Liver Dis.* **16**:201–210.

Lu, R., Kanai, N., Bao, Y., Wolkoff, A. W., and Schuster, V. L., 1996, Regulation of renal oatp mRNA expression by testosterone, *Am. J. Physiol.* **270**:F332–F337.

Ma, L., Krishnamachary, N., and Center, M. S., 1995, Phosphorylation of the multidrug resistance associated protein gene encoded protein P190, *Biochemistry* **34**:3338–3343.

Martel, F., Vetter, T., Russ, H., Gründemann, D., Azevedo, I., Koepsell, H., and Schömig, E., 1996, Transport of small organic cations in the rat liver. The role of the organic cation transporter OCT1, *Naunyn Schmiedebergs Arch. Pharmacol.* **354**:320–326.

Mauad, T. H., van Nieuwkerk, C. M. J., Dingemans, K. P., Smit, J. J. M., Schinkel, A. H., Notenboom, R. G. E., van den Bergh Weerman, M. A., Verkruisen, R. P., Groen, A. K.,

Oude Elferink, R. P. J., Van der Valk, M. A., Borst, P., and Offerhaus, G. J. A., 1994, Mice with homozygous disruption of the mdr2 P-glycoprotein gene. A novel animal model for studies of nonsuppurative inflammatory cholangitis and hepatocarcinogenesis, *Am. J. Pathol.* **145**:1237–1245.

Mayer, R., Kartenbeck, J., Büchler, M., Jedlitschky, G., Leier, I., and Keppler, D., 1995, Expression of the MRP gene-encoded conjugate export pump in liver and its selective absence from the canalicular membrane in transport-deficient mutant hepatocytes, *J. Cell Biol.* **131**:137–150.

Mayer, U., Wagenaar, E., Beijnen, J. H., Smit, J. W., Meijer, D. K. F., van Asperen, J., Borst, P., and Schinkel, A. H., 1996, Substantial excretion of digoxin via the intestinal mucosa and prevention of long-term digoxin accumulation in the brain by the mdr1a P-glycoprotein, *Br. J. Pharmacol.* **119**:1038–1044.

Meier, P. J., 1995, Molecular mechanisms of hepatic bile salt transport from sinusoidal blood into bile, *Am. J. Physiol.* **269**:G801–G812.

Meier, P. J., 1996, Hepatocellular transport systems: From carrier identification in membrane vesicles to cloned proteins, *J. Hepatol.* **24**:29–35.

Meier, P. J., Eckhardt, U., Schroeder, A., Hagenbuch, B., and Stieger, B., 1997, Substrate specificity of sinusoidal bile acid and organic anion uptake systems in rat and human liver, *Hepatology* **26**:1667–1677.

Meijer, D. K. F., 1989, Transport and metabolism in the hepatobiliary system, In: *Handbook of Physiology. The Gastrointestinal System* (J. Shulz, G. Forte, and B. B. Rauner, eds.), Oxford University Press, Oxford, Vol. III, pp. 717–758.

Meijer, D. K. F., and Groothuis, G. M. M., 1991, Hepatic transport of drugs and proteins, in: *Oxford Textbook of Clinical Hepatology* (N. McIntyre, J. P. Benhamou, J. Bircher, M. Rizzetto, and J. Rodes, eds.), Oxford University Press, Oxford, Vol. 1, pp. 40–78.

Meijer, D. K. F., and Ziegler, K., 1993, Mechanisms for the hepatic clearance of oligopeptides and proteins. Implications for rate of elimination, bioavailability and cell-specific drug delivery to the liver, in: *Biological Barriers to Protein Delivery* (K. L. Audus and T. J. Raub, eds.), Plenum Press, New York, pp. 339–408.

Meijer, D. K. F., Mol, W. E. M., Müller, M., and Kurz, G., 1990, Carrier-mediated transport in the hepatic distribution and elimination of drugs, with special reference to the category of organic cations, *J. Pharmacokinet. Biopharmacol.* **18**:35–70.

Meijer, D. K. F., Mol, W. E. M., Müller, M., Steen, H., and Kurz, G., 1991, Carrier-mediated transport in the hepatic distribution and elimination of organic cations, in: *Hepatic Metabolism and Disposition of Endo- and Xenobiotics* (K. W. Bock, S. Matern, W. Gerok, and R. Schmid, eds.), Kluwer Academic, Dordrecht, pp. 259–270.

Meijer, D. K. F., Smit, J. W., and Müller, M., 1997, Hepatobiliary elimination of cationic drugs: The role of P-glycoproteins and other ATP-dependent transporters, *Adv. Drug Deliv. Rev.* **25**:159–200.

Mol, W. E. M., Fokkema, G. N., Weert, B., and Meijer, D. K. F., 1988, Mechanisms for the hepatic uptake of organic cations. Studies with the muscle relaxant vecuronium in isolated rat hepatocytes, *J. Pharmacol. Exp. Ther.* **244**:268–275.

Mol, W. E. M., Müller, M., Kurz, G., and Meijer, D. K. F., 1992, Investigations on the hepatic uptake systems for organic cations with a photoaffinity probe of procainamide ethobromide, *Biochem. Pharmacol.* **43**:2217–2226.

Moriyama, Y., Amakatsu, K., and Futai, M., 1993, Uptake of the neurotoxin, 4-methyl-

phenylpyridinium, into chromaffin granules and synaptic vesicles: A proton gradient drives its uptake through monoamine transporter, *Arch. Biochem. Biophys.* **305:**271–277.
Moseley, R. H., and Van Dyke, R. W., 1995, Organic cation transport by rat liver lysosomes, *Am. J. Physiol.* **268:**G480–G486.
Moseley, R. H., Jarose, S. M., and Permoad, P., 1992, Organic cation transport by rat liver plasma membrane vesicles: Studies with tetraethylammonium, *Am. J. Physiol.* **263:**G775–G785.
Moseley, R. H., Smit, H., Van Solkema, B. G. H., Wang, W., and Meijer, D. K. F., 1996a, Mechanisms for the hepatic uptake and biliary excretion of tributylmethylammonium: Studies with rat liver plasma membrane vesicles, *J. Pharmacol. Exp. Ther.* **276:**561–567.
Moseley, R. H., Wang, W., Takeda, H., Lown, K., Shick, L., Ananthanarayanan, M., and Suchy, F. J., 1996b, Effect of endotoxin on bile acid transport in rat liver: A potential model for sepsis-associated cholestasis, *Am. J. Physiol.* **271:**G137–G146.
Moseley, R. H., Zugger, L. J., and Van Dyke, R. W., 1997, The neurotoxin 1-methyl-4-phenylpyridinium is a substrate for the canalicular organic cation/H^+ exchanger, *J. Pharmacol. Exp. Ther.* **281:**34–40.
Müller M., 1988, Membrantransport von organischen kationen in hepatocyten. Ein beitrag zur aufklärung von hepatischen transportvorgängen, Dissertation, Universität von Freiburg, Freiburg, Germany.
Müller, M., and Jansen, P. L. M., 1997, Molecular aspects of hepatobiliary transport, *Am. J. Physiol.* **272:**G1285–G1303.
Müller, M., and Jansen, P. L. M., 1998, The secretory function of the liver: New aspects of hepatobiliary transport, *J. Hepatol.* **28:**344–354.
Müller, M., Ishikawa, T., Berger, U., Klünemann, C., Lucka, L., Schreyer, A., Kannicht, C., Reutter, W., Kurz, G., and Keppler, D., 1991, ATP-dependent transport of taurocholate across the hepatocyte canalicular membrane mediated by a 110-kDa glycoprotein binding ATP and bile salt, *J. Biol. Chem.* **266:**18920–18926.
Müller, M., Mayer, R., Hero, U., and Keppler, D., 1994a, ATP-dependent transport of amphiphilic cations across the hepatocyte canalicular membrane mediated by mdr1 P-glycoprotein, *FEBS Lett.* **343:**168–172.
Müller, M., Meijer, C., Zaman, G. J. R., Borst, P., Scheper, R. J., Mulder, N. H., De Vries, E. G. E., and Jansen, P. L. M., 1994b, Overexpression of the gene encoding the multidrug resistance-associated protein results in increased ATP-dependent glutathione S-conjugate transport, *Proc. Natl. Acad. Sci. USA* **91:**13033–13037.
Müller, M., De Vries, E. G. E., and Jansen, P. L. M., 1996a, Role of multidrug resistance protein (MRP) in glutathione S-conjugate transport in mammalian cells, *J. Hepatol.* **24**(Suppl. 1):100–108.
Müller, M., Roelofsen, H., and Jansen, P. L. M., 1996b, Secretion of organic anions by hepatocytes: Involvement of homologues of the multidrug resistance protein, *Semin. Liver Dis.* **16:**211–220.
Müller, M., Vos, T. A., Koopen, N. R., Roelofsen, H., Kuipers, F., and Jansen, P. L. M., 1996c, Localization and regulation of a novel canalicular ABC-transporter, *Hepatology* **24** (No. 4, Pt. 2):368A (Abstract).
Müller, M., Vos, T. A., Roelofsen, H., Van Goor, H., Moshage, H., Kuipers, F., and Jansen,

P. L. M., 1997, Regulation of sister of P-glycoprotein, a major canalicular ABC-transporter, *J. Hepatol.* **26**(Suppl. 1):71.

Nagel, G., Volk, C., Friedrich, T., Ulzheimer, J. C., Bamberg, E., and Koepsell, H., 1997, A reevaluation of substrate specificity of the rat cation transporter rOCT1, *J. Biol. Chem.* **272**:31953–31956.

Neef, C., 1983, Structure–pharmacokinetics relationship of quaternary ammonium compounds, Thesis, Rijksuniversiteit Groningen, The Netherlands.

Neef, C., and Meijer, D. K. F., 1984, Structure–pharmacokinetics relationship of quaternary ammonium compounds. Correlation of physicochemical and pharmacokinetic parameters, *Naunyn Schmiedebergs Arch. Pharmacol.* **328**:111–118.

Neef, C., Oosting, R., and Meijer, D. K. F., 1984, Structure–pharmacokinetics relationship of quaternary ammonium compounds. Elimination and distribution characteristics, *Naunyn Schmiedebergs Arch. Pharmacol.* **328**:103–110.

Newton, A. C., 1995, Protein kinase C: Structure, function, and regulation, *J. Biol. Chem.* **270**:28495–28498.

Nies, A. T., Gatmaitan, Z., and Arias, I. M., 1996, ATP-dependent phosphatidylcholine translocation in rat liver canalicular plasma membrane vesicles, *J. Lipid Res.* **37**:1125–1136.

Nikaido, H., and Saier, M. H., Jr., 1992, Transport proteins in bacteria: Common themes in their design, *Science* **258**:936–942.

Nishida, T., Gatmaitan, Z., Che, M., and Arias, I. M., 1991, Rat liver canalicular membrane vesicles contain an ATP-dependent bile acid transport system, *Proc. Natl. Acad. Sci. USA* **88**:6590–6594.

Nishizuka, Y., 1988, The molecular heterogeneity of protein kinase C and its implications for cellular regulation, *Nature* **334**:661–665.

Nishizuka, Y., 1995, Protein kinases 5: Protein kinase C and lipid signaling for sustained cellular responses, *FASEB J.* **9**:484–496.

Noé, B., Hagenbuch, B., Stieger, B., and Meier, P. J., 1997, Isolation of a multispecific organic anion and cardiac glycoside transporter from rat brain, *Proc. Natl. Acad. Sci. USA* **94**:10346–10350.

Oike, M., Kitamura, K., and Kuriyama, H., 1993, Protein kinase C activates the non-selective cation channel in the rabbit portal vein, *Pflugers Arch.* **424**:159–164.

Oishi, K., Zheng, B., and Kuo, J. F., 1990, Inhibition of Na,K-ATPase and sodium pump by protein kinase C regulators sphingosine, lysophosphatidylcholine, and oleic acid, *J. Biol. Chem.* **265**:70–75.

Ortiz, D. F., St. Pierre, M. V., Abdulmessih, A., and Arias, I. M., 1997, A yeast ATP-binding cassette-type protein mediating ATP-dependent bile acid transport, *J. Biol. Chem.* **272**:15358–15365.

Oude Elferink, R. P. J., Meijer, D. K. F., Kuipers, F., Jansen, P. L. M., Groen, A. K., and Groothuis, G. M. M., 1995, Hepatobiliary secretion of organic compounds; molecular mechanisms of membrane transport, *Biochim. Biophys. Acta* **1241**:215–268.

Oude Elferink, R. P. J., Ottenhoff, R., Van Wijland, M., Frijters, C. M. G., Van Nieuwkerk, C., and Groen, A. K., 1996, Uncoupling of biliary phospholipid and cholesterol secretion in mice with reduced expression of mdr2 P-glycoprotein, *J. Lipid Res.* **37**:1065–1075.

Oude Elferink, R. P. J., Tytgat, G. N. J., and Groen, A. K., 1997, Hepatic canalicular membrane. 1. The role of mdr2 P-glycoprotein in hepatobiliary lipid transport, *FASEB J.* **11**:19–28.

Pan, B.-F., Dutt, A., and Nelson, J. A., 1994, Enhanced transepithelial flux of cimetidine by Madin–Darby canine kidney cells overexpressing human P-glycoprotein, *J. Pharmacol. Exp. Ther.* **270**:1–7.

Pan, M., and Stevens, B. R., 1995, Protein kinase C-dependent regulation of L-arginine transport activity in Caco-2 intestinal cells, *Biochim. Biophys. Acta* **1239**:27–32.

Paulusma, C. C., and Oude Elferink, R. P. J., 1997, The canalicular multispecific organic anion transporter and conjugated hyperbilirubinemia in rat and man, *J. Mol. Med.* **75**:420–428.

Paulusma, C. C., Bosma, P. J., Zaman, G. J. R., Bakker, C. T. M., Otter, M., Scheffer, G. L., Scheper, R. J., Borst, P., and Oude Elferink, R. P. J., 1996, Congenital jaundice in rats with a mutation in a multidrug resistance-associated protein gene, *Science* **271**:1126–1128.

Paulusma, C. C., Kool, M., Bosma, P. J., Scheffer, G. L., ter Borg, F., Scheper, R. J., Tytgat, G. N. J., Borst, P., Baas, F., and Oude Elferink, R. P. J., 1997, A mutation in the human canalicular multispecific organic anion transporter gene causes the Dubin–Johnson syndrome, *Hepatology* **25**:1539–1542.

Pfaller, W., Gstraunthaler, G., and Loidl, P., 1990, Morphology of the differentiation and maturation of LLC-PK$_1$ epithelia, *J. Cell. Physiol.* **142**:247–254.

Popescu, N. C., Silverman, J. A., and Thorgeirsson, S. S., 1993, Mapping of a rat multidrug resistance gene by fluorescence *in situ* hybridization, *Genomics* **15**:182–184.

Popkov, M., Lussier, I., Medvedkine, V., Esteve, P. O., Alakhov, V., and Mandeville, R., 1998, Multidrug-resistance drug-binding peptides generated by using a phage display library, *Eur. J. Biochem.* **251**:155–163.

Preisegger, K.-H., Stumptner, C., Riegelnegg, D., Brown, P. C., Silverman, J. A., Thorgeirsson, S. S., and Denk, H., 1996, Experimental Mallory body formation is accompanied by modulation of the expression of multidrug-resistance genes and their products, *Hepatology* **24**:248–252.

Pritchard, J. B., and Miller, D. S., 1997, Renal secretion of organic cations: A multistep process, *Adv. Drug Deliv. Rev.* **25**:231–242.

Rao, Y. P., Stravitz, R. T., Vlahcevic, Z. R., Gurley, E. C., Sando, J. J., and Hylemon, P. B., 1997, Activation of protein kinase C alpha and delta by bile acids: Correlation with bile acid structure and diacylglycerol formation, *J. Lipid Res.* **38**:2446–2454.

Raymond, M., and Gros, P., 1989, Mammalian multidrug-resistance gene: Correlation of exon organization with structural domains and duplication of an ancestral gene, *Proc. Natl. Acad. Sci. USA* **86**:6488–6492.

Raymond, M., Rose, E., Housman, D. E., and Gros, P., 1990, Physical mapping, amplification, and overexpression of the mouse mdr gene family in multidrug-resistant cells, *Mol. Cell. Biol.* **10**:1642–1651.

Reizer, J., Reizer, A., and Saier, M. H., Jr., 1994, A functional superfamily of sodium/solute symporters, *Biochim. Biophys. Acta* **1197**:133–166.

Roelofsen, H., Ottenhoff, R., Oude Elferink, R. P. J., and Jansen, P. L. M., 1991, Hepatocanalicular organic-anion transport is regulated by protein kinase C, *Biochem. J.* **278**:637–641.

Roelofsen, H., van der Veere, C. N., Ottenhoff, R., Schoemaker, B., Jansen, P. L. M., and

Oude Elferink, R. P. J., 1994, Decreased bilirubin transport in the perfused liver of endotoxemic rats, *Gastroenterology* **107**:1075–1084.

Roelofsen, H., Vos, T. A., Schippers, I. J., Kuipers, F., Koning, H., Moshage, H., Jansen, P. L. M., and Müller, M., 1997, Increased levels of the multidrug resistance protein in lateral membranes of proliferating hepatocyte-derived cells, *Gastroenterology* **112**:511–521.

Roninson, I. B., Chin, J. E., Choi, K. G., Gros, P., Housman, D. E., Fojo, A., Shen, D. W., Gottesman, M. M., and Pastan, I., 1986, Isolation of human mdr DNA sequences amplified in multidrug-resistant KB carcinoma cells, *Proc. Natl. Acad. Sci. USA* **83**: 4538–4542.

Ruetz, S., and Gros, P., 1994, Phosphatidylcholine translocase: A physiological role for the mdr2 gene, *Cell* **77**:1071–1081.

Ruetz, S., and Gros, P., 1995, Enhancement of Mdr2-mediated phosphatidylcholine translocation by the bile salt taurocholate. Implications for hepatic bile formation, *J. Biol. Chem.* **270**:25388–25395.

Ruetz, S., Brault, M., Dalton, W., and Gros, P., 1997, Functional interactions between synthetic alkyl phospholipds and the ABC transporters P-glycoprotein, Ste-6, MRP, and Pgh-1, *Biochemistry* **36**:8180–8188.

Schaub, T. P., Kartenbeck, J., Konig, J., Vogel, O., Witzgall, R., Kriz, W., and Keppler, D., 1997, Expression of the conjugate export pump encoded by the mrp2 gene in the apical membrane of kidney proximal tubules, *J. Am. Soc. Nephrol.* **8**:1213–1221.

Schinkel, A. H., Kemp, S., Dolle, M., Rudenko, G., and Wagenaar, E., 1993, N-glycosylation and deletion mutants of the human MDR1 P-glycoprotein, *J. Biol. Chem.* **268**:7474–7481.

Schinkel, A. H., Smit, J. J. M., van Tellingen, O., Beijnen, J. H., Wagenaar, E., Van Deemter, L., Mol, C. A. A. M., Van der Valk, M. A., Robanus-Maandag, E. C., te Riele, H. P. J., Berns, A. J. M., and Borst, P., 1994, Disruption of the mouse mdr1a P-glycoprotein gene leads to a deficiency in the blood–brain barrier and to increased sensitivity to drugs, *Cell* **77**:491–502.

Schinkel, A. H., Mol, C. A. A. M., Wagenaar, E., van Deemter, L., Smit, J. J. M., and Borst, P., 1995, Multidrug resistance and the role of P-glycoprotein knockout mice, *Eur. J. Cancer* **31A**:1295–1298.

Schinkel, A. H., Mayer, U., Wagenaar, E., Mol, C. A. A. M., Van Deemter, L., Smit, J. J. M., Van der Valk, M. A., Voordouw, A. C., Spits, H., van Tellingen, O., Zijlmans, J. M. J. M., Fibbe, W. E., and Borst, P., 1997, Normal viability and altered pharmacokinetics in mice lacking mdr1-type (drug-transporting) P-glycoproteins, *Proc. Natl. Acad. Sci. USA* **94**:4028–4033.

Schrenk, D., Gant, T. W., Preisegger, K.-H., Silverman, J. A., Marino, P. A., and Thorgeirsson, S. S., 1993, Induction of multidrug resistance gene expression during cholestasis in rats and nonhuman primates, *Hepatology* **17**:854–860.

Schuldiner, S., Shirvan, A., and Linial, M., 1995, Vesicular neurotransmitter transporters: From bacteria to humans, *Physiol. Rev.* **75**:369–392.

Senior, A. E., al-Shawi, M. K., and Urbatsch, I. L., 1995, The catalytic cycle of P-glycoprotein, *FEBS Lett.* **377**:285–289.

Shapiro, A. B., and Ling, V., 1994, ATPase activity of purified and reconstituted P-glycoprotein from Chinese hamster ovary cells, *J. Biol. Chem.* **269**:3745–3754.

Sharom, F. J., Yu, X., and Doige, C. A., 1993, Functional reconstitution of drug transport and ATPase activity in proteoliposomes containing partially purified P-glycoprotein, *J. Biol. Chem.* **268:**24197–24202.

Shen, D. W., Fojo, A., Chin, J. E., Roninson, I. B., Richert, N., Pastan, I., and Gottesman, M. M., 1986, Human multidrug-resistant cell lines: Increased mdr1 expression can precede gene amplification, *Science* **232:**643–645.

Shi, X. Y., Bai, S., Ford, A. C., Burk, R. D., Jacquemin, E., Hagenbuch, B., Meier, P. J., and Wolkoff, A. W., 1995, Stable inducible expression of a functional rat liver organic anion transport protein in HeLa cells, *J. Biol. Chem.* **270:**25591–25595.

Shneider, B. L., Dawson, P. A., Christie, D.-M., Hardikar, W., Wong, M. H., and Suchy, F. J., 1995, Cloning and molecular characterization of the ontogeny of a rat ileal sodium-dependent bile acid transporter, *J. Clin. Invest.* **95:**745–754.

Silverman, J. A., and Schrenk, D., 1997, Hepatic canalicular membrane. 4. Expression of the multidrug resistance genes in the liver, *FASEB J.* **11:**308–313.

Silverman, J. A., Raunio, H., Gant, T. W., and Thorgeirsson, S. S., 1991, Cloning and characterization of a member of the rat multidrug resistance (mdr) gene family, *Gene* **106:**229–236.

Simmons, N. L., Hunter, J., and Jepson, M. A., 1997, Renal secretion of xenobiotics mediated by P-glycoprotein: Importance to renal function in health and exploitation for targeted drug delivery to epithelial cysts in polycystic kidney disease, *Adv. Drug Deliv. Rev.* **25:**243–256.

Slapak, C. A., Martell, R. L., Terashima, M., and Levy, S. B., 1996, Increased efflux of vincristine, but not of daunorubicin, associated with the murine multidrug resistance protein (MRP), *Biochem. Pharmacol.* **52:**1569–1576.

Slovak, M. L., Ho, J. P., Bhardwaj, G., Kurz, E. U., Deeley, R. G., and Cole, S. P. C., 1993, Localization of a novel multidrug resistance-associated gene in the HT1080/DR4 and H69AR human tumor cell lines, *Cancer Res.* **53:**3221–3225.

Smit, J. J. M., Schinkel, A. H., Oude Elferink, R. P. J., Groen, A. K., Wagenaar, E., Van Deemter, L., Mol, C. A. A. M., Ottenhoff, R., Van der Lugt, N. M. T., van Roon, M. A., Van der Valk, M. A., Offerhaus, G. J. A., Berns, A. J. M., and Borst, P., 1993, Homozygous disruption of the murine mdr2 P-glycoprotein gene leads to a complete absence of phospholipid from bile and to liver disease, *Cell* **75:**451–462.

Smit, J. J. M., Schinkel, A. H., Mol, C. A. A. M., Majoor, D., Mooi, W. J., Jongsma, A. P. M., Lincke, C. R., and Borst, P., 1994, Tissue distribution of the human MDR3 P-glycoprotein, *Lab. Invest.* **71:**638–649.

Smit, J. W., Duin, E., Steen, H., Oosting, R., Roggeveld, J., and Meijer, D. K. F., 1998a, Interactions between P-glycoprotein substrates and other cationic drugs at the hepatic excretory level, *Br. J. Pharmacol.* **123:**361–370.

Smit, J. W., Schinkel, A. H., Müller, M., Weert, B., and Meijer, D. K. F., 1998b, Contribution of the murine mdr1a P-glycoprotein to hepato-biliary and intestinal elimination of cationic drugs as measured in mice with a mdr1a gene disruption, *Hepatology* **27:**1056–1063.

Smit, J. W., Schinkel, A. H., Weert, B., and Meijer, D. K. F., 1998c, Hepatobiliary and intestinal clearance of amphiphilic cationic drugs in mice in which both mdr1a and mdr1b genes have been disrupted, *Br. J. Pharmacol.* **124:**416–424.

Smit, J. W., Weert, B., Schinkel, A. H., and Meijer, D. K. F., 1998d, Heterologous expres-

sion of various P-glycoproteins in polarized epithelial cells induces directional transport of small (type1) and bulky (type2) cationic drugs, *J. Pharmacol. Exp. Ther.* **286**: 321–327.
Smith, A. J., Timmermans-Hereijgers, J. L. P. M., Roelofsen, B., Wirtz, K. W. A., van Blitterswijk, W. J., Smit, J. J. M., Schinkel, A. H., and Borst, P., 1994, The human MDR3 P-glycoprotein promotes translocation of phosphatidylcholine through the plasma membrane of fibroblasts from transgenic mice, *FEBS Lett.* **354**:263–266.
Smith, R. L., 1966, The biliary excretion and enterohepatic circulation of drugs and other organic compounds, *Fortschr. Arzneimittelforsch.* **9**:299–360.
Sohn, Y. J., Bencini, A. F., Scaf, A. H. J., Kersten, U. W., Gregoretti, S., and Agoston, S., 1982, Pharmacokinetics of vecuronium in man, *Anesthesiology* **57**(No. 3, Suppl.): A256 (Abstract).
Sparreboom, A., van Asperen, J., Mayer, U., Schinkel, A. H., Smit, J. W., Meijer, D. K. F., Borst, P., Nooijen, W. J., Beijnen, J. H., and van Tellingen, O., 1996, Limited oral bioavailability and active epithelial excretion of paclitaxel (Taxol) caused by P-glycoprotein in the intestine, *Proc. Natl. Acad. Sci. USA* **94**:2031–2035.
Speeg, K. V., and Maldonado, A. L., 1994, Effect of the nonimmunosuppressive cyclosporin analog SDZ PSC-833 on colchicine and doxorubicin biliary secretion by the rat in vivo, *Cancer Chemother. Pharmacol.* **34**:133–136.
Speeg, K. V., Maldonado, A. L., Liaci, J., and Muirhead, D., 1992, Effect of cyclosporine on colchicine secretion by a liver canalicular transporter studied in vivo, *Hepatology* **15**:899–903.
Steen, H., and Meijer, D. K. F., 1990, Mechanisms for uptake of the organic cation tri-butylmethyl-ammonium in the hepatocyte, *Eur. J. Pharmacol.* **183**:1358 (Abstract).
Steen, H., Oosting, R., and Meijer, D. K. F., 1991, Mechanisms for the uptake of cationic drugs by the liver: A study with tributylmethylammonium (TBuMA), *J. Pharmacol. Exp. Ther.* **258**:537–543.
Steen, H., Merema, M., and Meijer, D. K. F., 1992, A multispecific uptake system for taurocholate, cardiac glycosides and cationic drugs in the liver, *Biochem. Pharmacol.* **44**:2323–2331.
Steen, H., Smit, H., Nijholt, A., Merema, M., and Meijer, D. K. F., 1993, Modulators of the protein kinase C system influence biliary excretion of cationic drugs, *Hepatology* **18**:1208–1215.
Stieger, B., O'Neill, B., and Meier, P. J., 1992, ATP-dependent bile-salt transport in canalicular rat liver plasma-membrane vesicles, *Biochem. J.* **284**:67–74.
Stieger, B., Hagenbuch, B., Landmann, L., Höchli, M., Schroeder, A., and Meier, P. J., 1994, In situ localization of the hepatocytic Na^+/ taurocholate cotransporting polypeptide in rat liver, *Gastroenterology* **107**:1781–1787.
Strautnieks, S. S., Kagalwalla, A. F., Tanner, M. S., Knisely, A. S., Bull, L. N., Freimer, N. B., Kocoshis, S. A., Gardiner, R. M., and Thompson, R. J., 1997, Identification of a locus for progressive familiar intrahepatic cholestasis (PFIC2) on chromosome 2q24, *Am. J. Hum. Genet.* **61**:630–633.
Streich, S., Brüss, M., and Bönisch, H., 1996, Expression of the extraneuronal monoamine transporter (uptake$_2$) in human glioma cells, *Naunyn Schmiedebergs Arch. Pharmacol.* **353**:328–333.
Stride, B. D., Valdimarsson, G., Gerlach, J. H., Wilson, G. M., Cole, S. P. C., and Deeley,

R. G., 1996, Structure and expression of the messenger RNA encoding the murine multidrug resistance protein, an ATP-binding cassette transporter, *Mol. Pharmacol.* **49**:962–971.

Stride, B. D., Grant, C. E., Loe, D. W., Hipfner, D. R., Cole, S. P. C., and Deeley, R. G., 1997, Pharmacological characterization of the murine and human orthologs of multidrug-resistance protein in transfected human embryonic kidney cells, *Mol. Pharmacol.* **52**:344–353.

Taguchi, Y., Yoshida, A., Takada, Y., Komano, T., and Ueda, K., 1997, Anti-cancer drugs and glutathione stimulate vanadate-induced trapping of nucleotide in multidrug resistance-associated protein (MRP), *FEBS Lett.* **401**:11–14.

Takano, M., Kato, M., Takayama, A., Yasuhara, M., Inui, K.-I., and Hori, R., 1992, Transport of procainamide in a kidney epithelial cell line LLC-PK$_1$, *Biochim. Biophys. Acta* **1108**:133–139.

Takanishi, K., Miyazaki, M., Ohtsuka, M., and Nakajima, N., 1997, Inverse relationship between P-glycoprotein expression and its proliferative activity in hepatocellular carcinoma, *Oncology* **54**:231–237.

Taniguchi, K., Wada, M., Kohno, K., Nakamura, T., Kawabe, T., Kawakami, M., Kagotani, K., Okumura, K., Akiyama, S., and Kuwano, M., 1996, A human canalicular multispecific organic anion transporter (cMOAT) gene is overexpressed in cisplatin-resistant human cancer cell lines with decreased drug accumulation, *Cancer Res.* **56**:4124–4129.

Thiebaut, F., Tsuruo, T., Hamada, H., Gottesman, M. M., Pastan, I., and Willingham, M. C., 1987, Cellular localization of the multidrug-resistance gene product P-glycoprotein in normal human tissues, *Proc. Natl. Acad, Sci. U.S.A.* **84**:7735–7738.

Thiebaut, F., Tsuruo, T., Hamada, H., Gottesman, M. M., Pastan, I., and Willingham, M. C., 1989, Immunohistochemical localization in normal tissues of different epitopes in the multidrug transport protein P170: Evidence for localization in brain capillaries and crossreactivity of one antibody with a muscle protein, *J. Histochem. Cytochem.* **37**:159–164.

Trauner, M., Arrese, M., Soroka, C. J., Ananthanarayanan, M., Koeppel, T. A., Schlosser, S. F., Suchy, F. J., Keppler, D., and Boyer, J. L., 1997, The rat canalicular conjugate export pump (Mrp2) is down-regulated in intrahepatic and obstructive cholestasis, *Gastroenterology* **113**:255–264.

Tusnady, G. E., Bakos, E., Varadi, A., and Sarkadi, B., 1997, Membrane topology distinguishes a subfamily of the ATP-binding cassette (ABC) transporters, *FEBS Lett.* **402**:1–3.

Ullrich, K. J., 1997, Renal transporters for organic anions and organic cations. Structural requirements for substrates, *J. Membr. Biol.* **158**:95–107.

van Asperen, J., van Tellingen, O., Sparreboom, A., Schinkel, A. H., Borst, P., Nooijen, W. J., and Beijnen, J. H., 1997, Enhanced oral bioavailability of paclitaxel in mice treated with the P-glycoprotein blocker SDZ PSC B33, *Br. J. Cancer* **76**:1181–1183.

van der Bliek, A. M., Baas, F., Ten Houte de Lange, T., Kooiman, P. M., van der Velde-Koerts, T., and Borst, P., 1987, The human mdr3 gene encodes a novel P-glycoprotein homologue and gives rise to alternatively spliced mRNAs in liver, *EMBO J.* **6**:3325–3331.

van der Bliek, A. M., Kooiman, P. M., Schneider, C., and Borst, P., 1988, Sequence of mdr3 cDNA encoding a human P-glycoprotein, *Gene* **71**:401–411.

van Helvoort, A., Smith, A. J., Sprong, H., Fritzsche, I., Schinkel, A. H., Borst, P., and van Meer, G., 1996, MDR1 P-glycoprotein is a lipid translocase of broad specificity, while MDR3 P-glycoprotein specifically translocates phosphatidylcholine, *Cell* **87:**507–517.
Vos, T. A., Hooiveld, G. J. E. J., Koning, H., Childs, S., Meijer, D. K. F., Moshage, H., Jansen, P. L. M., and Müller, M. , 1998, Up-regulation of the multidrug resistance genes *mrp1* and *mdr1b* and down-regulation of the organic anion transporter, *mrp2*, and the bile salt transporter, *spgp*, in endotoxemic rat liver, *Hepatology* **28:**1637–1644.
Watanabe, T., Miyauchi, S., Sawada, Y., Iga, T., Hanano, M., Inaba, M., and Sugiyama, Y., 1992, Kinetic analysis of hepatobiliary transport of vincristine in perfused rat liver. Possible roles of P-glycoprotein in biliary excretion of vincristine, *J. Hepatol.* **16:**77–88.
Wijnholds, J., Evers, R., van Leusden, M. R., Mol, C. A. A. M., Zaman, G. J. R., Mayer, U., Beijnen, J. H., van der Valk, M., Krimpenfort, P., and Borst, P., 1997, Increased sensitivity to anticancer drugs and decreased inflammatory response in mice lacking the multidrug resistance-associated protein, *Nature Med.* **3:**1275–1279.
Wolkoff, A. W., 1996, Hepatocellular sinusoidal membrane organic anion transport and transporters, *Semin. Liver Dis.* **16:**121–127.
Wolters, H., Kuipers, F., Slooff, M. J. H., and Vonk, R. J., 1992, Adenosine triphosphate-dependent taurocholate transport in human liver plasma membranes, *J. Clin. Invest.* **90:**2321–2326.
Yang, J.-M., Chin, K.-V., and Hait, W. N., 1996, Interaction of P-glycoprotein with protein kinase C in human multidrug resistant carcinoma cells, *Cancer Res.* **56:**3490–3494.
Yelin, R., and Schuldiner, S., 1996, The pharmacological profile of the vesicular monoamine transporter resembles that of multidrug transporters, *FEBS Lett.* **377:**201–207.
Yousef, I. M., Mignault, D., Weber, A. M., and Tuchweber, B., 1990, Influence of dehydrocholic acid on the secretion of bile acids and biliary lipids in rats, *Digestion* **45:**40–51.
Zaman, G. J. R., Versantvoort, C. H. M., Smit, J. J. M., Eijdems, E. W. H. M., de Haas, M., Smith, A. J., Broxterman, H. J., Mulder, N. H., De Vries, E. G. E., Baas, F., and Borst, P., 1993, Analysis of the expression of MRP, the gene for a new putative transmembrane drug transporter, in human multidrug resistant lung cancer cell lines, *Cancer Res.* **53:**1747–1750.
Zaman, G. J. R., Lankelma, J., van Tellingen, O., Beijnen, J., Dekker, H., Paulusma, C., Oude Elferink, R. P. J., Baas, F., and Borst, P., 1995, Role of glutathione in the export of compounds from cells by the multidrug-resistance-associated protein, *Proc. Natl. Acad. Sci. USA* **92:**7690–7694.
Zaman, G. J. R., Flens, M. J., van Leusden, M. R., de Haas, M., Mulder H. S., Lankelma, J., Pinedo, H. M., Scheper, R. L., Baas, F., Broxterman, H. J., and Borst, P., 1994, The human multidrug resistance-associated protein MRP is a plasma membrane drug-efflux pump, *Proc. Natl. Acad. Sci. USA* **91:**8822–8826.
Zhang, L., Dresser, M. J., Gray, A. T., Yost, S. C., Terashita, S., and Giacomini, K. M., 1997, Cloning and functional expression of a human liver organic cation transporter, *Mol. Pharmacol.* **51:**913–921.
Zimonjic, D. B., Popescu, N. C., Brown, P. C., and Silverman, J. A., 1996, Localization of rat Pgy3 (mdr2) to the same region as Pgy2 (mdr1b) at 4q11–12, *Mamm. Genome* **7:**630–631.

5

Affinity of Drugs to the Different Renal Transporters for Organic Anions and Organic Cations

In Situ K_i Values

Karl Julius Ullrich

1. INTRODUCTION

Drugs are taken up into the body by different routes and are excreted by the kidney or by the liver, either directly or after metabolic transformation. Discriminating between renal and extrarenal clearance of drugs and their metabolites is fruitful, and tables containing quantitative clearance data are of value (Dettli, 1996). In the past two decades, renal drug excretion has been studied mainly on two levels: (1) via transporters for organic anions and for organic cations, and (2) localization of the transport processes at the contraluminal or luminal cell side of the proximal tubule. These results were achieved by kinetic transport studies using either the tubule *in situ* (Ullrich and Rumrich, 1990) or membrane vesicles from either side of the cell (Kinne-Saffran and Kinne, 1990). Our present knowledge of the respective transport mechanisms has been aided by the cloning and functional expression of the main transporters (Gründemann *et al.*,1994; Sekine *et al.*, 1997; Wolff *et al.*, 1997; Sweet *et al.*, 1997; Zhang *et al.*, 1997a) and by the evaluation of their substrate specificity. These topics are discussed in the proceedings

---

Karl Julius Ullrich • Max Planck Institute for Biophysics, 60596 Frankfurt am Main, Germany.
Membrane Transporters as Drug Targets, edited by Amidon and Sadée. Kluwer Academic/Plenum Publishers, New York, 1999.

of a recent symposium (Fleck *et al.*, 1998) and by Chapters 15 and 16 in this book. One is often faced with the question of whether a drug interacts with a certain transporter and what transport interactions with other drugs have to be expected. Fortunately, a large body of data is available to answer this question. A list of K_i values of drugs can be established, and the reciprocal of these K_i values can be considered as a reliable measure for the relative affinities of drugs for the respective transporters under *in situ* conditions. The data presented in this chapter were obtained on the rat proximal tubule *in situ* with the same micropuncture technique for the luminal as for the contraluminal cell side, so that all data can be compared quantitatively. For these transport studies appropriate test substances were chosen and the transporters named according to these test substances: para-aminohippurate (PAH), sulfate, and succinate as representative anions; N^1-methyl-nicotinamide ($NMeN^+$), tetraethylammonium (TEA^+), methyl-phenyl-pyridinium (MPP^+), and choline$^+$ as representative cations.

2. LOCATION OF THE TRANSPORT PROCESSES AND TRANSPORTERS IN THE PROXIMAL RENAL TUBULE

Figure 1 shows the location of the different transporters at either side of the proximal tubular cell. They are named according to the test substrate which they readily transport. For organic anions one has contraluminally the PAH/α-ketoglutarate exchanger, and luminally the urate/PAH exchanger; luminally the Na^+/sulfate cotransporter, and contraluminally the sulfate/oxalate exchanger; both luminally and contraluminally the Na^+/dicarboxylate cotransporters. For organic cations one has contraluminally the electrogenic TEA^+ ($NMeN^+$) transporter, which can be trans-stimulated by other organic cations, and luminally both the H^+/organic cation (MPP^+) exchanger and the electrogenic choline$^+$ transporter. We have investigated all these transporters extensively using micropuncture techniques in the kidney *in situ*. Applying initial flux inhibitory kinetics, we evaluated the affinity of the different groups of substrates against the different transporters (for references see Ullrich, 1997).

The main route for secretion of organic anions is via the contraluminal PAH/α-ketoglutarate exchanger, the precise mechanism of which was elucidated by Burckhardt (Shimada *et al.*, 1987) and Pritchard (1988), and luminally via the urate/PAH exchanger, which was mainly investigated by Aronson *et al.* (Kahn and Aronson 1983), Weinman *et al.* (1983), and Roch-Ramel *et al.* (1994, 1996, 1997). The contraluminal PAH transporters have been recently cloned by three groups independently, namely those of Endou (1998; Sekine *et al.*, 1997), Burckhardt (Wolff *et al.*, 1997; Burckhardt *et al.*, 1998), and Pritchard (Sweet *et al.*, 1997). The luminal urate/PAH exchanger has not yet been identified at the molecular level.

Affinity of Drugs to Renal Transporters for Organic Anions and Cations

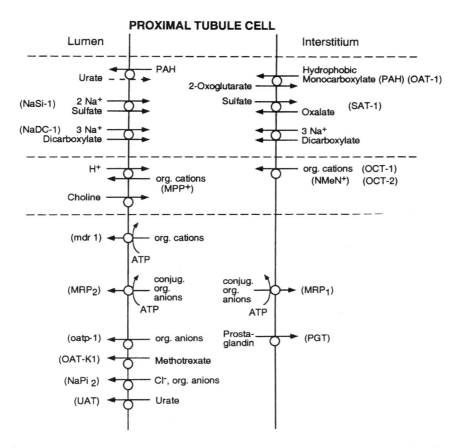

Figure 1. Sites of the transporters for organic anions and organic cations in the proximal renal tubule. The methotrexate transporter OAT-K1 was originally attributed to the contraluminal cell side (Saito et al., 1996), the OCT-2 transporter to the luminal cell side (Gründemann et al., 1997).

Both sulfate transporters, the luminal Na^+/sulfate cotransporter and the contraluminal sulfate/oxalate exchanger, have been cloned by the "Zürich" group under the leadership of Murer (Markovich et al., 1993, 1994). The luminal dicarboxylate transporter NaDC-1 was cloned by Pajor (1995) and by the group of Endou (Sekine et al. 1998), and the contraluminal dicarboxylate transporter (SDCT2, NaDC3) by the group of Hediger (Chen et al. 1999) and by Steffgen et al. (1999).

The main route for the secretion of organic cations occurs via the contraluminal TEA^+ ($NMeN^+$) transporter and the luminal H^+/organic cation (MPP^+) exchanger. The contraluminal $NMeN^+$ transporters OCT-1 and OCT-2 have been cloned by the group of Koepsell (1998; Gründemann et al., 1994), and was further investigated by Okuda et al. (1996) and Gründemann et al. (1997). Furthermore, there exists a luminal transport system for choline. The luminal H^+/organic cation (MPP^+) exchanger and the luminal choline transporter have not yet been identified on the molecular level. Candidates for this function could be the human pH-dependent oganic cation transporter OCTN1 (Tamai *et al.*, 1997), an alternatively spliced OCT1 isoform found in rat kidney cortex (Zhang *et al.* 1997a), or organic cation transporter-like proteins cloned from mouse kidney (Lopez-Nieto *et al.*, 1997; Mori *et al.*, 1997) Since the organic anion transporter OAT1 and the organic cation transporter OCT1 belong to the same gene family (Sekine *et al.*, 1997; Wolff *et al.*, 1997; Sweet *et al.*, 1997), the latter proteins might also be considered as candidates for organic anion transport.

The following transporters have not yet been investigated under *in situ* conditions nor sufficiently tested for their substrate specificity: the luminal ATP-driven multidrug-transporting P-glycoprotein mdr1 (Oude-Elferink *et al.*, 1998) and the luminal multidrug-related protein MRP-2 (Schaub *et al.*, 1997; Keppler *et al.*, 1998), as well as the luminal multispecific anion transporter oatp1 (Bergwerk *et al.*, 1996; Hagenbuch *et al.*, 1998), the related luminal methotrexate transporter oat-k1 (Masuda *et al.*, 1997), the luminal putative anion transporter NaPi-1 (Busch *et al.*, 1996; Lang *et al.*, 1998), and the luminal urate transporter/channel UAT (Leal-Pinto *et al.*, 1997). These luminal transporters may play a role in the transport of organic anions and cations from the cell into the tubular lumen. The following transporters were found at the basolateral cell side of the proximal tubule: the basolateral prostaglandin transporter PGT (Kanai *et al.*, 1995) and the multidrug-related protein MRP-1 (Evers *et al.*, 1996; Keppler *et al.*, 1998). The function of these transporters has not yet been defined *in situ*.

3. TRANSPORT PROCESSES FOR NET REABSORPTION AND NET SECRETION

During net secretion and net reabsorption, substrates have to cross two membrane barriers, the luminal and the contraluminal cell membranes of the proximal renal tubule. The first step, i.e., the entrance into the tubular cells, has been studied more extensively than the exit step. Thus, the K_i values of many drugs with transporters involved in secretion, i.e., the contraluminal organic acid (PAH) transporter, the contraluminal organic cation ($NMeN^+$, TEA^+) transporter, the less important contraluminal sulfate/oxalate exchanger, and the contraluminal Na^+ dicarboxylate cotransporter, are available and are listed in Table I. For the luminal exit step enough data are only available for the luminal H^+/organic cation (MPP^+) exchanger and are listed in Table I. For the luminal exit of organic anions only qualitative values are available (for references see Ullrich and Rumrich, 1997). The luminal steps for the reabsorption of filtered substrates occur usually by sodi-

Table I.
Alphabetical List of Drugs Which Interact with Transporters for Organic Anions and Organic Cations in the Proximal Renal Tubule[a]

Drug[b]	Apparent K_i (mmole/liter)					
	cl,PAH	cl,SO$_4^{2-}$	cl,succ	cl,NMeN$^+$	l,MPP$^+$	l,choline$^+$
Acetazolamide[6,19,22]	1.3	6.4	NS	NS	8.4	34.0
Acetylcholine[24]	—	—	—	2.46	NS	0.57
Acetylcysteine[11]	NS	NS	NS	—	—	—
Acetylcysteine–Hg complex[11]	1.7	1.6	NS	—	—	—
Acriflavine[17,21,24]	NS	—	—	0.28	0.12	0.24
Actinomycin D*	—	—	—	NS	—	—
Adenine*	NS	—	—	NS	—	—
Aflatoxin B$_1^{10}$	0.41	—	—	—	—	—
Agmatine*	—	—	—	3.1	—	—
d-Aldosterone[12,18,21,22]	0.18	NS	NS	0.9	0.87	NS
Alendronate[25]	NS	9.9	NS	—	—	—
ß-Amanitin[7]	0.7	—	—	—	—	—
Amiloride[17,21,22]	5.1	—	—	0.11	0.13	0.14
Amiloride-5-hexamethylene[17,21,22]	1.8	—	—	0.06	0.07	0.12
Ampicillin[6]	2.5	NS	NS	NS	—	—
Amphetamine[24]	—	—	—	0.7	—	5.6
Angiotensin II[7]	2.1	—	—	—	—	—
Antamanide[7]	1.7	—	—	—	—	—
Antifebrin[11]	2.0	NS	NS	—	—	—
Apalcillin[6]	0.02	NS	NS	NS	—	—
Arachidonate[10]	0.4	—	—	—	—	—
Aspirin[17]	1.1	NS	NS	—	—	—
Azidoclonidine[17]	3.3	—	—	0.5	—	—
Azidothymidine[15]	0.58	—	—	3.8	—	—
Bendroflumethiazide[19,22]	0.57	NS	—	7.2	2.0	22.1
Benzylamiloride[17,21,22]	1.3	—	—	0.05	0.024	0.025
Benzylpenicillin[6]	0.81	NS	15.8	NS	—	—
Betain*	—	—	—	NS	—	NS
8-Bromo-cGMP[9]	0.04	NS	NS	—	—	—
Bromodiphenylhydramine[17]	3.3	—	—	3.1	—	—
Brompheniramine[17,21]	NS	—	—	0.29	1.9	—
Bromosulfophthalein[2]	—	0.83	—	—	—	—
Bupranolol[17]	NS	—	—	0.9	—	—
Buspirone[12]	0.7	—	—	2.1	—	—
cAMP[9]	3.4	NS	NS	NS	—	—
Canrenoate[12]	0.22	NS	NS	—	—	—
Captopril[7]	1.5	—	—	—	—	—
Carbachol[24]	—	—	—	12.2	11.9	7.5
Carboplatin[16]	NS	NS	—	NS	—	—
Carnosine[7]	NS	—	—	—	—	—
Cefadroxil[6,18]	3.0	NS	NS	NS	—	—
Cefodicime[6]	0.22	4.7	NS	NS	—	—
Cefotaxime[6]	2.2	—	—	—	—	—

(continued)

Table I. (*Continued*)

Drug[b]	Apparent K_t (mmole/liter)					
	cl,PAH	cl,SO$_4^{2-}$	cl,succ	cl,NMeN$^+$	l,MPP$^+$	l,choline$^+$
Cefpirome[6]	NS	NS	NS	0.9	—	—
Cefsulodin[6]	1.9	7.2	NS	NS	—	—
Ceftriaxone[6]	1.5	12.7	NS	NS	—	—
Cephalexin[6]	2.3	NS	NS	NS	—	—
Cephaloridine[6,21]	NS	NS	NS	2.1	NS	—
cGMP[9]	0.27	NS	NS	NS	—	—
Chemotacticpeptide[7]	0.84	—	—	—	—	—
Chenodeoxycholate[26]	0.21	1.8	1.4	—	—	—
Chlorimipramine[17]	2.6	—	—	1.9	—	—
Chloroquine[17,21]	NS	—	—	0.25	0.12	—
Chlorpromazine[17,21]	1.4	—	—	1.3	3.5	—
Chlorpropamide[19]	0.24	23.6	—	NS	—	—
Chorthalidone[19,22]	NS	NS	—	8.0	0.87	1.8
Cholate[26]	0.29	5.3	—	NS	—	—
Choline[24]	—	—	—	1.4	21.2	0.18
Cimetidine[14,21]	1.7	—	—	0.16	0.07	—
Ciprofloxacin[18]	3.4	—	—	2.7	—	—
Cisplatin[16]	NS	—	—	NS	—	—
Citalopram[17,21]	4.5	—	—	0.82	1.4	—
Clodronate[25]	NS	1.5	NS	—	—	—
Clonazepam[17]	0.57	—	—	NS	—	—
Clonidine[17,21]	2.8	—	—	1.0	1.3	—
Cocaine*	4.8	—	—	2.0	6.0	6.6
Codeine[20]	3.0	—	—	0.72	5.1	16.4
Caffeine[9]	0.81	—	—	—	—	—
Colchicine*	—	—	—	—	2.0	4.7
Corticosterone[12,18,21]	0.23	NS	NS	0.2	1.1	—
Corticosteronesulfate[12]	0.17	3.2	NS	—	—	—
Cortisol[12,18,21]	0.25	NS	NS	0.36	0.50	—
Cortisolsulfate[12]	0.24	7.2	NS	—	—	—
Cortisone[12,18]	0.12	NS	NS	0.54	—	—
Cortisonesulfate[12]	0.37	NS	NS	—	—	—
Cyanine dye 863[24]	—	—	—	0.11	0.071	0.029
Cycloleucylglycine[7]	2.0	—	—	—	—	—
Cyclopenthiazide[19,22]	1.7	4.7	—	9.3	4.7	12.6
Cyclophosphamide[15]	4.8	—	—	5.0	—	—
Cycloplatam[16]	NS	—	—	NS	—	—
Daunomycine*	—	—	—	0.8	3.1	—
Deanol[24]	—	—	—	7.7	NS	0.18
Decynium[21,24]	—	—	—	0.42	0.013	0.045
(−)Deprenyl[23]	—	—	—	0.65	23,3	4.3
Dexamethasone[12,18]	0.20	NS	NS	NS	—	—
Dibutyryl cGMP[9]	0.05	NS	NS	—	—	—
Diclofenac[15]	0.18	18.5	—	—	—	—
H$_2$-DIDS[5]	0.02	0.4	1.8	—	—	—
Diethyldithiocarbamate[16]	1.1	NS	—	—	—	—

(*continued*)

Table I. (*Continued*)

Drug[b]	Apparent K_i (mmole/liter)					
	cl,PAH	cl,SO$_4^{2-}$	cl,succ	cl,NMeN$^+$	l,MPP$^+$	l,choline$^+$
Digoxine*	—	—	—	—	NS	—
Diltiazem[17]	1.6	—	—	2.7	—	—
Dimercaptopropanesulfonate[11]	0.56	NS	NS	—	—	—
O-Dimethyldopamine[13]	—	—	—	1.4	—	—
Dimidium[21,24]	—	—	—	0.07	0.04	0.05
Diphenylhydramine[17]	3.4	—	—	3.5	—	—
Dipivefrin[13]	—	—	—	0.8	—	—
(−)Disopyramide[23]	—	—	—	0.94	0.9	2.5
Dopamine[17,21,*]	NS	—	—	4.4	6.1	20.8
Doxorubicin*	—	—	—	1.0	7.0	—
Effortil[17]	NS	—	—	0.9	—	—
Enalapril[7]	1.1	—	—	—	—	—
Epinephrine[17]	NS	5.0	—	6.3	—	—
Enoxacin[18]	1.25	—	—	4.5	—	—
Enrofloxacin[18]	1.1	—	—	1.2	—	—
(1R,2S)-Ephedrine[23]	—	—	—	1.0	12.8	10.4
Estriolsulfate[8]	0.37	0.95	—	—	—	—
Ethacrynic acid[6,22]	0.12	4.7	NS	14.8	15.8	NS
Ethidiumbromide[17]	4.1	—	—	0.4	—	—
Etidronate[25]	NS	3.1	NS	—	—	—
Etozoline[22]	NS	—	—	—	17.4	NS
Euflavine[13]	—	—	—	0.28	—	—
Famotidine[17,21]	5.7	—	—	0.5	0.2	—
Fenoprofen[15]	0.09	6.9	—	—	—	—
Fleroxacin[18]	1.7	—	—	4.7	—	—
Flufenamic acid[15,19]	0.17	3.8	—	30.0	—	—
Fluphenazine[17,21]	0.7	—	—	13.1	13.4	—
Flurazepam[17]	1.8	—	—	1.7	—	—
Foscarnet[25]	NS	NS	NS	—	—	—
Furosemide[6,19,22]	0.04	0.87	5.1	11.5	1.7	17.5
Glibenclamide[19]	0.26	NS	—	NS	—	—
Glutathione[11]	NS	NS	10.6	—	—	—
Glutathione–Hg complex[11]	NS	1.5	18	—	—	—
Glycocholate[26]	0.62	5.8	NS	—	—	—
Glycylsarkosine[7]	NS	—	—	—	—	—
Guanfacin[17]	5.4	—	—	5.1	—	—
Guanidine[17,*]	NS	—	—	1.0	34.2	—
Haloperidol[17,21]	NS	—	—	3.5	8.9	—
Heroin[20]	2.11	—	—	0.58	1.2	6.6
Hippurate[8]	0.24	—	—	—	—	—
Hydrochorothiazide[19,22]	0.72	4.4	NS	5.0	18.5	14.3
Hydroxytriamterenesulfate[22]	0.56	NS	—	NS	8.3	NS
Hypoxanthine[9]	3.1	—	—	—	—	—
Ibuprofen[15]	0.16	NS	—	—	—	—
ICI 207.828[22]	0.75	—	—	NS	—	—
Idazoxan[17]	NS	—	—	0.4	—	—

(*continued*)

Table I. (*Continued*)

Drug[b]	Apparent K_i (mmole/liter)					
	cl,PAH	cl,SO$_4^{2-}$	cl,succ	cl,NMeN$^+$	l,MPP$^+$	l,choline$^+$
Ifosfamide[15]	NS	NS	—	3.2	—	—
Imipramine[17]	4.8	—	—	1.5	—	—
Indapamide[19,22]	1.0	NS	—	15.5	4.7	18.0
Indomethacin[15]	0.05	1.4	NS	—	—	—
Iprecynium[24]	—	—	—	0.15	0.02	0.056
Ipsapirone[17,21]	0.6	—	—	1.9	NS	—
Isobutylmethylxanthine[9]	3.1	—	—	—	—	—
Isoproterenol[17,21]	NS	—	—	3.0	3.9	—
Ketamine[17]	3.2	—	—	2.1	—	—
Ketanserin[17,21]	NS	—	—	6.9	16.7	—
α-Ketoglutarate[1,3]	0.03	NS	0.08	—	—	—
Ketoprofen[15]	0.09	7.8	—	—	—	—
Kö 1154[17]	NS	—	—	3.8	—	—
Leucotriene B$_4$[9]	0.20	—	—	—	—	—
Lysyl-acetylsalicylate[11]	2.0	NS	NS	—	—	—
Mefloquine[17]	1.7	—	—	9.0	—	—
Mefruside[19,22]	NS	6.5	—	NS	16.2	NS
Mercaptosuccinate[11]	0.77	6.1	0.28	—	—	—
Mercapturic acid[8]	NS	—	—	—	—	—
Mesna[11]	1.16	NS	NS	—	—	—
Methacholine[24]	—	—	—	7.6	5.0	9.4
O-Methylisoproterenol[17]	NS	—	—	3.3	—	—
N^1-Methylnicotinamide[13,21]	—	—	—	0.54	8.3	—
Methyl-phenylpyridinium[13,21,24]	—	—	—	0.12	0.21	1.3
Methylphenyltetrahydropyridine[13,21]	—	—	—	0.27	11.9	—
Metoclopramide[17]	5.3	—	—	5.8	—	—
Metronidazole[17]	NS	—	—	3.2	—	—
Morphine[20]	5.9	—	—	0.78	1.15	15.4
Muzolimine[22]	1.7	NS	—	11.3	NS	—
Nalidixic acid[18,21]	0.35	—	—	1.6	9.5	—
Naproxen[15]	0.1	NS	—	—	—	—
Nicotine[17]	NS	—	—	0.92	—	—
Niflumic acid[19]	0.21	2.7	—	22.0	—	—
Nitrazepam[17]	1.0	—	—	5.7	—	—
(1R,2S)-Norephedrine[23]	—	—	—	2.4	15.0	19.1
Norepinephrine[17,21]	NS	—	—	6.2	12.6	—
Norfloxacin[18]	1.9	—	—	2.3	—	—
(1S,2S)-Norpseudoephedrine[23]	—	—	—	2.3	8.7	14.8
Ochratoxin A[10]	0.02	—	—	—	—	—
Octopamine[17]	NS	—	—	4.3	—	—
Ofloxacin[18,21]	2.5	—	—	1.6	1.5	—
Orellanin[10]	0.13	—	—	—	—	—
Oxalate[5]	NS	1.1	NS	—	—	—
Ozolinone[22]	0.07	NS	—	—	—	—
Pamidronate[25]	NS	7.2	NS	—	—	—
Papaverine[9,13,21]	NS	—	—	0.73	0.41	—

(*continued*)

Affinity of Drugs to Renal Transporters for Organic Anions and Cations

Table I. (*Continued*)

Drug[b]	Apparent K_i (mmole/liter)					
	cl,PAH	cl,SO$_4^{2-}$	cl,succ	cl,NMeN$^+$	l,MPP$^+$	l,choline$^+$
Para-aminohippurate[5]	0.08	NS	1.8	—	—	—
Paracetamol[11]	2.8	NS	NS	—	—	—
Paracetamolsulfate[11]	1.7	5.4	—	—	—	—
Paraquat[13]	—	—	—	6.33	—	—
Pefloxacin[18]	1.2	—	—	2.1	—	—
Penicillamine[11]	NS	NS	NS	—	—	—
Pentagastrin[7]	0.3	—	—	—	—	—
Phaloidin[7]	NS	—	—	—	—	—
Phenacetin[11]	1.0	NS	NS	—	—	—
Phenetidine[11]	2.2	NS	NS	—	—	—
Phenolphthaleine[5]	0.24	NS	NS	—	—	—
Phenolsufonphthaleine[5]	0.04	1.7	NS	—	—	—
Phenoxymethylpenicillin[6]	0.7	NS	NS	NS	—	—
Pilocarpine[17,21,*]	2.7	—	—	0.33	2.8	7.4
Pinacidil[17]	NS	—	—	1.2	—	—
(−)Pindolol[23]	—	—	—	0.73	1.8	1.6
Pipemidic acid[18]	NS	—	—	4.1	—	—
Piretanide[6,19,22]	0.07	2.9	NS	9.1	10.9	16.7
Prednisolone[12,18]	0.27	NS	NS	0.54	—	—
Primaquine[17]	NS	—	—	2.1	—	—
Probenecid[6]	0.03	7.3	NS	NS	—	—
Proflavine[17,24]	NS	—	—	0.2	—	0.27
Prostaglandin E$_1$[9]	0.14	NS	NS	—	—	—
Prostaglandin E$_2$[9]	0.07	11.0	NS	—	—	—
Prostaglandin F$_{2a}$[9]	0.14	NS	NS	—	—	—
(1S,2S)-Pseudoephedrine[23]	—	—	—	0.91	6.4	12.2
Quinacrine[17,21]	3.1	—	—	0.36	0.17	—
(−)Quinine[21,23]	—	—	—	0.3	0.31	2.6
(+)Quinidine[23]	—	—	—	0.89	0.32	2.4
Quinoline[17]	4.5	—	—	2.3	—	—
Ramipril[7]	0.4	—	—	—	—	—
Ranitidine[17,21]	2.5	—	—	0.5	0.16	—
Rhodamine 123[21,24]	—	—	—	0.34	0.02	0.058
Rosoxacin[18]	0.5	—	—	3.2	—	—
Roxatidine[15]	NS	—	—	0.3	—	—
RU 28362-6[12]	0.46	NS	NS	—	—	—
Saccharin[6]	0.2	NS	NS	—	—	—
Salicylate[11]	0.93	NS	NS	—	—	—
Salicyluric acid[11]	0.03	NS	NS	—	—	—
Shionogi 6315S [6]	0.94	NS	NS	NS	—	—
Serotonin[21]	—	—	—	1.5	2.3	—
Sotalol[17]	NS	—	—	2.7	—	—
Spermine*	—	—	—	NS	19.9	—
Spermidine*	—	—	—	NS	14.5	—
Spironolactone[22]	0.07	NS	NS	—	—	—
Sulfinpyrazone[22]	0.17	—	—	3.8	3.5	11.9

(*continued*)

Table I. (*Continued*)

Drug[b]	Apparent K_i (mmole/liter)					
	cl,PAH	cl,SO$_4^{2-}$	cl,succ	cl,NMeN$^+$	l,MPP$^+$	l,choline
Sulpiride[17]	NS	—	—	3.7	—	—
2,4,5,T[6]	0.05	4.2	NS	—	—	—
Tacrine[17]	3.0	—	—	0.33	—	—
Terbutaline[17]	NS	—	—	1.8	—	—
Testosterone[18]	0.36	—	—	0.1	—	—
Tetracycline*	—	—	—	11.8	17.2	—
Tetraethylammonium[24]	—	—	—	0.07	1.85	15.6
Thebaine[20]	5.7	—	—	0.25	1.1	5.7
Theophylline[9]	0.61	—	—	—	—	—
Thromboxane B$_2$[9]	0.36	NS	NS	—	—	—
Ticarcillin[6]	0.84	8.2	NS	NS	—	—
Tienilic acid[6,22]	0.04	2.7	NS	20.5	7.7	NS
Tinidazole[17]	2.0	—	—	3.6	—	—
Tiopronin[11]	0.86	NS	NS	—	—	—
Tolbutamide[19]	0.22	17.9	—	17.7	—	—
Tolmetin[15]	0.14	NS	—	—	—	—
Torasemide[19,22]	0.16	11.4	—	11.2	9.0	18.5
(−)Tranylcypromine[23]	—	—	—	5.5	27.3	26.3
Trazodone[17]	0.9	—	—	3.0	—	—
Triamterene[22]	NS	—	—	NS	5.3	NS
Trichlorovinylacetylcysteine[7]	0.22	—	—	—	—	—
Trichlorovinylcysteine[7]	0.41	—	—	—	—	—
Trichlorovinylglutathione[7]	1.3	—	—	—	—	—
Triflocine[19,22]	0.15	4.1	NS	2.7	2.2	16.4
Trifluperazine[17,21]	1.4	—	—	6.4	15.3	—
Trifluperidol[17]	NS	—	—	2.2	—	—
Trofosfamide[15]	0.84	—	—	3.8	—	—
L-Tryptophan[7]	2.8	—	—	—	—	—
Tyramine[17,21]	NS	—	—	1.7	4.1	—
Tyrosine[7]	NS	—	—	—	—	—
Urate[5]	1.3	NS	NS	—	—	—
Ursodeoxycholate[26]	0.17	NS	NS	—	—	—
Valproate[4]	0.71	—	—	—	—	—
(−)Verapamil[17,21,23]	1.1	—	—	0.3	0.61	0.65
Vinblastine*	—	—	—	0.8	6.6	—
Vincristine*	—	—	—	1.6	3.8	—

[a] Listed are the apparent K_i values in mmole/liter. Roughly this gives the concentration of the drug by which half of the respective transporter is occupied. cl,PAH, apparent K_i against contraluminal PAH transport; cl,SO$_4^{2-}$, apparent K_i against contraluminal sulfate transport; cl,succ, apparent K_i against contraluminal dicarboxylate transport; cl,NMeN$^+$ (TEA$^+$), apparent K_i against contraluminal transport of organic cations; l,MPP$^+$, apparent K_i against luminal transport of organic cations; l,choline$^+$, apparent against luminal choline$^+$ transport. The K_m values for the test substances are as follows in mmole/liter: cl,PAH, 0.08; cl,SO$_4^{2-}$ 1.4; cl,succ, 0.09; cl,NMeN$^+$, 0.54; cl,TEA$^+$, 0.16; l,MPP$^+$, 0.21; l,choline$^+$, 0.18.

[b] Asterisk indicates unpublished results. Other references are as follows: 1, Ullrich *et al.* (1984); 2, Ullrich *et al.* (1985); 3, Ullrich *et al.* (1987a); 4, Ullrich *et al.* (1987b); 5, Ullrich and Rumrich (1988); 6, Ullrich *et al.* (1989a); 7, Ullrich *et al.* (1989b); 8, Ullrich *et al.* (1990); 9, Ullrich *et al.* (1991a); 10, Burckhardt *et al.* (1991); 11, Ullrich *et al.* (1991b); 12, Ullrich *et al.* (1991c); 13 Ullrich *et al.* (1992); 14, Ullrich and Rumrich (1992); 15, Ullrich *et al.* (1993a); 16, Ammer *et al.* (1993b); 17, Ullrich *et al.* (1993b); 18, Ullrich *et al.* (1993c); 19, Ullrich *et al.* (1994); 20, Ullrich and Rumrich (1995); 21, David *et al.* (1995); 22, Ullrich (1995); 23, Somogyi *et al.* (1996); 24, Ullrich and Rumrich (1996); 25, Ullrich *et al.* (1997); 26, Ullrich (1998).

um-driven cotransporters: the Na^+–sulfate cotransporter, the Na^+–dicarboxylate cotransporter, the Na^+–monocarboxylate (lactate) cotransporter, and the electrical potential-driven choline transporter. The latter may play a role in drug reabsorption. The respective K_i values for the choline transporter are also listed in Table I.

4. GENERAL INFORMATION ABOUT THE K_i VALUES GIVEN HERE

Table I contains K_i values for approximately 260 substances: drugs, test substances, and pharmacologically and toxicologically interesting substances. Each substance was tested on average against two or three transporters: approximately 70% of them against each the PAH transporter and the $NMeN^+$ transporter, one third against each the sulfate transporter and the MPP^+ transporter, and one fourth against the choline transporter. The K_i values can be compared with each other because they were obtained with comparable *in situ* techniques on the same rat strain. Data can deviate considerably if they are from different rat strains and/or were obtained with a different correction factor for initial flux rates (Brändle *et al.*, 1992; Ullrich *et al.*, 1992). Although in the given examples the deviation is by a fixed factor, the data of Brändle *et al.* (1992) are not included in Table I, though they are cited in Table III. The interested reader may consult the original publication. In Table I, NS means that transport inhibition was not significant at the applied dose, which was sometimes limited by the solubility of the applied substrates (for details see the original publications). In general NS means that the affinity is very low or absent. To get a better overview of the K_i values the following classification is used: $K_i < 0.3$ mmole/liter means good affinity, K_i values from 0.3 to 2.0 mmole/liter signify moderate affinity, K_i values from 2.0 to 10.0 mmole/liter represent low affinity, and K_i values >10 mmole/liter signify very low affinity. For practical evaluations, however, the K_i values should be related to the K_m values of the test substrates. These are listed in the first footnote to Table I; note that the K_m values of PAH, succinate, TEA^+, MPP^+, and $choline^+$ are between 0.08 and 0.21 mmole/liter, i.e., values corresponding to good-affinity substrates, while the K_m of $NMeN^+$ is 0.54 mmole/liter and that of sulfate is 1.4 mmole/liter, in the range of moderate-affinity substrates. The K_i values indicate only that the substrates tested interact with the respective transporters and inhibit the transport of other substrates. It remains open whether a substrate is indeed transported or only bound and not transported. However, as a rule it can be considered that high-affinity substrates are transported only at a reduced rate or not at all. The presence of a transporter for a certain substrate does not imply that the main route of transport of that substrate occurs via this transporter. If the substrate is lipid-soluble, the bulk flow might occur through the lipid bilayer of the cell membrane. That the transporter also plays a role in such a case is documented by the distribution of lipophilic cytotoxic drugs in cells that express multidrug-resistant transport proteins.

5. INTERACTION WITH CONTRALUMINAL PAH TRANSPORTER

The molecular features of substrates which interact with contraluminal PAH transport are reviewed by Ullrich (1997) and include the degree of hydrophobicity, the strength of negative ionic charge(s) (low pK_a values), and/or the strength of partial negative charges (electron-attracting side groups) which are able to form hydrogen bonds. Multiple hydrogen bonds are favorable; ionization is not necessary. Groups of compounds which have a good affinity to the PAH transporter (K_i < 0.3 mmole/liter) and examples of drugs are listed in Table II. Most drugs which interact with the PAH transporter have moderate affinity (K_i between 0.3 and 2.0 mmole/liter). This category includes penicillin and cephalosporin compounds (Ullrich et al., 1989a), angiotensin-converting enzyme inhibitors (Ullrich et al., 1989b), cysteine conjugates (Ullrich et al., 1989b), salicylate and aminobenzene analgesics (Ullrich et al., 1991b), analgesic aniline derivatives (Ullrich et al., 1993a), benzodiazepines and related compounds (Ullrich et al., 1993b), gyrase inhibitors (Ullrich et al., 1993c), some cyclophosphamide compounds (Ullrich et al., 1993c), and some thiazide diuretics (Ullrich et al., 1994).

Conjugation of substrates with sulfate, glucuronate, glycine, acetate, or benzoate creates or increases affinity of otherwise hydrophobic compounds toward the PAH transporter (Ullrich et al., 1989b, 1990). The same holds for conjugation with

Table II.
Classes of Substrates with a Good Affinity (K_i < 0.3 mmole/liter) to the Contraluminal PAH Transporter[a]

Group	Drug examples	References
Phenoxy derivatives	Ethacrynic acid, tienilic acid	Ullrich et al. (1989a)
Sulfamoylbenzoates	Probenecid, furosemide, piretanide	Ullrich et al. (1989a)
Eicosanoides	Prostaglandin E_2, iloprost, leukotriene B4	Ullrich et al. (1991a)
Cyclic GMP	Dibutyryl cGMP	Ullrich et al. (1991a)
Corticosteroids	Dexamethasone, spironolactone, canrenoate, d-aldosterone	Ullrich et al. (1991c)
Weak organic acid analgesics	Tolmetin, ibuprofen, naproxen, ketoprofen, fenoprofen, indomethacin	Ullrich et al. (1993a)
Sulfonylurea compounds	Tolbutamide, chlorpropamide, torasemide, glibenclamide	Ullrich et al. (1994)
Diphenylamine-2-carboxylates	Flufenamic acid, triflocine, niflumic acid, diclofenac	Ullrich et al. (1994)
Bile acids	Chenodeoxycholate, ursodeoxycholate, cholate, lithocholate-3-sulfate, lithocholate-3-glucuronide	Ullrich (1998)

[a]The contraluminal PAH and dicarboxylate transporters do not seem to have reactive SH groups, the occupancy of which inhibit drug interaction, but the contraluminal sulfate transporter has such groups (Ullrich et al., 1990).

glutathione and even more for consecutive processing to the respective L-cysteine and N-acetylcysteine compounds (Ullrich et al., 1989b, 1990).

6. INTERACTION WITH THE CONTRALUMINAL SULFATE AND DICARBOXYLATE TRANSPORTERS

The molecular features of substrates which interact with the contraluminal sulfate and dicarboxylate transporters were elaborated by Fritzsch et al. (1989). The important determinant for the sulfate transporter is an accumulation of negative charges and for the dicarboxylate transporter two negative charges at a favorable distance. Although there exists overlapping specificity between the PAH, sulfate, and dicarboxylate transporters, only a few of the substrates listed in Table I interact with all three transporters, such as furosemide, mercaptosuccinate, H_2-DIDS, and chenodeoxycholate. Mercaptosuccinate, like α-ketoglutarate, interacts with both the PAH and dicarboxylate transporters with high affinity. By serving the two transporters, α-ketoglutarate is able to drive the contraluminal uptake of PAH in the proximal tubular cell (Shimada et al., 1987; Pritchard, 1988). Among the substrates which interact with the contralumial sulfate transporter, the diphosphonates deserve special attention. The pharmacologically active diphosphonates alendronate, clodronate, etidronate, and pamidronate interact only with the contraluminal sulfate, but not with the other two organic anion transporters. It is likely that they are transported not only into proximal tubular cells, but also into bone cells by similar sulfate transporters (Ullrich et al., 1997; Hästbacka et al., 1994).

7. INTERACTION WITH THE THREE ORGANIC CATION TRANSPORTERS

The molecular features of substrates which interact with the organic cation transporters are similar to those of the contraluminal organic anion (PAH) transporter. However, instead of "electronegative" (electron attracting) one now has "electropositive" (electron donating), and instead of low pK_a, one has high pK_a. The necessary electropositive ionic charge or the electron-donating side group is usually connected with a nitrogen atom with not too low pK_a (as, for example, in aniline and analogs). Hydrophobic quaternary ammonium compounds have good affinity ($K_i < 0.3$ mmole/liter). Classes of substrates with good affinity to the contraluminal NMeN$^+$/TEA$^+$ transporter and examples of drugs are listed in Table III. Most NMeN$^+$/TEA$^+$ substrates also have an affinity to the luminal H$^+$/organic cation (MPP$^+$) exchanger and the luminal choline transporter (Table I). In general, the affinity of a substrate to the contraluminal NMeN$^+$ transporter is high-

Table III.
Classes of Substrates with a Good Affinity ($K_i < 0.3$ mmole/liter) to the Contraluminal $NMeN^+$, TEA^+ Transporter

Group	Drug examples	Reference
Pyridinium compounds	MPP^+, MPTP	Ullrich et al. (1992)
Catecholamines	Propylhexedrine, terbutaline, fenoterol, dopexamine	Brändle et al. (1992)
Aminoacridines	Proflavine, quinacrine, acriflavine, dimidium	Ullrich et al. (1992, 1993b)
Corticosteroides	Corticosterone, testosterone	Ullrich et al. (1993c)
Guanidines	Amiloride, benzylamiloride cimetidine	Ullrich et al. (1993b), David et al. (1995)
Morphine derivatives	Thebaine	Ullrich and Rumrich (1995)
Quinolines	Chloroquine, quinine	David et al. (1995)

$^a MPP^+$, Methyl-phenyl-pyridinium; MPTP, methylphenyltetrahydropyridine.

er than to the luminal H^+/organic cation (MPP^+) exchanger, and the affinity to the latter is higher than that to the luminal choline transporter (David et al., 1995; Ullrich and Rumrich, 1996). However, most substrates with a high affinity to the $NMeN^+$ transporter show the same or somewhat higher affinity to the MPP^+ transporter (David et al., 1995), while choline and related compounds have a higher affinity to the choline transporter than to the other two organic cation transporters (Ullrich and Rumrich, 1996).

8. BI- AND POLYSUBSTRATES: DRUGS WHICH INTERACT BOTH WITH TRANSPORTERS FOR ORGANIC ANIONS AND TRANSPORTERS FOR ORGANIC CATIONS

Molecules containing side groups which enable them to interact with both organic anion and organic cation transporters usually interact with one type strongly and the other one weakly (Ullrich et al., 1993a–c; Ullrich, 1995; Ullrich and Rumrich, 1995). Such substrates can be zwitterionic, anionic, cationic, or with no ionic charge at all, but having the respective electron-attracting and electron-donating side groups. The bisubstrate behavior is of special interest, as it turns out that the contraluminal organic anion (PAH) transporter (OAT1) and the contraluminal organic cation ($NMeN^+/TEA^+$) transporter (OCT1) belong to the same gene family (Sekine et al., 1997; Wolff et al., 1997; Sweet et al., 1997). For some bi- or polysubstrates, such as cimetidine, amiloride, and ranitidine, it was shown that not only do they interact, but they are also transported by the two types of transporters (Ullrich et al., 1993b).

9. CONCLUSIONS AND PERSPECTIVES

Drug interaction as quantified by K_i values refers to the interaction of drugs with the respective transporters under standard conditions, i.e., interaction on the molecular level. It does not exactly predict to what degree drug interaction occurs at the cellular or higher level, i.e., whether the interacting drug is transported itself or to what degree it might inhibit transtubular net transport of other substrates. In the situation of net transtubular transport, a substrate has to cross both cell sides via at least two transporters with the possibility of cis-inhibition, trans-stimulation, or trans-inhibition by interfering substrates. In experiments on the kidney *in situ*, using fluorescent substrates, such interactions can be readily measured and have been documented. Thus, the net secretion of the anionic sulfofluorescein was cis-inhibited at the contraluminal cell side by probenecid and apalcillin and transstimulated by glutarate (Ammer *et al.*, 1993a), while the net reabsorption of the cationic dimethylaminostyryl-*N*-methyl-pyridinium was trans-stimulated at the luminal cell side by amiloride and cimetidine (Pietruck and Ullrich, 1995). The K_i values listed in Table I refer to six transporters in the kidney, as well as to drug transport in other organs where the same transporters are present, namely liver, intestine, bone, central nervous system, and others (see Chapters 15 and 16 in this book).

Today structural requirements for substrates of the organic anion and cation transporters are known, and the main transporters have been cloned. Thus, it might be possible to produce two- or three-dimensional crystals from transporter proteins and to determine the functional structure by electron or X-ray crystallography (Kühlbrandt 1997; Ostermeier *et al.*, 1995). Then substrate stucture/transporter structure profiles for drugs could also be established with the use of computer-based drug design.

ACKNOWLEDGMENTS

I thank Profs. Gerhard Burckhardt, Göttingen, Kathleen M. Giacomini, San Francisco, and Joseph Pfeilschifter, Frankfurt, for valuable suggestions.

REFERENCES

Ammer, U., Natochin, Y., and Ullrich, K. J., 1993a, Tissue concentration and urinary excretion pattern of sulfofluorescein by the rat kidney, *J. Am. Soc. Nephrol.* **3:**1474–1487.

Ammer, U., Natochin, Y., David, C. Rumrich, G., and Ullrich, K. J., 1993b, Cisplatin nephrotoxicity: Site of functional disturbance and correlation to loss of body weight, *Renal Physiol. Biochem.* **16:**131–145.

Bergwerk, A. J., Shi, X., Ford, A. C., Kanai, N., Jacquemin, E., Burk, R. D., Bai, S., Novikoff, P. M., Stieger, B., Meier, P. J., Schuster, V. L., and Wolkoff, A. W., 1996, Immunologic distribution of an organic anion transport protein in rat liver and kidney, *Am. J. Physiol.* **271**:G231–G238.

Brändle, E., Fritzsch, G., and Greven, J., 1992, Affinity of different local anesthetic drugs and catecholamines for the contraluminal transport system for organic cations in proximal tubules of rat kidneys, *J. Pharmacol. Exp. Ther.* **260**:734–741.

Burckhardt, G, Schmitt, Ch., and Ullrich, K. J., 1991, p-Aminohippurate uptake across the basolateral membrane of rat proximal tubule cells: Specificity and mode of energetization, in: *Proceedings of the 11th International Congress of Neophrology* (M. Hatano et al., eds.), Springer-Verlag, Tokyo, vol. 2, pp. 1380–1390.

Burckhardt, G., Porth, J., and Wolff, N. A., 1998, Functional and molecular characterizaton of the renal transporter for p-aminohippurate, *Nova Acta Leopoldina NF* **78**:35–40.

Busch, A. E., Schuster, A., Waldegger, S., Wagner, C. A., Zempel, G., Broer, S., Biber, J., Murer, H., and Lang, F., 1996, Expression of a renal type I sodium/phosphate transporter (NaPi-1) induces a conductance in *Xenopus* oocytes permeable for organic and inorganic anions, *Proc. Natl. Acad. Sci. USA* **93**:5347–5351.

Chen, X., Tsukaguchi, H., Chen, X.-Z., Berger, U. V., and Hediger, M. A., 1999, Molecular and functional analysis of SCDT2, a novel rat sodium-dependent dicarboxylate transporter, *J. Clin. Invest.* **103**:1159–1168.

David, C., Rumrich, G., and Ullrich, K. J., 1995, Luminal transport system for H^+/organic cations in the rat proximal tubule: Kinetics, dependence on pH; specificity as compared with the contraluminal organic cation-transport system, *Pfluegers Arch.* **430**:477–492.

Dettli, L., 1996, Grundlagen der Arzneimitteltherapie, 14th edition, Documed, Basel, in: "Allgemeine und spezielle Pharmakologie und Toxikologie". 7th edition 1998, W. Forth, D. Henschler W. Rummel, and K. Starke eds). Spektrum Akademischer Verlag Heidelberg, Berlin, Oxford, pp. 94–101.

Evers, R., Zaman, G. J. R., Van Deemter, L., Jansen, H., Calafat, J., Oomen L. C. J. M., Oude-Elferink, R. P. J., Borst, P., and Schinkerl A. H., 1996, Basolateral localization and export activity of the human multidrug resistance-associated protein in polarized pig kidney cells, *J. Clin. Invest.* **97**:1211–1218.

Fleck, Ch., 1998, Renal and hepatic transport, *Nova Acta Leopoldina* **78**: 371 pages.

Fritzsch, G., Rumrich, G., and Ullrich, K. J., 1989, Anion transport through the contraluminal cell membrane of renal proximal tubule. The influence of hydrophobicity and molecular charge distribution on the inhibitory activity of organic anions, *Biochim. Biophys. Acta* **978**:249–256.

Gründemann, D., Gorboulev, V., Gambarian, S., Veyhl, M., and Koepsell, H., 1994, Drug excretion by a new prototype of polyspecific transporter, *Nature* **372**:549–552.

Gründemann, D., Babin-Ebell, J., Martel, F., Örding, N., Schmitt, A., and Schömig E., 1997, Primary structure and functional expression of the apical organic cation transporter from rat kidney epithelial LLC-PK_1 cells, *J. Biol. Chem.* **272**:10408–10413.

Hagenbuch, B., Echhardt, U., Noé, B., Stieger, B., and Meier, P. 1998, Multispecific hepatocellular bile acid and organic anion uptake systems, *Nova Acta Leopoldina NF* **78**:129–133.

Hästbacka, J., de la Chapelle, A., Mahtani, M. M., Clines, G., Reeve-Daly, M. P., Daly, M., Hamilton, B. A., Kusumi, K., Trivedi, B., Weaver, A., Coloma, A., Lovett, M. Buckler, A., Kaitila, I., and Lander E. S., 1994, The diastrophic dysplasia gene encodes a novel sulfate transporter: Positional cloning by fine-structure linkage disequilibrium mapping, *Cell* **78:**1073–1087.

Kahn, A. M., and Aronson P. S., 1983, Urate transport via anion exchange in dog renal microvillus membrane vesicles, *Am. J. Physiol.* **244:**F56–F63.

Kanai, N., Lu, R., Satriano, J. A., Bao, Y., Wolkoff, A. W., and Schuster V. L., 1995, Identification and characterization of a prostaglandin transporter, *Science* **268:**866–869.

Keppler, D., König, J., Schaub, T., and Leier, I., 1998, Molecular basis of ATP-dependent transport of anionic conjugates in kidney and liver, *Nova Acta Leopoldina NF* **78:**213–221.

Kinne-Saffran, E., and Kinne, R. K. H., 1990, Isolation of lumenal and contralumenal plasma membrane vesicles from kidney, in: *Methods in Enzymology, Vol. 191, Biomembranes, Part V, Cellular and Subcellular Transport, Epithelial Cells* (S. Fleischer and B. Fleischer, eds.), Academic Press, New York, pp. 450–469.

Koepsell, H., 1998, Organic cation transporters in intestine, kidney, liver, and brain, *Annu. Rev. Physiol.* **60:**243–266.

Kühlbrandt, W., 1997, Pumping ions, *Nature Struct. Biol.* **4:**773

Lang, F., Waldegger, S., Murer, H., Foster, I., Koepsell, H., Palacin, M., Bröer, S., Capasso, G., Matskevich, J., Wagner, C., and Busch, A. E., 1998, Electrophysiology of proximal renal transporters, *Nova Acta Leopoldina NF* **78:**73–78.

Leal-Pinto, E., Tao, W., Rappaport J., Richardson, M., Knorr, B. A., and Abramson, R. G., 1997, Molecular cloning and functional reconstitution of a urate transporter/channel, *J. Biol. Chem.* **272:**617–625.

Lopez-Nieto, C. E., You, G., Bush, K. T. Barros, E. J. G., Beier, D. R., and Nigam, S. K., 1997, Molecular cloning and characterization of NKT, a gene product related to the organic cation transporter family that is almost exclusively expressed in the kidney, *J. Biol. Chem.* **272:**6471–6478.

Markovich, D., Forgo, J., Stange, G., Biber, J., and Murer, H., 1993, Expression cloning of rat renal Na^+/SO_4^{2-}-cotransport, *Proc. Natl. Acad. Sci. USA* **90:**8073–8077.

Markovich, D., Bissig, M., Sorribas, V., Hagenbuch, B., Meier, P. J., and Murer, H., 1994, Expression of rat renal sulfate transport systems in *Xenopus laevis* oocytes, *J. Biol. Chem.* **269:**3022–3026.

Masuda, S., Saito, H., Nonoguchi, H., Tomita, K., and Inui, K., 1997, mRNA distribution and membrane localization of the OAT-K1 organic anion transporter in rat renal tubules, *FEBS Lett.* **407:**127–131.

Mori, K., Ogawa, Y., Ebihara, K., Aoki, T., Tamura, N., Sugawara, A., Kuwahara, T., Ozaki, S., Mukoyama, M., Tashiro, K., Tanaka, I., and Nakao, K., 1997, Kidney-specific expression of a novel mouse organic cation transporter-like protein, *FEBS Lett.* **417:**371–374.

Okuda, M., Saito, H., Urakami, Y., Takano, M., and Inui, K.-I., 1996, cDNA cloning and functional expression of a novel rat kidney organic cation transporter, OCT2, *Biochem. Biophys. Res. Commun.* **224:**500–507.

Ostermeier, C., Iwata, S., Ludwig, B., and Michel H., 1995, F_v fragment mediated crystallization of the membrane protein bacterial cytochrome c oxidase, *Nature Struct. Biol.* **2**:842–846.

Oude-Elferink, R. P. J., Paulusma, C., de Vree, M., Frijters C. M. G., and Groen, B. K., 1998, The role of ATP-transporters in hepatobiliary excretion, *Nova Acta Leopoldina NF* **78**:157–163.

Pajor, A. M., 1995, Sequence and functional characterization of renal sodium/dicarboxylate cotransporter, *J. Biol. Chem.* **270**:5779–5785.

Pietruck, F., and Ullrich, K. J., 1995, Transport interactions of different organic cations during their excretion by the intact rat kidney, *Kidney Int.* **47**:1647–1657.

Pritchard, J. B., 1988, Coupled transport of *p*-aminohippurate by rat kidney basolateral membrane vesicles, *Am. J. Physiol.* **255**:F597–F604.

Roch-Ramel, F., Werner, D., and Guisan, B., 1994, Urate transport in brush-border membrane of human kidney, *Am. J. Physiol.* **266**:F797–F805.

Roch-Ramel, F., Guisan, B., Jaeger, Ph., and Diezi, J., 1996, Transport of urate and other organic anions by anion exchange in human renal brush border membrane vesicles, *Cell. Physiol. Biochem.* **6**:60–71.

Roch-Ramel, F., Guisan, B., and Diezi, J., 1997, Effect of uricosuric and antiuricosuric agents on urate transport in human brush border membrane vesicles, *J. Pharmacol. Exp. Ther.* **280**:839–845.

Saito, H., Masuda, S., and Inui, K., 1996, Cloning and functional characterization of a novel rat organic anion transporter mediating basolateral uptake of methotrexate in the kidney, *J. Biol. Chem.* **271**:20719–20725.

Schaub, T. P., Kartenbeck, J., König J., Vogel, O., Witzgall, R., Kriz, W., and Keppler, D., 1997, Expression of the conjugate export pump encoded by the *mrp-2* gene in the apical membrane of kidney proximal tubules, *J. Am. Soc. Nephrol.* **8**:1213–1221.

Sekine, T., Watanabe, N., Hosoyamada, M., Kanai, Y., and Endou, H., 1997, Expression cloning and characterization of a novel multispeciific organic anion transporter, *J. Biol. Chem.* **272**:18526–18529.

Sekine, T., Watanabe, N., Hosoyamada, M., Kanai, Y., and Endou, H., 1998, A novel renal multispecific organic anion transporter (PAH transporter): Its structure and functional characteristics, *Nova Acta Leopoldina, NF* **78**:119–126.

Shimada, H., Moewes, B., and Burckhardt, G., 1987, Indirect coupling to Na^+ of *p*-aminohippuric acid uptake into rat renal basolateral membrane vesicles, *Am. J. Physiol.* **253**:F795–F801.

Somogyi, A. A., Rumrich, G., Fritzsch, G., and Ullrich, K. J., 1996, Stereospecificity in contraluminal and luminal transporters of organic cations in the rat renal proximal tubules, *J. Pharmacol. Exp. Ther.* **278**:31–36.

Steffgen, J., Burckhardt, B. C., Langenberg, C., Kühne, L., Müller, G. A., Burckhardt, G., and Wolff, N. A., 1999, Expression cloning and characterization of a novel sodium-dicarboxylate cotransporter from winter flounder kidney. *J. Biol. Chem.* **274**:20191–20196.

Sweet, H. D., Wolff, N. A., and Pritchard, J. B., 1997, Expression cloning and characterization of ROAT1. The basolateral organic anion transporter in rat kidney, *J. Biol. Chem.* **272**:30088–30095.

Tamai, I., Yabuuchi, H., Nezu, J.-I., Sai, Y., Oku, A., Shimane, M., and Tsuji, A., 1997, Cloning and characterization of a novel pH dependent organic cation transporter, *FEBS Lett.* **419**:107–111.

Ullrich, K. J., 1995, Interaction of diuretics with transport systems in the proximal renal tubule, in: *Handbook of Experimental Pharmacology: Diuretics* (R. Greger, H. Knauf, and E. Mutschler, eds.), Springer, Berlin, pp. 201–219.

Ullrich, K. J., 1997, Renal transport of organic anions and organic cations. Structural requirements for substrates, *J. Membr. Biol.* **158**:95–107.

Ullrich, K. J., 1998, Features of substrates for interaction with renal transporters of organic anions and cations, *Nova Acta Leopoldina NF* **78**:23–34.

Ullrich, K. J., and Rumrich, G., 1988, Contraluminal transport systems in the proximal renal tubule involved in secretion of organic anion, *Am. J. Physiol.* **254**:F453–F462.

Ullrich, K. J., and Rumrich, G., 1990, Kidney: Microperfusion-double perfused tubule *in situ*, in: *Methods in Enzymology, Vol. 191, Biomembranes, Part V, Cellular and Subcellular Transport, Epithelial Cells* (S. Fleischer, and B. Fleischer, eds.), Academic Press, New York, pp. 98–107.

Ullrich, K. J., and Rumrich, G., 1992, Renal contraluminal transport systems for organic anions (*para*-aminohippurate, PAH) and organic cations (N^1-methyl-nicotinamide, NMeN) do not see the degree of substrate ionization, *Pfluegers Arch,* **421**:286–288.

Ullrich, K. J., and Rumrich, G., 1995, Morphine analogues: Relationship between chemical structure and interaction with proximal tubular transporters: Contraluminal organic cation and anion transporter, luminal H^+/organic cation exchanger, and luminal choline transporter, *Cell. Physiol. Biochem.* **5**:290–298.

Ullrich, K. J., and Rumrich, G., 1996, Luminal transport system for choline$^+$ in relation to other organic cation transport systems in the rat proximal tubule. Kinetics, specificity: Alkyl/arylamines, alkylamines with OH, O, SH, NH_2, ROCO, RSCO and H_2PO_4-groups, methylaminostyryl, rhodamine, acridine, phenanthrene and cyanine compounds, *Pfluegers Arch.* **432**:471–485.

Ullrich, K. J., and Rumrich, G., 1997, Luminal transport step of *para*-aminohippurate (PAH): Transport from PAH loaded proximal tubular cells into the tubular lumen of rat kidney *in vivo*, *Pfluegers Arch.* **433**:735–743.

Ullrich, K. J., Fasold, H., Rumrich, G., and Klöss, S., 1984, Secretion and contraluminal uptake of dicarboxylic acids in the proximal convolution of the rat kidney, *Pfluegers Arch.* **400**:241–249.

Ullrich, K. J., Rumrich, G., Klöss, S., and Lang, H.-J., 1985, Contraluminal sulfate transport in the proximal tubule of the rat kidney. V. Specificity: Phenolphthaleins, sulfonphthaleins, and other sulfo dyes, sulfamoyl-compounds and diphenylamine-2-carboxylates, *Pfluegers Arch.* **404**:311–318.

Ullrich, K. J., Rumrich, G., Fritzsch, G., and Klöss, S., 1987a, Contraluminal *para*-aminohippurate transport in the proximal tubule of the rat kidney. II. Specificity: Aliphatic dicarboxylic acid, *Pfluegers Arch.* **408**:38–45.

Ullrich, K. J., Rumrich, G., and Klöss, S., 1987b, Contraluminal *para*-aminohippurate transport in the proximal tubule of the rat kidney. III. Specificity: Monocarboxylic acids, *Pfluegers Arch.* **409**:547–554.

Ullrich, K. J., Rumrich, G., and Klöss, S., 1989a, Contraluminal organic anion and cation transport in the proximal renal tubule: V. Interaction with sulfamoyl- and phenoxy diuretics, and β-lactam antibiotics, *Kidney Int.* **36**:78–88.

Ullrich, K. J., Rumrich, G., Wieland, T., and Dekant, W., 1989b, Contraluminal *para*-aminohippurate (PAH) transport in the proximal tubule of the rat kidney. VI: Amino acids, their *N*-methyl-, *N*-acetyl- and *N*-benzoylderivatives; glutathion- and cysteine conjugates, di- and oligopeptides, *Pfluegers Arch.* **415**:342–350.

Ullrich, K. J., Rumrich, G., Gemborys M., and Dekant, W., 1990, Transformation and transport: How does metabolic transformation change the affinity of substrates for renal contraluminal anion and cation transporters? *Toxicology Lett.* **53**:19–27.

Ullrich, K. J., Rumrich, G. Papavassiliou, F., Klöss, S., and Fritzsch, G., 1991a, Contraluminal *para*-aminohippurate transport in the proximal tubule of the rat kidney. VII. Specificity: Cyclic nucleotides, eicosanoids, *Pfluegers Arch.* **418**:360–370.

Ullrich, K. J., Rumrich, G., Gemborys, M. W., and Dekant, W., 1991b, Renal transport and nephrotoxicity, in: *Nephrotoxity* (P. H. Bach, N. J. Gregg, M. F. Wilks, and L. Delacruz, eds.), Marcel Dekker, New York, pp. 1–8.

Ullrich, K. J., Rumrich, G., Papavassiliou, F., Hierholzer, K., 1991c, Contraluminal *p*-aminohippurate transport in the proximal tubule of the rat kidney. VIII. Transport of corticosteroids, *Pfluegers Arch.* **418**:371–382.

Ullrich, K. J., Rumrich, G., Neiteler, K., and Fritzsch, G., 1992, Contraluminal transport of organic cations in the proximal tubule of the rat kidney: II. Specificity: Anilines, phenylalkylamines (catecholamines), heterocyclic compounds (pyridines, quinolines, acridines), *Pfluegers Arch.* **420**:29–38.

Ullrich, K. J., Rumrich, G., David, C., and Fritzsch, G., 1993a, Interaction of xenobiotics with organic anion and cation transport systems in renal proximal tubule cells, in: *Renal Disposition and Nephrotoxicity of Xenobiotics* (M. W. Anders, W. Dekant, D. Henschler, H. Oberleithner, and S. Silbernagl, eds.), Academic Press, San Diego, California, pp. 97–115.

Ullrich, K. J., Rumrich, G., David, C., and Fritzsch, G., 1993b, Bisubstrates: Substances that interact with renal contraluminal organic anion and organic cation transport systems: I. Amines, piperidines, piperazines, azepines, pyridines, quinolines, imidazoles, thiazoles, guanidines, and hydrazines, *Pfluegers Arch.* **425**:280–299.

Ullrich, K. J., Rumrich, G., David, C., and Fritzsch, G., 1993c, Bisubstrates: substances that interact with both renal contraluminal organic anion and organic cation transport systems: II. *Zwitterionic substrates:* Dipeptides, cephalosporins, quinolon-carboxylate gyrase inhibitors, and phosphamide thiazine carboxylates. *Nonionizable substates:* Steroid hormones and cyclophosphamides, *Pfluegers Arch.* **425**:300–312.

Ullrich, K. J., Fritzsch, G., Rumrich, G., and David, C., 1994, Polysubstrates that interact with renal contraluminal PAH, sulfate, and NMeN transport: Sulfamoyl-, sulfonylurea-, thiazide-, benzeneamino-carboxylate (nicotinate) compounds, *J. Pharmacol. Exp. Ther.* **269**:684–692.

Ullrich, K. J., Rumrich, G., Burke, R. T., Shirazi-Beechey, S. P., and Lang, H.-J., 1997, Interaction of alkyl/aryl-phosphonates, phosphonocarboxylates and diphosphonates with different anion transport systems in the proximal renal tubule, *J. Pharmacol. Exp. Ther.* **283**:1223–1229.

Weinman, E. J., Sansom, S. C., Bennett, S., and Kahn, A. M., 1983, Effect of anion exchange inhibitors and *para*-aminohippurate on the transport of urate in the rat proximal tubule, *Kidney Int.* **23**:832–837.

Wolff, N. A., Werner, A., Burkhardt, S., and Burckhardt, G., 1997, Expression cloning and characterization of a renal organic anion transporter from Winter Flounder, *FEBS Lett.* **417**:287–291.

Zhang, L., Dresser, M. J., Chun, J. K., Babbitt, P. C., and Giacomini, K. M., 1997a, Cloning and functional characterization of a rat renal organic cation transporter isoform (rOCT1A), *J. Biol. Chem.* **272**:16548–16554.

Zhang, L., Dresser, M. J., Gray, A. T., Yost, S. C., Terashita, S., and Giacomini, K. M., 1997b, Cloning and functional expression of a human liver organic cation transporter, *Mol. Pharmacol.* **51**:913–921.

6

Drug Disposition and Targeting

Transport across the Blood–Brain Barrier

Bertrand Rochat and Kenneth L. Audus

1. INTRODUCTION

The concept of the blood–brain barrier (BBB) was developed late in the last century on the basis that certain dyes did not accumulate in the brain, in contrast to other organs. Just after the middle of this century, research demonstrated that the BBB resides in the endothelium of the brain microvessels, represented a significant surface area of exchange between blood and brain (around 12 m^2), and was, however, the primary barrier to drug and peptide entry into the brain (Pardridge, 1994). The modern view of the BBB may be regarded as a dynamic barrier where many of its characteristics potentially may be up- or downregulated in normal as well as in disease states.

Several specific characteristics of the brain endothelium participate in formation of this barrier, including anatomical or morphological features illustrated in Fig. 1. In contrast to peripheral endothelial microvessels, the endothelium comprising the BBB has no fenestrations and a reduced number of pinocytic vesicles, suggesting that bulk transfer and transcytotic mechanisms for moving macromolecules are not well elaborated at the BBB. However, as reviewed below, specific, unidirectional movement of a limited number of macromolecules does occur at the BBB (Broadwell and Banks, 1993). The cells also exhibit the presence of unusu-

---

Bertrand Rochat and Kenneth L. Audus • Department of Pharmaceutical Chemistry, School of Pharmacy, University of Kansas, Lawrence, Kansas 66047.

Membrane Transporters as Drug Targets, edited by Amidon and Sadée. Kluwer Academic/Plenum Publishers, New York, 1999.

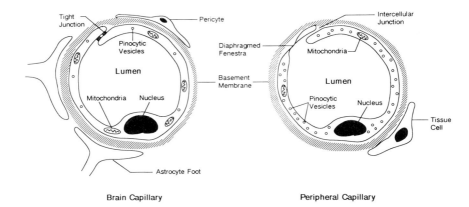

Figure 1. Anatomical and morphological characteristics of endothelium of the blood–brain barrier and the peripheral vasculature. Reprinted with permission from Audus and Borchardt (1991). Copyright 1991 Springer-Verlag GmbH & Co. KG, Heidelberg.

ally tight intercellular junctions (Reese and Karnovsky, 1967; Brightman and Reese, 1969) which effectively "seal" the apposing intercellular membranes of the endothelial cells together and confer a high membrane resistance (Crone and Olesen, 1982). Thus, the passive movement of macromolecules and hydrophilic compounds paracellularly is effectively blocked and forces substances to undergo transcytotic transport. The brain endothelial cells also contain many mitochondria, indicating that the barrier between blood and brain can contribute to biotransformation of xenobiotics (Pardridge, 1983, Ghersi-Egea *et al.*, 1995; Minn *et al.*, 1991). Cells surrounding the brain endothelium (i.e., neurons, astrocytes, and pericytes), as well as the basal lamina, seem to contribute to the stability and the barrier characteristics of the brain endothelial cells. These cell–cell interactions appear to be via the secretion of trophic factors rather than direct physical contact (Joo, 1996; Gragera *et al.*, 1993; Claudio *et al.*, 1995; Ment *et al.*, 1995).

In order to understand the transport mechanisms of the BBB, *in vivo, in situ,* and *in vitro* systems have been developed and are now in use to explore the biochemical and molecular basis of the BBB. While *in vivo* or *in situ* studies can provide an indication of the delivery potential and pharmacological consequences in animal models, the *in vitro* systems (i.e., isolated brain capillaries, cell cultures, etc.) offer promising tools in order to characterize properties specific to brain endothelial cell transport and metabolism at the cellular, biochemical, and molecular levels (Audus *et al.*, 1996). Addressing the application, advantages, and disadvantages of different experimental systems is beyond the scope of this chapter and the interested reader is directed to recent reviews (Audus *et al.*, 1996; Smith, 1996).

Further discussions in this chapter will survey the membrane transport processes that are currently known to allow drugs and macromolecules to permeate the BBB. A number of transcellular transport systems in brain endothelia exist for the transport of vital nutrients and waste products between the blood and the central nervous. The presence of an efflux pump, the multidrug-resistant gene product 1 (MDR1), an overexpressed P-glycoprotein, compounds the picture of the BBB and is included here as a significant problem that must be overcome in many drug delivery strategies. We also include mention of several less well-characterized systems which might contribute to transporter-targeted delivery strategies in the future. We hope this discussion might stimulate researchers to develop a better knowledge of natural transport systems at the BBB and underscore more strongly their possible primordial role in drug delivery and therapeutic efficacy.

2. PERMEABILITY OF THE BLOOD–BRAIN BARRIER

2.1. Carrier-Mediated Transport

Carrier-mediated transport may be described as either facilitative, which is saturable, stereoselective, and energy independent, or active, which is saturable, stereoselective, and directly dependent on energy expenditure by the cell. Several carrier mechanisms have been described at the BBB and now include both active and facilitative types as summarized in Table I. Only a few of these carriers have been extensively characterized and cloned. For some, such as the neutral amino acid carrier, their utility in drug delivery to the brain is well known.

2.1.1. CARRIERS AND EFFLUX SYSTEMS FOR LOW-MOLECULAR-WEIGHT MOLECULES

At least five D-glucose transporters have been described (Thorens, 1996). From the earliest studies, it appeared that D-glucose crossed the BBB easily and stereospecifically, thanks to a high density of facilitative-type transporters. Discrepant results have been published concerning the reported affinity constants of glucose for "its" transporter, GLUT-1 (Audus *et al.*, 1992), and could arise from the overlapping specificities of the various transporters. A recent study has described a sodium-dependent, active transport of glucose at the BBB that apparently functions in parallel to the facilitative GLUT-1 transporter (Nishizaki *et al.*, 1995).

The GLUT-1 transporter is inhibited by barbiturates (Honkanen *et al.*, 1995) and its gene expression can be controlled by cerebrolysin (Boado, 1995). Although there is a relatively low affinity of the substrate D-glucose for the GLUT-1 carrier

Table I.
Blood–Brain Barrier Transport Systems for Nutrients and Low-Molecular-Weight Organic Molecules

Carrier	Type	Substrate	Direction	Reference
A amino acid	Active	Alanine	Brain-to-blood	Komura et al. (1996)
Acidic amino acid	Facilitated	Glutamate	Blood-to-brain	Benrabh and Lefauconnier, (1996)
Amine	Active	Choline	—	Galea and Estrada (1992)
Basic amino acid	Facilitated	Arginine	—	Oldendorf (1971)
Basic organic molecules	—	Mepyramine	—	Yamazaki et al. (1994)
Biotin	Facilitated	Biotin	Bidirectional	Shi et al. (1993)
Carnitine	Facilitated	Carnitine	Brain-to-blood	Mroczkowska et al. (1997)
Medium-chain fatty acid	—	Valproic acid	—	Adkison and Shen (1996)
Hexose	Facilitated	Glucose	Bidirectional	Pardridge et al. (1990)
Monocarboxylic acid	Active	Lactate	Bidirectional	Conn et al. (1983)
Large neutral amino acid (LNAA)	Facilitated	Phenylalanine	Bidirectional	Oldendorf and Szabo (1976)
Pantothenic acid	Facilitated	Pantothenic acid	Blood-to-brain	Spector et al. (1986)
Purines	Facilitated	Adenine	—	Cornford and Oldendorf (1975)
Pyrimidines	Facilitated	Thymidine	—	Thomas and Segal (1997)
Taurine	Active	Taurine	Bidirectional	Tamai et al. (1995)
Thiamine	Facilitated	Thiamine	—	Greenwood et al. (1982)
Thyroid hormone	Facilitated	Thyroid hormone	—	Pardridge, (1979)

(K_m in the mM range), no drugs have been shown to be competing substrates. There are a few indications that GLUT-1 can transport substrates other than D-glucose. As examples, glucosylation of few peptides has been shown to increase their transport across the BBB and the evidence seemed to indicate that GLUT-1 was responsible for this transport (Polt et al., 1994). Additionally, the oxidized form of ascorbic acid has been shown to be transported by the GLUT-1 carrier (Agus et al., 1997). Attempts have also been made to couple chemotherapeutics with glucose (e.g., glucose–chlorambucil derivatives) only to find that, while the agents inhibit the GLUT-1 transporter, the agents are not transported (Halmos et al., 1996). Despite some positive observations, GLUT-1 does not appear to be an obvious carrier of choice for the drug delivery into the brain.

Saturable and bidirectional carriers for monocarboxilic acids have been decribed at the BBB and are responsible of the transport of L-lactate, L-acetate, and L-pyruvate, valproic acid, salicylic acid, and saturated fatty acids (Conn et al., 1983; Adkison and Shen, 1996; Gerhart et al., 1997). In general, these carriers have been more studied in intestinal cells like Caco-2 cells and are better known in that cell type (Tamai et al., 1995; Takanaga et al., 1994). Drugs such as valproic acid inhibit the uptake of short-chain monocarboxylic acids such as lactate, acetate, and pyruvate at the BBB. However, the uptake of valproic acid is not inhibited by short-chain monocarboxylic acids. Rather, the passage of valproic acid across the BBB appears to occur through a medium-chain fatty acid transporter (Adkinson and Shen, 1996). Therefore, it appears that multiple carriers for monocarboxylic acid compound are probably expressed at the BBB and show overlaps in substrate selectivity. Specific antibodies as well as molecular probes may allow a better characterization of the transporters of monocarboxylic acid compounds and eventually demonstrate their importance in the delivery of drug across the BBB.

Neutral amino acid carriers have proven to be capable of transporting many amino acids, drugs, and endogenous compounds with similar stuctures. Leucine, cysteine, serine, alanine, phenylalanine, L-dopa, L-tryptophan, the alkylating agent melphalan, the antiepileptic drug gabapentin, the muscle relaxant baclofen, and the neurotoxin beta-N-methylamino-L-alanine are all examples of agents shown to be substrates of neutral amino acid carriers (Begley, 1996, De Boer and Breimer, 1994; Smith et al., 1992; Oldendorf and Szabo, 1976). Various forms of neutral amino acid carriers have been described (Sanchez del Pino et al., 1995; Zerangue and Kavanaugh, 1996). For instance, the carrier for large neutral amino acids (LNAA) mainly transports phenylalanine with a high affinity (K_m about 10 μM) and is sodium independent. Since LNAA carrier is saturated under normal conditions by neutral amino acids present in blood, substrate competition does occur *in vivo* and does impair delivery of lower affinity substrates (Shulkin et al., 1995). The large neutral amino acid carriers, located on the luminal and abluminal membranes of the brain endothelial cells, appear to be subject to regulation by endogenous compounds such as oxoproline (W. J. Lee *et al.,* 1996) and vasopressin (Reichel *et al.,* 1996).

In contrast to dopamine, L-dopa is transported across the BBB by the LNAA carrier and is readily biotransformed into the brain to dopamine. At present, the prodrug L-dopa represents the best example of drug delivery mediated by a facilitated amino acid carrier at the BBB. Alkylating agents, mephalan and the nitrogen mustard, DL-NAM, have also shown to be substrates of LNAA carrier at the BBB (Begley, 1996; Takada *et al.,* 1992). The affinity of DL-NAM for the LNAA carrier is 100 times higher than mephalan because the addition of a naphthoic side chain. LNAA carriers show stereospecific activity for the S-isoforms of phenylalanine derivatives prepared as NMDA receptor antagonists. As an example, L-4-chlorokynurenine might be a useful prodrug for brain delivery of glycine–NMDA receptor antagonist since it is rapidly transported across the BBB by the LNAA carrier and enzymatically converted in brain parenchyma to the therapeutic form, 7-chlorokynurenic acid (Hokari *et al.,* 1996). The cytotoxic agent acivicin causes neurotoxicity due to its broad penetration into the brain, largely mediated by LNAA carriers. Thus, acivicin derivatives showing a lower affinity for LNAA carriers could exhibit a lower CNS toxicity (Chikhale *et al.,* 1995a).

Other neutral amino acid carriers (i.e, types A, ASC, and N) showing more selectivity for small neutral amino acid and a sodium dependence have been described at the BBB (Zerangue and Kavanaugh, 1996; Sanchez del Pino *et al.,* 1995). These amino acid carriers do have overlaps in amino acid recognition. For instance, the BBB transport of L-glutamine, mainly found in the cerebrospinal fluid, has been investigated and there has been an indication that two neutral amino acid carriers were involved in the L-glutamine transport: a Na-independent system L and a Na-dependent system N (Keep and Xiang, 1995). Taking into consideration all of these examples, we see that the neutral amino acid carriers continue to have the potential for brain delivery of therapeutic agents. Nevertheless, progress has yet to be made regarding the definition of the broad functions, affinities, and the specificities of all of BBB neutral amino acid carriers.

Basic amino acid and acidic amino acid transporters have been shown to transport asparagine, lysine, and aspartate and glutamate across the BBB (Oldendorf, 1971; Benrabh and Lefauconnier, 1996). Amine transporters are able to transport choline across the BBB (Galea and Estrada, 1992). The existence of nucleoside transporters has also been demonstrated (Joo, 1996). These latter carriers are able to transport the purines adenosine, guanosine, and inosine and the pyrimidine uridine across the BBB (Cornford and Oldendorf, 1975; Thomas and Segal, 1997). Thiamine and thyroid transporters are able to transport thiamine, vitamins such as folates and ascorbic acid, and T3 across the BBB, respectively (Greewood *et al.,* 1982; Joo, 1996). Once more, molecular and biochemical characterizations of these transporters will allow a better understanding of their potential for exploitation in drug delivery.

Many drugs have been shown to cross the brain endothelial cells poorly in spite of their favorable coefficient partitions (i.e., relatively lipophilic compounds). In fact, many drugs and peptides are pumped outside the cells by active transporters which are present in many cell types (Lum and Gosland, 1995). These transmembrane pumps contribute to the resistance to anticancer agents and probably psychotropic drugs in both normal and tumor tissues. P-glycoprotein (P-gp) is actually the best-documented efflux pump in resistant cell lines as well as in normal tissues (Stein, 1997; Lum and Gosland, 1995) and is responsible for the multidrug-resistance type 1 (MDR1) phenotype *in vitro* as well as *in vivo* (Lum and Gosland 1995; Huang *et al.,* 1997). Carcinogens and various xenobiotics are capable of coinducing both the expression of P-gp as well as some cytochrome P-450 isozymes in normal tissues as well as in cancer-resistant cell lines (Burt and Thorgeirsson, 1988; Thorgeirsson *et al.,* 1991). Inhibition of the efflux mechanism by the so-called reversing agents allows an increased accumulation of P-gp substrates inside the cells (Wigler, 1996). The P-gp of the BBB has been shown to play a sometimes dominant role in the efficacy of many drugs in the central nervous system (Jette *et al.,* 1995; Schinkel *et al.,* 1994, 1996). In humans, the activity of the brain P-gp is about one-fourth that expressed in the gut (Silverman and Schrenk, 1997).

The number and diversity of drugs shown to be substrates of the P-gp suggest that, like cytochrome P-450, the specificity of P-gp is very broad (Gottesman and Pastan, 1993). In the near future, other drug efflux systems (MDR-like pumps) are likely to be discovered and characterized at the BBB. Clearly, not all drugs have the same affinity for MDR1 present at the BBB. MDR1 systems do vary somewhat among species (Schinkel *et al.,* 1994, 1996), and there are some indications that MDR1 may vary in localization at the blood–brain barrier, particularly in the human, where there is conflicting evidence of MDR1 localization from the typical luminal membrane (Stewart *et al.,* 1996) to an abluminal and astrocytic foot location (Pardridge *et al.,* 1997). The multidrug-resistance protein MRP, which has been described in cancer-resistant cell lines as well as normal tissues (Kavallaris, 1997; Berger *et al.,* 1997; Barrand *et al.,* 1997), is expressed and functional at the BBB (Miller *et al.,* 1997). In cancer cell lines, other MDR transporters, different from the MDR1 and MRP, have already been identified (Izquierdo *et al.,* 1996; Huang *et al.,* 1997; S. E. Lee *et al.,* 1997).

It appears that in most instances lipophilic drugs are substrates of an ATP-dependent efflux pump, MDR1, an overexpressed P-glycoprotein (P-gp) localized at luminal membranes of the brain endothelial cells. Structurally diverse compounds are able to inhibit, sometimes in a competitive manner, the P-gp activity. Known substrates include vinblastine, reserpine, verapamil, trifluoperazine, amiodarone, daunomycin, progesterone, propafenone, and quinidine (Drion *et al.,* 1996; Wigler, 1996; Kavallaris, 1997; Stein, 1997).

As transporters, drug efflux systems remain important BBB targets that must be neutralized to realize delivery of a wide variety of therapeutic agents. Cyclosporine A is able to inhibit completely the P-gp activity (Wigler, 1996) and the cyclosporine D derivative SDZ PSC 833 has also been used *in vivo* and *in vitro* and shows a potent P-gp inhibition. The pipecolinate derivative of cyclosporine D, VX-710, which is able to inhibit P-gp, showed a stronger inhibition of the MRP activity effect than either verapamil or cyclosporine A (Germann *et al.*, 1997). The ATPase portion of the P-gP can be inhibited by bafilomycin A1. P-gp ATPase activity can be stimulated by colchicine, verapamil, and trifluoperazine (Sharom, 1995). The phospholipids surrounding the P-gp ATPases have been identified and could play an important role in their activity. It has been suggested that the inhibition of the P-gp activity by so-called fluidizers could occur by perturbing the lipid intregrity around the P-gp and its ATPase (Drori *et al.*, 1995).

2.1.2. PEPTIDE CARRIERS

A limited but growing body of research has been focused on saturable carriers for selected peptides at the BBB. These carriers are both unidirectional, either transporting peptides out of the brain or into the brain, and bidirectional (Banks and Kastin, 1996). Carrier systems for small peptides at the BBB are summarized in the Table II. At present, there appear to be no carriers of peptides larger than 10 amino acids at the BBB and those larger peptides would be expected to undergo transcytosis (Broadwell and Banks, 1993). At present, many of these carrier mechanisms lack sufficient characterization to propose precise roles for the mechanisms in drug delivery schemes.

Several small peptides have been shown to be apparent substrates of P-gp efflux at the BBB (Sharma *et al.*, 1992; Sarkadi *et al.*, 1994; Chikhale *et al.*, 1995b). However, independent of P-gp, the unidirectional transport of peptides such as corticotropin-releasing hormone (CRH) has been shown to be mediated by an active transporter and dependent on calcium channels (Martins *et al.*, 1996, 1997). Similary, a number of other small peptides also have distinct, independent transport systems to move the peptides from the brain into the systemic circulation. Among these are small tyrosinated peptides, Met-enkephalin and Tyr-MIF-1 (Banks and Kastin, 1997; Banks *et al.*, 1994). The transporter for these latter peptides is notable for being sensitive to ethanol, which can enhance the brain Met-enkephalin concentration (Plotkin *et al.*, 1997). A separate efflux system has also been described for RC-160, a somatostatin analog (Banks *et al.*, 1994) and arginine vasopressin (Begley, 1996).

In *in vivo* experiments, D-[Ala^1]-peptide T amide is transported from blood to brain by a saturable system (Barrera *et al.*, 1987). Using intraventricular injections of peptide T, the transport out of the brain was not saturable, indicating a specific localization of the carrier at the membrane of the endothelial cells. Studies

Table II.
Blood–Brain Barrier Peptide Transport Systems

Carrier	Type	Substrate	Direction	Reference
Corticotropin-releasing hormone (CRH)	Active	CRH	Brain-to-blood	Martins et al. (1996)
D-[Ala¹]-peptide T-amide	Facilitated	D-[Ala¹]-peptide T-amide	Blood-to-brain	Barrera et al. (1987)
Deltorphin	Facilitated	Deltorphin	Blood-to-brain	Fiori et al. (1997)
[D-Penicillamine^{2,5}]-enkephalin (DPDPE)	—	DPDPE	Blood-to-brain	Thomas et al. (1997)
Glutathione	Active	Glutathione	Bidirectional	Kannan et al. (1996)
Cytokines	—	Interleukin-1	Bidirectional	Banks et al. (1989)
Leucine-enkephalin	—	Leucine-enkephalin	—	Zlokovic et al. (1987)
Luteinizing hormone-releasing hormone (LHRH)	Facilitated	LHRH	Bidirectional	Barrera et al. (1991)
Neurotensin	Facilitated	Neurotensin	—	Banks et al. (1995)
Somatostatin	—	Somatostatin	Brain-to-blood	Banks et al. (1994)
Tyrosinated peptides	Active	Methionine-enkephalin	Brain-to-blood	Banks and Kastin (1997)
Vasopressin	Facilitated	Vasopressin	Brain-to-blood	Begley (1996); Banks et al. (1987)

employing intravenous injection of radiolabeled leptin revealed the presence of a saturable blood-to-brain transporter (Banks *et al.,* 1996). Similar to peptide T, no saturable transport of radiolabeled leptin out of the brain was observed. Neurotensin is another representative small peptide that also has shown only unidirectional transport into the brain (Banks *et al.,* 1995).

Some of the opioid peptides, including deltorphins, cross the BBB rather effectively and experiments performed with isolated bovine brain microvessels indicate the presence of a naloxone-sensitive carrier (Fiori *et al.,* 1997). Preloading the microvessel preparations with L-glutamine can also transiently stimulate deltorphin uptake, suggesting some interrelationships with amino acid transport at the BBB. The delta opioid receptor-selective [D-penicillamine-2,5] enkephalin (DPDPE) has been shown to penetrate the brain by a saturable system (Thomas *et al.,* 1997). The uptake system for DPDPE, which does not involve the large amino acid transporter, has yet to be fully characterized. Transport of leucine-enkephalin across the BBB has also been demonstrated *in vivo* and exhibited a saturable mechanism (Zlokovic *et al.,* 1987). The potential relationships among the carriers for opioids at the BBB have not been resolved. The possibility obviously exists, however, for the design and development of opioid peptide therapeutics that target BBB carriers and achieve improved brain delivery (Thomas *et al.,* 1997).

The last group of peptide carriers to be mentioned facilitate peptide distribution across the BBB in both directions. Bidirectional and saturable transporters exist for peptides such as luteinizing hormone-releasing hormone (LHRH; Barrera *et al.,* 1991) and interleukin-1 alpha (Banks *et al.,* 1989). Different transporters, sodium dependent or independent, have also been identified for glutathione (Kannan *et al.,* 1996).

2.2. Transcytosis

Macromolecules may cross the brain microvessel endothelial cells by transcytosic mechanisms mediated by luminal surface receptors (Audus *et al.,* 1992; Broadwell and Banks, 1993). While the role of plasmalemmal vesicles as transcytotic carriers through the endothelium was postulated in the 1960s, the use of a combination of a nontransportable moiety with a natural macromolecule, a chimeric drug delivery system, recognized by membrane receptors mediating transcytosis at the BBB was not proposed until the late 1980s (Pardridge *et al.,* 1987). Subsequently, several receptor systems mediating transcytotic processing through the BBB have been described and are summarized in the Table III. In several instances, these transcytotic systems have been explored as mechanisms to target and facilitate the delivery of therapeutic entities to brain capillary endothelial cells and eventually the brain.

Table III.
Blood–Brain Barrier Macromolecule Endocytic and Transcytotic Systems

Carrier	Type	Reference
Amyloid β-protein	Receptor	Poduslo et al. (1997)
Basic peptides	Adsorptive	Tamai et al. (1997)
Histone	Adsorptive	Pardridge et al. (1989)
Immunoglobulin G albumin	Adsorptive (?)	Poduslo et al. (1994)
Insulin	Receptor	Duffy and Pardridge (1987)
Insulin-like growth factor	Receptor	Reinhardt and Bondy (1994)
Low-density lipoprotein	Receptor	Dehouck et al. (1997)
Lectins	Adsorptive	Villegas and Broadwell (1993)
Leptin	Receptors	Banks et al. (1996)
Modified albumins	Adsorptive	Kumagai et al. (1987)
Transferrin	Receptor	Broadwell et al. (1996), Bradbury (1997)*

A high rate of transcytosis of insulin in brain capillary endothelial has been reported *in vivo* and suggests that the insulin receptor could be an effective target for drug delivery into the brain (Duffy and Pardridge, 1987; Podulso et al., 1994). Related peptides, insulin-like growth factors I and II, also show significant brain levels following injection into the carotid artery of rats (Rheinhardt and Bondy, 1994). Native insulin has been used as a carrier of horseradish peroxidase, a protein which crosses the BBB very slowly (Fukuta et al., 1994). The difficulty in using native insulin as carrier is the associated hypoglycemia. For this reason, insulin fragments have also been examined as carriers and have shown some promise (Fukuta et al., 1994).

Transferrin is another natural polypeptide that like, insulin, crosses the BBB by a receptor-mediated system (Broadwell et al., 1996). While native transferrin has also been considered as a potential chimeric carrier, most current efforts focus on the use of monoclonal antibodies directed at specific receptor systems. Monoclonal antibodies generated against transferrin or insulin receptors have been shown to cross the BBB (Friden et al., 1996; Pardridge et al., 1995b). Due to its rapid transcytosis the insulin receptor monoclonal antibody has been proposed as an effective drug delivery vector and potential diagnostic tool and has been demonstrated in a primate model (Wu et al., 1997). The murine OX26 monoclonal antibody to rat transferrin receptor was also successfully used as vector in order to increase the delivery of brain-derived neurotrophic factor and polyamide nucleic acids across the BBB (Pardridge et al., 1994, 1995a). On comparison, horseradish peroxidase and native transferrin complexes do cross the BBB more efficiently than horseradish peroxidase and OX26 antibody complexes. After intravenous injection in rat, some differences in intracellular and extracellular distributions in brain were observed indicating that fates in the central nervous system of delivery systems employing native transferrin and monoclonal antibodies to the transferrin receptor will differ (Broadwell *et al.*, 1996). Indications are that both insulin and transferrin receptor targeting monoclonal antibodies could be useful for noninvasive delivery of therapeutic entities to the brain.

Other endogenous substances appear to undergo receptor-mediated transcytosis at the BBB and their receptors may be considered potential delivery targets. Low-density lipoprotein (LDL) is specifically transcytosed across brain capillary endothelial cells (Dehouck *et al.*, 1997). LDL is transcytosed by a receptor-mediated system which can be upregulated by the depletion of cholesterol in the astrocyte:endothelial cell cocultures. A receptor-mediated transport process has also been reported for the amyloid beta-protein suggesting that sources outside the nervous system could contribute, at least partially, to the cerebral A beta-amyloid deposits seen in Alzheimer's patients (Podulso *et al.*, 1997). Finally, leptin, a protein synthesized by adipose tissue as a signal of satiety, penetrates the BBB via trancytosis mediated by specific receptors (Golden *et al.*, 1997).

Several proteins have been shown to undergo adsorptive transcytosis at the BBB; the precise specificity, however, and therefore targeting potential is not well developed for this group. A recent report indicates that the C-terminal structure and basicity of peptides seem to be important for the saturable uptake by an adsorptive-mediated endocytosis system at the BBB (Tamai *et al.*, 1997). The endogenous cationic protein histone is also capable of crossing the BBB via an adsorptive transcytotic system (Pardridge *et al.*, 1989). Natural proteins such as immunoglobulin G and albumin have been shown to cross the BBB with low permeability surface area products, approximately about 25 and 330 times less than the BBB permeability of transferrin and insulin, respectively (Podulso *et al.*, 1994). Although albumin transcytosis apparently does not involve a receptor-mediated system (Vorbrodt and Trowbridge, 1991), the more rapid rate of immunoglobulin G transfer suggests the potential of a transport mechanism (Podulso *et al.*, 1994). Cationized albumin, with a greater pI than the native albumin, enters more readily into the brain and seems to be a tool to study the adsorptive-mediated endocytosis at the brain endothelial cells (Kumagai *et al.*, 1987). Selected lectins undergo transcytosis at the BBB. Complexes of horseradish peroxidase and wheat germ agglutinin, for example, penetrate the brain 10 times higher than does horseradish peroxidase alone via a probable nonspecific adsorptive (Villegas and Broadwell, 1993; Banks and Broadwell, 1994).

It appears that the complexing of a therapeutic agent with a carrier recognized by brain capillary-specific endothelial cell receptors can provide a choice for some drug delivery strategies since the complex may in some instances, such as transferrin, selectively target the BBB (Friden, 1994). Due to the probable differing fates, structures, and capacities for complexation of the carrier molecules, a single carrier system will not be practical. Thus, the precise determination of chemical structures recognized by BBB receptors mediating transcytosis should continue to be a fruitful area of research.

3. SUMMARY AND FUTURE PERSPECTIVES

The restrictive permeability properties of the BBB remain a challenge to providing sufficient drug delivery into the central nervous system. Basic research in the past few decades surveyed here has revealed, however, that the BBB exhibits an array of specific transport systems for both conventional low-molecular-weight nutrients and hormones, as well as a limited selection of systems for natural peptides and proteins. Future success in exploiting these transport processes is now dependent, to some degree, on incorporating the advances in cellular, biochemical, and molecular biology into thinking about targeting natural transporters to facilitate drug delivery into the central nervous system.

ACKNOWLEDGMENTS

The authors gratefully acknowledge the support of the University of Kansas and the Fonds National Suisse de la Recherche Scientifique (BR) for funding research on this subject.

REFERENCES

Adkison, K. D., and Shen, D. D., 1996, Uptake of valproic acid into rat brain is mediated by a medium-chain fatty acid transporter, *J. Pharmacol. Exp. Ther.* **276**:1189–1200.

Agus, D. B., Gambhir, S. S., Pardridge, W. M., Spielholz, C., Baselga, J., Vera, J. C., and Golde, D. W., 1997, Vitamin C crosses the blood–brain barrier in the oxidized form through the glucose transporters, *J. Clin. Invest.* **100**:2842–2848.

Audus, K. L., and Borchardt, R. T., 1991, Transport of macromolecules across the capillary endothelium, *Handbook Exp. Pharmacol.* **100**:43–70.

Audus, K. L., Chikhale, P. J., Miller, D. W., Thompson, S. E., and Borchardt, R. T., 1992, Brain uptake of drugs: The influence of chemical and biological factors, *Adv. Drug Res.* **23**:1–64.

Audus, K. L., Ng, L., Wang, W., and Borchardt, R. T., 1996, Brain microvessel endothelial cell culture systems, *Pharmaceut. Biotechnol.* **8**:239–258.

Banks, W. A., and Broadwell, R. D., 1994, Blood to brain and brain to blood passage of native horseradish peroxidase, wheat germ agglutinin, and albumin: Pharmacokinetic and morphological assessments, *J. Neurochem.* **62**:2404–2419.

Banks, W. A., and Kastin, A. J., 1996, Passage of peptides across the blood–brain barrier: Pathophysiological perspectives, *Life Sci.* **59**:1923–1943.

Banks, W. A., and Kastin, A. J., 1997, The role of the blood–brain barrier transporter PTS-1 in regulating concentrations of methionine enkephalin in blood and brain, *Alcohol* **14**:237–245.

Banks, W. A., Kastin, A. J., Horvath, A., and Michals, E. A., 1987, Carrier-mediated transport of vasopressin across the blood–brain barrier of the mouse, *J. Neurosci. Res.* **18**:326–332.

Banks, W. A., Kastin, A. J., and Durham, D. A., 1989, Bidirectional transport of interleukin-1 across the blood–brain barrier, *Brain Res. Bull.* **23**:433–437.

Banks, W. A., Kastin, A. J., Sam, H. M., Cao, V. T., King, B., Maness, L. M., and Schally, A. V., 1994, Saturable efflux of the peptides RC-160 and Tyr-MIF-1 by different parts of the blood–brain barrier, *Brain Res. Bull.* **35**:179–182.

Banks, W. A., Wustrow, D. J., Cody, W. L., Davis, M. D., and Kastin, A. J., 1995, Permeability of the blood–brain barrier to the neurotensin8–13 analog NT1, *Brain Res.* **695**:59–63.

Banks, W. A., Kastin, A. J., Huang, W., Jaspan, J. B., and Maness, L. M., 1996, Leptin enters the brain by a saturable system independent of insulin, *Peptides* **17**:305–311.

Barrand, M. A., Bagrij, T., and Neo, S. Y., 1997, Multidrug resistance-associated protein: A protein distinct from P-glycoprotein involved in cytotoxic drug expulsion, *Gen. Pharmacol.* **28**:639–645.

Barrera, C. M., Kastin, A. J., and Banks, W. A., 1987, D-[Ala^1]-peptide T-amide is transported from blood to brain by a saturable system, *Brain Res. Bull.* **19**:629–633.

Barrera, C. M., Kastin, A. J., Fasold, M. B., and Banks, W. A., 1991, Bidirectional saturable transport of LHRH across the blood–brain barrier, *Am. J. Physiol.* **261**:E312–E318.

Begley, D. J., 1996, The blood–brain barrier: Principles for targeting peptides and drugs to the central nervous system, *J. Pharm. Pharmacol.* **48**:136–146.

Benrabh, H., and Lefauconnier, J. M., 1996, Glutamate is transported across the rat blood–brain barrier by a sodium-independent system, *Neurosci. Lett.* **210**:9–12.

Berger, W., Hauptmann, E., Elbling, L., Vetterlein, M., Kokoschka, E. M., and Micksche, M., 1997, Possible role of the multidrug resistance-associated protein (MRP) in chemoresistance of human melanoma cells, *Int. J. Cancer* **71**:108–115.

Boado, R. J., 1995, Brain-derived peptides regulate the steady state levels and increase stability of the blood–brain barrier GLUT1 glucose transporter mRNA, *Neurosci. Lett.* **197**:179–182.

Bradbury, M. W., 1997, Transport of iron in the blood–brain–cerebrospinal fluid system, *J. Neurochem.* **69**:443–454.

Brightman M. W., and Reese, T. S., 1969, Junctions between intimately apposed cell membranes in the vertebrate brain, *J. Cell. Biol.* **40**:648–677.

Broadwell, R. D., and Banks, W. A., 1993, A cell biological perspective for the transcytosis of peptides and proteins through the mammalian blood–brain fluid barriers, in: *The Blood–Brain Barrier: Cellular and Molecular Biology* (W. M. Pardridge, ed.), Raven Press, New York, pp. 165–199.

Broadwell, R. D., Baker-Cairns, B. J., Friden, P. M., Oliver, C., and Villegas, J. C., 1996, Transcytosis of protein through the mammalian cerebral epithelium and endothelium. III. Receptor-mediated transcytosis through the blood–brain barrier of blood-borne transferrin and antibody against the transferrin receptor, *Exp. Neurol.* **142**:47–65.

Burt, R. K., and Thorgeirsson, S. S., 1988, Coinduction of MDR-1 multidrug-resistance and cytochrome P-450 genes in rat liver by xenobiotics, *J. Natl. Cancer. Inst.* **80**:1383–1386.

Chikhale, E. G., Chikhale, P. J., and Borchardt, R. T., 1995a, Carrier-mediated transport of the antitumor agent acivicin across the blood–brain barrier, *Biochem. Pharmacol.* **49**:941–945.

Chikhale, E. G., Burton, P. S. and Borchardt, R. T., 1995b, The effect of verapamil on the transport of peptides across the blood–brain barrier in rats: Kinetic evidence for an apically polarized efflux mechanism, *J. Pharmacol. Exp. Ther.* **273**:298–303.

Claudio, L., Raine, C. S., and Brosnan, C. F., 1995, Evidence of persistent blood–brain barrier abnormalities in chronic-progressive multiple sclerosis, *Acta Neuropathol. Berl.* **90**:228–38.

Conn, A. R., Fell, D. I., and Steele, R. D., 1983, Characterization of alpha-keto acid transport across blood–brain barrier in rats, *Am. J. Physiol.* **245**:E253–E260.

Cornford, E. M., and Oldendorf, W. H., 1975, Independent blood–brain barrier transport systems for nucleic acid precursors, *Biochim. Biophys. Acta* **394**:211–219.

Crone, C., and Olesen, S. P., 1982, Electrical resistance of brain microvascular endothelium, *Brain Res.* **241**:49–55.

De Boer, A. G., and Breimer, D. D., 1994, The bloodbrain barrier: Clinical implications for drug delivery to the brain, *J. R. Coll. Physicians* **28**:502–506.

Dehouck, B., Fenart, L., Dehouck, M. P., Pierce, A., Torpier, G., and Cecchelli, R., 1997, A new function for the LDL receptor: Transcytosis of LDL across the blood–brain barrier, *J. Cell Biol.* **138**:877–889.

Drion, N., Lemaire, M., Lefauconnier, J. M., and Scherrmann, J. M., 1996, Role of P-glycoprotein in the blood–brain transport of colchicine and vinblastine, *J. Neurochem.* **67**:1688–1693.

Drori, S., Eytan, G. D., and Assaraf, Y. G., 1995, Potentiation of anticancer-drug cytotoxicity by multidrug-resistance chemosensitizers involves alterations in membrane fluidity leading to increased membrane permeability, *Eur. J. Biochem.* **228**:1020–1029.

Duffy, K. R., and Pardridge, W. M., 1987, Blood–brain barrier transcytosis of insulin in developing rabbits, *Brain Res.* **420**:32–38.

Fiori, A., Cardelli, P., Negri, L., Savi, M. R., Strom, R., and Erspamer, V., 1997, Deltorphin transport across the blood–brain barrier, *Proc. Natl. Acad. Sci. USA* **94**:9469–9474.

Friden, P. M., 1994, Receptor-mediated transport of therapeutics across the blood-brain barrier, *Neurosurgery* **35**:294–298.

Friden, P. M., Olson, T. S., Obar, R., Walus, L. R., and Putney, S. D., 1996, Characterization, receptor mapping and blood–brain barrier transcytosis of antibodies to the human transferrin receptor, *J. Pharmacol. Exp. Ther.* **278**:1491–1498.

Fukuta, M., Okada, H., Iinuma, S., Yanai, S., and Toguchi, H., 1994, Insulin fragments as a carrier for peptide delivery across the blood–brain barrier, *Pharmaceut. Res.* **11**:1681–1688.

Galea, E., and Estrada, C., 1992, Ouabain-sensitive choline transport system in capillaries isolated from bovine brain, *J. Neurochem.* **59**:936–941.

Gerhart, D. Z., Enerson, B. E., Zhdankina, O. Y., Leino, R. L. and Drewes, L. R., 1997, Expression of monocarboxylate transporter MCT1 by brain endothelium and glia in adult and suckling rats, *Am. J. Physiol.* **273**:E207–E213.

Germann, U. A., Ford, P. J., Shlyakhter, D., Mason, V. S., and Harding, M. W., 1997, Chemosensitization and drug accumulation effects of VX-710, verapamil, cyclosporin A, MS-209 and GF120918 in multidrug resistant HL60/ADR cells expressing the multidrug resistance-associated protein MRP, *Anticancer Drugs* **8**:141–155.

Ghersi-Egea, J. F., Leininger-Muller, B., Cecchelli, R., and Fenstermacher, J. D., 1995, Blood–brain interfaces: Relevance to cerebral drug metabolism, *Toxicol. Lett.* 82–83:645–653.

Golden, P. L., Maccagnan, T. J., and Pardridge, W. M., 1997, Human blood–brain barrier leptin receptor. Binding and endocytosis in isolated human brain microvessels, *J. Clin. Invest.* 99:14–18.

Gottesman, M. M., and Pastan, I., 1993, Biochemistry of multidrug resistance mediated by the multidrug transporter, *Annu. Rev. Biochem.* 62:385–427.

Gragera, R. R., Muniz, E., and Martinez-Rodriguez, R., 1993, Molecular and ultrastructural basis of the blood–brain barrier function. Immunohistochemical demonstration of Na+/K+ ATPase, alpha-actin, phosphocreatine and clathrin in the capillary wall and its microenvironment, *Cell. Mol. Biol.* 39:819–828.

Greenwood, J., Love, E. R., and Pratt, O. E., 1982, Kinetics of thiamine transport across the blood–brain barrier in the rat, *J. Physiol. (London)* 327:95–103.

Halmos, T., Santarromana, M., Antonakis, K., and Scherman, D., 1996, Synthesis of glucose-chlorambucil derivatives and their recognition by the human GLUT1 glucose transporter, *Eur. J. Pharmacol.* 318:477–484.

Hokari, M., Wu, H. Q., Schwarcz, R., and Smith, Q. R., 1996, Facilitated brain uptake of 4-chlorokynurenine and conversion to 7-chlorokynurenic acid, *NeuroReport* 8:15–18.

Honkanen, R. A., McBath, H., Kushmerick, C., Callender, G. E., Scarlata, S. F., Fenstermacher, J. D., and Haspel, H. C., 1995, Barbiturates inhibit hexose transport in cultured mammalian cells and human erythrocytes and interact directly with purified GLUT-1, *Biochemistry* 34:535–544.

Huang, Y., Ibrado, A. M., Reed, J. C., Bullock, G., Ray, S., Tang, C., and Bhalla, K., 1997, Coexpression of several molecular mechanisms of multidrug resistance and their significance for paclitaxel cytotoxicity in human AML HL-60 cells, *Leukemia* 11:253–257.

Izquierdo, M. A., Neefjes, J. J., Mathari, A. E., Flens, M. J., Scheffer, G. L., and Scheper, R. J., 1996, Overexpression of the ABC transporter TAP in multidrug-resistant human cancer cell lines, *Br. J. Cancer* 74:1961–1967.

Jette, L., Murphy, G. F., Leclerc, J. M., and Beliveau, R., 1995, Interaction of drugs with P-glycoprotein in brain capillaries, *Biochem. Pharmacol.* 50:1701–1709.

Joo, F., 1996, Endothelial cells of the brain and other organ systems: Some similarities and differences, *Prog. Neurobiol.* 48:255–273.

Kannan, R., Yi, J. R., Tang, D., Li, Y., Zlokovic, B. V., and Kaplowitz, N., 1996, Evidence for the existence of a sodium-dependent glutathione (GSH) transporter, *J. Biol. Chem.* 271:9754–9758.

Kavallaris, M., 1997, The role of multidrug resistance-associated protein (MRP) expression in multidrug resistance, *Anticancer Drugs* 8:17–25.

Keep, R. F., and Xiang, J., 1995, N-system amino acid transport at the blood–CSF barrier, *J. Neurochem.* 65:2571–2576.

Komura, J., Tamai, I., Senmaru, M., Terasaki, T., Sai, Y., and Tsuji, A., 1996, Sodium and chloride-dependent transport of beta-alanine across the blood–brain barrier, *J. Neurochem.* 67:330–335.

Kumagai, A. K., Eisenberg, J. B., and Pardridge, W. M., 1987, Adsorptive-mediated endocytosis of cationized albumin and a β-endorphin-cationized albumin chimeric peptide by isolated brain capillaries, *J. Biol. Chem.* **262:**15214–15219.

Lee, J. S., Scala, S., Matsumoto, Y., Dickstein, B., Robey, R., Zhan, Z., Altenberg, G., and Bates, S. E., 1997, Reduced drug accumulation and multidrug resistance in human breast cancer cells without associated P-glycoprotein or MRP overexpression, *J. Cell. Biochem.* **65:**513–526.

Lee, W. J., Hawkins, R. A., Peterson, D. R., and Vina, J. R., 1996, Role of oxoproline in the regulation of neutral amino acid transport across the blood–brain barrier, *J. Biol. Chem.* **271:**19129–19133.

Lum, B. L., and Gosland, M. P., 1995, MDR expression in normal tissues. Pharmacologic implications for the clinical use of P-glycoprotein inhibitors, *Hematol. Oncol. Clin.* **9:**319–336.

Martins, J. M., Kastin, A. J., and Banks, W. A., 1996, Unidirectional specific and modulated brain to blood transport of corticotropin-releasing hormone, *Neuroendocrinology* **63:**338–343.

Martins, J. M., Banks, W. A., and Kastin, A. J., 1997, Acute modulation of active carrier-mediated brain-to-blood transport of corticotropin-releasing hormone, *Am. J. Physiol.* **272:**E312–E319.

Ment, L. R., Stewart, W. B., Ardito, T. A., and Madri, J. A., 1995, Germinal matrix microvascular maturation correlates inversely with the risk period for neonatal intraventricular hemorrhage, *Brain Res. Dev. Brain Res.* **84:**142–149.

Miller, D. W., Han, H. Y., and Carney, D., 1997, Is the probenecid-sensitive transporter in the blood–brain barrier multidrug resistance associated protein (MRP)? *Pharmaceut. Res.* **14:**332.

Minn, A., Ghersi-Egea, J. F., Perrin, R., Leininger, B., and Siest, G., 1991, Drug metabolizing enzymes in the brain and cerebral microvessels, *Brain Res.* **16:**65–82.

Mroczkowska, J. E., Galla, H. J., Nalecz, M. J., and Nalec, K. A., 1997, Evidence for an asymmetrical uptake of L-carnitine in the blood–brain barrier *in vitro*, *Biochem. Biophys. Res. Commun.* **241:**127–131.

Nishizaki, T., Kammesheidt, A., Sumikawa, K., Asada, T., and Okada, Y., 1995, A sodium- and energy-dependent glucose transporter with similarities to SGLT1–2 is expressed in bovine cortical vessels, *Neurosci. Res.* **22:**13–22.

Oldendorf, W. H., 1971, Brain uptake of radiolabeled amino acids, amines, and hexoses after arterial injection, *Am. J. Physiol.* **221:**1629–1639.

Oldendorf, W. H., and Szabo, J., 1976, Amino acid assignment to one of three blood–brain carrier amino acid carriers, *Am. J. Physiol.* **230:**94–98.

Oldendorf, W. H., Hyman, S., Braun, L. D., and Oldendorf, S. Z., 1972, Blood–brain barrier: Penetration of morphine, codeine, heroin, and methadone after carotid injection, *Science* **178:**984.

Pardridge, W. M., 1979, Carrier mediated transport of thyroid hormones through the blood–brain barrier. Primary role of albumin bound hormone, *Endocrinology* **105:**605–612.

Pardridge, W. M, 1983, Brain metabolism: A perspective from the blood–brain barrier, *Physiol. Rev.* **63:**1481–1535.

Pardridge, W. M., 1994, New approaches to drug delivery through the blood–brain barrier, *Trends Biotechnol.* **12:**239–245.

Pardridge, W. M., Kumagai, A. K., and Eisenberg, J. B., 1987, Chimeric peptides as a vehicle for peptide pharmaceutical delivery through the blood–brain barrier, *Biochem. Biophys. Res. Commun.* **146**:307–313.

Pardridge, W. M., Triguero, D., and Buciak, J., 1989, Transport of histone through the blood–brain barrier, *J. Pharmacol. Exp. Ther.* **251**:821–826.

Pardridge, W. M., Boado, R. J., and Farrell, C. R., 1990, Brain-type glucose transporter (GLUT-1) is selectively localized to the blood–brain barrier. Studies with quantitative western blotting and *in situ* hybridization, *J. Biol. Chem.* **265**:18035–18040.

Pardridge, W. M., Kang, Y. S., and Buciak, J. L., 1994, Transport of human recombinant brain-derived neurotrophic factor (BDNF) through the rat blood–brain barrier in vivo using vector-mediated peptide drug delivery, *Pharmaceut. Res.* **11**:738–746.

Pardridge, W. M., Boado, R. J., and Kang, Y. S., 1995a, Vector-mediated delivery of a polyamide ("peptide") nucleic acid analogue through the blood–brain barrier *in vivo*, *Proc. Natl. Acad. Sci. USA* **92**:5592–5596.

Pardridge, W. M., Kang, Y. S., Buciak, J. L., and Yang, J., 1995b, Human insulin receptor monoclonal antibody undergoes high affinity binding to human brain capillaries *in vitro* and rapid transcytosis through the blood–brain barrier *in vivo* in the primate, *Pharmaceut. Res.* **12**:807–816.

Pardridge, W. M., Golden, P. L., Kang, Y. S., and Bickel, U., 1997, Brain microvascular and astrocyte localization of P-glycoprotein, *J. Neurochem.* **68**:1278–1285.

Plotkin, S. R., Banks, W. A., Waguespack, P. J., and Kastin, A. J., 1997, Ethanol alters the concentration of Met-enkephalin in brain by affecting peptide transport system-1 independent of preproenkephalin mRNA, *J. Neurosci. Res.* **48**:273–280.

Poduslo, J. F., Curran, G. L., and Berg, C. T., 1994, Macromolecular permeability across the blood–nerve and blood–brain barriers, *Proc. Natl. Acad. Sci. USA* **91**:5705–5709.

Poduslo, J. F., Curran, G. L., Haggard, J. J., Biere, A. L., and Selkoe, D. J., 1997, Permeability and residual plasma volume of human, Dutch variant, and rat amyloid beta-protein 1–40 at the blood–brain barrier, *Neurobiol. Dis.* **4**:27–34.

Polt, R., Porreca, F., Szabo, L. Z., Bilsky, E. J., Davis, P., Abbruscato, T. J., Davis, T. P., Harvath, R., Yamamura, H. I., and Hruby V. J., 1994, Glycopeptide enkephalin analogues produce analgesia in mice: Evidence for penetration of the blood–brain barrier, *Proc. Natl. Acad. Sci. USA* **91**:7114–7118.

Reese, T. S., and Karnovsky, M. J., 1967, Fine structural localization of a blood–brain barrier to exogenous peroxidase, *J. Cell. Biol.* **34**:207–217.

Reichel, A., Begley, D. J., and Ermisch, A., 1996, Arginine vasopressin reduces the blood–brain barrier transfer of L-tyrosine and L-valine: Further evidence of the effect of the peptide on the L-system transporter at the blood–brain barrier, *Brain Res.* **713**:232–239.

Reinhardt, R. R., and Bondy, C. A., 1994, Insulin-like growth factors cross the blood–brain barrier, *Endocrinology* **135**:1753–1761.

Sanchez del Pino, M. M., Peterson, D. R., and Hawkins, R. A., 1995, Neutral amino acid transport characterization of isolated luminal and abluminal membranes of the blood–brain barrier, *J. Biol. Chem.* **270**:14913–14918.

Sarkadi, B., Muller, M., Homolya, L., Hollo, Z., Seprodi, J., Germann, U. A., Gottesman, M. M., Price, E. M., and Boucher, R. C., 1994, Interaction of bioactive hydrophobic peptides with the human multidrug transporter, *FASEB J.* **8**:766–770.

Schinkel, A. H., Smit, J. J. M., van Tellingen, O., Breijnen, J. H., Wagenaar, E., Van Deemter, L., Mol, C. A. A. M., van der Valk, M. A., Robanus-Maanday, E. C., te Riele, H. P. J., *et al.*, 1994, Disruption of the mouse mdr1a P-glycoprotein gene leads to a deficiency in the blood–brain barrier and to increased sensitivity to drugs, *Cell* **77**:491–502.

Schinkel, A. H., Wagenaar, E., Mol, C. A., and van Deemter, L., 1996, P-Glycoprotein in the blood–brain barrier of mice influences the brain penetration and pharmacological activity of many drugs, *J. Clin. Invest.* **97**:2517–2524.

Sharma, R. C., Inoue, S., Roitelman, J., Schimke, R. T., and Simoni, R. D., 1992, Peptide transport by the multidrug resistance pump, *J. Biol. Chem.* **267**:5731–5734.

Sharom, F. J., Yu, X., Chu, J. W., and Doige, C. A., 1995, Characterization of the ATPase activity of P-glycoprotein from multidrug-resistant Chinese hamster ovary cells, *Biochem. J.* **308**:381–390.

Shi, F., Bailey, C., Malick, A. W., and Audus, K. L., 1993, Biotin uptake and transport across bovine brain microvessel endothelial cell monolayers, *Pharmaceut. Res.* **10**:282–288.

Shulkin, B. L., Betz, A. L., Koeppe, R. A., and Agranoff, B. W., 1995, Inhibition of neutral amino acid transport across the human blood–brain barrier by phenylalanine, *J. Neurochem.* **64**:1252–1257.

Silverman, J. A., and Schrenk, D., 1997, Hepatic cannicular membrane 4: Expression of the multidrug resistance genes in the liver, *FASEB J.* **11**:308–313.

Smith, Q. R., 1996, Brain perfusion systems for studies of drug uptake and metabolism in the central nervous system, *Pharmaceut. Biotechnol.* **8**:285–308.

Smith, Q. R., Nagura, H., Takada, Y., and Duncan, M. W., 1992, Facilitated transport of the neurotoxin, beta-*N*-methylamino-L-alanine, across the blood–brain barrier, *J. Neurochem.* **58**:1330–1337.

Spector, R., Sivesind, C., and Kinzenbaw, D., 1986, Pantothenic acid transport through the blood–brain barrier, *J. Neurochem.* **47**:966–971.

Stein, W. D., 1997, Kinetics of the multidrug transporter (P-glycoprotein) and its reversal, *Physiol. Rev.* **77**:545–590.

Stewart, P. A., Beliveau, R., and Rogers, K. A., 1996, Cellular localization of P-glycoprotein in brain versus gonadal capillaries, *J. Histochem. Cytochem.* **44**:679–685.

Takada, Y., Vistica, D. T., Greig, N. H., Purdon, D., Rapoport, S. I., and Smith, Q. R., 1992, Rapid high-affinity transport of a chemotherapeutic amino acid across the blood–brain barrier, *Cancer Res.* **52**:2191–2196.

Takanaga, H., Tamai, I., and Tsuji, A., 1994, pH-dependent and carrier-mediated transport of salicylic acid across Caco-2 cells, *J. Pharm. Pharmacol.* **46**:567–570.

Tamai, I., Senmaru, M., Terasaki, T., and Tsuji, A., 1995, Na(+)- and Cl(−)-dependent transport or taurine at the blood–brain barrier, *Biochem. Pharmacol.* **50**:1783–1793.

Tamai, I., Sai, Y., Kobayashi, H., Kamata, M., Wakamiya, T., and Tsuji, A., 1997, Structure–internalization relationship for adsorptive-mediated endocytosis of basic peptides at the blood–brain barrier, *J. Pharmacol. Exp. Ther.* **280**:410–415.

Thomas, S. A., and Segal, M. B., 1997, The passage of azidodeoxythymidine into and within the central nervous system: Does it follow the parent compound, thymidine, *J. Pharmacol. Exp. Ther.* **281**:1211–1218.

Thomas, S. A., Abbruscato, T. J., Hruby, V. J., and Davis, T. P., 1997, The entry of [D-penicillamine^{2,5}]enkephalin into the central nervous system: Saturation kinetics and specificity, *J. Pharmacol. Exp. Ther.* **280**:1235–1240.

Thorens, B., 1996, Glucose transporters in the regulation of intestinal, renal, and liver glucose fluxes, *Am. J. Physiol.* **270**:G541–G553.

Thorgeirsson, S. S., Silverman, J. A., Gant, T. W., and Marino, P. A., 1991, Multidrug resistance gene family and chemical carcinogens, *Pharmacol. Ther.* **49**:283–292.

Villegas, J. C., and Broadwell, R. D., 1993, Transcytosis of protein through the mammalian cerebral epithelium and endothelium: II. Adsorptive transcytosis of WGA-HRP and the blood–brain barriers, *J. Neurocytol.* **22**:67–80.

Vorbrodt, A. W., and Trowbridge, R. S., 1991, Ultrastructural study of transcellular transport of native and cationized albumin in cultured sheep brain microvascular endothelium, *J. Neurocytol.* **20**:998–1006.

Wigler, P. W., 1996, Cellular drug efflux and reversal therapy of cancer, *J. Bioenerg. Biomembr.* **28**:279–284.

Wu, D., Yang, J., and Pardridge, W. M., 1997, Drug targeting of a peptide radiopharmaceutical through the primate blood–brain barrier *in vivo* with a monoclonal antibody to the human insulin receptor, *J. Clin. Invest.* **100**:1804–1812.

Yamazaki, M., Fukuoka, H., Nagata, O., Kato, H., Ito, Y., Terasaki, T., and Tsjui, A., 1994, Transport mechanism of an H1-antagonist at the blood–brain barrier: Transport mechanism of mepyramine using the carotid injection technique, *Biol. Pharmaceut. Bull.* **17**:676–679.

Zerangue, N., and Kavanaugh, M. P., 1996, Interaction of L-cysteine with a human excitatory amino acid transporter, *J. Physiol.* **493**:419–423.

Zlokovic, B. V., Lipovac, M. N., Begley, D. J., Davson, H., and Rakic, L. J., 1987, Transport of leucine-enkephalin across the blood–brain barrier in the perfused guinea pig brain, *J. Neurochem.* **49**:300–305.

7

The Mammalian Facilitative Glucose Transporter (GLUT) Family

Michael J. Seatter and Gwyn W. Gould

1. INTRODUCTION

Glucose is one of the predominant sources of energy utilized by mammalian cells. The majority of cells acquire this resource via protein-dependent movement of glucose across the plasma membrane, down its chemical gradient. This process is bidirectional, specific for the D-enantiomer of glucose, and independent of any energy-requiring components, such as ATP hydrolysis or ion gradients. As will be discussed in this chapter, a family of proteins have been identified to be responsible for this process and have been labeled the GLUTs. These facilitative glucose transporters are functionally and genetically distinct from the Na^+-dependent transporters which actively accumulate glucose. The current world-wide scientific interest in facilitative transporters was ignited by the realization that glucose transport into adipose and muscle tissues of higher mammals is regulated in both acute and chronic fashions by circulating hormones, and hence defects in this transport system may underlie diseases such as diabetes mellitus or hypertension. Today we know of the existence of a family of related transport proteins which are the products of distinct genes, possess subtly different kinetic properties, and are expressed in a highly regulated, tissue-specific fashion. This rigorous control of multiple glucose transport proteins, expressed in different tissues, implies that each member is likely to play a distinct role in the regulation of whole-body glu-

---

Michael J. Seatter and Gwyn W. Gould • Division of Biochemistry and Molecular Biology, Institute of Biomedical and Life Sciences, University of Glasgow, Glasgow G12 8QQ, Scotland.
Membrane Transporters as Drug Targets, edited by Amidon and Sadée. Kluwer Academic/Plenum Publishers, New York, 1999.

cose homeostasis, and transporter dysfunction could potentially be an underlying defect in various diseases.

2. BACKGROUND

The observations that sugar transport by erythrocytes is saturable at high sugar concentrations (Widdas, 1952) and that various sugars display competitive effects with each other (reviewed in LeFevre, 1961) suggested the presence of specific transport proteins which conform to modified Michaelis–Menten kinetics. This behavior is analogous to that of simple enzymes, assuming that D-glucose is both the substrate prior to and the product after transport. The kinetics of glucose uptake was also found to vary subtly among tissues, predicting the existence of more than one facilitative transporter. That this glucose transport moiety was indeed a protein was demonstrated by Jung, who showed that the transport capacity of the erythrocyte was retained by erythrocyte ghosts, but not by erythrocyte lipids alone (Jung, 1971). This finding was reinforced by the observation that very low concentrations of two compounds, cytochalasin B and phloretin, are capable of inhibiting transport very efficiently with K_i values of 140 and 200 nM, respectively (LeFevre, 1961; Bloch, 1973; Zoccoli *et al.,* 1978).

Although protein-mediated uptake of sugar had been detected in many tissue types, all the early kinetic studies were performed on human erythrocytes, since they could be obtained easily in great quantity and be easily manipulated. The transporter protein content of human erythrocytes has since been determined to be 5% of the total membrane protein (Allard and Lienhard, 1985). The functional purification of the erythrocyte transporter was a substantial step forward and was achieved independently by two groups. The first group based their purification on the ability of reconstituted fractionated erythrocyte membrane proteins to transport glucose (Kasahara and Hinkle, 1977). The second group used the specific binding of the transport inhibitor cytochalasin B as a basis for the purification procedure (S. A. Baldwin *et al.,* 1979; J. M. Baldwin *et al.,* 1981). Both studies revealed a heterologously glycosylated integral membrane protein which migrated on sodium dodecyl sulfate–polyacrylamide gels (SDS–PAGE) as a broad band with an approximate molecular mass of 55 kDa, which was reduced to a mass of 46 kDa upon treatment with endoglycosidase H. When this protein was reconstituted into phospholipid vesicles its kinetic behavior was identical to the native erythrocyte protein (Wheeler and Hinkle, 1981). Additionally, when present in vesicles this protein binds cytochalasin B at a stoichiometric ratio of 1:1 (J. M. Baldwin *et al.,* 1981). The purification of the erythrocyte transporter led to the production of antibody probes and its partial protein sequencing. This in turn led to cloning and sequencing of the transporter from human hepatoma G2 (HepG2) cells (Mueckler, 1985). This glucose transporter has been named GLUT1. Other mem-

Mammalian Facilitative Glucose Transporter Family

bers of the facilitative glucose transporter family have been named GLUTs 2–5, in the chronological order of the isolation of cDNAs (see Section 6 below).

3. THE STRUCTURE OF THE GLUTs

3.1. Predicted Secondary Structure of GLUT1

Hydropathy analysis of the primary sequence of GLUT1 suggests the formation of 12 membrane-spanning α-helices. Five of these are predicted to be amphipathic—helices 3, 5, 7, 8, and 11—and could form hydrophilic binding regions through the membrane bilayer allowing the passage of D-glucose. The hydropathy analyses of subsequently cloned transporters display identical patterns. The amino acid sequences of the putative transmembrane helices are more highly conserved than those of the extramembranous loops. The two largest loops and the N- and C-termini are least conserved (see Fig. 1).

The N- and C-termini and the large loop between transmembrane helices 6 and 7 are predicted to lie on the cytoplasmic side of the membrane and the other large loop, between helices 1 and 2, on the extracellular side (Fig. 1). The other cytoplasmic loops are uniformly short (eight residues long) and probably serve to

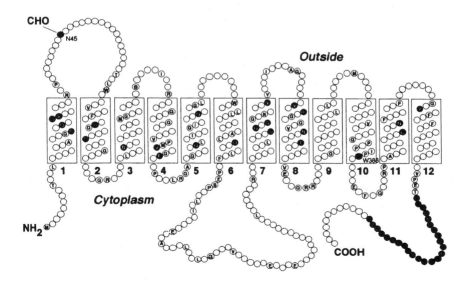

Figure 1. Schematic arrangement of the 12 membrane-spanning domain structure of the GLUTs. Residues conserved between GLUTs 1–5 are indicated by single-letter codes. Darker regions have been predicted to be important functional domains for GLUT1 (see also Table III).

restrict the movement of the transmembrane helices. The extracellular loops are not as conserved in size and sequence as those on the cytoplasmic side. This may allow the protein some flexibility of movement during transport. Constraints imposed by the short length of the cytoplasmic loops suggest that the transporter tertiary structure may be composed of two groups of six helices in a bilobular structure as observed with low-resolution electron microscopic images of *Escherichia coli* lactose permease (Kaback, 1996).

3.2. Evidence in Favor of the Secondary Structure Model

Site-specific tryptic digestion of GLUT1 yields peptide fragments corresponding to the C-terminal tail and the large central loop only when erythrocytes are permeabilized (Cairns *et al.,* 1987). Similarly, antibodies raised against peptides representing the C-terminal tail have also been found to bind to GLUT1 in inside-out vesicles (Cairns *et al.,* 1987; Davies *et al.,* 1987, 1990) confirming that these regions are intracellularly disposed (Fig. 1). Moreover, the loop linking helices 1 and 2 contains the only site of *N*-linked glycosylation in the primary sequence, asparagine-45 (Cairns *et al.,* 1987). Since the protein is heavily glycosylated, this residue must lie in an extracellular location. Substitution of this residue abolishes *N*-linked glycosylation of GLUT1 (Asano *et al.,* 1991). The use of impermeant labels, coupled with tryptic digestion, has also confirmed the external accessibility of certain predicted extracellular residues.

Perhaps the most elegant evidence in favor of the 12 membrane-spanning domain model has come from the use of so-called "glycosylation scanning mutagenesis." In this method, the sequential introduction of a sequence encoding the GLUT4 glycosylation site into each of the short loops of a GLUT1 mutant lacking its native glycosylation sequence, followed by analysis of their glycosylation status in *Xenopus* oocytes, provided a system by which Mueckler and colleagues could confirm the location of each loop (Hresko *et al.,* 1994). A change of mobility on SDS–PAGE after glycosidase digestion indicated that the inserted sequence was presented to the lumen of the endoplasmic reticulum and Golgi apparatus. Insertions into each predicted exofacial loop always produced proteins which were glycosylated, and insertions into predicted endofacial loops in each case yielded nonglycosylated proteins, thus confirming the predicted 12-transmembrane-helices model.

3.3. Biophysical Investigation of GLUT1

Infrared (IR) spectroscopy is a valuable method for investigation of protein secondary structure, assessing the relative proportions of α-helical, β-strand, and random coil conformations. Fourier-transforming infrared (FTIR) studies, which allow the analysis of protein structure in dilute aqueous media, have shown that GLUT1 protein (in lipid bilayers) contains predominantly α-helical structure, but also a proportion of β-sheet, β-turns, and random coil conformation (Alvarez *et al.*, 1987). Polarized FTIR spectroscopy results suggest that the α-helices are preferentially oriented perpendicular to the plane of the lipid bilayer, with a tilt of <38° from the membrane normal (Chin *et al.*, 1986). Circular dichroism (CD) spectroscopy detects 82% α-helix, 10% β-turns, and 8% random coil structure, but no β-sheet structure (Chin *et al.*, 1986). Hydrogen–deuterium exchange studies of purified GLUT1 in lipid bilayers (Jung *et al.*, 1986; Alvarez *et al.*, 1987) demonstrate that 80% of the polypeptide backbone of GLUT1 is readily accessible to solvent. The residual portion of the protein which is exchanged only slowly includes α-helical structure, which presumably exists in a hydrophobic environment. This suggests the presence of a hydrophilic porelike structure, possibly surrounded by amphipathic transmembrane helices.

All the above data are consistent with the 12-transmembrane α-helix structure and further suggest that D-glucose is transported across the lipid bilayer via a hydrophilic pore formed by bundles of helices which span the bilayer.

4. THE DYNAMICS OF GLUCOSE TRANSPORT

The majority of kinetic, thermodynamic, and ligand-binding studies have been carried out investigating GLUT1, due to its availability from human erythrocytes and its functional purification. Many properties measured with GLUT1 undoubtedly apply to all the members of the glucose transporter family since they all share very similar structures, although sugar specificities and kinetic parameters (Table I), targeting patterns, ligand binding, and both acute and chronic hormonal regulation of expression is found to vary among different transporter isoforms.

GLUT1 sugar transport exhibits simple Michaelis–Menton kinetics when measured under a range of different conditions measuring, for example, equilibrium exchange transport or zero-*trans* entry or exit. However, measurements made under these different conditions yield distinct values for the K_m and V_{max} of transport. For example, in erythrocytes, the K_m and V_{max} values measured for zero-*trans* efflux of glucose are some 10-fold greater than values for zero-*trans* influx. Hence, in intact erythrocytes, the transporter is said to be asymmetric. GLUT1 also displays the phenomenon of *trans*-acceleration. This describes the ability of unlabeled glucose on one side of the membrane to stimulate the transport of radiolabeled glucose from the other (*trans*) side. This is demonstrated by differences in the rates of glucose transport when measured under equilibrium exchange or zero-*trans* conditions. These properties may be important for GLUT1's function as the "housekeeping" transporter (see below).

Table I.
Kinetic Parameters and Substrate Specificities of the Human GLUTs[a]

Transporter	K_m (3-O-MG) (mM)	$T_{1/2}$ (min^{-1})	K_i Cyto B (μM)	K_m (D-Gal) (mM)	K_m (D-Fruct)	Other substrates?
GLUT1	~20	10,000–27,000	0.1–0.2	~17	>5M	Mannose
GLUT2	~42	High	~2	>50	66 mM	Mannose
GLUT3	~10	51,000	~0.4	8.5	—	Mannose
GLUT4	~2	~25,000	0.1 - 0.2	>50	—	Mannose
GLUT5	—	—	Not sensitive	—	6–14 mM	—

[a]Shown are experimentally derived values for affinity for 3-O-methylglucose, D-galactose, and D-fructose (where appropriate), together with estimates of the K_i for cytochalasin B and turnover numbers ($T_{1/2}$) where available. (See Gould and Holman, 1993; Baldwin, 1993 for details.)

4.1. The Alternating Conformation Model

The most popular kinetic model for transport is the alternating conformation model (Figs. 2 and 3). This model predicts a single glucose-binding site which is exposed either extracellularly (T_o) or intracellularly (T_i), but never both simultaneously. Both the unloaded (T_o, T_i) and loaded (T_oS, T_iS) sites can reorient to face the opposite side of the plasma membrane. It is the reorientation of T_oS to T_iS which allows the movement of glucose into the cell. Note that this mechanism applies to substrate influx and efflux, but net flow will be determined by the substrate concentration. Numerous studies have suggested that this reorientation of the substrate-binding site occurs as a consequence of a conformational change within the protein, and this has been measured using a number of biophysical approaches such as fluorescence changes (see below). In the scheme shown, the rate constants which govern substrate binding to T_o or T_i differ, as do the rates of transporter reorientation, giving rise to the observed kinetic asymmetry. Furthermore, the rate constants which govern the reorientation of the empty transporter are less than those which govern the reorientation of the loaded transporter, accounting for the phenomenon of *trans*-acceleration.

Additional evidence in favor of this simple model has come from pre-steady-state analyses of glucose uptake in erythrocytes (Lowe and Walmsley, 1986; Walmsley and Lowe, 1987; Lowe et al., 1991). These experiments use specific inhibitors of the transport to follow a single half-turnover of the transport. So, for example, incubation of cells with the membrane-impermeant substrate analogue maltose should result in this nontransported disaccharide binding to the outward-facing substrate-binding site, which, according to the single-site model, should then recruit all the transporters to the T_o conformation as the outward-facing substrate-binding sites are filled. If one subsequently dilutes away the inhibitor in the

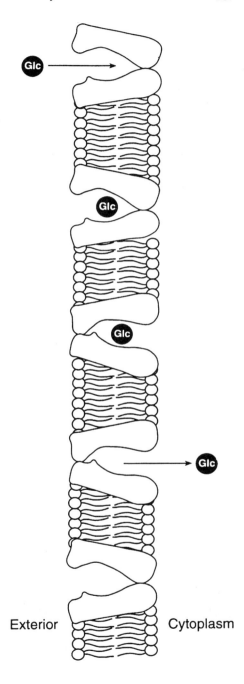

Figure 2. Schematic model of the alternating conformational model of GLUT function. For explanation, see text.

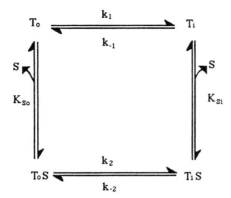

Figure 3. Simple kinetic scheme for GLUT1. Here T_o and T_i represent the unloaded transporter in an outward-facing conformation and an inward-facing conformation, respectively. T_oS and T_iS represent the transporter with bound substrate (D-glucose) in an outward-facing conformation and inward-facing conformation, respectively. K_{So} and K_{Si} represent the dissociation constants for substrate (D-glucose) at the exofacial and endofacial binding sites, respectively. Rate constants for the reorientation transporter are denoted as $k_1, k_{-1}, k_2,$ and k_{-2}. Values of these constants for GLUT1 at 0°C have been determined (see text). The rate of reorientation of loaded transporter from an external conformation to an internal one is the fastest step, $k_2 = 1113 \pm 498$ sec^{-1} compared to $k_{-2} = 90.3 \pm 3.47$ sec^{-1}. The slowest step is the reorientation of the unloaded carrier from inward-facing to outward-facing $k_{-1} = 0.726 \pm 0.498$ sec^{-1}. The rate of movement of unloaded transporter from an external conformation to an internal one is $k_1 = 12.1 \pm 0.98$ sec^{-1}. It should be noted that the rates of substrate binding and dissociation greatly exceed the reorientation rates. Note also that the observed asymmetry is reduced somewhat at room temperature. (For details see Allard and Lierhard, 1985.)

presence of glucose, there is a rapid single half-turnover of the transporters to the inward-facing conformation. Thus, before steady state is achieved there is a burst of uptake the magnitude of which reflects the numbers of functional transporters in the cells. This rate of reaction can be measured as a shift in the fluorescence absorption spectra.

4.2. Oligomerization of GLUT1

Several studies of GLUT1 in its native erythrocyte environment apparently demonstrated the simultaneous occupancy of both inward- and outward-facing substrate-binding sites. Clearly, this situation is at odds with the alternating conformational model described above, a central tenet of which is that there is only a single binding site which exists in one of two mutually exclusive conformations (see, for example, Fig. 2). This situation was further complicated by the inability to replicate results obtained in erythrocytes when using purified recombinant GLUT1 (Gorga and Lienhard, 1982; Appleman and Lienhard, 1989). One clue to this apparent dichotomy was suggested by the finding that the application of size-exclusion chromatography and sucrose gradient ultracentrifugation to cholate-solubilized GLUT1, and reconstitution of the resulting complexes into proteoliposomes, leads to the formation of transporter complexes which display D-glucose-inhibitable cytochalasin B binding (Hebert and Carruthers, 1991, 1992). The sizes of these particles indicate that GLUT1 can exist as a tetramer or a dimer under different reducing conditions. Further hydrodynamic studies, chemical cross-linking, and use of conformational-specific antibodies suggest that the tetramer binds 1 mole of cytochalasin B/2 moles GLUT1, presenting only two binding sites per tetramer to D-glucose; and the dimer binds 1 mole of cytochalasin B/mole GLUT1, able to present all binding sites to D-glucose simultaneously (Hebert and Carruthers, 1991, 1992).

Carruthers has suggested that the tetramer is the native form of transporter in erythrocytes, which is formed of two dimers stabilized by intramolecular disulfide bonds (Hebert and Carruthers, 1991, 1992). He suggests that the tetrameric transporter isoform is two- to eightfold catalytically more efficient than dimeric forms, and that discrepancies in previous kinetic studies with GLUT1 arise from the fact that these have been performed on both erythrocytes (tetrameric form) and detergent-treated reconstituted transporters (dimeric form). Carruthers proposes a model to explain this difference in efficiency between different forms. In the dimeric form, each subunit is independent of the actions of the other, and thus it displays no allosteric interactions, each subunit behaving in such a manner as explained by the single-site alternating conformation model. On the other hand, the tetramer is composed of two dimers which are bound to each other in such a way that they adopt a pseudo-D2 symmetry, and the active region of each subunit is conformationally constrained by the isomerization of the neighbor such that the binding sites are arranged in an antiparallel manner. Thus each "dimer" in the tetrameric structure presents one exofacial and one endofacial binding site, and the tetramer displays characteristics of a multisite, allosteric transporter which is consistent with the two-site fixed-site model. Carruthers speculates that the tetramer is intrinsically more efficient since the conformation change induced by transport of glucose by one subunit causes the reorientation of the neighboring unloaded subunit, overcoming the large energy barrier required for this step (Lowe and Walmsley, 1986) in dimers and monomers. Presumably this rate-limiting reorientation step would not occur under equilibrium exchange conditions, so that the steady-state kinetics observed for both tetramers and dimers would be identical.

5. TRANSPORTER PHOTOAFFINITY LABELING

The location of the D-glucose-binding sites is unknown, but various studies have used photoaffinity ligands whose binding both inhibits glucose transport and is inhibited by the presence of glucose. Given that the site of photoaffinity labeling does not merely sterically hinder glucose binding, whether directly or by affecting transporter conformation, and assuming that the site of photoaffinity binding and the site of reversible inhibition of glucose transport are the same, it has proved possible to use limited proteolysis to locate the approximate location of the glucose-binding site.

One such ligand is the fungal metabolite cytochalasin B, which binds GLUT1 in a D-glucose-sensitive manner at the ratio of one molecule per polypeptide chain (S. A. Baldwin and Lienhard, 1989) with a K_d of 120 nM (Zoccoli *et al.,* 1978). It binds GLUT2 with relatively weak affinity (K_d of ~1 μM) (Axelrod and Pilch, 1983), and does not appreciably bind human GLUT5 (Burant *et al.,* 1992). Cytochalasin B binds GLUT1 at the intracellular surface of the membrane. Photolabeling with [^{3}H]cytochalasin B after tryptic cleavage yields a labeled fragment corresponding to residues 270–456, which comprises most of the C-terminal half of the protein (Cairns *et al.,* 1987). The site of photoaffinity labeling has been proposed to lie within residues 389–412, which comprises the loop connecting helices 10 and 11, and helix 11 (Saravolac and Holman, 1997). Recent studies suggest that Trp^{388} is also included in this sequence (Inukai *et al.,* 1994). This sequence is believed to contain the endofacial glucose-binding site. The D-glucose-inhibitable binding of cytochalasin B and forskolin derivatives to the GLUT1 (and other bacterial transporters) endofacial binding site can be envisioned when one realizes that the oxygen atoms of these molecules are superimposable with those of D-glucose, suggesting these inhibitors might interact with the binding site via hydrogen bonding. Forskolin has been found to inhibit D-glucose transport by GLUT1. It is also found to bind to the intracellular surface of the membrane and bind GLUT1 in a cytochalasin B- and D-glucose-inhibitable manner (reviewed in Baldwin, 1993). Therefore it binds close to the cytochalasin B-binding site. Proteolytic digestion and labeling with the photoactivatable derivative [^{3}H]IAPS-forskolin 3-iodo-4-azidopherethylamido-7-0-succinyldeacetyl followed by proteolytic and chemical cleavage has suggested that it binds within helix 10, that is, residues 369–389, which may also comprise part of the endofacial binding site.

The requirement for a membrane-impermeant affinity label to participate in the quantitation of cell surface glucose transporters led Holman and colleagues to develop a series of bis-hexoses with the rationale that these should bind the exofacial D-glucose-binding site. The bis-mannose derivative 1,3-bis-(D-mannos-4-yloxy)-2-propylamine (BMPA), a hydrophilic impermeable compound consisting of two mannose moieties, bisected by an amino group which can be linked to a photoactivatable group, was found to be useful. Although derivatives of this were found to bind to GLUT1 with lower affinity than cytochalasin B or IAPS-forskolin, they bound with greater affinity than D-mannose itself. The derivative 2-*N*-(4-azidosalicoyl)-1,3-bis(D-mannos-4-yloxy)propyl-2-amine (ASA-BMPA), which possesses an azidosalicoyl photoactivatable group, was found to label selectively the extracellular surface of GLUT1. This compound was found to bind to GLUT1 at the region between residues 347–388, probably at the extracellular end of helix 9, which was proposed to be the exofacial glucose-binding site (Gould and Holman, 1993).

Mammalian Facilitative Glucose Transporter Family

However, ASA-BMPA was found not to label selectively plasma membrane glucose transporters when they were in low abundance. Therefore a new photoaffinity probe, 2-N-[(4-azi-2,2,2-trifluoroethyl) benzoyl]-1,3-bis-(D-mannos-4-yloxy)-2-propylamine (ATB-BMPA), was developed. This compound, which incorporates an azitrifluoro-ethylbenzoyl reactive group, was found to exhibit better selectivity than its predecessor, allowing its use even with membrane fractions. The site of ATB-BMPA labeling has been localized to residues 301–330, within transmembrane helix 8 (reviewed in Gould and Holman, 1993).

Thus, the site of the exofacial binding site is proposed to include helices 8 and 9, whereas the endofacial binding site is proposed to include helices 10 and 11. Therefore there seems to be some structural separation between the endofacial and exofacial binding sites. The results of several mutagenesis studies have supported the proposed position of the endofacial binding site, and further suggested that helix 7 is also included as part of the exofacial binding site.

6. THE TISSUE-SPECIFIC DISTRIBUTION OF GLUCOSE TRANSPORTERS

Members of the GLUT family exhibit tissue-specific patterns of expression which can be explained in part on the basis of their functional properties (Table II). Below, we briefly summarize the salient features of each member and outline their proposed function in whole-body glucose homeostasis. A comparison of the sequences of GLUTs 1–5 is presented in Table III.

6.1. GLUT1

The purification of this protein led to the generation of antibodies which, together with partial sequence information, resulted in the isolation of a cDNA clone for the transporter in 1985. This was the first eukaryotic membrane transporter to be cloned (Mueckler *et al.*, 1985; Birnbaum *et al.*, 1986). Utilizing both cDNA and antibody probes, many subsequent studies have demonstrated that both the GLUT1 protein and mRNA are present in many tissues and cells (Flier *et al.* 1987a,b; Gould and Bell, 1990). In humans it is expressed at highest levels in brain and erythrocytes, but is also enriched in the cells of the blood-tissue barriers such as the blood–brain/nerve barrier, the placenta, and the retina (Calderhead and Lienhard, 1988; *Harik et al.*, 1990a,b; Pardridge *et al.*, 1990; Farrell and Pardridge, 1991; Farrell *et al.*, 1992). GLUT1 protein is found at low levels in muscle and fat, but these tissues express much higher levels of GLUT4, which is probably responsible for the bulk of glucose transport into these tissues after insulin stimula-

Table II.
Tissue Distribution of the Human GLUTs

GLUT1	Blood–tissue barriers (placenta, retina, blood–brain barrier) erythrocytes, low levels in many tissues
GLUT2	Liver, pancreatic beta cells, adsorptive epithelia
GLUT3	Brain, nerve (high levels)
GLUT4	Fat, muscle and heart (insulin-responsive tissues)
GLUT5	Spermatozoa, small intestine epithelia, low levels other tissues (such as fat and muscle)

tion (see Section 6.4). GLUT1 is found only at very low levels in the liver, which instead has high levels of GLUT2. Hence, the tissues responsible for maintaining whole-body glucose homeostasis all contain high levels of transporter isoforms other than GLUT1; in these tissues, the expression of GLUT1 is probably of secondary importance. The role of GLUT1 in peripheral tissues most likely involves a housekeeping function, providing fat and muscle with a constant low level of glucose required for resting cellular homeostasis. This function is discussed below, but it is probably the asymmetry of transport by GLUT1 enabling the effective unidirectional transport of glucose under hypoglycemic conditions (see Sections 3 and 4 above).

It is also interesting to note that GLUT1 levels are increased after transformation, in cellular growth, and in response to stress. It is well established that transformation of cell culture lines results in a pronounced elevation of GLUT1 protein and mRNA levels, and that this general phenomenon is observed for all cell culture lines (Mueckler *et al.*, 1985; Birnbaum *et al.*, 1987; Flier *et al.*, 1987a,b; Hiraki *et al.*, 1989). Additionally, many mitogens stimulate GLUT1 transcription, and glucose starvation can also stimulate GLUT1 expression (Mueckler *et al.*, 1985; Birnbaum *et al.*, 1987; Flier *et al.*, 1987a,b; Williams and Birnbaum, 1988; Hiraki *et al.*, 1989; Merrall *et al.*, 1993; Gould *et al.*, 1995).

6.2. GLUT2

Initial studies investigating glucose transport in hepatocytes measured kinetic parameters which were radically different from those of erythrocytes, having much higher K_m and V_{max} values. The discovery that these cells contained very little GLUT1 suggested the presence of a distinct transporter (Craik and Elliott, 1979; Axelrod and Pilch, 1983; Elliott and Craik, 1983).

The isolation of the GLUT1 cDNA led to its use as a homologous probe by two independent groups to screen hepatocyte libraries under conditions of low stringency (Fukumoto *et al.*, 1988; Thorens *et al.*, 1988). Both groups isolated a

Table III.
Sequence alignments of GLUTs 1–5[a]

gt1	. . . . M	EPSSKKLTGR	LMLAVGGAVL	G.SLQFGYNT	GVINAP...Q	KVIEEFYNQT
gt2	. . . .	.MTEDKVTGT	LVFTVITAVL	G.SFQFGYDI	GVINAP...Q	QVIISHYRHV
gt3	. . . .	.MGTQKVTPA	LIFAITVATI	G.SFQFGYNT	GVINAP...E	KIIKEFINKT
gt4	. . . . MPS	EPPQQRVTGT	LVLAVFSAVL	G.SLQFGYNI	GVINAP...Q	KVIEQSYNET
gt5	. . . . MEQQDQ	SMKEGRLTLV	LALATLIAAF	GSSFQYGYNV	AAVNSP...A	LLMQQFYNET
				←—— TM1 ——→		
gt1	WVHRYG....	....	PKPTPWAEEE	ESILPTTLTT	LWSLSVAIFS	VGGMIGSFSV
gt2	LGVPLDDRKA	INNYVINSTD	....	TVAAAQLITM	LWSLSVSSFA	VGGMTASFFG
gt3	LTDKGN....	....	....	APPSEVLLTS	LWSLSVAIFS	VGGMIGSFSV
gt4	WLGRQG....	....	....	SSIPPGTLTT	LWALSVAIFS	VGGMISSFLI
gt5	YYGRTG....	....	.....PEGP	EFMEDFPLTL	LWSVTVSMFP	FGGFIGSLLV
					←——— TM2 ———→	
gt1	GLFVNRFGRR	NSMLMMNLLA	FVSAVLMGFS	KLGKSFEMLI	LGRFIIGVYC	GLTTGFVPMY	VGEVSPTAFR
gt2	GWLGDTLGRI	KAMLVANILS	LVGALLMGFS	KLGPSHILII	AGRSISGLYC	GLISGLVPMY	IGEIAPTALR
gt3	GLFVNRFGRR	NSMLIVNLLA	VTGGCFMGLC	KVAKSVEMLI	LGRLVIGLFC	GLCTGFVPMY	IGEISPTALR
gt4	GIISQWLGRK	RAMLVNNVLA	VLGGSLMGLA	NAAASYEMLI	LGRFLIGAYS	GLTSGLVPMY	VGEIAPTHLR
gt5	GPLVNKFGRK	GALLFNNIFS	IVPAILMGCS	RVATSFELII	ISRLLVGICA	GVSSNVVPMY	LGELAPKNLR
		←—— TM3 ——→				
gt1	GALGTLHQLG	IVVGILIAQV	FGLDSIMGNK	DLWPLLLSII	FIPALLQCIV	LPFCPESPRF	LLINRNEENR
gt2	GALGTFHQLA	IVTGILISQI	IGLEFILGNY	DLWHILLGLS	GVRAILQSLL	LFFCPESPRY	LYIKLDEEVK
gt3	GAFGTLNQLG	IVVGILVAQI	FGLEFILGSE	ELWPLLLGFT	ILPAILQSAA	LPFCPESPRF	LLINRKEEEN
gt4	GALGTLNQLA	IVIGILIAQV	LGLESLLGTA	SLWPLLLGLT	VLPALLQLVL	LPFCPESPRY	LYIIQNLEGP
gt5	GALGVVPQLF	ITVGILVAQI	FGLRNLLANV	DGWPILLGLT	GVPAALQLLL	LPFFPESPRY	LLIQKKDEAA
		←—— TM5 ——→		←—— TM6 ——→		
gt1	AKSVLKKLRG	TADVTHDLQE	MKEESRQMMR	EKKVTILELF	RSPAYRQPIL	IAVVLQLSQQ	LSGINAVFYY
gt2	AKQSLKRLRG	YDDVTKDINE	MRKEREEASS	EQKVSIIQLF	TNSSYRQPIL	VALMHVAQQ	FSGINGIFYY

(continued)

Table III. (Continued)
Sequence alignments of GLUTs 1–5[a]

gt3	AKQILQRLWG	TQDVSQDIQE	MKDESARMSQ	EKQVTVLELF	RVSSYRQPII	ISIVLQLSQQ	LSGINAVFYY
gt4	ARKSLKRLTG	WADVSGVLAE	LKDEKRKLER	ERPLSLLQLL	GSRTHRQPLI	IAVVLQLSQQ	LSGINAVFYY
gt5	AKKALQTLRG	WDSVDREVAE	IRQEDEAEKA	AGFISVLKLF	RMRSLRWQLL	SIIVLMGGQQ	LSGVNAIYYY
						←——— TM7 ———→	gt1
	STSIFEKAGV				QQ...PVYAT	IGSGIVNTAF	
	TVVSLFVVER				AGRRTLHLIG	LAGMAGCAIL	
	MTIALALLEQ						
gt2	STSIFQTAGI	SK...PVYAT	IGVGAVNMVF	TAVSVFLVEK	AGRRSLFLIG	MSGMFVCAIF	MSVGLVLLNK
gt3	STGIFKDAGV	QE...PIYAT	IGAGVVNTIF	TVVSLFLVER	AGRRTLHMIG	LGGMAFCSTL	MTVSLLLKDN
gt4	STSIFETAGV	GQ...PAYAT	IGAGVVNTVF	TLVSVLLVER	AGRRTLHLLG	LAGMCGCAIL	MTVALLLER
gt5	ADQIYLSAGV	PE.EHVQYVT	AGTGAVNVVM	TFCAVFVVEL	LGRRLLLLG	FSICLIACCV	LTAALALQDT
			←—— TM8 ——→			←—— TM9 ——	
gt1	LPWMSYLSIV	AIFGFVAFFE	VGPGPIPWFI	VAELFSQGPR	PAAIAVAGFS	NWTSNFIVGM	CFQYVEQLCG
gt2	FSWMSYVSMI	AIFLFVSFFE	IGPGPIPWFM	VAEFFSQGPR	PAALAIAAFS	NWTCNFIVAL	CFQYIADFCG
gt3	YNGMSFVCIG	AILVFVAFFE	IGPGPIPWFI	VAELFSQGPR	PAAMAVAGCS	NWTSNFLVGL	LFPSAAHYLG
gt4	VPAMSYVSIV	AIFGFVAFFE	IGPGPIPWFI	VAELFSQGPR	PAAMAVAGFS	NWTSNFIIGM	GFQYVAEAMG
gt5	VSWMPYISIV	CVISYVIGHA	LGPSPIPALL	ITEIFLQSSR	PSAFMVGGSV	HWLSNFTVGL	IFPFIQEGLG
		←—— TM10 ——→			←——— TM11 ————→		
gt1	PYVFIIFTVL	LVLFFIFTYF	KVPETKGRTF	DEIASGF..R	QGGASQ.SDK	TPEELFHPLG	ADSQV.....
gt2	PYVFFLFAGV	LLAFTLFTFF	KVPETKGKSF	EEIAAEFQKK	SGSAHR.P..	KAAVEMKFLG	ATETV.....
gt3	AYVFIIFTGF	LITFLAFTFF	KVPETRGRTF	EDITRAFEGQ	AHGADR.SGK	DGVMEMNSIE	PAKETTTNV..
gt4	PYVFLLFAVL	LLGFFIFTFL	RVPETRGRTF	DQISAAFHRT	PSLLEQ.EVK	.PSTELEYLG	PDEND.....
gt5	PYSFIVFAVI	CLLTTIYIFL	IVPETKAKTF	IEINQIFTKM	NKVSEVYPEK	EELKELPPVT	SEQ.......
	←——— TM12 ———→						

[a] Shown is a protein sequence alignment of GLUTs 1–5 (gt1 to gt5, respectively). The approximate position of the transmembrane helices is illustrated. Residues identified as being important for transporter function are highlighted in bold. (For further details see Baldwin et al., 1993 for example).

cDNA whose predicted amino acid sequence exhibited 55% identity and 80% homology to GLUT1. Furthermore, hydropathy plots of the predicted GLUT1 and GLUT2 proteins are virtually superimposable, suggesting that the transporters share the same overall structure.

Immunocytochemical and Northern analyses have shown that GLUT2 is expressed at highest levels in liver, in the β-cells of the islets of Langerhans of the pancreas, and in the adsorptive epithelia of the small intestine and kidney proximal tubule (see Fig. 4). Functional studies of GLUT2 heterologously expressed in a range of cells have subsequently shown that this protein has a supraphysiological K_m for glucose, has a high transport capacity, and exhibits the ability to transport D-fructose in addition to D-glucose (a unique feature of this isoform; see Table I).

In the liver, GLUT2 is found in the sinusoidal plasma membrane of the hepatocytes, but there is a gradient of expression with highest levels observed in the periportal area compared to the perivenous regions (Thorens et al., 1990). The high capacity of GLUT2 for glucose should allow it to transport glucose efficiently across the sinusoidal plasma membranes in either direction under varying blood glucose concentrations, allowing the liver to respond quickly and effectively to altered glycemic conditions.

The expression of GLUT2 in the pancreatic β-cells, but not the surrounding α- or δ-cells (Orci et al., 1989), suggests that it has a role in the secretion of insulin. Elevated blood glucose levels lead to an increase in the intracellular concentration of some metabolite of glucose within the b-cells which triggers the secretion of stored insulin. The specific expression of GLUT2 in these cells suggests its importance, as the high transport capacity is crucial for the rapid equilibration of the cytosol with glucose from the extracellular space, a vital property for glucose-sensing cells. The importance of GLUT2 has been further identified by the demonstration of reconstitution of glucose-stimulated insulin secretion from a cell line expressing GLUT2 but not GLUT1 (Hughes et al., 1992). However, this is the subject of much debate, centered around the relative importance of GLUT2 and the unique hexokinase isoform expressed in β-cells, glucokinase (see, for example, Unger, 1991; Milburn et al., 1993).

GLUT2 is also found at high levels in the absorptive epithelia of the small intestine and kidney, but its expression is confined to the basolateral membranes (Thorens et al., 1990; Miyamoto et al., 1992), suggesting that this protein acts to transport glucose and fructose down their concentration gradients into the bloodstream (Fig. 4). This is possibly due to the concentration of glucose in these cells by active transepithelial uptake via the sodium-dependent glucose transporter SGLT1 (Haase et al., 1990; Hwang et al., 1991). Again, the high transport capacity of GLUT2 is an important characteristic, allowing rapid efflux of the accumulated glucose into the blood.

6.3. GLUT3

The low-stringency hybridization method detected another novel transporter-like cDNA, GLUT3, in a human fetal muscle library, which was subsequently cloned (Kayano *et al.,* 1990; Nagamatsu, 1992, 1993). However, Northern blot analysis revealed that in humans this transporter was barely detectable in adult skeletal muscle, but was rather expressed at high levels in the brain and neural tissue, and at lower levels in fat, kidney, liver muscle tissue, and the placenta. Analysis of the sites of expression of GLUT3 mRNA for this isoform in monkey tissues suggested that the transcript was widely expressed (Yano *et al.,* 1991). In contrast, in rodents the transcript was found only in neural tissue.

Antipeptide antibodies specific for mouse GLUT3 have demonstrated that the expression of GLUT3 is restricted to brain and neural cell lines and is not immunologically detectable in highly purified mouse muscle, liver, or fat membranes (Gould *et al.,* 1992; Maher *et al.,* 1992; Shepherd *et al.,* 1992b; Maher, 1995). Immunological analysis of human tissues revealed the presence of GLUT3 at high levels in the brain, with much lower amounts present in the placenta, liver, heart, and kidney, but not from three different muscle groups, soleus, vastus lateralis, and psoas major (Shepherd *et al.,* 1992b), although GLUT3 mRNA levels in these tissues are relatively abundant. This disparity could be due to significant neural contamination of the tissue sections used to prepare the mRNA for the Northern analysis, or these tissues may be exhibiting posttranscriptional regulation of GLUT3.

GLUT3 protein expression levels are highest in brain and neural tissue, both of which exhibit a high glucose demand. Indeed, the brain is dependent upon glucose as its energy source. For glucose to be used by the CNS, it must first be transported across the walls of the cerebral blood vessels before encountering the plasma membranes of glial cells and neurons. In the peripheral nervous tissues, glucose must cross the paranodal ion-channel and diffusion barriers by traversing the Schwann cells which envelop the axons (Sima *et al.,* 1986). It is well established that the major glucose transporter expressed at the blood–nerve and blood–brain barrier is GLUT1. Studies of rat peripheral nerve, for example, have demonstrated that GLUT1 is expressed in the paranodal region and incisures of Schwann cells in rat peripheral nerve, suggesting that GLUT1 is responsible for transport across the Schwann cells from the endoneurial (Magnani *et al.,* 1996). GLUT3 is expressed in the plasma membranes of glial cells and neurons in both the central and peripheral nervous systems (Maher *et al.,* 1994; Maher, 1995; Magnani *et al.,* 1996).

In brain, under normal conditions the capacity of hexokinase for glucose is considerably greater than the capacity of the glucose transport systems in this tis-

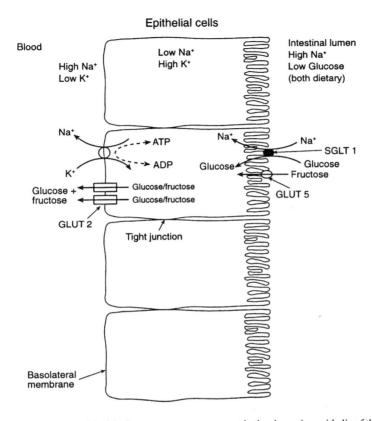

Figure 4. Schematic model of the hexose transporters present in the absorptive epithelia of the small intestine. Note the polarized distribution of the sodium-dependent transporter (SGLT1), GLUT2, and GLUT5.

sue. However, under conditions either of high glucose demand or hypoglycemia, the expression of a transporter in the brain with a low K_m for glucose may be required to utilize efficiently low concentrations of blood glucose. The measured K_m for both 2-deoxyglucose and 3-O-methyl-D-glucose for GLUT3 is relatively low (Gould et al., 1991; Asano et al., 1992; Colville et al., 1993; Maher et al., 1996), and it has been estimated that the V_{max} of this isoform is about 50% that of GLUT1 when expressed in CHO cells. Hence the function of GLUT3 may be more relevant under hypoglycemic conditions, whereas the GLUT1 may be dominant under "normal" conditions. It is likely that both GLUT1 and GLUT3 play an important role in the maintenance of glucose supply to the nervous system.

6.4. GLUT4

It has been demonstrated that glucose transport by rat adipocytes increases 30-fold in the presence of insulin. This change was found to be accompanied by a massive redistribution of cytochalasin B-binding sites from unidentified intracellular membranes to the plasma membrane fraction, representing a translocation of glucose transporters in response to insulin (Cushman and Wardzala, 1980; Suzuki and Kono, 1980). This effect was reversible upon insulin removal. Initially, this transporter was believed to be GLUT1, but the "translocation" of immunologically detectable GLUT1 to the plasma membrane was too low to account for the large increase in insulin-stimulated glucose transport, suggesting the possibility that the insulin response was mediated by another, unidentified transporter (Calderhead and Lienhard, 1988; James et al., 1988; Oka et al., 1988).

The isolation of a cDNA encoding a fourth distinct glucose transporter-like protein was reported independently by several laboratories (Birnbaum, 1989; Charron et al., 1989; Fukumoto et al., 1989; James et al., 1989; Kaestner et al., 1989). GLUT4 was subsequently shown to be expressed at high levels in adipocytes and both skeletal and cardiac muscle, those tissues which display insulin-dependent regulation of glucose transport.

In human adipose cells the response to insulin is much smaller than in rat adipocytes, approximately two- to fourfold as opposed to 30-fold (Kozka et al., 1995). Insulin has been shown to increase glucose transport activity in rat muscle by sevenfold but only by twofold in human muscle (Dohm et al., 1988; Kozka et al., 1995).

Kinetic studies (Ploug et al., 1987; Dohm et al., 1988; Palfreyman et al., 1992; Ploug et al., 1993; Kozka et al., 1995) have shown that the K_m of this transporter expressed in adipocytes is of the order of $\sim$2–5 mM and that the major effect of insulin is to increase the V_{max} of glucose uptake, which represents the increased number of transporters at the cell surface. The relatively low K_m will ensure that it operates close to the V_{max} over the normal range of blood glucose concentrations, ensuring the rapid disposal of postprandial blood glucose into the primary peripheral glucose-storing tissues (muscle and fat).

The crucial feature of GLUT4 is that in the absence of insulin, this transporter is sequestered into a subcellular compartment(s). In brown adipocytes, for example, less than 2% of the total cellular GLUT4 is present at the cell surface (Slot et al., 1991a,b, 1997). In response to insulin, there is a rapid and large mobilization of the sequestered transporter to the plasma membrane, resulting in large increases in the number of functional GLUT4 molecules at the cell surface. Defects in this translocation event are proposed to underlie insulin resistance (Kahn 1992a,b; Garvey et al., 1993; Livingstone et al., 1995). GLUT4 has been extensively studied in terms of its regulation by insulin and terms of its subcellular trafficking. The interested reader is referred to recent reviews on this subject (Kahn, 1992a,b; Gould and Holman, 1993; James and Piper, 1993, 1994; Holman and Cushman, 1994; Holman *et al.*, 1994).

6.5. GLUT5

Yet another member of the human GLUT family was isolated in 1990. GLUT5 cDNA encodes a protein of 501 amino acids which is the most divergent of all the transporter isoforms cloned to date, being only 40% identical to the other known isoforms (Kayano *et al.*, 1990). Northern analysis located highest levels of mRNA in the duodenal and jejunal regions in small intestine, and at lower levels in ileum and colon (Davidson *et al.*, 1992). Peptide-specific antibodies localized the protein to the apical membranes of the absorptive epithelia (Davidson *et al.*, 1992) (Fig. 4). Further immunoblot analyses have demonstrated that this protein is also expressed in a range of other human tissues, including muscle (soleus, rectus abdominus, psoas major, and vastus lateralis), brain and adipose tissue, and kidney and testis (Burant *et al.*, 1992; Shepherd *et al.*, 1992b).

In rats and rabbits, GLUT5 mRNA and protein are expressed in the small intestine, with highest levels in the proximal regions, but not in testis, muscle, or adipose tissue (Rand *et al.*, 1993; Miyamoto *et al.*, 1994)

Expression of human GLUT5 cRNA in oocytes revealed that the protein was incapable of measurable glucose transport, but was found instead to be a high-affinity D-fructose transporter ($K_m \sim 6$ mM). Hence perhaps its role in the apical membrane of absorptive epithelia is uptake of dietary fructose from the lumen of the small intestine (Burant *et al.*, 1992; Kane *et al.*, 1997). Additionally, expression of GLUT5 in adipose and muscle may allow fructose transport into these cells.

However, although human GLUT5 does not transport glucose and is not inhibited by either D-glucose or cytochalasin B (Burant *et al.*, 1992; Kane *et al.*, 1997), both rat and rabbit isoforms are subtly different in nature. The rat clone exhibits a significant degree of glucose transport, and fructose transport by this isoform can be blocked by D-glucose. The fructose transport property of this isoform is insensitive to cytochalasin B, but glucose transport mediated by this isoform is inhibited (Rand *et al.*, 1993). The rabbit GLUT5 isoform transported fructose and is clearly inhibited by glucose when expressed in oocytes (Miyamoto *et al.*, 1994). Differences are further suggested by the demonstration that both human and rabbit GLUT5 can be photolabeled by cytochalasin B, yet the functional properties of the protein are unaffected by this inhibitor (Shepherd *et al.*, 1992b; Miyamoto *et al.*, 1994).

These data suggest distinct and overlapping substrate-binding sites for glucose and fructose by GLUT5, which are differentially sensitive to cytochalasin B, but more experimental analysis is required to confirm this.

7. RECENT WORK AND FUTURE DIRECTIONS

This chapter should serve as an introduction to the sugar transporter family only, describing the basic models for transport and briefly introducing the members of the family. There is, of course, a wealth of data which has only been briefly touched upon.

Extensive mutagenesis has been carried out on the various transporters, including substituting individual residues and by construction of chimeras between the various transporter isoforms (Katagiri et al., 1991; Hashiramoto et al., 1992; Arbuckle et al., 1996; Dauterive et al., 1996; Saravolac and Holman, 1997; Seatter et al., 1997, 1998). Analysis of the kinetic profiles of these mutants, as well as their interactions with the above-mentioned ligands, has provided valuable insights into the mechanism of sugar transport.

Additionally, the trafficking and translocation of GLUT4 has been mentioned only in passing. The investigation of the intracellular vesicular components and protein machinery involved in this process is extensive, as is the regulation of this process by insulin. The intracellular signaling pathways which translate insulin/receptor binding to the translocation of glucose transporters to the cell surface is also the subject of great scrutiny (Lienhard, 1994; Shepherd et al., 1996).

Other subjects which have also been studied in detail include the genetic organization of the transporters and their transcriptional regulation, both by mitogens and nutrients, the production of transgenic animals, and, of course the relevance to related research investigating the more clinical aspects of glucose transporter dysfunction (Gould and Seatter, 1997).

The challenges which confront those of us interested in transporter biology include primarily the elucidation of the structure of the proteins at high resolution. Such information will allow us to compare the substrate-binding sites of the GLUT family members, and perhaps employ these tissue-specific proteins as vehicles for drug delivery to defined cell populations. Such studies utilizing a range of approaches are underway in many laboratories worldwide.

ACKNOWLEDGMENTS

Work in G.W.G.'s laboratory is supported by the Medical Research Trust, the Wellcome Trust, the British Diabetic Association, Tenovus (Scotland), the Scottish Hospitals Endowments Research Trust, the Sir Jules Thorne Charitable Trust, the British Heart Foundation, Smith Kline Beecham Pharmaceuticals, Pfizer Central Research, and the Lister Institute of Preventive Medicine. G.W.G. is a Lister Fellow.

REFERENCES

Allard, W. J., and Lienhard, G. E., 1985, Monoclonal antibodies to the glucose transporter from human erythrocytes: Identification of the transporter as a Mr 55,000 protein. *J. Biol. Chem.* **260**:8668–8675.

Alvarez, J., Lee, D. C., Baldwin, S. A., and Chapman, D., 1987, Fourier transform infrared spectroscopic study of the structure and conformational changes of the human erythrocyte glucose transporter, *J. Biol. Chem.* **262**:3502–3509.

Appleman, J. E., and Lienhard, G. E., 1989, Kinetics of the purified glucose transporter. Direct measurement of the rates of interconversion of transporter conformers, *Biochemistry* **28**:8221–8227.

Arbuckle, M. I., Kane, S., Porter, L. M., Seatter, M. J., and Gould, G. W., 1996, Structure–function analysis of the liver-type (GLUT2) and brain-type (GLUT3) glucose transporters: Expression of chimeric transporters in *Xenopus* oocytes suggests an important role for putative transmembrane helix VII in determining substrate selectivity, *Biochemistry* **35**:16519–16527.

Asano, A., Katagiri, H., Takata, K., Lin, J.-L., Ishihara, H., Inukai, K., Tsukuda, K., Kikuchi, M., Hirano, H., Yazaki, Y., and Oka, Y., 1991, The role of N-glycosylation of GLUT1 for glucose transport activity, *J. Biol. Chem.* **266**:24632–24636.

Asano, T., Katagiri, H., Takata, K., Tsukuda, K., Lin, J.-L., Ishihara, H., Inukaio, K., Hirano, H., Yazaki, Y., and Oka, Y., 1992, Characterization of GLUT3 protein expressed in Chinese hamster ovary cells, *Biochem. J.* **288**:189–193.

Axelrod, J. D., and Pilch, P. F., 1983, Unique cytochalasin B binding characteristics of the hepatic glucose carrier, *Biochemistry* **22**:2222–2227.

Baldwin, S. A., 1993, Mammalian passive glucose transports: Members of a ubiquitous family of active and passive transport proteins. *Biochim et Biophys Acta* **1154**:17–49.

Baldwin, J. M., Gorga, J. C., and Lienhard, G. E., 1981, The monosaccharide transporter of the human erythrocyte, *J. Biol. Chem.* **256**:3685–3689.

Baldwin, S. A., and Lienhard, G. E., 1989, Purification and reconstitution of glucose transporter from human erythrocytes, *Meth. Enzymol.* **174**:39–50.

Baldwin, S. A., Baldwin, J. M., Gorga, F. R., and Lienhard, G. E., 1979, Purification of the cytochalasin B binding component of the human erythrocyte monosaccharide transport system, *Biochim. Biophys. Acta* **552**:183–188.

Birnbaum, M. J., 1989, Identification of a novel gene encoding an insulin-responsive glucose transporter protein, *Cell* **57**:305–315.

Birnbaum, M. J., Haspel, H. C., and Rosen, O. M., 1986, Cloning and characterization of a cDNA encoding the rat brain glucose-transporter protein, *Proc. Natl. Acad. Sci. USA* **83**:5784–5788.

Birnbaum, M. J., Haspel, H. C., and Rosen, O. M., 1987, Transforamation of rat fibroblasts by FSV rapidly increases glucose transporter gene transcription, *Science* **235**:1495–1498.

Bloch, R., 1973, Inhibition of glucose transport in the human erythrocyte by cytochalasin B, *Biochemistry* **12:**4799–4801.

Burant, C. F., Takeda, J., Brot-Laroche, E., Bell, G. I., and Davidson, N. O., 1992, Fructose transporter in human spermatozoa and small intestine is GLUT5, *J. Biol. Chem.* **267:**14523–14526.

Cairns, M. T., Alvarez, J., Panico, M., Gibbs, A. F., Morris, H. R., Chapman D., and Baldwin, S.A., 1987, Investigation of the structure and function of the human erythrocyte glucose transporter by proteolytic dissection, *Biochim. Biophys. Acta* **905:**295–310.

Calderhead, D. M., and Lienhard, G. E., 1988, Labeling of glucose transporters at the cell surface in 3T3-L1 adipocytes, *J. Biol. Chem.* **263:**12171–12174.

Charron, M. J., Brosius III, F. C., Alper, S. L., and Lodish, H. F., 1989, A glucose transport protein expressed predominantly in insulin-responsive tissues, *Proc. Natl. Acad. Sci. USA* **86:**2535–2539.

Chin, J. J., Jung, E. K. Y., and Jung, C. Y., 1986, Structural basis of human erythrocyte glucose transporter function in reconstituted vesicles, *J. Biol. Chem.* **261:**7101–7104.

Colville, C. A., Seatter, M. J., Jess, T. J., Gould, G. W., and Thomas, H. M., 1993, Kinetic analysis of the liver-type (GLUT 2) and brain-type (GLUT 3) glucose transporters expressed in oocytes: Substrate specificities and effects of transport inhibitors, *Biochem. J.* **290:**701–706.

Craik, J. D., and Elliott, K. R. F., 1979, Kinetics of 3-*O*-methyl-D-glucose transport in isolated rat hepatocytes, *Biochem. J.* **182:**503–508.

Cushman, S. W., and Wardzala, L. J., 1980, Potential mechanism of insulin action on glucose transport in the isolated rat adipose cell. Apparant translocation of intracellular transport systems to the plasma membrane, *J. Biol. Chem.* **255:**4758–4762.

Dauterive, R., Laroux, S., Bunn, S. C., Chaisson, A., Sanson, T., and Reed, B. C., 1996, C-terminal mutations that alter the turnover number for 3-*O*-methylglucose transport by GLUT1 and GLUT4, *J. Biol. Chem.* **271:**11414–11421.

Davidson, N. O., Hausman, A. M. L., Ifkovits, C. A., Buse, J. B., Gould, G. W., Burant, C. F., and Bell, G. I., 1992, Human intestinal glucose transporter expression and localization of GLUT5, *Am. J. Physiol.* **262:**C795-C800.

Davies, A., Meeran, K., Cairns, M. T., and Baldwin, S. A., 1987, Peptide specific antibodies as probes of the orientation of the glucose transporter in the human erythrocyte membrane, *J. Biol. Chem.* **262:**9347–9352.

Davies, A., Ciardelli, T. L., Lienhard, G. E., Boyle, J. M., Wheton, A. D., and Baldwin, S. A., 1990, Site-specific antibodies as probes of the topology and function of the human erythrocyte glucose transporter, *Biochem. J.* **266:**799–808.

Dohm, G. L., Tapscott, E. B., Pories, W. J., Flickinger, E. G., Meelheim, D., Fushiki, T., Atkinson, S. M., Elton, C. W., and Caro, J. F., 1988, An *in vitro* human muscle preparation suitable for metabolic studies, *J. Clin. Invest.* **82:**486–494.

Elliott, K. R. F., and Craik, J. D., 1983, Sugar transport across the hepatocyte plasma membrane, *Biochem. Soc. Trans.* **10:**12–13.

Farrell, C. L., and Pardridge, W. M., 1991, Blood–brain barrier glucose transporter is asymmetrically distributed on brain capillary endothelial lumenal and ablumenal membranes: An electron miscroscopic immunogold study, *Proc. Natl. Acad. Sci. USA* **88:**5770–5783.

Farrell, C. L., Yang, J., and Pardridge, W. M., 1992, GLUT-1 glucose transporter is present

within apical and basolateral membranes of brain epithelial interfaces and in microvascular endothelia with and without tight junctions, *J. Histochem. Cytochem.* **40**:193–199.
Flier, J. S., Mueckler, M., McCall, A. L., and Lodish, H. F., 1987a, Distribution of glucose transport messenger RNA transcripts in tissues of rat and man, *J. Clin. Invest.* **79**:657–661.
Flier, J. S., Mueckler, M. M., Usher, P., and Lodish, H. F., 1987b, Elevated levels of glucose transport and transporter messenger RNA are induced by *ras* or *src* oncogenes, *Science* **235**:1492–1495.
Fukumoto, H., Seino, S., Imura, H., Seino, Y., Eddy, R. L., Fukushima, Y., Byers, M. B., Shows, T. B., and Bell, G. I., 1988, Sequence, tissue distributioin, and chromosomal localization of mRNA encoding a human glucose transporter-like protein, *Proc. Natl. Acad. Sci. USA* **85**:5434–5438.
Fukumoto, H., Kayano, T., Buse, J. B., Edwards, Y., Pilch, P. F., Bell, G. I., and Seino, S., 1989, Cloning and characterizationof the major insulin-responsive glucose transporter expressed in human skeletal muscle and other insulin-responsive tissues, *J. Biol. Chem.* **264**:7776–7779.
Garvey, W. T., Maianu, L., Zhu, J.-H., Hancock, J. A., and Golichowski, A. M., 1993, Multiple defects in the adipocyte glucose transport system cause cellular insulin resistance in gestational diabetes, *Diabetes* **42**:1773–1785.
Gorga, F. R., and Lienhard, G. E., 1982, Changes in the intrinsic fluorescence of the human erythrocyte monosaccharide transporter upon ligand binding, *Biochemistry* **21**:1905–1908.
Gould, G. W., and Bell, G. I., 1990, Facilitative glucose transporters—An expanding family, *Trends Biochem. Sci.* **15**:18–23.
Gould, G. W., and Holman, G. D., 1993, The glucose transporter family: Structure, function and tissue-specific expression, *Biochem. J.* **295**:329–341.
Gould, G. W., and Seatter, M. J., 1997, Introduction to the facilitative glucose transporter family, in: *Facilitative Glucose Transporters.* (G. W. Gould, ed.), Landes, Austin, Texas, pp. 1–38.
Gould, G. W., Thomas, H. M., Jess, T. J., and Bell, G. I., 1991, Expression of human glucose transporters in *Xenopus* oocytes: Kinetic characterisation and substrate specificities of the erythrocyte, liver, and brain isoforms, *Biochemistry* **30**:5139–5145.
Gould, G. W., Brant, A. M., Kahn, B. B., Shepherd, P. R., McCoid, S. C., and Gibbs, E. M., 1992, Expression of the brain-type glucose transporter (GLUT 3) is restricted to brain and neuronal cells in mice, *Diabetelogia* **35**:304–309.
Gould, G. W., Cuenda, A., Thomson, F. J., and Cohen, P., 1995, The activation of distinct mitogen-activated protein kinase cascades is required for the stimulation of 2-deoxyglucose uptake by interleukin-I and insulin-like growth factor-I in KB cells, *Biochem. J.* **311**:735–738.
Haase, W., Heitman, K., Friese, W., Ollig, D., and Koepsell, H., 1990, Charcaterisation and histochemical localisation of the rat intestinal sodium-dependent glucose transporter by monoclonal antibodies, *Eur. J. Cell. Biol.* **52**:297–309.
Harik, R. J., Kalaria, R. N., Andersson, L., Lundahl, P., and Perry, G., 1990a, Immunocytochemical localisation of the erythroid glucose transporter: Abundance in tissues with barrier functions, *J. Neurosci.* **10**:3862–3867.

Harik, R. J., Kalaria, R. N., Whitney, P. M., Andersson, L., Lundahl, P., Ledbetter, S. R., and Perry, G., 1990b, Glucose transporters are abundant in cells with occluding junctions at the blood–eye barriers, *Proc. Natl. Acad. Sci. USA* **87**:4261–4265.

Hashiramoto, M., Kadowaki, T., Clark, A. E., Muraoka, A., Momomura, K., Sakura, H., Tobe, K., Akanuma, Y., Yazaki, Y., Holman, G. D., and Kasuga, M., 1992, Site-directed mutagenesis of GLUT1 in helix 7 residue 282 results in perturbation of exofacial ligand binding, *J. Biol. Chem.* **267**:17502–17507.

Hebert, D. N., and Carruthers, A., 1991, Cholate-solubilised erythrocyte glucose transporters exist as a mixture of homodimers and homotetramers, *Biochemistry* **30**:4654–4658.

Hebert, D. N., and Carruthers, A., 1992, Glucose transporter oligomeric structure determines transporter function, *J. Biol. Chem.* **267**:23829–23838.

Hiraki, Y., Garcia de Herreros, A., and Birnbaum, M. J., 1989, Transformation stimulation glucose transporter gene expression in the absence of protein kinase C, *Proc. Natl. Acad. Sci. USA* **86**:8252–8256.

Holman, G. D., and Cushman, S. W., 1994, Subcellular localisation and trafficking of the GLUT4 glucose transporter isoform in insulin-responsive cells, *BioEssays* **16**:753–759.

Holman, G. D., Leggio, L. L., and Cushman, S. W., 1994, Insulin-stimulated GLUT4 glucose transporter recycling: A problem in membrane protein subcellular trafficking through multiple pools, *J. Biol. Chem.* **269**:17516–17524.

Hresko, R. C., Kruse, M., Strube, M., and Mueckler, M., 1994, Topology of the GLUT1 glucose transporter deduced from glycoylation scanning mutagenesis, *J. Biol. Chem.* **269**:20482–20488.

Hughes, S. D., Johnson, J. H., Quaade, C., and Newgard, C. B., 1992, Engineering of glucose-stimulated insulin secretion and biosynthesis in non-islet cells, *Proc. Natl Acad. Sci. USA* **89**:677–692.

Hwang, E. S., Hiraayama, B. A., and Wright, E. M., 1991, Distribution of SGLT1 sodium/glucose co-transporter and mRNA along the crypt/villus axis of rabbit small intestine, *Biochem. Biophys. Res. Commun.* **181**:1208–1217.

Inukai, K., Asano, T., Katagiri, H., Anai, M., Funaki, M., Ishihara, H., Tsukuda, K., Kikuchi, M., Yazaki, Y., and Oka, Y., 1994, Replacement of both tryptophan residues at 388 and 412 completely abolished cytochalasin B photolabelling of the GLUT1 glucose transporter, *Biochem. J.* **302**:355–361.

James, D. E., and Piper, R. C., 1993, Targeting of mammalian glucose transporters, *J. Cell Sci.* **104**:607–612.

James, D. E., and Piper, R. C., 1994, Insulin resistance, diabetes, and the insulin-regulated trafficking of GLUT4, *J. Cell. Biol.* **126**:1123–1126.

James, D. E., Brown, R., Navarro, J., and Pilch, P. F., 1988, Insulin-regulatable tissues express a unique insulin-sensitive glucose transport protein, *Nature* **333**:183–185.

James, D. E., Strube, M., and Mueckler, M., 1989, Molecular cloning and characterization of an insulin-regulatable glucose transporter, *Nature* **338**:83–87.

Jung, C. Y., 1971, Permeability of bimolecular membranes made from lipid extracts of human red blood cell ghosts to sugars, *J. Membr. Biol.* **5**:200–214.

Jung, E. K. Y., Chin, J. J., and Jung, C. Y., 1986, Structural basis of human erythrocyte glucose transporter function in reconstituted system, *J. Biol. Chem.* **261**:9155–9160.

Kaback, H. R. (1996). The Lactose Permease of *Escherichia coli:* An Update, in: *Molecu-*

lar Biology of Membrane Transport Disorders (S. G. Shultz, *et al.*) New York. Plenum Press, pp. 111–125.

Kaestner, K. H., Christy, R. J., McLenithan, J. C., Braiterman, L. T., Cornelius, P., Pekla, P. H., and Lane, M. D., 1989, Sequence, tissue distribution, and differential expression of mRNA for a putative insulin-responsive glucose transporter in mouse 3T3-L1 adipocytes, *Proc. Natl. Acad. Sci. USA* **86**:3150–3154.

Kahn, B. B., 1992a, Alterations in glucose transporter expression and function in diabetes: Mechanisms for insulin resistance, *J. Cell. Biochem.* **48**:122–128.

Kahn, B. B., 1992b, Facilitative glucose transporters: regulatory mechanisms and dysregulation in diabetes, *J. Clin. Invest.* **89**:1367–1374.

Kane, S., Seatter, M. J., and Gould, G. W., 1997, Functional studies of human GLUT5: Effect of pH on substrate selection and an analysis of substrate interactions, *Biochem. Biophys. Res. Commun.* **238**:503–505.

Kasahara, M., and Hinkle, P. C., 1977, Reconstitution and purification of the D-glucose transporter from human erythrocytes, *J. Biol. Chem.* **252**:7384–7390.

Katagiri, H., Asano, T., Shibasaki, Y., Lin, J.-L., Tsukuda, K., Ishihara, H., Akanuma, Y., Takaku, F., and Oka, Y., 1991, Substitution of luecine for tryptophan 412 does not abolish cytochalasin B labelling but markedly decreases the intrinsic activity of GlUT1 glucose transporter, *J. Biol. Chem.* **266**:7769–7773.

Kayano, T., Burrant, C. F., Fukumoto, H., Gould, G. W., Fan, Y.-S., Eddy, R. L., Byers, M. G., Shows, T. B., Seino, S., and Bell, G. I., 1990, Human facilitative glucose transporters: Isolation, functional characterisation and gene localisation of cDNAs encoding an isoform expressed in small intestine, kidney, muscle and adipose tissue (GLUT 5) and an unusual glucose transporter pseudogene-like sequence (GLUT 6), *J. Biol. Chem.* **265**:13276–13282.

Kozka, I. J., Gould, G. W., Reckless, J. P. D., Cushman, S. W., and Holman, G. D., 1995, The effects of insulin on the levels and activities of the glucose transporter isoforms present in human adipose cells, *Diabetelogia* **38**:661–666.

LeFevre, P. G., 1961, Sugar transport in the red blood cell: Structure–activity relationships in substrates and antagonists, *Pharmacol. Rev* **13**:39–70.

Lienhard, G. E., 1994, Life without the IRS, *Nature* **372**:128–129.

Livingstone, C., Dominiczak, A. F., Campbell, I. W., and Gould, G. W., 1995, Insulin resistance, hypertension and the insulin-responsive glucose transporter, GLUT4, *Clin. Sci.* **89**:109–116.

Lowe, A. G., and Walmsley, A. R., 1986, The kinetics of glucose transport in human red blood cells, *Biochim. Biophys. Acta* **857**:146–154.

Lowe, A. G., Critchley, A. J., and Brass, A., 1991, Inhibition of glucose transport in human erythrocytes by ubiquinone Q_o, *Biochim. Biophys. Acta* **1069**:223–228.

Magnani, P., Cherian, P. V., Gould, D. A., Greene, G. W., Sima, A. A. F., and Brosius, F. C., 1996, Glucose transporters in rat peripheral nerve: Paranodal expression of GLUT1 and GLUT3, *Metab. Clin. Exp.* **45**:1466–1473.

Maher, F., 1995, Immunolocalization of GLUT1 and GLUT3 glucose transporters in primary cultured neurons and glia, *J. Neurosci. Res.* **42**:459–469.

Maher, F., Vannucci, S. J., Takeda, J., and Simpson, I. A., 1992, Expression of mouse-GLUT3 and human-GLUT3 glucose transporter proteins in brain, *Biochem. Biophys. Res. Commun.* **182**:703–711.

Maher, F., Vannucci, S. J., and Simpson, I. A., 1994, Glucose transporter proteins in the brain, *FASEB J.* **8**:1003–1011.

Maher, F., Davies-Hill, T. M., and Simpson, I. A., 1996, Substrate specificity and kinetic parameters of GLUT3 in rat cerebellar granule neurons, *Biochem. J.* **315**:827–831.

Merrall, N. M., Plevin, R. J., and Gould, G. W., 1993, Mitogens, growth factors, oncogenes and the regulation of glucose transport, *Cell. Signalling* **5**:667–675.

Milburn, J. L., Ohneda, M., Johnson, J. H., and Unger, R. H., 1993, Beta-cell GLUT2 loss and NIDDM: Current status of the hypothesis, *Diabet. Metab. Rev.* **9**:231–236.

Miyamoto, K.-I., Takago, T., Fujii, T., Matsubara, T., Hase, K., Taketani, Y., Oka, T., Minami, H., and Nakabou, Y., 1992, Role of liver-type glucose transporter (GLUT2) in transport across the basolateral membrane in rat jejunum, *FEBS Lett.* **314**:466–470.

Miyamoto, K., Tatsumi, S., Morimoto, A., Minami, H., Yamamoto, Y., Sone, K., Taketani, E., Nakabou, Y., Oka, T., and Takeda, E., 1994, Characterisation of the rabbit intestinal fructose transporter (GLUT5), *Biochem. J.* **303**:877–883.

Mueckler, M., Caruso, C., Badlwin, S. A., Panico, M., Blench, I., Morris, H. R., Allard, W. J., Leinahrd, G. E., and Lodish, H. F., 1985, Sequence and structure of a human glucose transporter, *Science* **229**:941–945.

Nagamatsu, S., Kornhauser, J. M., Burant, C. F., Seino, S., Mayo, K. E., and Bell, G. I., 1992, Glucose transporter expression in brain. cDNA sequence of mouse GLUT3, the brain facilitative glucose transporter isoform, and identification of sites of expression by *in situ* hybridization, *J. Biol. Chem.* **267**:467–472.

Nagamatsu, S., Sawa, H., Kamada, K., Nakamichi, Y., Yoshitomo, K., and Hoshino, T., 1993, Neuron-specific glucose transporter: CNA distribution of GLUT3 rat glucose transporter in rat central neurons, *FEBS Lett.* **334**:289–295.

Oka, Y., Asano, T., Shibasaki, Y., Kasuga, M., Kanazawa, Y., and Takaku, F., 1988, Studies with antipeptide antibody suggest the presence of at least two types of glucose transporter in rat brain and adipocytes, *J. Biol. Chem.* **263**:13432–13439.

Orci, L., Thorens, B., Ravazzola, M., and Lodish, H. F., 1989, Localisation of the pancreatic beta cell glucose transporter to specific plasma membrane domains, *Science* **245**:295–297.

Palfreyman, R. W., Clark, A. E., Denton, R. M., and Holman, G. D., 1992, Kinetic resolution of the separate GLUT1 and GLUT4 glucose transport activities in 3T3-L1 cells, *Biochem. J.* **284**:275–281.

Pardridge, W. M., Boado, R. J., and Farrell, C. R., 1990, Brain-type glucose transporter (GLUT1) is selectively localised to the blood–brain barrier. Studies with quantitative Western blotting and *in situ* hybridization, *J. Biol. Chem.* **265**:18035–18041.

Ploug, T., Galbo, H., Vinten, J., Jorgensen, M., and Richter, E. A., 1987, Kinetics of glucose transport in rat muscle: Effects of insulin and contractions, *Am. J. Physiol. (Endocrinol. Metab.)* **253** (16):E12–E20.

Ploug, T., Wojtaszewski, J., Kristiansen, S., Hespel, P., Galbo, H., and Richter, E. A., 1993, Glucose transport and transporters in muscle giant vesicles: Differential effects of insulin and contractions, *Am. J. Physiol.* **264**:E270–E278.

Rand, E. B., Depaoli, A. M., Davidson, N. O., Bell, G. I., and Burant, C. F., 1993, Sequence, tissue distribution, and functional characterization of the rat fructose transporter GLUT5, *Am. J. Physiol.* **264**:G1169–G1176.

Saravolac, E. G., and Holman, G. D., 1997, Glucose transport: Probing the structure/function relationship, in: *Facilitative Glucose Transporters* (G. W. Gould, ed.), Landes, Austin, Texas, pp. 39–61.

Seatter, M., Kane, S., Porter, L., Melvin, D. R., and Gould, G. W., 1997, Structure–function studies of the brain-type glucose transporter GLUT3: Alanine-scanning mutagenesis of putative transmembrane helix VIII, and an investigation of the role of proline residues in transport catalysis, *Biochemistry* **36**:6401–6407.

Seatter, M. J., De La Rue, S. A., Porter, L. M., and Gould, G. W., 1998, The QLS motif in transmembrane helix VII of the glucose transporter (GLUT) family interacts with the C-1 position of D-glucose and is involved in substrate selection at the exofacial binding site, *Biochemistry* **37**:1322–1326.

Shepherd, P. R., Gibbs, E. M., Wesslau, C., Gould, G. W., and Kahn, B. B., 1992a, The human small intestine facilitative fructose/glucose transporter (GLUT 5) is also present in insulin-responsive tissues and brain: Investigation of biochemical characteristics and translocation, *Diabetes* **41**:1360–1365.

Shepherd, P. R., Gould, G. W., Colville, C. A., McCoid, S. C., Gibbs, E. M., and Kahn, B. B., 1992b, Distribution of GLUT 3 glucose transporter in human tissues, *Biochem. Biophys. Res. Commun.* **188**:149–154.

Shepherd, P. R., Navé, B. T., and O'Rahilly, S., 1996, The role of phosphoinositide 3-kinase in insulin signalling, *J. Mol. Endocrinol.* **17**:175–184.

Sima, A. A. F., Lattimer, S. A., Yagihashi, S., and Greene, D. A., 1986, "Axo-glial dysjunction": A novel structural lesion that accounts for poorly reversible slowing of nerve conduction in the spontaneously diabetic BB-rat, *J. Clin. Invest.* **77**:414–425.

Slot, J. W., Geuze, H. J., Gigengack, S., James, D. E., and Lienhard, G. E., 1991a, Translocation of the glucose transporter GLUT4 in cardiac myocytes of the rat, *Proc. Natl. Acad. Sci. USA* **88**:7815–7819.

Slot, J. W., Geuze, H. J., Gigengack, S., Lienhard, G. E., and James, D. E., 1991b, Immunolocalization of the insulin regulatable glucose transporter in brown adipose tissue of the rat, *J. Cell Biol.* **113**:123–135.

Slot, J. W., Garruti, G., Martin, S., Oorschot, V., Posthuma, G., Kraegen, E. W., Laybutt, R., Thibault, G., and James, D. E., 1997, Glucose transporter (GLUT-4) is targeted to secretory granules in rat atrial cardiomyocytes, *J. Cell Biol.* **137**:1243–1254.

Suzuki, K., and Kono, T., 1980, Evidence that insulin causes translocation of glucose transport activity to the plasma membrane from an intracellular storage site, *Proc. Natl. Acad. Sci. USA* **77**:2542–2545.

Thorens, B., Sarkar, H. K., Kaback, H. R., and Lodish, H. F., 1988, Cloning and functional expression in bacteria of a novel glucose transporter present in liver, intestine, kidney and b-pancreatic islet cells, *Cell* **55**:281–290.

Thorens, B., Cheng, Z.-Q., Brown, D., and Lodish, H. F., 1990, Liver glucose transporter: A basolateral protein in hepatocytes and intestine and kidney cells, *Am. J. Physiol.* **260**:C279–C285.

Unger, R. H., 1991, Diabetic hyperglycemia: Link to impaired glucose transport in pancreatic b cells, *Science* **251**:1200–1205.

Walmsley, A. R., and Lowe, A. G., 1987, Comparison of the kinetics and thermodynamics of the carrier systems for glucose and leucine in human red blood cells, *Biochim. Biophys. Acta* **901**:229–238.

Wheeler, T. J., and Hinkle, P. C., 1981, Kinetic properties of the reconstituted glucose transporter from human erythrocytes, *J. Biol. Chem.* **256**:8907–8914.

Widdas, W. F., 1952, Inability of diffusion to account for placental glucose transport in the sheep and consideration of the kinetics of a possible carrier transfer, *J. Physiol.* **118**:23–39.

Williams, S. A., and Birnbaum, M. J., 1988, The rat facilitated glucose transporter gene. Transformation and serum-stimulated transcription initiate from identical sites, *J. Biol. Chem.* **263**:19513–19518.

Yano, H., Seino, Y., Inagaki, N., Hinokio, Y., Yamamoto, T., Yasuda, K., Masuda, K., Someya, Y., and Imura, H., 1991, Tissue distribution and species difference of the brain type glucose transporter (GLUT3), *Biochem. Biophys. Res. Commun.* **174**:470–477.

Zoccoli, M. A., Baldwin, S. A., and Lienhard, G. E., 1978, The monosaccharide transport system of the human erythrocyte, *J. Biol. Chem.* **253**:6923–6930.

8

Cationic Amino Acid Transporters (CATs)

Targets for the Manipulation
of NO-Synthase Activity?

Ellen I. Closs and Petra Gräf

1. INTRODUCTION

Cationic amino acids are not only building blocks of proteins, but also substrates for many metabolic pathways (Fig. 1). L-Lysine is an essential amino acid that is exclusively used for protein synthesis. L-Arginine is considered a *"semi*essential" amino acid. It can be synthesized from L-ornithine via L-citrulline by the enzymes of the urea cycle. Accordingly L-ornithine can be produced from L-arginine by the action of arginases, but it can also be synthesized from L-glutamate. The *de novo* synthesis of L-arginine/L-ornithine seems yet not to be sufficient to sustain an adequate supply of these cationic amino acids, especially when there is a high need, e.g., during growth and wound healing (for review see Jenkinson *et al.,* 1996). Besides being a component of proteins and the starting point for urea and ornithine synthesis, L-arginine also serves as substrate for the synthesis of nitric oxide (NO), a radical involved in such divergent actions as smooth muscle relaxation, host defense, and learning (for review see W. C. Sessa, 1994; Förstermann *et al.,* 1995,b; Michel and Feron, 1997). L-Ornithine is not proteinogenic. It can be metabolized to the proteinogenic amino acid L-proline and to polyamines, important regulators

---

Ellen I. Closs and Petra Gräf • Department of Pharmacology, Johannes Gutenberg University, D-55101 Mainz, Germany.
Membrane Transporters as Drug Targets, edited by Amidon and Sadée. Kluwer Academic/Plenum Publishers, New York, 1999.

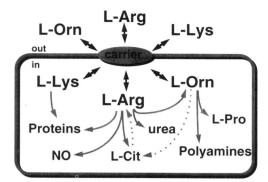

Figure 1. Cationic amino acids: transmembrane transport and major metabolic pathways. Transmembrane transport (black arrows): Cationic amino acids share the same carrier systems for transport into and out of mammalian cells. The carriers are energy-independent in most cells and mediate an exchange of their substrate between the extracellular and intracellular space according to the electrochemical equilibrium (facilitated diffusion). Some of these carriers also transport neutral amino acids (not shown). Major metabolic pathways (gray): The essential cationic amino acid L-lysine and the semiessential L-arginine are building blocks of proteins. In addition, L-arginine serves as the substrate for the synthesis of NO and L-citrulline by NO synthases and it can be metabolized to urea and L-ornithine by the action of arginases. L-Ornithine is not a component of proteins, but it is a precursor for the proteinogenic amino acid L-proline and for polyamines. In some cells L-arginine can be resynthesized from L-ornithine and/or L-citrulline by the enzymes of the urea cycle (dashed arrows).

of cell growth and differentiation (for review see Calandra et al., 1996; A. Sessa and Perin, 1997). Given the multitude of processes that involve cationic amino acids, it is evident that basically every mammalian cell must be capable of exchanging cationic amino acids with its environment. This exchange is mediated by specialized carrier proteins that recognize either exclusively cationic or also neutral amino acids. In most cells, transport of cationic amino acids is energy-independent, leading to an exchange of substrate between the extracellular and intracellular spaces according to the electrochemical equilibrium (facilitated diffusion). The carrier proteins mediating the transport of cationic amino acids through the plasma membrane might well be targets for the manipulation of processes such as NO or polyamine synthesis. We are only beginning to know the proteins involved in cationic amino acid transport and to learn about their physiological roles. Here, we review a family of related carrier proteins for cationic amino acids (CAT)* and focus on their possible role in supplying substrate for NO synthesis.

*Abbreviations used: 4F2hc, heavy-chain cell-surface antigen; AITV, α-amino-δ-isothioureidovaleric acid; CAT, cationic amino acid transporter (m, mouse; h, human; r, rat); IL-1β, interleukin-1β; INF-γ, interferon-γ; K_M, Michaelis–Menten constant; L-NMA, N^G-methyl-L-arginine; L-SDMA, symmetrical N^G,N^G-dimethyl-L-arginine; L-ADMA, asymmetrical N^G,N'^G-dimethyl-L-arginine; L-NIO, L-N^5-(1-iminoethyl)-ornithine; L-NNA, N^G-nitro-L-arginine; LPS, bacterial lipopolysaccharide; rBAT, related to b$^{o,+}$ transporter; TM, transmembrane domain; TNF-α, tumor necrosis factor-α; V_{max}, maximal velocity.

Cationic Amino Acid Transporters (CATs)

2. IDENTIFICATION OF FOUR RELATED CARRIER PROTEINS FOR CATIONIC AMINO ACIDS IN MAMMALIAN CELLS

2.1. Murine CAT-Proteins (mCATs)

In recent years four distinct isoforms of a family of carrier proteins that catalyze the transport of cationic amino acids through the plasma membrane have been identified. A cDNA encoding the first member of this family was cloned when Albritton and co-workers identified the host cell protein responsible for susceptibility to infection by murine ecotropic leukemia viruses (MuLV). Introduction of genomic DNA from mouse cells (which are infectable by ecotropic MuLVs) into human cells (which are not infectable) led to a human cell line susceptible for ecotropic MuLV infection. This cell line allowed the isolation of a cDNA that codes for the receptor of ecotropic MuLVs (Albritton *et al.*, 1989). The receptor contains 622 amino acids and has a predicted molecular mass of 67 kDa. Based on hydrophobicity plots, Albritton and co-workers suggested the receptor is an integral membrane protein that spans the membrane 14 times and has intracellular N- and C-termini (Fig. 2A). Although the receptor had no sequence homology with

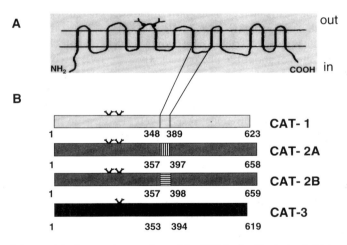

Figure 2. Scheme of the four murine members of the CAT family. The hydrophobicity plots of all known members of the CAT family are almost identical. (A) A model postulating 14 transmembrane domains (TMs) and intracellular N- and C-termini. (B) Schematic alignment of the four murine CAT proteins. CATs have two conserved sites for N-linked glycosylation, except for mCAT-3, which has only one site. The protein domains where mCAT-2A and mCAT-2B differ and the corresponding regions in mCAT-1 and mCAT-3 are boxed. According to the 14-TM model, this domain lies in the intracellular loop connecting TM VIII and IX.

proteins known at the time, similarities between its secondary structure and the L-histidine and the L-arginine permeases from *Saccharomyces cerevisiae* suggested that the receptor might also function as a solute transporter (Kim *et al.*, 1991). The expression of the MuLV receptor in *Xenopus* oocytes and subsequent transport studies demonstrated that the receptor mediates the Na^+-independent transport of cationic amino acids (Kim *et al.*, 1991; Wang *et al.*, 1991). Therefore, the receptor has been renamed mCAT-1 (mouse cationic amino acid transporter). mCAT-1 mRNA has been shown to be ubiquitously expressed with the exception of the liver. As a result of mutational analysis the viral binding site of mCAT-1 has been mapped to the third extracellular loop (between TM V and TM VI) (Albritton *et al.*, 1993). This domain also contains two consensus sites for N-linked glycosylation. The mutation of the respective Asn^{223} and Asn^{229} to histidine demonstrated that both Asn residues are glycosylated when mCAT-1 is expressed in *Xenopus* oocytes and suggested that they are the unique glycosylation sites of this protein (Kim and Cunningham, 1993).

The second member of the CAT transporter family was also identified serendipitously. MacLeod and co-workers isolated a cDNA from the murine T-lymphoma cell line SL 12.4. which they initially named Tea ("T-cell early activation"; MacLeod *et al.*, 1990). Translation of the open reading frame demonstrated a significant amino acid sequence identity to the ecotropic MuLV receptor, but the encoded protein was truncated at the N-terminus. Later, full-length cDNAs were isolated by the same group (Reizer *et al.*, 1993) as well as others (Closs *et al.*, 1993a,b; Kavanaugh *et al.*, 1994). The coding sequence of all cDNAs is identical except for a stretch of about 120 nucleotides where the liver-derived cDNAs (Closs *et al.*, 1993a; Kavanaugh *et al.*, 1994) differ from the cDNAs obtained from SL 12.2. cells or bacterial lipopolysaccharide (LPS)-activated mouse macrophages encoding for Tea (Closs *et al.*, 1993b; Reizer *et al.*, 1993). The cDNAs encode proteins of 657 and 658 amino acids, respectively. Expression in *Xenopus* oocytes demonstrated that both proteins have the same substrate specificity for cationic amino acids as mCAT-1. To reflect their cellular function they were renamed mCAT-2A (liver-derived isoform) and mCAT-2B (T-cell/macrophage-derived isoform). The two proteins most likely result from differential splicing of transcripts from the same gene (mCAT-2). They are identical except for the stretch of 42 amino acids encoded by the 120 divergent nucleotides, where they differ in 20 amino acids. This protein domain is located within the predicted fourth intracellular loop (Fig. 2). In an optimal alignment, the amino acid sequences of the two mCAT-2 isoforms are 60% identical with mCAT-1.

Recently Hosokawa and co-workers identified the fourth isoform of the CAT family (Hosokawa *et al.*, 1997). By screening a rat brain cDNA library with a mCAT-1 hybridization probe they isolated a cDNA encoding a protein with 619 amino acids (rCAT-3) that is 53–58% identical to the other three CAT isoforms. Interestingly, at the same time a cDNA that codes for mouse CAT-3 was isolated

Cationic Amino Acid Transporters (CATs)   233

by Ito and Groudine when identifying germ-layer-specific mRNA transcripts from mouse embryos (Ito and Groudine, 1997). In mid-streak-stage embryos mCAT-3 expression has only been detected in the mesoderm. However, at a later fetal stage mCAT-3 has been found to be expressed in a variety of tissues. In adult rats and mice CAT-3 seems to be exclusively expressed in brain (Hosokawa et al., 1997; Ito and Groudine, 1997).

The hydrophobicity plots of all mCAT proteins are very similar suggesting that the number of transmembrane domains (TMs) is the same for each family member. Alternative to the 14-TM model, MacLeod and co-workers proposed a model with 12 TMs. In addition, some programs predict 13 TMs with the C-terminus residing outside the cell.

2.2. Human CAT-Proteins (hCATs)

Two groups isolated cDNAs encoding the human homologues of mCAT-1 (Yoshimoto et al., 1991; Albritton et al., 1992). In addition, cDNAs encoding rat CAT-1 have been cloned (Wu et al., 1994; Puppi and Henning, 1995). An alignment of the deduced amino acid sequences shows that CAT-1 proteins are highly conserved among different mammalian species. Mouse and human proteins are 86% identical, and mouse and rat proteins are 95% identical. We isolated cDNAs encoding the human homologues of mCAT-2A and mCAT-2B (Closs et al., 1997b). The open reading frames code for proteins of 657 and 658 amino acids, respectively. The human CAT-2 proteins differ also only by 21 amino acids in a stretch of 42 amino acids corresponding to the same protein domain divergent in the mCAT-2 proteins. A sequence comparison with the mouse CAT-2 proteins reveals 90% identity. The divergent regions of hCAT-2A and hCAT-2B deviate in only one residue from their respective murine homologues. In contrast, in the corresponding region the sequences of the human and mouse CAT-1 proteins differ in seven residues.

3. TRANSPORT PROPERTIES OF THE CAT PROTEINS

Most studies on the transport properties of the different CAT isoforms have been performed in *Xenopus* oocytes, where each carrier can be expressed individually against a low endogenous background. The transport activity can be assayed either using radiolabeled amino acids or by measuring amino acid-induced currents. The carrier-mediated transport activity is determined by subtracting the values obtained with water-injected oocytes from those with cRNA-injected oocytes encoding for a specific carrier. CAT activity has also been measured in mammalian cells transfected with expression vectors encoding for various CAT isoforms (Kakuda *et al.,* 1993; Kavanaugh *et al.,* 1994; Hosokawa *et al.,* 1997). However, even under the control of a strong promoter (e.g., SV40 or cytomegalovirus [CMV]) only a moderate expression of the CAT proteins is seen in mammalian cells. In addition, all tissue culture cells so far investigated express at least CAT-1. Therefore, it is not always possible to distinguish the exogenous and endogenous transport activities unambiguously. In the following, we will focus on the transport properties of the four mCAT isoforms obtained by expression in *Xenopus* oocytes.

3.1. Substrate Specificity of the mCAT Proteins

All four CAT isoforms share the same substrate specificity for cationic amino acids. Furthermore, CAT-mediated transport has been shown to be stereoselective for L-amino acids and independent of the presence of Na^+ ions. Figure 3 shows a typical experiment where uptake of 13 different amino acids into oocytes injected with mCAT-2A cRNA has been assayed. These oocytes took up about 10–15 times

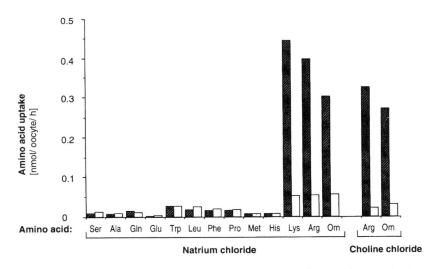

Figure 3. Substrate specificity and Na^+ independence of mCAT-2A. Oocytes injected with 50 ng of mCAT-2A cRNA (dark bars) or with water (white bars) were incubated in an isotonic salt solution containing either 100 mM NaCl or 100 mM choline chloride and the indicated ^{3}H- or ^{14}C-labeled L-amino acids (0.1 mM, 1 μCi/ ml, 1 hr, 20°C). After incubation, oocytes were washed, lysed in 1% sodium dodecyl sulfate, and the radioactivity was determined by liquid scintillation counting. Columns represent means, $n = 3-5$.

more of the cationic amino acids L-arginine, L-lysine, and L-ornithine than water-injected controls. In contrast, the uptake of neutral and acidic amino acids did not differ significantly between cRNA-injected oocytes and controls. Exchange of the Na^+ ions in the incubation buffer by choline did not alter the transport activity of the oocytes. Uptake studies have shown similar results for mCAT-1 (Kim et al., 1991) and mCAT-2B (Closs et al., 1993b) (for review see Closs, 1996). mCAT-1 and mCAT-2B seem to also recognize neutral amino acids, but with a very low affinity and only in the presence of Na^+ ions (Wang et al., 1991; Kakuda et al., 1993). mCAT-3-mediated L-arginine transport has been shown to be competitively inhibited not only by other cationic L-amino acids, but also by D-arginine, L-methionine, L-cysteine, L-histidine, and even L-aspartate and L-glutamate (Ito and Groudine, 1997). In contrast to mCAT-1, recognition of the neutral amino acids was Na^+-independent. However, an excess of about 10^4–10^6 in the concentration of the inhibitor over that of the substrate used in these studies inhibited the transport rate only by 50–75%, putting into question the specificity of the interaction. In our hands, an excess of 30- to 100-fold L-lysine is sufficient to block L-arginine uptake mediated by mCAT-1 or -2B almost completely.

3.2. Kinetics of mCAT-Mediated Transport

The V_{max} values for L-arginine uptake obtained for different CAT proteins expressed in *Xenopus* oocytes differ significantly (Table I). Whereas the values for mCAT-1, -2A, and -2B range from about 1 to 8 nmole/oocyte/hr, the V_{max} of mCAT-3 is only 0.02–0.06 nmole/oocyte/hr.

According to their substrate affinities, CATs can be classified into two groups: the high-affinity transporters CAT-1, -2B, and -3 with K_M values for cationic amino acids in the range of 40–400 μM, and the low-affinity transporter CAT-2A with a K_M of 2–5 mM (Table I). The apparent K_M values for L-arginine reported in the literature range from 0.07 to 0.25 mM for mCAT-1, from 0.04 to 0.38 mM for mCAT-2B, and from 0.04 to 0.12 mM for mCAT-3. The apparent K_M values of the CAT-proteins for L-lysine and L-ornithine are similar to those for L-arginine, suggesting that these substrates are recognized with a similar affinity. Kavanaugh described a voltage dependence of mCAT-1-mediated transport (Kavanaugh, 1993): Hyperpolarization increases the V_{max} and decreases the apparent K_M for influx and it decreases the V_{max} and increases the apparent K_M for efflux. Variations in K_M values might therefore result from different membrane potentials in oocytes under different experimental conditions.

The CAT proteins differ also in their sensitivity to trans-stimulation. While the activities of mCAT-1 and -2B are stimulated by physiological concentrations of substrate at the trans-side of the membrane (250 μM), transport mediated by

Table I.
Comparison of the Transport Properties of System y^+ and the CAT Proteins[a]

	Expression system	K_M (mM L-Arg)	V_{max} (nmole L-Arg/oocyte/hr)	Transstimulation[b]
System y^+	Human fibroblasts	0.03–0.2	—	7.3
mCAT-1	*Xenopus* oocytes	0.14–0.25	1.1–1.6	8.3
mCAT-2A	*Xenopus* oocytes	2.10–5.20	3.9–7.1	1.5
mCAT-2B	*Xenopus* oocytes	0.25–0.38	1.1–3.4	2.9
mCAT-3	*Xenopus* oocytes	0.04–0.10	0.02–0.06	2.0
hCAT-1	*Xenopus* oocytes	0.11–0.16	1.6–1.8	9.8
hCAT-2A	*Xenopus* oocytes	3.36–3.90	2.2–8.4	1.1
hCAT-2B	*Xenopus* oocytes	0.32–0.73	1.2–4.0	1.8

[a] Data for system y^+ are from White (1985). Data for K_M and V_{max} of L-arginine and trans-stimulation of transporters expressed in *Xenopus* oocytes are from Closs *et al.* (1997b) for hCAT proteins, from Closs *et al.* (1993b) for mCAT-1, -2A and -2B, and from Ito and Groudine (1997) for mCAT-3.
[b] Trans-stimulation: Fold transport activity at high (CAT-1, -2A and -2B: 0.25 mM; mCAT-3: 10 mM) versus zero trans-substrate (L-arginine).

mCAT-2A is largely independent of the presence of trans-substrate (Closs *et al.*, 1993a,b). The most pronounced trans-stimulation effect has been observed for mCAT-1 (Table I). For mCAT-3 an about twofold higher activity has been observed at a trans-substrate concentration of 10 mM compared to zero trans-substrate, but no data are available for more physiological concentrations (Ito and Groudine, 1997).

It is noticeable that mCAT-2A and mCAT-2B isoforms diverge in their apparent substrate affinity and sensitivity to trans-stimulation although their amino acid sequences are 97% identical and differ only in a stretch of 42 amino acids. Replacement of the corresponding region of mCAT-1 by that of mCAT-2A or -2B and vice versa led to chimeric proteins with transport properties of the donor of that region (including the apparent affinity for L-arginine and sensitivity to trans-stimulation) (Closs *et al.*, 1993b). These data suggest that the transport properties of the CAT proteins are determined by that region located either at the inside (14-TM model) or at the outside (12-TM model) of the membrane.

3.3. hCAT-Mediated Transport

The transport properties of the hCAT proteins determined using *Xenopus* oocytes are very similar to those obtained with the mouse proteins (Table I). However, unlike the mouse homologues, the K_M values of hCAT-1 and hCAT-2B differ significantly. In addition, hCAT-1 is more sensitive to trans-stimulation than hCAT-2B and its transport activity is independent of the extracellular pH in a range from 5.5–8.0, whereas at pH 5.5 both CAT-2 proteins exhibit only half maximal activity.

4. CATs AND KNOWN TRANSPORT SYSTEMS FOR CATIONIC AMINO ACIDS

The transport activities measured in intact cells or membrane vesicles using radioactive labeled amino acids have been referred to as "transport systems" (for review see Christensen, 1990). This term implies that each transport activity might be mediated by a complex of different proteins rather than a single carrier protein. System y^+ has been described as the major Na^+-independent transport activity for cationic amino acids in mammalian cells. It is almost ubiquitous with the exception of the liver (Christensen, 1964; Christensen and Antonioli, 1969; White and Christensen, 1982a,b; White et al., 1982). Cationic amino acid transport by system y^+ is pH-independent, stimulated by substrate at the trans-side of the membrane, and has a K_M of 0.025–0.2 mM (Christensen and Antonioli, 1969; White and Christensen, 1982a). The activity of system y^+ is increased by hyperpolarization (Bussolati et al., 1993). In the presence of Na^+ neutral amino acids can also interfere with y^+-mediated transport.

The expression pattern of CAT-1 as well as the transport activity induced by expression of CAT-1 in *Xenopus* oocytes or mammalian cells are consistent with system y^+. However, in the overexpression experiments we cannot exclude that CAT-1 might be associated with one or more proteins endogenous to the host cell. The answer to the question of whether CAT-1 is one and the same as system y^+ or only one of several compounds of system y^+ will have to await reconstitution of the pure CAT-1 protein in artificial membranes.

Recently, Perkins et al. introduced a germ-line null mutation into the mouse CAT-1 gene by targeted mutagenesis of embryonic stem cells (Perkins et al., 1997). The heterozygous animals of two mouse strains carrying the CAT-1 knockout are essentially normal. In contrast, the homozygous mutant mice die on day 1 after birth and show a 25% reduction in size compared to the wild-type animals. The homozygous mice suffer from severe anemia suggesting a substantial role of mCAT-1 in hematopoesis and growth control during mouse development. However, no other abnormality has been found in these mice. The surprising result that the knockout of such a major carrier for cationic amino acids has no effect on other cells or organs than the erythroid lineage suggests that the loss of CAT-1 function might be compensated by other carrier(s).

CAT-2B and CAT-3 demonstrate very similar transport properties to CAT-1, although they are less sensitive to trans-stimulation and (at least CAT-2B) more sensitive to pH changes. Their expression patterns are much more restricted than that of CAT-1 [for review on CAT-2B see Macleod and Kakuda (1996)]. The expression of CAT-2B and probably CAT-3 is found in cells that also express CAT-1. As the activities of the three different isoforms can hardly be distinguished, it is puzzling to ask about the specific physiological role of each "y^+ carrier."

In primary hepatocytes and liver slices White and Christensen observed the absence of system y^+ and described a transport activity with very low affinity for cationic amino acids (White and Christensen, 1982a). No specific name has been assigned to this activity. They assumed that a lack of the high-affinity system y^+ (CAT-1) prevents hydrolysis of the plasma L-arginine by hepatocyte arginase. CAT-2A seems to be the low-affinity carrier that allows hepatocytes to rapidly take up cationic amino acids at high plasma concentrations, while leaving a sufficient amount in the circulation for cells expressing high-affinity carriers. The independence of the CAT-2A activity of the presence of trans-substrate allows an efficient uptake of cationic amino acids in hepatocytes that have very low concentrations of these amino acids.

Besides system y^+, four different Na^+-independent transport systems for cationic amino acids have been described. These systems differ from y^+ and from each other by their interaction with neutral amino acids. Because of the very weak interaction of CAT-1 with neutral amino acids it has been proposed that the CAT proteins might represent systems b_1^+ or b_2^+, two transport activities that can not be inhibited by neutral amino acids even in the presence of Na^+ (Van Winkel and Campione, 1990). However, the two systems demonstrate a pronounced preference for L-arginine compared to L-lysine that is not observed with any of the CAT proteins.

CAT-mediated transport activity can clearly be distinguished from that of system y^+L and $b^{0,+}$. The first is a very high affinity system for cationic amino acids (K_M for L-lysine is about 10 μM) that has been described in human erythrocytes (Deves et al., 1992), intestine, and placenta (Harvey et al., 1993; Fei et al., 1995). In the presence of Na^+ or Li^+ some neutral amino acids are recognized with similar affinity as cationic amino acids. System $b^{0,+}$ transports both cationic and neutral amino acids Na^+-independently and with similar affinity. Its activity was initially described in early mouse embryos, but it has also been observed in the epithelia of small intestine and renal tubules (Van Winkel et al., 1988; Magagnin et al., 1992). Two distantly related proteins (4F2hc and rBAT) that mediate y^+L and $b^{0,+}$ activity, respectively, when expressed in Xenopus oocytes have been described (for review see Palacin, 1994). They are predicted to span the membrane only one or four times and to have large intra- and extracellular domains, a structure not typical for carrier proteins. Both proteins seem to form heterodimers with yet-unidentified proteins that might represent the actual carriers.

Only one Na^+-dependent transport system for cationic amino acids ($B^{0,+}$) has been described (Van Winkel et al., 1985). Its activity has been found in the epithelia of small intestine and renal tubules, tissues with a strong capability to concentrate solutes. In most other cells, the negative membrane potential seems to be sufficient to accumulate enough intracellular cationic amino acids.

5. EXPRESSION OF THE CAT PROTEINS IN NO-PRODUCING CELLS

In most NO-producing cells (endothelial cells, smooth muscle cells, macrophages, and neuronal cells) transport of L-arginine and cationic NO synthase (NOS) inhibitors has been shown to be mediated predominantly by an Na^+-independent L-arginine transporter classified as system y^+ (Mann et al., 1989; Bogle et al., 1991, 1992b, 1994, 1996; Bussolati et al., 1993; Greene et al., 1993; Wu and Meininger, 1993; Schmidlin and Wiesinger, 1994; Block et al., 1995; Sobrevia et al., 1995; Sobrevia et al., 1996). However, one or more of the other transport systems for cationic amino acids described above might also be present in some of these cells yet not identified.

In various NO-producing cells, y^+ activity has been found to increase in response to exogenous stimuli such as LPS, interleukin-1β (IL-1β), tumor necrosis factor-α (TNF-α), insulin, angiotensin II, and bradykinin (Bogle et al., 1991, 1992a; Lind et al., 1993; Durante et al., 1995; Pan et al., 1995; Schmidlin and Wiesinger, 1995; Wileman et al., 1995; Cendan et al., 1996; Durante et al., 1996). Recent results suggest that this is due at least in part to a differential induction of CAT gene expression in these cells (Closs et al., 1993b; Gill et al., 1996; Simmons et al., 1996a,b; Stevens et al., 1996). In these studies, the expression of CAT-1, -2A, and -2B (but not CAT-3) mRNA has been examined (Table II). Consistent with system y^+, CAT-1 expression has been found in all cells investigated. CAT-2B is expressed in LPS-activated macrophages and glia cells, IL-1-β- and TNFα-induced vascular smooth muscle cells, and in IL-1β- and interferon-γ-(INF-γ)-induced cardiac myocytes and cardiac microvascular endothelial cells (CMEC). In all cases a significant expression of CAT-2B is only found after cytokine or LPS treatment that leads either to a concomitant increase of CAT-1 expression or leaves the CAT-1 expression unchanged. Therefore, CAT-2B seems to be an inducible carrier, always coexpressed with the more constitutive CAT-1. The expression of CAT-2B is often induced together with NOS II, the inducible isoform of NO synthase, suggesting that CAT-2B might have a specific role in delivering substrate to NOS II. Surprisingly, the expression of the low-affinity carrier CAT-2A in addition to CAT-1 and -2B has been found in cardiomyocytes and CMEC stimulated with IL-1β and INF-γ. The physiological role of a carrier with a K_M for cationic amino acids ≥2 mM in cells that are normally exposed to plasma with a cationic amino acid concentration of about 0.2 mM is not evident. In all the cell systems investigated, the observed increase in CAT mRNA expression is accompanied by a much

Table II.
CAT Expression in NO-Producing Cells[a]

Cell type	Treatment	Stimulation of y^+ Activity	Regulation of CAT expression			Reference
			CAT-1	CAT-2A	CAT-2B	
RAW 264.7	LPS	2×↑	↔	—	10×↑	Closs et al. (1993b)
Astroglia cells	LPS + INF-γ	4×↑	↔	n.d.	10×↑	Stevens et al. (1996)
Neurons	LPS + INF-γ	n.d.	↑	n.d.	↔	Stevens et al. (1996)
VSMC	Angiotensin II	2×↑	10×↑	—	—	Gill et al. (1996)
	IL1-β + TNFα	2×↑	↔	↑	—	
Cardiomyocytes	Insulin	2×↑	5×↑	↔	↔	Simmons et al. (1996a)
	IL1-β + INF-γ	2×↑	10×↑	10×↑	5×↑	
	Insulin + IL1-β + INF-γ	4×↑	50×↑	10×↑	5×↑	
CMEC	IL1-β + INF-γ	2×↑	2×↑	10×↑	10×↑	Simmons et al. (1996b)
	Dexamethasone + IL1-β + INF-γ	↔	↔	↔	↔	

[a]RAW 264.7, murine macrophage cell line; astroglia cells and neurons, cells from 1-day-old rats; VSMC, vascular smooth muscle cells from adult rats; cardiomyocytes, ventricular myocytes from adult and neonatal rats; CMEC, cardiac microvascular endothelial cells from adult rats. —, no expression; ↔, no change in expression; ↑, upregulation; ↓, downregulation; n.d., not determined. The numbers given for ×-fold up- or down-regulation are estimates from the respective references. For VSMC no distinction has been made between CAT-2A and -2B expression.

more moderate increase in y^+ activity, suggesting either that the mRNA is not translated into protein or that a large portion of the newly synthesized carriers are not translocated to the plasma membrane. To answer this question, the expression and subcellular localization of the CAT proteins have to be studied.

6. L-ARGININE TRANSPORT AND NO SYNTHESIS

The regulation of L-arginine transport combined with the differential expression of the CAT proteins suggests that the CATs play an important role in regulating L-arginine homeostasis in NO-producing cells. In addition, many commonly used NOS inhibitors are L-arginine analogues. Competition assays using intact cells suggest that some L-arginine analogues such as N^G-methyl-L-arginine (L-NMA) utilize the y^+ transport system, whereas others such as N^G-nitro-L-arginine (L-NNA) are transported by different carrier systems (Bogle et al., 1992b; Schmidt et al., 1993, 1994, 1995; Baydoun and Mann, 1994; Forray et al., 1995). We have recently shown that the substrate specificity of hCAT-2B for L-arginine analogues differs markedly from that of the inducible NOS II (Closs et al., 1997a). Thus there are some NOS inhibitors that are recognized and transported by hCAT-2B [e.g., L-NMA, L-N^5-(1-iminoethyl)-ornithine (L-NIO), or asymmetrical N^G,N'^G-dimethyl-L-arginine (L-ADMA)], while others are not (L-NNA). Conversely, some L-arginine analogues and cationic amino acids are transported by hCAT-2B [e.g., symmetrical N^G,N^G-dimethyl-Larginine (L-SDMA), L-lysine, or α-amino-δ-isothioureidovaleric acid (AITV)], but are not recognized by the NOS protein and do not inhibit its activity directly. The inhibitory effect of L-NMA, L-NIO, L-ADMA, and L-NNA on macrophage-derived NOS activity was confirmed by in vitro measurements of the NOS activity. In contrast, even high concentrations (1 mM) of L-SDMA and AITV did not interfere with NOS activity in vitro (Fig. 4A). Using Xenopus oocytes expressing hCAT-2B, we demonstrated directly that the NOS inhibitors L-NMA, L-NIO, and L-ADMA interfere with L-arginine transport by the y^+ carrier hCAT-2B (Fig. 4B). However, L-NNA inhibited the hCAT-2B-mediated L-arginine transport only slightly and at very high concentrations (10 mM). Because a similar degree of inhibition was observed for Glut-1-mediated D-glucose transport, this inhibition seems to be nonspecific (Closs et al., 1997a).

The different behavior of the arginine derivatives with respect to CAT-mediated transport is most likely due to differences in the charge of their side chain. On the other hand, NOS enzymes seem to require at least one free guanidino nitrogen for substrate/inhibitor recognition, because L-ADMA (in which both methyl groups are on the same guanidino nitrogen) is an inhibitor of the enzyme, whereas L-SDMA (which has one methyl group on each guanidino nitrogen) and AITV

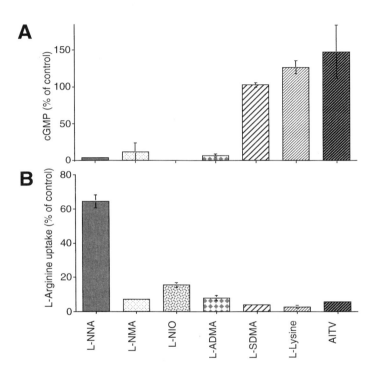

Figure 4. Differences between NOS II and CAT-2B in the recognition of L-arginine analogues. Effect of different L-arginine analogues and L-lysine on NOS II activity (A) and CAT-2B-mediated L-arginine transport (B). Data are from Closs et al. (1997a). The chemical structures of the L-arginine analogues are shown on top of the respective columns in panel A (b, L-arginine backbone, identical in all compounds); the names of the compounds are indicated at the bottom of panel B (L-NNA: N^G-nitro-L-arginine; L-NMA: N^G-methyl-L-arginine; L-NIO: L-N^5-(1-iminoethyl)-ornithine; L-ADMA: asymmetrical N^G,N'^G-dimethyl-L-arginine; L-SDMA: symmetrical N^G,N^G-dimethyl-L-arginine; AITV: α-amino-δ-isothioureidovaleric acid). (A) Lysates from 10^6 LPS/INF-γ-induced RAW 269.7 macrophages (as the source of NOS II) were incubated for 3 min on 10^6 RFL-6 fibroblasts in the presence of the indicated L-arginine analogues or L-lysine. The cGMP content of the RFL-6 cells at the end of the 3 min was determined by radioimmunoassay. The basal cGMP content of the RFL-6 cells (0.6 ± 0.1 pmole/ 10^6 RFL-6 cells) was subtracted from that of the test samples. Data are expressed as

(in which the guanidino nitrogen at C5 of L-arginine has been replaced with a sulfur) are not.

Because of a trans-stimulation effect L-arginine analogues that are CAT substrates not only compete with L-arginine for transport, but also drive out the intracellular L-arginine. A high-pressure liquid chromatography analysis demonstrated that SDMA is a good substrate for hCAT-2B and that intracellular L-arginine is exchanged for extracellular L-SDMA (Closs *et al.*, 1997a). In contrast, the intracellular L-arginine concentration changed to a much lesser extent after a 30-min incubation with 5 mM AITV. Therefore, only little AITV seems to be transported by hCAT-2B. New L-arginine analogues that compete with L-arginine for the substrate-binding site of the CAT proteins but are not transported could probably be developed on the structural basis of AITV. Such compounds would not change the intracellular L-arginine pool and could therefore be used to elucidate the role of CATs in the actual transport processes that provide NOS enzymes with substrate.

An indirect inhibitory effect on the NOS activity in intact cells may be mediated by compounds like L-SDMA or AITV that do not interfere with the enzyme directly, but limit substrate availability. Also, in addition to their direct inhibition of NOS, inhibitors such as L-NMA, L-NIO, and L-ADMA may have an indirect effect on the enzyme in intact cells by depletion of L-arginine. This may be especially relevant in cells like macrophages where NOS activity is largely dependent on extracellular L-arginine (Baydoun and Mann, 1994).

7. CONCLUSIONS

When the transport properties of mCAT-1 were described in 1991, *the* y^+ carrier and major transporter for cationic amino acids seemed to be discovered. Today, we know that there are at least three different CAT isoforms that mediate y^+ activity and the family might be growing. In addition, transport systems for cationic amino acids other than y^+ have been described and proteins that induce the respective transport activities have been identified. Consequently, the transport of cationic amino acids appears to be a complex process involving many proteins— carriers and possibly also regulatory proteins—whose expression is cell-specific

---

Figure 4. (*continued*) percent of the cGMP content measured in RFL-6 cells incubated with RAW 264.7 lysate alone (11.9 ± 2.3/10^6 RFL-6 cells). Columns represent means ± SEM, $n = 5-8$. (B) L-Arginine uptake was measured in oocytes injected with 25 ng of cRNA encoding hCAT-2B and 3 days later incubated for 30 min with uptake solution containing 50 μM [^{3}H]-L-arginine and 10 mM of L-arginine analogues. Data are expressed as percent of the uptake measured in control oocytes incubated with 50 μM [^{3}H]-L-arginine alone. One hundred percent is equivalent to 0.2 ± 0.05 nmole L-arginine/oocyte/30 min. Symbols represent means ± SEM, $n = 5-10$.

and dependent on a variety of external stimuli. The multitude of different carrier proteins possibly offers an opportunity to inhibit specifically transport processes in a given cell without interfering with the overall transport of cationic amino acids. A functional knockout of individual proteins would lead to a better understanding of the physiological role of each protein. To date no specific inhibitors are available to block the function of the CAT proteins or other transporters for cationic amino acids. The discovery of such inhibitors not only could provide scientific tools, but could also lead to the development of new drugs targeting specific carrier proteins. It would, for instance, be desirable to inhibit CAT-2B in cells expressing the inducible NOS without interfering with CAT-1 activity. If CAT-2B played a crucial role in delivering substrate to NOS II, NO production could be reduced without interfering with other processes depending on cationic amino acids supply by CAT-1.

ACKNOWLEDGMENTS

This work was supported by the Deutsche Forschungsgemeinschaft, Bonn, Germany, by Grants C1 100/ 3-2 and SFB 553/Project B4 (to E.I.C.).

REFERENCES

Albritton, L. M., Tseng, L., Scadden, D., and Cunningham, J. M., 1989, A putative murine ectropic retrovirus receptor gene encodes a multiple membrane-spanning protein and confers susceptibility to virus infection, *Cell* **57**:659–666.

Albritton, L. M., Bowcock, A. M., Eddy, R. L., Morton, C. C., Tseng, L., Farrer, L. A., Cavalli, S. L., Shows, T. B., and Cunningham, J. M., 1992, The human cationic amino acid transporter (ATRC1): Physical and genetic mapping to 13q12–q14, *Genomics* **12**:430–434.

Albritton, L. M., Kim, J. W., Tseng, L., and Cunningham, J., 1993, Envelope binding domain in the cationic amino acid transporter determines the host range of ecotropic murine retroviruses, *J. Virol.* **67**:2091–2096.

Baydoun, A. R., and Mann, G. E., 1994, Selective targeting of nitric oxide synthase inhibitors to system y^+ in activated macrophages, *Biochem. Biophys. Res. Commun.* **200**:726–731.

Block, E. R., Herrera, H., and Couch, M., 1995, Hypoxia inhibits L-arginine uptake by pulmonary artery endothelial cells, *Am. J. Physiol.* **269**:L574–L580.

Bogle, R. G., Coade, S. B., Moncada, S., Pearson, J. D., and Mann, G. E., 1991, Bradykinin and ATP stimulate L-arginine uptake and nitric oxide release in vascular endothelial cells, *Biochem. Biophys. Res. Commun.* **180**:926–932.

Bogle, R. G., Baydoun, A. R., Pearson, J. D., Moncada, S., and Mann, G. E., 1992a, L-Arginine transport is increased in macrophages generating nitric oxide, *Biochem. J.* **284**:15–18.

Bogle, R. G., Moncada, S., Pearson, J. D., and Mann, G. E., 1992b, Identification of inhibitors of nitric oxide synthase that do not interact with the endothelial cell L-arginine transporter, *Br. J. Pharmacol.* **105**:768–770.
Bogle, R. G., Mann, G. E., Pearson, J. D., and Morgan, D. M. L., 1994, Endothelial polyamine uptake—Selective stimulation by L-arginine deprivation or polyamine depletion, *Am. J. Physiol.* **266**:C776–C783.
Bogle, R. G., Baydoun, A. R., Pearson, J. D., and Mann, G. E., 1996, Regulation of L-arginine transport and nitric oxide release in superfused porcine aortic endothelial cells, *J. Physiol.* **490**:229–241.
Bussolati, O., Sala, R., Astorri, A., Rotoli, B. M., Dall'Asta, V., and Gazzola, G. C., 1993, Characterization of amino acid transport in human endothelial cells, *Am. J. Physiol.* **265**:C1006–C1014.
Calandra, R. S., Rulli, S. B., Frungieri, M. B., Suescun, M. O., and Gonzalez Calvar, S. I., 1996, Polyamines in the male reproductive system, *Acta Physiol. Pharmacol. Ther. Latinoam.* **46**:209–222.
Cendan, J. C., Topping, D. L., Pruitt, J., Snowdy, S., Copeland, E. M., and Lind, D. S., 1996, inflammatory mediators stimulate arginine transport and arginine-derived nitric oxide production in a murine breast cancer cell line, *J. Surg. Res.* **60**:284–288.
Christensen, H. N., 1964, A transport system serving for mono- and diamino acids. *Proc. Natl. Acad. Sci. USA* **51**:337–344.
Christensen, H. N., 1990, Role of amino acid transport and countertransport in nutrition and metabolism, *Physiol. Rev.* **70**:43–77.
Christensen, H. N., and Antonioli, J. A., 1969, Cationic amino acid transport in the rabbit reticulocyte, *J. Biol. Chem.* **244**:1497–1504.
Closs, E. I., 1996, CATs, a family of three distinct cationic amino acid transporters, *Amino Acids* **11**:193–208.
Closs, E. I., Albritton, L. M., Kim, J. W., and Cunningham, J. M., 1993a, Identification of a low affinity, high capacity transporter of cationic amino acids in mouse liver, *J. Biol. Chem.* **268**:7538–7544.
Closs, E. I., Lyons, C. R., Kelly, C., and Cunningham, J. M., 1993b, Characterization of the third member of the MCAT family of cationic amino acid transporters—Identification of a domain that determines the transport properties of the MCAT proteins, *J. Biol. Chem.* **268**:20796–20800.
Closs, E. I., Basha, F. Z., Habermeier, A., and Förstermann, E., 1997a, Interference of L-arginine analogues with L-arginine transport mediated by the y^+ carrier hCAT-2B, *Nitric Oxide Biol. Chem.* **1**:67–73.
Closs, E. I., Gräf, P., Habermeier, A., Cunningham, J. M., and Förstermann, U., 1997b, The human cationic amino acid transporters hCAT-1, hCAT-2A and hCAT-2B: Three related carriers with distinct transport properties, *Biochemistry* **36**:6462–6468.
Deves, R., Chavez, P., and Boyd, C. A., 1992, Identification of a new transport system (y+L) in human erythrocytes that recognizes lysine and leucine with high affinity, *J. Physiol. (Lond.)* **454**:491–501.
Durante, W., Liao, L., and Schafer, A. I., 1995, Differential regulation of L-arginine transport and inducible NOS in cultured vascular smooth muscle cells, *Am. J. Physiol.* **37**:H1158–H1164.
Durante, W., Liao, L., Iftikhar, I., Obrien, W. E., and Schafer, A. I., 1996, Differential reg-

ulation of L-arginine transport and nitric oxide production by vascular smooth muscle and endothelium, *Circ. Res.* **78**:1075–1082.
Fei, Y. J., Prasad, P. D., Leibach, F. H., and Ganapathy, V., 1995, The amino acid transport system y(+)L induced in *Xenopus laevis* oocytes by human choriocarcinoma cell (JAR) mRNA is functionally related to the heavy chain of the 4F2 cell surface antigen, *Biochemistry* **34**:8744–8751.
Forray, M. I., Angelo, S., Boyd, C. A. R., and Deves, R., 1995, Transport of nitric oxide synthase inhibitors through cationic amino acid carriers in human erythrocytes, *Biochem. Pharmacol.* **50**:1963–1968.
Förstermann, U., Gath, I., Schwarz, P., Closs, E. I., and Kleinert, H., 1995a, Isoforms of nitric oxide synthase: Properties, cellular distribution and expressional control, *Biochem. Pharmacol.* **50**:1321–1332.
Förstermann, U., Kleinert, H., Gath, I., Schwarz, P., Closs, E. I., and Dun, N. J., 1995b, Expression and expressional control of nitric oxide synthases in various cell types, in: *Nitric Oxide—Biochemistry, Molecular Biology and Therapeutic Implications* (L. Ignarro and F. Murad, eds), Academic Press, San Diego, California, pp. 171–186.
Gill, D. J., Low, B. C., and Grigor, M. R., 1996, Interleukin-1β and tumor necrosis factor-α stimulate the cat-2 gene of L-arginine transporter in cultured vascular smooth muscle cells, *J. Biol. Chem.* **271**:11280–11283.
Greene, B., Pacitti, A. J., and Souba, W. W., 1993, Characterization of L-arginine transport by pulmonary artery endothelial cells, *Am. J. Physiol.* **264**:L351–L356.
Harvey, C. M., Muzyka, W. R., Yao, S. Y. M., Cheeseman, C. I., and Young, J. D., 1993, Expression of rat intestinal L-lysine transport systems in isolated oocytes of *Xenopus laevis, Am. J. Physiol.* **265**:G99–G106.
Hosokawa, H., Sawamura, T., Kobayashi, S., Ninomiya, H., Miwa, S., and Masaki, T., 1997, Cloning and characterization of a brain-specific cationic amino acid transporter, *J. Biol. Chem.* **272**:8717–8722.
Ito, K., and Groudine, M., 1997, A new member of the cationic amino acid transporter family is preferentially expressed in adult mouse brain, *J. Biol. Chem.* **272**:26780–26786.
Jenkinson, C. P., Grody, W. W., and Cederbaum, S. D., 1996, Comparative properties of arginases, *Comp. Biochem. Physiol.* **114**:107–132.
Kakuda, D. K., Finley, K. D., Dionne, V. E., and MacLeod, C. L., 1993, Two distinct gene products mediate y^+ type cationic amino acid transport in *Xenopus* oocytes and show different tissue expression patterns, *Transgene* **1**:91–101.
Kavanaugh, M. P., 1993, Voltage dependence of facilitated arginine flux mediated by the system y^+ basic amino acid transporter, *Biochemistry* **32**:5781–5785.
Kavanaugh, M. P., Wang, H., Zhang, Z., Zhang, W., and Wu, Y. N., 1994, Control of cationic amino acid transport and retroviral receptor functions in a membrane protein family, *J. Biol. Chem.* **269**:15445–15450.
Kim, J. W., and Cunningham, J. M., 1993, N-linked glycosylation of the receptor for murine ecotropic retroviruses is altered in virus-infected cells, *J. Biol. Chem.* **268**:16316–16320.
Kim, J. W., Closs, E. I., Albritton, L. M., and Cunningham, J. M., 1991, Transport of cationic amino acids by the mouse ecotropic retrovirus receptor, *Nature* **352**:725–728.
Lind, D. S., Copeland, E. M., Souba, W. W., Baker, C. C., and Billiar, T., 1993, Endotoxin stimulates arginine transport in pulmonary artery endothelial cells, *Surgery* **114**:199–205.

Macleod, C. L., and Kakuda, D. K., 1996, Regulation of CAT: Cationic amino acid transporter gene expression, *Amino Acids* **11**:171–191.
MacLeod, C. L., Finley, K., Kakuda, D., Kozak, C. A., and Wilkinson, M. F., 1990, Activated T cells express a novel gene on chromosome 8 that is closely related to the murine ectropic retroviral receptor, *Mol. Cell. Biol.* **10**:3663–3674.
Magagnin, S., Bertran, J., Werner, A., Markovich, D., Biber, J., Palacin, M., and Murer, H., 1992, Poly(A)$^+$ RNA from rabbit intestinal mucosa induces $b^{0,+}$ and y^+ amino acid transport activities in *Xenopus laevis* oocytes, *J. Biol. Chem.* **267**:15384–15390.
Mann, G. E., Pearson, J. D., Sheriff, C. J., and Toothill, V. J., 1989, Expression of amino acid transport systems in cultured human umbilical vein endothelial cells, *J. Physiol. (Lond.)* **410**:325–339.
Michel, T., and Feron, E., 1997, Nitric oxide synthases: Which, where, how, and why? *J. Clin. Invest.* **100**:2146–2152.
Palacin, M., 1994, A new family of proteins (rBAT and 4F2hc) involved in cationic and zwitterionic amino acid transport: A tale of two proteins in search of a transport function. *J. Exp. Biol.* **196**:123–137.
Pan, M., Wasa, M., Lind, D. S., Gertler, J., Abbott, W., and Souba, W. W., 1995, TNF-stimulated arginine transport by human vascular endothelium requires activation of protein kinase C, *Ann. Surg.* **221**:590–601.
Perkins, C. P., Mar, V., Shutter, J. R., Castillo del, J., Danilenko, D. M., Medlock, E. S., Ponting, I. L., Graham, M., Stark, K. L., Zuo, Y., Cunningham, J. M., and Bossekmann, R. A., 1997, Anemia and perinatal death result from loss of the murine ectropic retrovirus receptor mCAT-1, *Genes Dev.* **11**:914–925.
Puppi, M., and Henning, S. J., 1995, Cloning of the rat ecotropic retroviral receptor and studies of its expression in intestinal tissues, *Proc. Soc. Exp. Biol. Med.* **209**:38–45.
Reizer, J., Finley, K., Kakuda, D., MacLeod, C., Reizer, A., and Saier, M., 1993, Mammalian integral membrane receptors are homologous to facilitators and antiporters of yeast, fungi and eubacteria, *Protein Sci.* **2**:20–30.
Schmidlin, A., and Wiesinger, H., 1994, Transport of L-arginine in cultured glial cells, *Glia* **11**:262–268.
Schmidlin, A., and Wiesinger, H., 1995, Stimulation of arginine transport and nitric oxide production by lipopolysaccharide is mediated by different signaling pathways in astrocytes, *J. Neurochem.* **65**:590–594.
Schmidt, K., Klatt, P., and Mayer, B., 1993, Characterization of endothelial cell amino acid transport systems involved in the actions of nitric oxide synthase inhibitors, *Mol. Pharmacol.* **44**:615–621.
Schmidt, K., Klatt, P., and Mayer, B., 1994, Uptake of nitric oxide synthase inhibitors by macrophage RAW 264.7 cells, *Biochem. J.* **301**:313–316.
Schmidt, K., List, B. M., Klatt, P., and Mayer, B., 1995, Characterization of neuronal amino acid transporters: Uptake of nitric oxide synthase inhibitors and implication for their biological effects, *J. Neurochem.* **64**:1469–1475.
Sessa, A., and Perin, A., 1997, Ethanol and polyamine metabolism: Physiologic and pathologic implications: A review, *Alcohol Clin. Exp. Res.* **21**:318–325.
Sessa, W. C., 1994, The nitric oxide synthase family of proteins, *J. Vasc. Res.* **31**:131–143.
Simmons, W. W., Closs, E. I., Cunningham, J. M., Smith, T. W., and Kelly, R. A., 1996a,

Cytokines and insulin induce cationic amino acid transporter (CAT) expression in cardiac myocytes. Regulation of L-arginine transport and NO production by CAT-1, CAT-2A, and CAT-2B, *J. Biol. Chem.* **271**:11694–11702.

Simmons, W. W., Ungureanu-Longrois, D., Smith, G. K., Smith, T. W., and Kelly, R. A., 1996b, Glucocorticoids regulate inducible nitric oxide synthase by inhibiting tetrahydrobiopterin synthesis and L-arginine transport, *J. Biol. Chem.* **271**:23928–23937.

Sobrevia, L., Cesare, P., Yudilevich, D. L., and Mann, G. E., 1995, Diabetes-induced activation of system y(+) and nitric oxide synthase in human endothelial cells: Association with membrane hyperpolarization, *J. Physiol.* **489**:183–192.

Sobrevia, L., Nadal, A., Yudilevich, D. L., and Mann, G. E., 1996, Activation of L-arginine transport (system y(+)) and nitric oxide synthase by elevated glucose and insulin in human endothelial cells—Short paper given rapid review, *J. Physiol.* **490**:775–781.

Stevens, B. R., Kakuda, D. K., Yu, K., Waters, M., Vo, C. B., and Raizada, M. K., 1996, Induced nitric oxide synthesis is dependent on induced alternatively spliced CAT-2 encoding L-arginine transport in brain astrocytes, *J. Biol. Chem.* **271**:24017–24022.

Van Winkel, L. J., Christensen, H. N., and Campione, A. L., 1985, Na^+-dependent transport of basic, zwitterionic, and bicyclic amino acids by a broad scope system in mouse blastocytes, *J. Biol. Chem.* **260**:12118–12123.

Van Winkel, L. J., Campione, A. L., and Gorman, J. M., 1988, Na^+-independent transport of basic and zwitterionic amino acids in mouse blastocytes by a shared system and by processes which distinguish between these substrates, *J. Biol. Chem.* **263**:3150–3163.

Van Winkel, L. J., and Campione, A. L., 1990, Functional changes in cation-preferring amino acid transport during development of preimplantation mouse conceptuses, *Biochim. Biophys. Acta* **1028**:165–173.

Wang, H., Kavanaugh, M. P., North, R. A., and Kabat, D., 1991, Cell-surface receptor for ecotropic murine retroviruses is a basic amino-acid transporter, *Nature* **352**:729–731.

White, M. F., 1985, The transport of cationic amino acids across the plasma membrane of mammalian cells, *Biochim. Biophys. Acta* **822**:355–374.

White, M. F., and Christensen, H. N., 1982a, Cationic amino acid transport into cultured animal cells. II Transport system barely perceptible in ordinary hepatocytes, but active in hepatoma cell lines, *J. Biol. Chem.* **257**:4450–4457.

White, M. F., and Christensen, H. N., 1982b, The two-way flux of cationic amino acids across the plasma membrane of mammalian cells is largely explained by a single transport system, *J. Biol. Chem.* **257**:10069–10080.

White, M. F., Gazzola, G. C., and Christensen, H. N., 1982, Cationic amino acid transport into cultured animal cells. I. Influx into cultured human fibroblasts, *J. Biol. Chem.* **257**:4443–4449.

Wileman, S. M., Mann, G. E., and Baydoun, A. R., 1995, Induction of L-arginine transport and nitric oxide synthase in vascular smooth muscle cells: Synergistic actions of proinflammatory cytokines and bacterial lipopolysaccharide, *Br. J. Pharmacol.* **116**:3243–3250.

Wu, G. Y., and Meininger, C. J., 1993, Regulation of L-arginine synthesis from L-citrulline by L-glutamine in endothelial cells, *Am. J. Physiol.* **265**:H1965–H1971.

Wu, J. Y., Robinson, D., Kung, H. J., and Hatzoglou, M., 1994, Hormonal regulation of the gene for the type C ecotropic retrovirus receptor in rat liver cells, *J. Virol.* **68**:1615–1623.

Yoshimoto, T., Yoshimoto, E., and Meruelo, D., 1991, Molecular cloning and characterization of a novel human gene homologous to the murine ecotropic retroviral receptor, *Virology* **185**:10–15.

9

Electrophysiological Analysis of Renal Na$^+$-Coupled Divalent Anion Transporters

Ian Forster, Jürg Biber, and Heini Murer

1. INTRODUCTION

The homeostasis of inorganic phosphate (P_i) and sulfate (S_i) in mammals is largely achieved by reabsorption of these anions in the renal proximal tubule and the small intestine. Under physiological conditions, the kidney plays the major role in maintaining the extracellular concentration of both anions. In both cases transepithelial transport is performed by apically located Na$^+$-dependent transport systems (Na$^+$/P_i and Na$^+$/S_i cotransporters) and basolaterally localized anion exchange mechanisms (Murer and Biber 1992; Murer *et al.*, 1994). *In vivo* microperfusion studies as well as *in vitro* studies on tubules and brush border membrane vesicles (BBMVs) obtained from appropriately pretreated animals suggest that the brush border membrane entry step is the most probable final target for physiological and pathophysiological regulation of both proximal tubular P_i reabsorption (for review, see Biber *et al.*, 1996; Murer and Biber; 1997) and S_i reabsorption (Markovich *et al.*, 1998; Fernandes *et al.*, 1997; for review see Murer *et al.*, 1994).

Although the steady-state kinetics of both apical Na$^+$-coupled anion co-

---

Ian Forster, Jürg Biber, and Heini Murer • Physiological Institute, University of Zurich, CH-8057 Zurich, Switzerland.

Membrane Transporters as Drug Targets, edited by Amidon and Sadée. Kluwer Academic/Plenum Publishers, New York, 1999.

transporter systems has been extensively characterized by isotope tracer flux studies using the BBMV preparation isolated from the kidney cortex and small intestine from a number of species as well as microperfusion studies (for review, see Murer et al., 1994), a significant advance in our understanding of the mechanisms of Na^+/P_i and Na^+/S_i cotransport systems has been gained through the application of expression cloning techniques. Here the specific cotransporter proteins have been isolated and their function studied essentially in isolation using, for example, the *Xenopus laevis* oocyte expression system. Results from tracer flux studies have confirmed that the cloned transporters for both P_i and S_i display all the characteristics previously established using BBMVs and microperfusion techniques (Magnanin et al., 1993; Markovich et al., 1993). Moreover, the recent finding that these cotransporters are electrogenic, whereby a net charge transfer occurs per transport cycle, has allowed the use of electrophysiological techniques to characterize their kinetics by means of the *Xenopus* oocyte expression system (Busch et al., 1994a,b, 1995; Hartmann et al., 1995; Forster et al., 1997a, 1998; for review see Busch et al., 1996).

In the case of renal P_i apical reabsorption, two Na^+/P_i cotransporters have been cloned, termed type I and type II, which are distinguished on both the molecular and functional levels (for review, see Murer and Biber, 1996, 1997; Busch et al., 1996). The type II Na^+/P_i cotransporter, of which to date seven species-specific isoforms have so far been identified, has been most extensively characterized (Murer and Biber, 1997). Similarly, a renal sulfate transporter was cloned from rat renal cortex (NaSi-1) (Markovich et al., 1993; for review, see Murer et al., 1994) and its expression in *Xenopus* oocytes revealed properties consistent with Na^+-coupled transport (Busch et al., 1994b).

The location of both Na^+-coupled anion transporter systems on the renal proximal tubule brush border membrane makes them potential targets for drugs which would be freely filtered by the glomerulus and which could potentially regulate the reabsorption process of the respective anions. Moreover, serum levels of drugs could be maintained through reabsorption at the apical membrane by the cotransporter. In this context, the oocyte expression system in combination with electrophysiology provides a versatile means to investigate and quantify the action of specific substances on renal cotransport kinetics and the cotransport properties of other potential substrates.

In this chapter we focus on the type II Na^+/P_i cotransporter by describing the steady-state and pre-steady-state kinetic properties with emphasis on both the methodology and analysis aspects. This has allowed the determination of substrate binding order, identification of voltage-dependent steps in the transport cycle, and development of a kinetic model for Na^+/P_i cotransport based on detailed biophysical studies of two cotransporter isoforms (Forster et al., 1997a, 1998). This material can serve as a basis for characterizing other electrogenic renal Na^+-coupled transporters.

2. STEADY-STATE ELECTROPHYSIOLOGICAL CHARACTERISTICS

2.1. Recording P_i-Induced Currents from the *Xenopus* Oocyte

The most common means of investigating the electrogenic transport by cotransporters expressed in *Xenopus laevis* oocytes utilizes the two-electrode voltage clamp as depicted in Fig. 1A [for general review of *Xenopus* oocyte methods

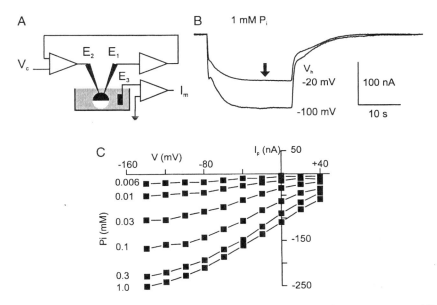

Figure 1. Electrogenic behavior of type II Na^+/P_i cotransporter expressed in *Xenopus* oocytes. (A) Basic scheme of two-electrode voltage-clamp recording system. The oocyte is placed in a small recording chamber and impaled with two glass microelectrodes filled with 3 M KCl (E_1, E_2). Electrode E_1 measures the transmembrane potential with respect to a bath electrode E_3 which is held at signal ground by a virtual ground bath amplifier. The membrane potential is amplified and compared to a command signal (V_c). Current is then supplied to the cell via electrode E_2 to maintain the membrane potential at the desired value. The complete arrangement, including the cell, comprises a negative-feedback control system. The current I_m is measured at the output of the bath amplifier. (B) When P_i is applied to an oocyte expressing the type II Na^+/P_i cotransporter, NaPi-2, an inward current is induced which is dependent on the holding potential V_h of the cell. Arrow indicates the time at which steady-state conditions have been reached, and a voltage staircase is applied to determine the $I-V$ response (see text). The baselines of the two responses in the control solution preceding application of P_i have been superimposed. (C) Typical $I-V$ data from an oocyte determined by applying a staircase voltage from −140 to +40 mV at the P_i indicated and subtracting the response of the control solution to give the P_i-induced current I_p. Data points have been joined for visualization purposes only. Data redrawn and adapted from Forster *et al.* (1998).

and voltage clamping, see appropriate articles in Rudy and Iverson 1992)]. This technique allows direct measurement of the electrical response of the cell to different superfusates and the preparation can be maintained stably for up to several hours without deterioration. As shown in Fig. 1B, under voltage clamp conditions, oocytes injected with cRNA coding for the type II Na^+/P_i cotransporter and superfused with P_i (1 mM) in the presence of Na^+ (100 mM) display an inward current, the magnitude of which depends on the holding potential V_h. Such behavior is expected for electrogenic cotransport and indicates a net inward movement of positive charge. If this experiment is repeated with different P_i concentrations and over a range of membrane potentials, a set of current–voltage (I–V) curves can be directly obtained as shown in Fig. 1C. In practice this is achieved easily by applying a staircase voltage which covers the range of potentials of interest for each P_i to the cell (Forster et al., 1998). The P_i-dependent response at each V and P_i is then obtained by subtracting a record made under control conditions (i.e., with 0 mM P_i) to give the I–V relation. These data indicate that the response saturates at sufficiently large P_i, characteristic for membrane transporter phenomena. Moreover, for any given P_i, the I–V relation is nonlinear at extreme hyperpolarizing and depolarizing potentials, which suggests the presence of voltage-independent rate-limiting steps in the transport cycle.

2.2. Dose and Voltage Dependence of Steady-State Kinetics

At a given membrane potential, dose–response relationships can then be obtained from the I–V data as illustrated in Fig. 2. These data are quantitated by fitting to the modified Hill equation given by $I_p = I_{p,\max}[S]^n/\{[S]^n + (K_m^s)^n\}$, where [S] is the substrate concentration, $I_{p,\max}$ is the extrapolated maximum current, K_m^s is the concentration of substrate S which gives a half-maximum response or apparent affinity constant, and n is the Hill coefficient. The latter parameter provides a measure of the degree of cooperativity of the substrate-binding reaction (e.g., Weiss, 1997) and under conditions of high positive cooperativity it can be taken as an index of the stoichiometry of the reaction.

For P_i as the variable substrate (Fig. 2A), the dose–response at $V_h = -50$ mV has a rectangular hyperbolic form and the maximum P_i-induced current is dependent on the external Na^+ concentration. Moreover, fits to these data indicate that the half-maximum concentration or apparent affinity coefficient for P_i (K_m^{Pi}) is also dependent on Na^+. Of particular significance is the finding that K_m^{Pi} is a function of V_h, showing a markedly stronger voltage dependence at reduced Na^+ (Fig. 2B). Furthermore, the fits to these data give a Hill coefficient n close to unity for both Na^+ concentrations, suggesting that each transport cycle involves the binding of only one P_i molecule. For the complementary dose–response with Na^+ as the variable substrate, type II Na^+/P_i cotransporters typically do not show saturation up

Renal Na⁺-Coupled Divalent Anion Transporters

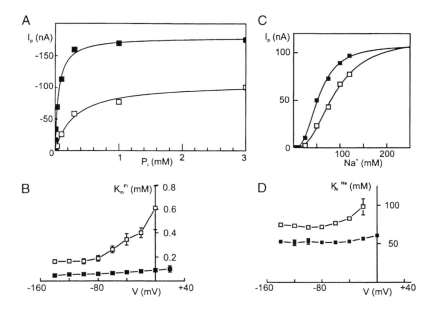

Figure 2. Steady-state dose–response characteristics of type II Na⁺P$_i$ cotransporter NaPi2. Both the apparent K_m for P$_i$ (K_m^{Pi}) and that for Na⁺ (K_m^{Na}) are voltage dependent and also dependent on the concentrations of the other substrate (Na⁺ or P$_i$, respectively) indicating that these conditions should always be specified when comparing steady-state kinetic data. (A) Typical dose–response data for the same oocyte with respect to P$_i$ for two Na⁺ concentrations: 96 mM (filled symbols) and 50 mM (open symbols). Continuous lines are fits to the modified Hill equation (see text). For this cell the fit parameters for 96 (50) mM Na⁺ were n = 0.8 (0.9), K_m^{Pi} = 0.054 (0.26) mM, and $I_{p,max}$ = 74 (50) nA. All data are at a holding potential of −50 mV. (B) Voltage dependence of K_m^{Pi} for two Na⁺ concentrations: 96 mM (filled symbols) and 50 mM (open symbols), pooled from four oocytes. (C) Typical dose–response data for the same oocyte with respect to Na⁺ for two P$_i$ concentrations: 1.0 mM (filled symbols) and 0.1 mM (open symbols). Continuous lines are fits to the modified Hill equation as in panel A. For this cell the fit parameters at 1.0 (0.1) mM P$_i$ were n = 2.6 (2.6), K_m^{Na} = 89 (50) mM, and $I_{p,max}$ = 114 (109) nA. All data were obtained at a holding potential of −50 mV. (D) Voltage dependence of K_m^{Na} for two P$_i$ concentrations: 1.0 mM (filled symbols) and 0.1 mM (open symbols), pooled from four oocytes. Data redrawn and adapted from Forster et al. (1998).

to the maximum external Na⁺ concentration permitted by the intact oocyte preparation. Nevertheless, as indicated in Fig. 2C, free fits to the Hill equation strongly suggest that the maximum P$_i$-induced current is independent of P$_i$. The corresponding K_m^{Na} (Fig. 2D) shows a weak voltage dependence at saturating P$_i$ (1 mM) and a significant voltage dependence for P$_i$ close to K_m^{Pi} (0.1 mM). Moreover, the Hill coefficient predicted from these fits lies close to 3, indicating that more than one Na⁺ ion binds per transport cycle. Under conditions of strong positive cooperativity a Hill coefficient >2 is expected since at neutral pH, P$_i$ is predominantly divalent and at least three Na⁺ ions would be required to give the observed inward current in the presence of P_i. These conclusions have now been confirmed by direct measurement of substrate flux and charge movement under voltage clamp conditions from the same oocyte (Forster *et al.*, 1999a).

2.3. Sodium Slippage in the Absence of P_i

Recent studies of the rat isoform NaPi-2 (Forster *et al.*, 1998) have also revealed that type II Na^+/P_i cotransporters have a Na^+ "leak" or slippage, similar to that first described by Umbach *et al.* (1990) for the sodium–glucose cotransporter SGLT1, whereby Na^+ ions can translocate in the absence of P_i. For NaPi-2, the slippage current shows a saturating dose dependence with a Hill coefficient close to unity and accounts for about 10% of maximum P_i-induced current. When characterizing cotransporter steady-state kinetics, slippage should be considered because its presence can affect the accuracy of the kinetic parameters, since they are derived on the assumption the currents measured result from Na^+-coupled transport only.

2.4. The Order of Substrate Binding and Voltage Dependence Based on Steady-State Kinetics

The steady-state dose–response data are consistent with an ordered binding scheme in which P_i binding precedes Na^+ binding before the fully loaded carrier translocates, based on our finding that $I_{p,\max}$ is dependent on Na^+. Furthermore, we have interpreted the presence of slippage as being a strong indicator that Na^+ is the first substrate to bind to the empty carrier, and based on our analyses we have concluded that this step could involve one Na^+ ion. This first Na^+-binding step contributes to the voltage dependence of the apparent affinity constants for both substrates, whereas the weaker voltage dependence of K_m^{Na} at saturating P_i suggests that the final Na^+-binding step is most likely voltage independent.

3. PRE-STEADY-STATE ELECTROPHYSIOLOGICAL CHARACTERISTICS

3.1. General Properties of Pre-Steady-State Relaxations

Pre-steady-state relaxations, resulting from the application of voltage steps to the voltage-clamped cell, are a general feature of many electrogenic cation-coupled cotransporters, having been first reported by Birnir *et al.* (1991) for the cloned $Na^+/$

glucose cotransporter (SGLT1). Subsequent detailed kinetic studies of the SGLT family by Wright and colleagues (e.g., Hazama et al., 1997; Loo et al., 1993; Parent et al., 1992a,b) and studies of other cation-coupled cotransporters [e.g., $Na^+/Cl^-/$ GABA transporter (GAT-1; Mager et al., 1993), Na^+/myo-inositol transporter (SMIT; Hager et al., 1995), Na^+/glutamate transporter (EEAT-2, Wadiche et al., 1995), and Na^+/I^- transporter (NIS, Eskandari et al., 1997)] have established the usefulness of this technique for the quantification of kinetic differences between cotransporter isoforms and identification of partial reactions in the transport cycle. For example, our detailed analysis of the pre-steady-state kinetics of the flounder (NaPi-5) and rat (NaPi-2) isoforms (Forster et al., 1997a, 1998) revealed species-related differences in the rate constants associated with the empty carrier, and Na^+ binding which could not be determined directly from steady-state analysis.

Figure 3 shows a typical example of pre-steady-state relaxations recorded

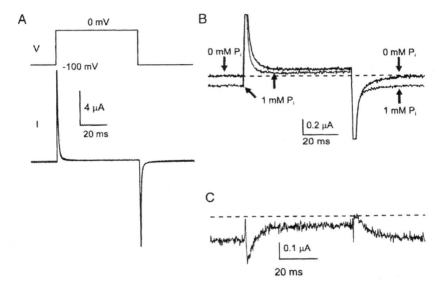

Figure 3. Voltage steps elicit pre-steady-state charge movements recorded from an oocyte expressing NaPi-2. (A) Voltage step between -100 and 0 mV applied to oocyte and corresponding membrane currents superimposed for 96 mM Na^+ with and without 1 mM P_i. The small displacement in the baseline preceding the step represents the steady-state induced current at -100 mV. (B) Magnified view of current records in panel A showing a clear difference in the relaxations depending on the presence or absence of substrate. Thin trace is for 1 mM P_i, bold trace is for 0 mM P_i. (C) Difference between records in panel B, showing the relaxations superimposed upon the P_i-induced steady-state currents. Note that the rate of relaxation for the forward step (-100 to 0 mV) is faster than the return relaxation, indicating that the relaxation time constant depends on the target potential, which is a characteristic of voltage-dependent charge movements. The area under each relaxation, determined by numerical integration, is a measure of the apparent charge translocated and should be equal for the depolarizing and hyperpolarizing steps.

from an oocyte expressing the type II Na^+/P_i cotransporter isoform NaPi-2. When a voltage step is applied to the oocyte with normal Na^+ (96 mM) in the presence and absence of saturating P_i, the resulting current transient is primarily due to the charging of the oocyte capacitance as shown in Fig. 3A. However, at higher gain (Fig. 3B), the relaxation to the steady state is significantly slower in the absence of P_i. This additional component (termed the pre-steady-state relaxation), superimposed upon the endogenous oocyte charging transient is not observed in control oocytes under the same superfusion conditions. The magnitude of these relaxations can be directly correlated with the P_i-induced cotransport and they can be shown to have properties consistent with the translocation of charged entities within the transmembrane electric field.

To quantitate such relaxations, a reasonable signal bandwidth should be used (typically 2–3 kHz) and signal averaging is routinely applied to improve the signal-to-noise ratio unless high levels of expression are obtained. Relaxations are then fit using an exponential curve-fitting routine. In general, two exponentials are sufficient to describe the complete relaxation, the faster component (in the range 400–1000 μsec, depending on the speed of the voltage clamp) represents the charging of the endogenous oocyte capacitance and the slower component, in the millisecond range, represents the cotransporter relaxation. Often the endogenous transient contains a small additional tail component, arising from inhomogeneous charging of the oocyte membrane, which can potentially contaminate the pre-steady-state relaxations. In this case we usually fit a single exponential to the main relaxation, commencing about 5 msec after the step. Alternatively, the oocyte endogenous components can be first eliminated by subtracting the response from a saturating P_i application as shown in Fig. 3C (Forster *et al.,* 1998) under the assumption that all relaxations have been suppressed. At present there is no blocker with high specificity for type II Na^+/P_i cotransporters which fully suppresses cotransport function, analogous to the action of phlorizin on Na^+/glucose cotransporters (e.g., Loo *et al.,* 1993; Chen *et al.,* 1996).

3.2. Voltage Dependence of Pre-Steady-State Relaxations

The voltage dependence of the time constant τ and equivalent charge Q associated with the relaxation (equal to the integral of the relaxation) are then obtained by analyzing the relaxations recorded from a series of voltage jumps from the same holding potential to voltages in the range -140 to $+80$ mV. Typical results for the NaPi-2 isoform are shown in Figs. 4 and 5. The inset of Fig. 4A shows a typical set of relaxations for voltage steps from $V_h = -100$ mV, to which single exponentials were fit. The resulting Q–V data for three V_h values show saturation at extreme hyperpolarizing and depolarizing potentials and the data at each V_h are simply displaced along the voltage axis as would be expected for a fixed number

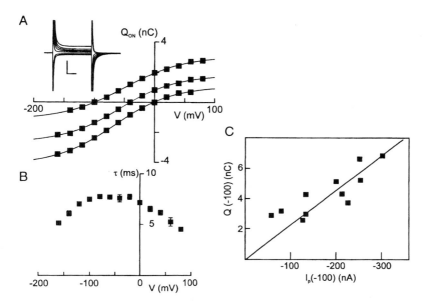

Figure 4. General characteristics of pre-steady-state relaxations for NaPi-2. (A) Charge translocation Q as a function of voltage derived from single-exponential fits to pre-steady-state relaxations from three V_h values ($-100, -40, 0$ mV) to test potentials between -140 and $+80$ mV commencing 5 msec after the step onset. Inset shows a typical family of records for $V_h = -100$ mV (scale: vertical 1 μA, horizontal 10 msec). Continuous lines are fits to the Boltzmann equation (see text). For this particular cell the fits to the Q–V for each V_h gave typically $z = 0.5e^-$ and $V_{0.5} = -40$ mV. (B) Voltage dependence of relaxation time constant τ obtained from a single-exponential fits. Data show mean±SEM for records corresponding to the same target potential for the same V_h values. (C) Correlation between charge available for translocation at -100 mV and steady-state P_i-induced current at -100 mV for 11 oocytes. Straight line is a linear regression line forced through the origin (slope = 46 sec^{-1}), which can be used to estimate the cotransporter turnover rate (see text). Data redrawn and adapted from Forster et al. (1988).

of translocatable charges. Figure 4B shows the corresponding τ–V data, which have a "bell"-shaped form typical for relaxations associated with membrane-bound charge displacements. The Q–V data can be further quantified by fitting to the data a Boltzmann equation, which describes a simple two-state model for voltage–dependent charge movement: $Q = Q_{hyp} + Q_{max}/\{1 + \exp(-ze(V - V_{0.5})/kT)\}$, where Q_{max} is the maximum charge translocated, Q_{hyp} is the charge translocated at the hyperpolarizing limit, $V_{0.5}$ is the voltage at which the charge is distributed equally between the two states, e is the electronic charge, k is Boltzmann's constant, and T is the absolute temperature. Fits using this equation give estimates of the apparent valence for the charge movement z, the potential at which half the charge has been displaced $V_{0.5}$, and the total charge movement Q_{max}. These parameters are useful for comparing results from different cotransporters and for the same cotransporter under different superfusion conditions. Their physical significance should be treated with caution since the fitting procedure is based on the validity of the Boltzmann model. For example, whereas the $V_{0.5}$ for the rat isoform NaPi-2 shown here is approximately -30 mV, that of the flounder isoform (NaPi-5) lies close to 0 mV (Forster *et al.*, 1997a) illustrating how species differences can be reflected in pre-steady-state data. Such differences have also been reported for the SGLT family (Mackenzie *et al.*, 1996).

3.3. Estimation of Transporter Number and Turnover

The Boltzmann parameters Q_{max} and z can be used to estimate the total number of cotransporters N_t, from $N_t = Q_{max}/ze$. Furthermore, if Q_{max} is plotted as a function of the corresponding I_p at saturating P_i for oocytes having different levels of expression, a linear correlation is obtained which confirms that the relaxations are directly related to functional cotransporters as shown in Fig. 4C. The slope of this relation can be used to estimate the transporter turnover ϕ from $\phi = I_{p,max} ze/Q_{max}$. For the NaPi-2 isoform we have estimated for ϕ at -100 mV a value of 25 sec^{-1}, which agrees with the estimates reported for other Na$^+$-coupled cotransporters (e.g., Loo *et al.*, 1993; Mager *et al.*, 1993; Wadiche *et al.*, 1995). It should be noted that the estimates of N_t and ϕ assume that (i) the charge translocation involves a single step, (ii) the same number of transporters contribute to $I_{p,max}$ in the presence of saturating P_i as contribute to the maximum charge movement in the absence of P_i, and (iii) the cotransporter does not comprise multiple subunits each having an apparent charge z (Zamphigi *et al.*, 1995).

Despite these limitations, the estimation of N_t from the pre-steady-state analysis provides a useful index of cotransporter numbers when studying transporter regulation and modulation, since from steady-state measurements alone a change in $I_{p,max}$ can result from either changes in N_t or cotransporter turnover, or both. For example, in the case of the Na$^+$/glucose cotransporter SGLT-1, Hirsch *et al.* (1996) combined both steady-state and pre-steady-state measurements together with estimates of oocyte membrane area obtained from the capacitive charging transient to study the retrieval and insertion of SGLT-1 under the regulation of phosphokinases A and C. Similar findings have recently been obtained for type II Na$^+$/P$_i$ cotransporters expressed in *Xenopus* oocytes (Forster *et al.*, 1999b).

3.4. The Effects of Substrates on Pre-Steady-State Kinetics

Further insight into the kinetic behavior of the cotransporter is obtained by measuring the pre-steady-state relaxations under conditions of varying substrates as shown in Fig. 5. In the absence of P_i, a reduction in external Na$^+$ shifts the $V_{0.5}$ toward negative potentials (Fig. 5A), whereas Q_{max} and z remain essentially constant. This behavior is consistent with a simple model for voltage-dependent ion binding in which Na$^+$ ions moving to a binding site within the transmembrane

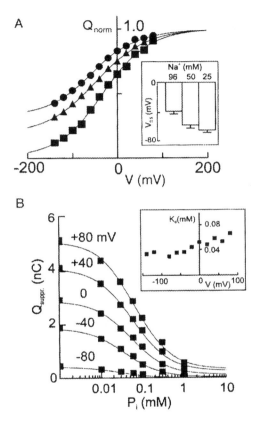

Figure 5. Pre-steady-state relaxations are modified by changing the external substrate. (A) The effect of varying external Na$^+$ in the absence of P_i on pre-steady-state relaxations. As the external Na$^+$ is reduced, the midpoint voltage $V_{0.5}$ of the Q–V shifts to more hyperpolarizing potentials (inset, showing pooled data, mean ± SEM; for four oocytes), which is consistent with an "ion-well" model in which Na$^+$ binding occurs within the transmembrane field. Continuous lines are fits to the Boltzmann equation. The z predicted from the fits was approximately 0.5 e^- and the Q_{max} decreased by approximately 20% with a fourfold change in Na$^+$. Symbols: (■) 96 mM Na$^+$, (▲) 50 mM Na$^+$, (●) 25 mM Na$^+$. (B) The effect of varying the external P_i in the presence of 96 mM Na$^+$ on pre-steady-state relaxations. As P_i increases, the relaxations are suppressed in a dose-dependent manner. Fitting the data to a competition function (continuous lines) gives the voltage dependence of the apparent K_d which is similar to the K_m^{Pi} predicted from the steady-state response (see Fig. 2). Data redrawn and adapted from Forster et al. (1998).

field, a so-called "ion-well." However, this model predicts changes in the $\tau-V$ data which are not observed and indicates that at least one additional voltage-dependent step contributes to the overall $Q-V$ data. Moreover, in the complete absence of external Na^+, pre-steady-state relaxations are still observed, indicating that the recorded charge movements arise from both the empty carrier and the binding of Na^+. With P_i as the variable substrate, as expected, pre-steady-state relaxations are suppressed in a dose-dependent manner which shows similar voltage dependence (Fig. 5B) to the steady-state K_m^{Pi}. Furthermore, changes in P_i do not alter the $\tau-V$ data nor result in shifts in $V_{0.5}$, which supports the notion that the P_i-binding step is not intrinsically voltage dependent.

4. A KINETIC SCHEME FOR Na^+-COUPLED P_i COTRANSPORT

4.1. The Sequential, Alternating Access Model for Type II Na^+/P_i Cotransport

Based on the steady-state and pre-steady-state characterization of the two type II Na^+/P_i cotransporter isoforms NaPi-2 (Forster et al., 1998) and NaPi-5 (Forster et al., 1997a), we have proposed a simple kinetic scheme which can account for the observed behavior as shown in Fig. 6. This scheme is similar to the sequential alternating access scheme proposed for the cloned Na^+/glucose cotransporter (SGLT1) by Parent et al. (1992b), but includes a second Na^+-binding step to account for the dependence of $I_{p,max}$ on Na^+ in the steady state. We also assumed that, like the P_i-binding site, the second Na^+-binding site is external to the transmembrane field, so that no contribution to pre-steady-state charge movements is made by this reaction, also consistent with our finding of a weak voltage dependence of K_m^{Na} at saturating P_i (Fig. 2D).

The two voltage-dependent partial reactions identified by pre-steady-state experiments (translocation of the unloaded carrier. $6 \Leftrightarrow 1$; binding of the first Na^+, $1 \Leftrightarrow 2$) are shown shaded. The empty cotransporter is assumed to have a valence of -1, but this apparent charge only translocates through a fraction of the transmembrane field to account for an apparent valence <1 obtained from fits to the Boltzmann $Q-V$ equation. Both the slippage ($2 \Leftrightarrow 5^*$) pathway and translocation of the fully loaded carrier ($4 \Leftrightarrow 5$) are assumed to be electroneutral. Except for slippage, the order of binding/release of substrates on the cytosolic side cannot be determined using the intact oocyte preparation, and therefore the cytosolic release of cotransported P_i together with two Na^+ ions is lumped as one reaction ($5 \Leftrightarrow 5^*$). For $V < 0$, substrate binding is facilitated on the *cis* side due to a net outward movement of negative charge associated with transition $6 \Rightarrow 1$. This allows one Na^+ ion to reach its binding site within the transmembrane field, leading to an increasing the affinity of the transporter for P_i, which then binds ($2 \Leftrightarrow 3$)

Renal Na$^+$-Coupled Divalent Anion Transporters

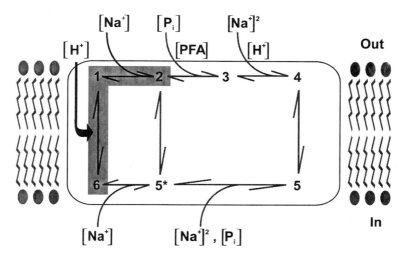

Figure 6. An ordered kinetic model for type II Na$^+$/P$_i$ cotransport. This scheme shows the partial reactions which have been identified by the steady-state and pre-steady-state analyses. The shaded reactions represent the two voltage-dependent steps which contribute to the pre-steady-state charge movements. Substrate concentrations are assumed to modify the binding rate constants only. Protons modulate the voltage-dependent rate constant associated with the empty carrier as well as the final Na$^+$-binding step, whereas PFA most likely competes directly with P$_i$ binding. The binding/debinding steps on the cytosolic side have not yet been characterized.

in a voltage-independent manner. A further voltage-independent step involving the binding of two Na$^+$ ions (3 ⇔ 4) results in an electroneutral fully loaded carrier, which then translocates (4 ⇔ 5), releases substrates on the *trans* side, and returns to state 6. A net inward charge movement of +1 charge occurs per cycle, which is manifested as I_p. By assigning rate constants to the voltage-dependent partial reactions identified from the pre-steady-state analysis we have been able to predict the general features of the both the pre-steady-state and steady-state findings (Forster *et al.*, 1997a, 1998). Simulations using this model have also predicted the existence of a second, very fast relaxation ($\tau < 1000$ μsec) related to the Na$^+$-binding transition, which is not normally resolvable with the two-electrode voltage clamp.

4.2. Site of Interaction of Protons and Foscarnet with Type II Na$^+$/P$_i$ Cotransporters

Simulations using the above model have been used to indicate which substrate-binding steps are accessible by protons. In the steady state, decreasing external pH leads to a significant reduction in the maximum transport rate. From studies on BBMVs (Amstutz et al., 1985) and type II Na^+/P_i cotransporters expressed in Xenopus oocytes (Busch et al., 1995), the most likely explanation is that H^+ ions compete for Na^+ binding. Our studies of the effects of pH on pre-steady-state relaxations indicate that protons cause a significant shift in the pre-steady-state $Q-V$ relation and a concomitant shift in the voltage dependence of the relaxation time constant (Forster et al., 1997b, 1998) which cannot be explained in terms of Na^+ competition alone. Simulations using the above model have revealed that these effects can be best accounted for in terms of a proton modulation of the empty carrier kinetics, whereas the steady-state observations require an additional interaction of protons with the final Na^+-binding step as indicated in Fig. 6.

The antiviral agent foscarnet (phosphonoformic acid, PFA) is a known inhibitor of Na^+/P_i cotransport both in the kidney and small intestine (for review, see Kempson, 1988). In the intestine, it is most likely absorbed via a P_i-dependent, carrier-mediated system (for review, see Tamai and Tsuji, 1996) and in the kidney it is also associated with nephrotoxic effects (Deray et al., 1989). Based on isotope flux on studies in BBMVs, opossum kidney cells (Loghman-Adham and Dousa, 1992), and electrophysiological studies on the human Na^+/P_i cotransporter clone NaPi-3, expressed in Xenopus oocytes (Busch et al., 1995), it appears that PFA competes for P_i binding resulting in a shift in K_m^{Pi} without changing the maximum transport rate. The apparent binding affinity for PFA is in the millimolar range and from oocyte experiments, PFA appears not to be transported. Pre-steady-state measurements have confirmed that at a concentration of 3 mM PFA, at which the P_i-induced current (0.3 mM P_i) is suppressed by over 70%, relaxations in the absence of P_i were suppressed just as for the saturating P_i case, consistent with PFA competing at the P_i-binding site (Forster et al., 1998).

5. CONCLUSIONS

Characterization of the electrophysiological properties of electrogenic cotransporter function in terms of both steady-state and pre-steady-state kinetics provides essential information for both model generation and identification of partial reactions in the transport cycle which play a critical role in determining kinetic parameters. Furthermore, the voltage dependence of the apparent affinity constants obtained from the steady state serves to emphasize the importance of defining membrane potential when specifying kinetic parameters. In the context of studying the interaction of drugs with cotransporters and the potential application of cotransporters as drug delivery systems, steady-state kinetic characterization provides estimates of the maximum transport rates, apparent affinity coefficients, and stoichiometry for substrates as well as allowing quantification of the interaction between substrates. On the other hand, pre-steady-state characterization gives additional information which can be related to partial reactions in the cotransport cycle and thereby enables identification of the site(s) of interaction of substrates and functional modulators with the cotransporter.

ACKNOWLEDGMENTS

The work was supported by grants to H.M. from the Swiss National Science Foundation (SNF: 31-46523), the Hartmann Müller-Stiftung (Zurich), the Olgar Mayenfisch-Stiftung (Zurich), and the Schweizerischer Bankgesellschaft (Zurich) (Bu 704/7-1).

REFERENCES

Amstutz, M., Mohrmann, M., Gmaj, P., and Murer, H. 1985, Effect of pH on phosphate transport in rat renal brush border vesicles, *Am. J. Physiol.* **248**:F705–F710.

Biber, J., Custer, M., Magagnin, S., Hayes, G., Werner, A., Lötscher, M., Kaissling, B., and Murer, H., 1996, Renal Na/Pi-cotransporters, *Kidney Int.* **49**:981–985.

Birnir, B., Loo, D. D. F., and Wright, E. M., 1991, Voltage-clamp studies of the Na^+/glucose cotransporter cloned from rabbit small intestine, *Pflügers Arch.* **418**:78–95.

Busch, A. E., Waldegger, S., Herzer, T., Biber, J., Markovich, D., Hayes, G., Murer, H., and Lang, F., 1994a, Electrophysiological analysis of Na^+/P_i cotransport mediated by a transporter cloned from rat kidney in *Xenopus* oocytes, *Proc. Natl. Acad. Sci. USA* **91**:8205–8208.

Busch, A. E., Waldegger, S., Herzer, T., Biber, J., Markovich, D., Murer, H., and Lang, F., 1994b, Electrogenic cotransport of Na^+ and sulfate in *Xenopus* oocytes expressing the cloned Na^+/SO_4^{2-} transport protein NaSi-1, *J. Biol. Chem.* **269**:12407–12409.

Busch, A. E., Wagner, C. A., Schuster, A., Waldegger, S., Biber, J., Murer, H., and Lang, F., 1995, Properties of electrogenic P_i transport by NaPi-3, a human renal brush border Na^+/P_i transporter, *J. Am. Soc. Nephrol.* **6**:1547–1551.

Busch, A. E., Biber, J., Murer, H., and Lang, F., 1996, Electrophysiological insights of type I and II Na/P_i transporters, *Kidney Int.* **49**:986–987.

Chen, X. Z., Coady, M. J., and Lapointe, J.-Y., 1996, Fast voltage clamp discloses a new component of presteady-state currents from the Na^+-glucose cotransporter, *Biophys. J.* **71**:2544–2552.

Deray, G., Martinez, F., Katlama, C., Levatier, B., Beaufils, H., Danis, M., Rozenheim, M., Baumelou, A., Dohin, E., Gentilini, M., and Jacobs, C., 1989, Foscarnet nephrotoxicity: Mechanism, incidence, and prevention, *Am. J. Nephrol.* **9**:316–321.

Eskandari, S., Loo, D. D. F., Dai, G., Levy, O., Wright E. M., and Carrasco, N., 1997, Thyroid Na^+/I^- symporter, *J. Biol. Chem.* **43**:27230–27238.

Fernandes, I., Hampson, G., Cahours, X., Morin, P., Coureau, C., Couette, S., Prie, D.,

Biber, J., Murer, H., Friedlander, G., and Silve, C., 1997, Abnormal sulfate metabolism in vitamin D-deficient rats, *J. Clin. Invest.* **100**:2196–2203.

Forster, I. C., Wagner, C. A., Busch, A. E., Lang, F., Biber, J., Hernando, N., Murer, H., and Werner, A., 1997a, Electrophysiological characterization of the flounder type II Na^+/P_i cotransporter (NaPi-5) expressed in *Xenopus laevis* oocytes, *J. Membr. Biol.* **160**:9–25.

Forster, I. C., Biber, J., and Murer, H., 1997b, Modulation of the voltage-dependent kinetics of renal type II Na^+/P_i cotransporters by external pH, *J. Am. Soc. Nephrol.* **8**:A2611 (Abstract).

Forster, I. C., Loo, D. D. F., and Eskandari, S., 1999a, Stoichiometry and Na^+ binding co-operativity of rat and flounder renal type II Na^+–P_i cotransporters. *Am. J. Physiol. Renal* **276**:F644–F649.

Forster, I. C., Traebert, M., Jankowski, M., Strange, G., Biber, J., and Murer, H., 1999b, Protein kinase C activators induce membrane retrieval of type II Na^+–phosphate cotransporters expressed in *Xenopus* oocytes. *J. Physiol. (Lond.)* **517**:327–340.

Forster, I. C., Hernando, N., Biber, J., and Murer, H., 1998, The voltage dependence of a cloned mammalian renal type II Na^+–P_i cotransporter (NaPi-2), *J. Gen. Physiol.* **112**:1–18.

Hager, K., Hazama, A., Kwon, H. M., Loo, D. D. F., Handler, J. S., and Wright, E. M., 1995, Kinetics and specificity of the renal Na^+/Myo-inositol cotransporter expressed in *Xenopus* oocytes. *J. Membrane Biology* **143**:103–113.

Hartmann, C. M., Wagner, C. A., Busch, A. E., Markovich, D., Biber, J., Lang, F., and Murer, H., 1995, Transport characteristics of a murine renal Na/P_i-cotransporter, *Pfluegers Arch.* **430**:830–836.

Hazama, A., Loo, D. D. F., and Wright, E. M., 1997, Presteady-state currents of the Na^+/glucose cotransporter (SGLT1), *J. Membr. Biol.* **155**:175–186.

Hirsch, J. R., Loo, D. D. F., and Wright, E. M., 1996, Regulation of Na^+/glucose cotransporter expression by protein kinases in *Xenopus laevis* oocytes, *J. Biol. Chem.* **271**: 14740–14746.

Kempson, S. A., 1988, Novel specific inhibitors of epithelial phosphate transport, *NIPS* **3**:154–157.

Loghman-Adham, M., and Dousa, T. P., 1992, Dual action of phosphonoformic acid on Na^+–P_i cotransport in opossum kidney cells, *Am. J. Physiol.* **263**:F301–F310.

Loo, D. D. F., Hazama, A., Supplisson, S., Turk, E., and Wright, E. M., 1993, Relaxation kinetics of the Na^+/glucose cotransporter, *Proc. Natl. Acad. Sci. USA* **90**:5767–5771.

Mackenzie, B., Loo, D. D. F., Panayotova-Heiermann, M., and Wright, E. M., 1996, Biophysical characteristics of the pig kidney Na^+/glucose cotransporter SGLT2 reveal a common mechanism for SGLT1 and SGLT2, *J. Biol. Chem.* **271**:32678–32683.

Magagnin, S., Werner, A., Markovich, D., Sorribas, V., Stange, G., Biber, J., and Murer, H., 1993, Expression cloning of human and rat renal cortex Na/Pi cotransport, *Proc. Natl. Acad. Sci. USA* **90**:5979–5983.

Mager, S., Naeve, J., Quick, M., Labarca, C., Davidson, N., and Lester, H. A., 1993, Steady states, charge movements, and rates for a cloned GABA transporter expressed in *Xenopus* oocytes, *Neuron* **10**:177–188.

Markovich, D., Forgo, J., Stange, G., Biber, J., and Murer, H., 1993, Expression cloning of rat renal Na^+/SO_4^{--} cotransport, *Proc. Natl. Acad. Sci. USA* **90**:9073–8077.

Markovich, D., Murer, H., Biber, J., Sakhaee, K., Pak, C., and Levi, M., 1998, Dietary sulfate regulates the expression of the renal brush border Na/S_i-cotransporter NaS_i-1, *J. Am Soc. Nephrol.* **9**(9):1568–1573.

Murer, H., and Biber, J., 1992, Renal tubular phosphate transport. Cellular mechanisms, in: *The Kidney. Physiology and Pathophysiology,* 2nd ed. (D. W. Seldin and G. Giebisch, eds.), Raven Press, New York, pp. 2481–2509.

Murer, H., and Biber, J., 1996, Molecular mechanisms of renal apical Na phosphate cotransport, *Annu. Rev. Physiol.* **58**:607–618.

Murer, H., and Biber, J., 1997, A molecular view of proximal tubular inorganic phosphate (P_i) reabsorption and of its regulation, *Pflügers Arch.* **433**:379–389.

Murer, H., Markovich, D., and Biber, J., 1994, Renal and small intestinal sodium-dependent symporters of phosphate and sulphate, *J. Exp. Biol.* **196**:167–181.

Parent, L., Supplisson, S., Loo, D. D. F., and Wright, E. M., 1992a, Electrogenic properties of the cloned Na^+/glucose cotransporter: I. Voltage clamp studies. *J. Membr. Biol.* **125**:49–62.

Parent, L., Supplisson, S., Loo, D. D. F., and Wright, E. M., 1992b, Electrogenic properties of the cloned Na^+/glucose cotransporter: II. A transport model under nonrapid equilibrium conditions, *J. Membr. Biol.* **125**:63–79.

Rudy, B., and Iverson, L. E., 1992, Ion channels, in: *Methods in Enzymology,* Vol. 207, Academic Press, New York, pp. 1–917.

Tamai, I., and Tsuji, A., 1996, Carrier-mediated approaches for oral drug delivery, *Adv. Drug Delivery Rev.* **20**:5–32.

Umbach, J. A., Coady, M. J., and Wright, E. M., 1990, Intestinal Na^+/glucose cotransporter expressed in *Xenopus* oocytes is electrogenic, *Biophys. J.* **57**:1217–1224.

Wadiche, J. I., Arriza, J. L., Amara, S. G., and Kavanaugh, M. P., 1995, Kinetics of a human glutamate transporter, *Neuron* **14**:1019–1027.

Weiss, J. N., 1997, The Hill equation revisited: Uses and misuses, *FASEB J.* **11**:835–841.

Zampighi, G. A., Kreman, M., Boorer, K. J., Loo, D. D. F., Bezanilla, F., Chandy, G., Hall, J. E., and Wright, E. M., 1995, A method for determining the unitary functional capacity of cloned channels and transporters expressed in *Xenopus laevis* oocytes, *J. Membr. Biol.* **148**:65–78.

10

Dipeptide Transporters

Ken-ichi Inui and Tomohiro Terada

1. INTRODUCTION

Transport of di- or tripeptides across plasma membranes in the small intestine and the kidney proximal tubules plays a pivotal role in efficient absorption of protein digestion products. The absorption process is mediated actively by H^+-coupled peptide transporters localized in the brush-border membranes of these epithelia. As there are 20 amino acids that comprise small peptides, there may be 400 dipeptides and 8000 tripeptides with various charges and molecular sizes. Therefore, peptide transporters could have a broad range of substrate specificity. Consequently, various foreign compounds structurally related to these small peptides such as β-lactam antibiotics and bestatin, a dipeptide-like anticancer agent, are recognized by the peptide transporters, and the intestinal and renal absorption of these drugs is mediated by them (Fig. 1).

The identification of two peptide transporters, PEPT1 and PEPT2, by Leibach's and Hediger's groups represented a major step forward toward molecular understanding of the physiological and pharmacological significance of peptide transporters. There are several recent reviews documenting the historical development and recent progress in our understanding of the peptide transporters (Leibach and Ganapathy, 1996; Adibi, 1997; Daniel and Herget, 1997). In this chapter we will focus mainly on drug transport mediated by the peptide transporters.

Ken-ichi Inui and Tomohiro Terada • Department of Pharmacy, Kyoto University Hospital, Sakyo-ku, Kyoto 606-8507, Japan.
Membrane Transporters as Drug Targets, edited by Amidon and Sadée. Kluwer Academic/Plenum Publishers, New York, 1999.

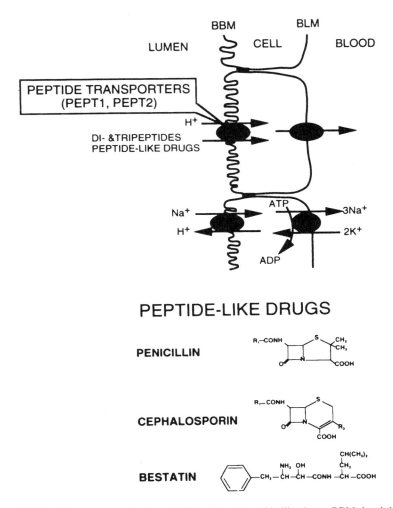

Figure 1. Peptide transporters in the epithelial cells and peptide-like drugs. BBM, brush-border membrane; BLM, basolateral membrane.

2. BRUSH-BORDER MEMBRANE TRANSPORT

2.1. Small Peptides

Dietary protein is broken down to many small peptides as well as amino acids in the lumen of the small intestine. These products are then taken up from the intestinal epithelial cells and are delivered into the circulation. Numerous studies have indicated that digestion products of protein are absorbed in the form of small peptides (Matthews, 1975). In addition to the small intestine, the renal proximal tubular cells efficiently reabsorb the filtrated small peptides to conserve amino acid nitrogen (Silbernagl et al., 1987). The mechanism of the transport system for small peptides is known as the peptide transporter, and the transport process across the brush-border membranes has been well characterized using intestinal and renal brush-border membrane vesicles. The peptide transporter was elegantly demonstrated to be an electrogenic H^+-coupled transport system driven by an inward H^+ gradient and stimulated by a negative membrane potential (V. Ganapathy and Leibach, 1983; Takuwa et al., 1985). Intact small peptides consisting of two or three amino acids can be accepted as substrates for the peptide transporter, while free amino acids and peptides consisting of four or more amino acids cannot.

2.2. Peptide-like Drugs

β-Lactam antibiotics for oral administration are absorbed efficiently from the small intestine even though they are ionized at physiological pH and have low lipid solubilities. Using the *in situ* perfusion technique, the loop technique, and the *in vitro* everted sac technique, it had been suggested that amino-β-lactam antibiotics are absorbed by a carrier-mediated system (Dixon and Mizen, 1977; Kimura et al., 1978; Tsuji et al., 1981), and that some of these molecules might share a common transport system with peptide transporter due to their structural similarities with small peptides (Kimura et al., 1983; Nakashima et al., 1984; Nakashima and Tsuji, 1985). Direct evidence for the absorption mechanism of orally active β-lactam antibiotics via H^+-coupled peptide transporter was obtained using intestinal brush-border membrane vesicles (Okano et al., 1986a, b). As shown in Fig. 2, a clear overshoot phenomenon was observed for cephradine (aminocephalosporin) uptake by rabbit intestinal brush-border membrane vesicles in the presence of an inward H^+ gradient, and this uptake was completely inhibited by the protonophore FCCP, carbonylcyanide p-trifluoromethoxyphenylhydrazone. Cephradine uptake was markedly inhibited by dipeptides, but not by amino acids. These results represent the first demonstration that foreign compounds as well as native small peptides are recognized and transported by the intestinal H^+-coupled peptide transporter.

As the native small peptides have an α-amino group in their structure, the transport of β-lactam antibiotics by the peptide transporter had generally been believed to apply only to β-lactam antibiotic analogues with an α-amino group in their side chains. However, cephalosporin antibiotics lacking an α-amino group such as ceftibuten and cefixime were shown to be transported by the peptide transporter (Tsuji et al., 1987; Inui et al., 1988; Muranushi et al., 1989). This discovery encouraged research into the many peptide-like drugs without an α-amino

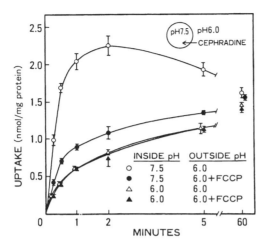

Figure 2. Effects of FCCP on H$^+$ gradient-dependent cephradine uptake by intestinal brush-border membrane vesicles. Reprinted with permission from Okano et al. (1986a).

group as substrates for the peptide transporter. As a result, bestatin (Tomita et al., 1990), renin inhibitors (Kramer et al., 1990), and angiotensin-converting enzyme inhibitors (Kitagawa et al., 1997) were demonstrated to be recognized by the peptide transporter. Recent progress in the delivery of peptide-like drugs using peptide transporter is described in Section 5.

By using renal brush-border membranes vesicles, several studies on the transport of peptide-like drugs in the kidney have also been carried out. The aminocephalosporin cephalexin was the first peptide-like drug whose transport was demonstrated to be mediated via the peptide transporter in the kidney (Inui et al., 1983, 1984). Thereafter, such peptide-like drugs as bestatin (Hori et al., 1993), cefadroxil (Ries et al., 1994), and ceftibuten (Naasani et al., 1995) were reported to be transported by peptide transporter(s). However, unlike the intestine, it was suggested that the transport of cefadroxil and ceftibuten in the renal proximal tubules was mediated by at least two distinct peptide transporters; namely, high-affinity, low capacity and low-affinity, high-capacity transport systems (Ries et al., 1994; Naasani et al., 1995).

3. TRANSCELLULAR TRANSPORT

3.1. Caco-2 Cell Monolayers

The absorption of peptide-like drugs through the intestine requires crossing of two distinct membranes, i.e., uptake by epithelial cells from the lumen across the brush-border membranes, followed by transfer to the blood across the basolateral membranes. The peptide transporter in the brush-border membranes has been well characterized as described in Section 2, but the transport of small peptides and peptide-like drugs across the basolateral membrane of epithelial cells has yet to be fully elucidated. The human colon adenocarcinoma cell line Caco-2 has been reported to be a useful *in vitro* model for studying the intestinal absorption mechanisms of a variety of drugs, including orally active β-lactam antibiotics (Dantzig and Bergin, 1990). These cells spontaneously differentiate in culture into polarized cell monolayers possessing microvilli and many enterocyte-like properties. Using Caco-2 cells grown on microporous membrane filters, we attempted to clarify the cellular mechanism of the transepithelial transport of peptide-like drugs (e.g. cephalosporins and bestatin) (Inui *et al.*, 1992; Saito and Inui, 1993; Matsumoto *et al.*, 1994).

3.2. Peptide Transporter in Apical Membranes

The cephalosporin antibiotics available for oral administration such as cephradine, ceftibuten (Inui *et al.*, 1992; Matsumoto *et al.*, 1994), and bestatin (Saito and Inui, 1993) were accumulated concentratively in Caco-2 monolayers from the apical side, but cefotiam, a parenteral agent, was not. The accumulation of these drugs was pH-dependent, saturable, and affected by FCCP. Various dipeptides and peptide-like drugs inhibited the accumulation of these drugs, but glycine did not. These observations are similar to those obtained in studies using intestinal brush-border membrane vesicles. The apical H^+/peptide cotransport system was regulated by cell growth and/or differentiation in the Caco-2 cells (Matsumoto *et al.*, 1995). Therefore, the Caco-2 cell monolayers appear to provide a useful model for studying the intestinal absorption mechanisms of peptide-like drugs.

3.3. Peptide Transporter in Basolateral Membranes

In 1992, we demonstrated that a specific transport system was involved in the efflux of cephradine across the basolateral membranes of Caco-2 cells (Inui *et al.*, 1992). Thereafter, the transport of bestatin across the basolateral membranes was shown to be less sensitive to the medium pH, and it did not occur against a concentration gradient, although the basolateral transport of bestatin was inhibited by glycyl-L-leucine (Fig. 3) (Saito and Inui, 1993). In addition, FCCP had little effect on the uptake of ceftibuten in the basolateral membranes (Matsumoto *et al.*, 1994) and the accumulation of these drugs via basolateral membranes was inhibited by various dipeptides and peptide-like drugs, but not by glycine (Saito and Inui, 1993; Matsumoto *et al.*, 1994). These results suggest that a facilitative peptide transport

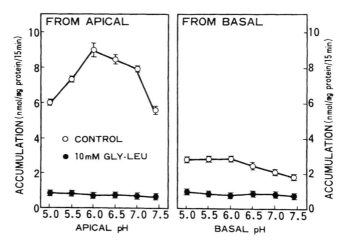

Figure 3. The pH dependence of bestatin accumulation by Caco-2 monolayers. Reprinted with permission from Saito and Inui (1993).

system which is distinct from the apical H^+-coupled peptide transporter is localized in the basolateral membranes of the Caco-2 cells. In contrast, there have been a few reports that the peptide transporter in the basolateral membranes is H^+-dependent (Dyer *et al.*, 1990; Thwaites *et al.*, 1993).

Although the peptide transporters in the brush-border and basolateral membranes have been suggested to be distinct from each other, the question remains as to which peptide transporter determines the absorption rate of peptide-like drugs. We compared the apparent absorption rates of various cephalosporins estimated by disappearance from the intestinal loop with initial uptake rates by the intestinal brush-border membrane vesicles (Hori *et al.*, 1988). As shown in Fig. 4, a good correlation was found between *in situ* intestinal absorption and the brush-border membrane transport of cephalosporins, suggesting that transport process across the brush-border membranes is the rate-limiting step for the intestinal absorption of these antibiotics.

4. CLONING OF PEPTIDE TRANSPORTERS

4.1. Structure

Although considerable progress has been made in delineation of the functional aspects of the peptide transporters over the last decade, their structural

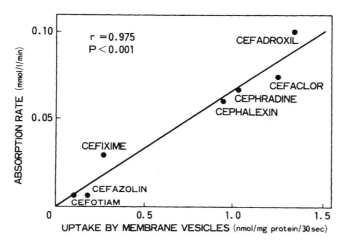

Figure 4. Correlation between intestinal absorption and brush-border membrane transport of cephalosporin antibiotics. Reprinted with permission from Hori *et al.* (1988).

properties at the molecular level have not yet been determined. However, using expression cloning strategy with *Xenopus laevis* oocytes, a cDNA encoding the oligopeptide transporter (PEPT1) in the rabbit small intestine was isolated by Leibach's and Hediger's groups (Fei *et al.*, 1994). Subsequently, they also succeeded in cloning the human PEPT1 (Liang *et al.*, 1995). Using a polymerase chain reaction technique based on the amino acid sequence of rabbit PEPT1, we isolated the rat PEPT1 cDNA from a rat kidney cDNA library (Saito *et al.*, 1995). The rat PEPT1 cDNA encodes a 710-amino acid protein with 77% and 83% identity to rabbit and human PEPT1, respectively. Hydropathy analysis of the rat PEPT1 suggested that there are 12 putative membrane-spanning domains with a large extracellular loop between transmembrane domains 9 and 10. An isoform of intestinal peptide transporter PEPT1, the renal peptide transporter PEPT2 cDNA was isolated from various species (Liu *et al.*, 1995; Boll *et al.*, 1996; Saito *et al.*, 1996). Rat PEPT2 consists of 729 amino acids, and shows 48% identity and similar membrane topology with the rat PEPT1. Both rat PEPT1 and PEPT2 have a number of putative N-glycosylation sites in the large extracellular loop, and potential protein kinase A and C phosphorylation sites in the intracellular domains. Recently, a novel protein which showed a significant homology with the NH_2-terminal sequence of human PEPT1 was identified from the human duodenum cDNA library (Saito *et al.*, 1997). Although the physiological function and significance of this protein are not clear, it may be involved in the regulation of peptide transporter function.

4.2. Tissue Distribution

The message for rat PEPT1 was ~2.9 kb and found to be expressed mainly in the small intestine and slightly in the kidney (Saito *et al.,* 1995; K. Miyamoto *et al.,* 1996). On Western blot analysis, the rat PEPT1 protein with a molecular mass of 75 kDa was detected in the crude membrane fractions from the small intestine (duodenum, jejunum, and ileum) and kidney (Fig. 5A) (Ogihara *et al.,* 1996). Immunohistochemical studies revealed that rat PEPT1 was localized at the brush border of absorptive epithelial cells in the villi (Fig. 5B). The esophagus, stomach, colon, and rectum were negative for immunostaining. On the other hand, Northern blot analysis revealed a ~4-kb band corresponding to rat PEPT2 mRNA in the kidney, especially in the medulla, but not in the small intestine (Saito *et al.,* 1996). This finding that two peptide transporters, PEPT1 and PEPT2, are expressed in the kidney is consistent with the results of vesicle studies indicating that at least two peptide transporters exist in the kidney (Y. Miyamoto *et al.,* 1988; Daniel *et al.,* 1991; Takahashi *et al.,* 1998). Rat PEPT2 mRNA was also detected in the brain, lung, and spleen, and therefore PEPT2 may participate in as-yet-undetermined physiological function(s) related to the active transport of small peptides across plasma membranes of these tissues. It has been reported that di- and tripeptides can be transported across the epithelium of the lung, suggesting that PEPT2 may play an important role in lung homeostasis (Meredith and Boyd, 1995).

4.3. Substrate Specificity and Recognition

PEPT1 and PEPT2 have been functionally expressed in various heterologous expression systems and the transport of small peptides and peptide-like drugs by PEPT1 and PEPT2 has been investigated. Both PEPT1 and PEPT2 can transport di- and tripeptides, but not free amino acids and tetrapeptides (Fei *et al.,* 1994; Boll *et al.,* 1996). Peptide-like drugs such as β-lactam antibiotics and bestatin were also demonstrated to be transported by PEPT1 and PEPT2 (M. E. Ganapathy *et al.,* 1995; Saito *et al.,* 1995, 1996; Wenzel *et al.,* 1996; Tamai *et al.,* 1997).

Kinetic analysis of the transport of chemically diverse dipeptides via human PEPT1 and PEPT2 revealed that PEPT1 is a low-affinity transporter, whereas PEPT2 is a high-affinity transporter (Ramamoorthy *et al.,* 1995). Following the studies of substrate specificity of native small peptides by PEPT1 and PEPT2, the recognition of the β-lactam antibiotics by both transporters was examined. M. E. Ganapathy *et al.* (1995) evaluated the inhibition constants K_i of cefadroxil (aminocephalosporin) and cyclacillin (aminopenicillin) for glycylsarcosine uptake by human PEPT1 or PEPT2, and suggested the differential recognition of these an-

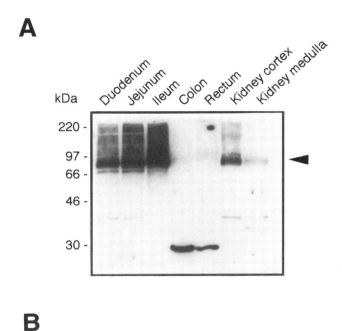

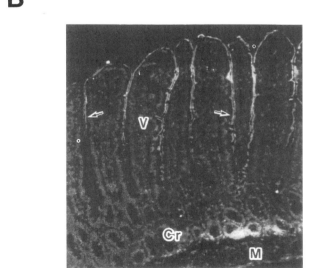

Figure 5. (A) Immunoblot analysis of the rat digestive tract and kidney with anti-PEPT1 serum. (B) Immunofluorescence localization of PEPT1 in the jejunum. Intestinal villi (V) were strongly stained for PEPT1 (arrows). Cr, crypt; M, muscle coat. Reprinted with permission from Ogihara *et al.* (1996).

tibiotics by the two peptide transporters: PEPT1 had a much higher affinity for cyclacillin than for cefadroxil, whereas PEPT2 preferred cefadroxil to cyclacillin. Using LLC-PK$_1$ cells stably transfected with either rat PEPT1 or PEPT2 cDNA, we also examined the recognition of β-lactam antibiotics and bestatin by PEPT1 and PEPT2 (Terada et al., 1997a, b). By comparing the K_i values of these drugs for glycylsarcosine transport between PEPT1 and PEPT2, we found that β-lactam antibiotics (except ceftibuten and cefixime) and bestatin showed much more potent inhibition of glycylsarcosine uptake via PEPT2 than via PEPT1 (Table I). These results suggested that rat PEPT2 has a much higher affinity for β-lactam antibiotics with an α-amino group than rat PEPT1 does, whereas rat PEPT1 preferred β-lactam antibiotics without an α-amino group.

Daniel and Adibi (1993) reported that increasing the hydrophobicity of the NH$_2$-terminal side chain increased the affinities of aminocephalosporins and aminopenicillins to the renal H$^+$/peptide cotransporter. In comparing the K_i values for cefadroxil and cephalexin or for amoxicillin and ampicillin, cefadroxil and amoxicillin had much higher affinities for both PEPT1 and PEPT2 than did cephalexin and ampicillin, respectively (Table I). Cefadroxil and cephalexin (aminocephalosporins) or amoxicillin and ampicillin (aminopenicillins) have very similar structures; the former two have a p-hydroxyphenyl group and the latter two have a phenyl group in the side chain in the position 7 or 6 of the cephalosporin and penicillin structure, respectively. Because the p-hydroxyphenyl group is much less hydrophobic than the phenyl group, the hydroxyphenyl group at the NH$_2$-terminal side chain of some β-lactams would make a stronger interaction with the peptide transporter than the hydrophobic interaction. These findings suggest that not only hydrophobicity, but also the hydroxyphenyl group at the side chain of these antibiotics are involved in substrate recognition by both rat PEPT1 and PEPT2.

4.4. Structural Features

Use of brush-border membrane vesicles treated with diethylpyrocarbone (DEPC), a histidine-residue modifier, suggested that the histidine residues were essential for the transport activity of the H$^+$-coupled peptide transporter (Y. Miyamoto et al., 1986; Kato et al., 1989). As histidine residues located in the putative transmembranes are considered to play an important role, histidine residues 57 and 121 of rat PEPT1, which are located in the predicted transmembrane domains 2 and 4, respectively, were replaced by glutamine residues using site-directed mutagenesis (Figure 6A). When these histidine-mutant proteins were expressed in Xenopus oocytes, they could not transport glycylsarcosine (Fig. 6B), although their insertion in the plasma membranes was not impaired (Terada et al.,

Table I.
Configurations and Inhibition Constants K_i for β-Lactam Antibiotics[a]

Drug	R_1	R_2	K_i (mM) PEPT1	K_i (mM) PEPT2
Ampicillin	phenyl-CH(NH₂)-		47,800	669
Amoxicillin	HO-phenyl-CH(NH₂)-		13,000	179
Cyclacillin	cyclohexyl-NH₂		168	27
Cephalexin	phenyl-CH(NH₂)-	CH₃	4,500	49
Cefadroxil	HO-phenyl-CH(NH₂)-	CH₃	2,170	3
Cephradine	cyclohexadienyl-CH(NH₂)-	CH₃	8,540	47
Cefdinir	H₂N-thiazolyl-C(=NOH)-	CH=CH₂	11,900	20,100
Ceftibuten	H₂N-thiazolyl-C(=CHCOOH)H-	H	597	1,340
Cefixime	H₂N-thiazolyl-C(=NOCH₂COOH)-	CH=CH₂	6,920	11,900
Bestatin			505	20

[a]Reproduced with permission from Terada et al. (1997b).
[b]Chemical structures illustrated as R_1 and R_2 represent substituents in penicillin or cephalosporin shown in Fig. 1. The apparent K_i values were calculated from the inhibition plots based on the transformed Michaelis–Menten equation using nonlinear least square regression analysis.

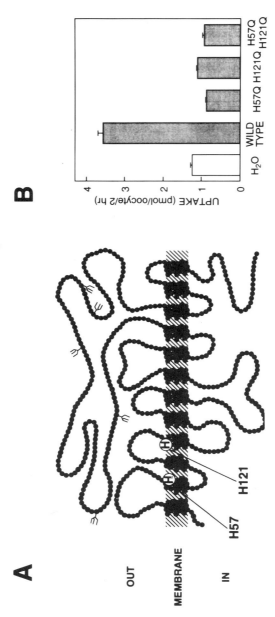

Figure 6. (A) Conserved histidine residues in rat, rabbit, and human peptide transporters PEPT1 and PEPT2. The numbers of histidine residues were derived from rat PEPT1. (B) [^{14}C]Glycylsarcosine uptake by oocytes expressing wild-type and histidine-mutant rat PEPT1. H, histidine; Q, glutamine (Terada et al., 1996).

1996). Interestingly, the corresponding histidine residues are also conserved in the human, rabbit, and rat PEPT2. These histidine residues of human PEPT2 were reported to play a significant role in the maintenance of transport activity (Fei et al., 1997). Although the essential histidine residues have been identified, the functional roles of histidine residues have yet to be fully elucidated. Recent studies revealed that the α-amino group of dipeptides and aminocephalosporins interacts with the DEPC-sensitive histidine residue of PEPT1 and PEPT2 (Terada et al., 1998). Not only histidine residues, but also sulfhydryl groups were reported to be essential components of the peptide transporters (Y. Miyamoto et al., 1986; Saito and Inui, 1993).

To identify the structural domains responsible for the functional properties of transporters, it is useful to generate functional chimeras between closely related family members. Döring et al. (1996) constructed a chimeric peptide transporter between rabbit PEPT1 and PEPT2 and compared their functions with those of the parent transporters. They concluded that the phenotypic characteristics of PEPT2 were determined by its amino-terminal region. This finding is consistent with the above observations that the essential histidine residues are present in the amino-terminal region of transmembrane domains 2 and 4. Further investigations to identify the essential amino acid residues and functional domains are needed to understand the transport mechanisms of peptide transporters.

Based on our data, the characteristics of rat H^+/peptide cotransporters PEPT1 and PEPT2 are summarized in Table II.

5. APPLICATION TO DRUG DELIVERY

Utilization of peptide transporters for drug delivery has been considered a promising strategy. L-α-Methyldopa is a poorly absorbed antihypertensive agent and amino acid analogue. When converted to dipeptidyl derivatives, these were more efficiently absorbed from the intestine than L-α-methyldopa itself because dipeptidyl derivatives serve as substrates for the intestinal peptide transporter, whereas L-α-methyldopa does not (Hu et al., 1989; Tsuji et al., 1990). These observations indicate that the peptide transporter is a possible route for improving the intestinal absorption of pharmacologically active amino acid analogues. In addition, recent studies have revealed that various peptide-like compounds are recognized by peptide transporter, even though they do not have a peptide bond or free amino and/or carboxy terminal. As examples of compounds without a peptide bond, the aminopeptidase inhibitor arphamenine (Daniel and Adibi, 1994), the antiviral agent valacyclovir (Han and Amidon, 1996), and 4-aminophenylacetic acid (Temple et al., 1998) were reported to be substrates. As compounds without a free

Table II.
Characteristics of Rat H^+/Peptide Cotransporters PEPT1 and PEPT2[a]

	PEPT1	PEPT2
mRNA size, (kb)	3	4
Structure		
Amino acid residues	710	729
Molecular mass of immunodetectable protein (kDa)	75	105
Predicted membrane spannings	12	12
N-Glycosylation site	5	4
Homology (%)	48	
Tissue distribution	Small intestine > kidney cortex	Kidney medulla > kidney cortex > brain, lung, spleen
Localization	Brush-border membrane	Brush-border membrane
Function		
Driving force	Proton gradient, membrane potential	Proton gradient, membrane potential
Substrate	Di-, tripeptides, β-lactam antibiotics, bestatin, angiotensin-converting enzyme inhibitors	Di-, tripeptides, β-lactam antibiotics, Bestatin
Substrate affinity		
Apparent K_m (Gly-Sar) (mM)	1.1	0.11
Essential residues	His57, His121	His87, His142

[a] Data from Saito et al. (1995 and 1996), Ogihara et al. (1996), and Terada et al. (1996 and 1997a,b).

amino and/or carboxy terminal, cyclic dipeptides (Hidalgo et al., 1995) and nephrotoxin ochratoxin A (Zingerle et al., 1997) were also shown to be substrates (Fig. 7). Valacyclovir, which is the L-valyl ester of acyclovir, has been used as an oral prodrug of acyclovir and reported to show three- to fourfold bioavailability increase relative to acyclovir. Thus, as the number of possible substrates for the peptide transporter is increasing, more drug targeting via peptide transporters will be exploited in the future. Moreover, a dipeptide transport system which is similar but not identical to peptide transporters PEPT1 and PEPT2 exists in fibroblast-derived tumor cells, but not in normal fibroblasts (Nakanishi et al., 1997). This finding could be the basis of a novel strategy for the specific delivery of peptide-like anticancer drugs into tumor cells, although it is necessary to confirm the selective expression of this transport system in various tumor cells, but not in normal cells.

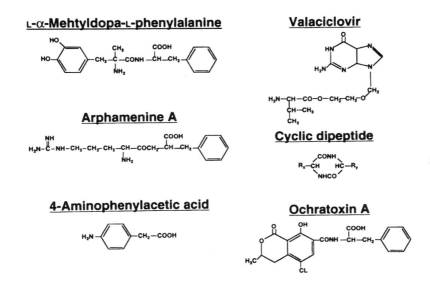

Figure 7. The chemical structures of cyclic dipeptide and peptide-like compounds.

6. SUMMARY AND PERSPECTIVE

Peptide transporters PEPT1 and PEPT2 have been shown to have important physiological, pharmacological, and nutritional roles. Molecular understanding of the drug recognition mechanisms by PEPT1 and PEPT2 will provide useful information for drug design and delivery systems to improve the efficiency of chemotherapy. From the pharmaceutical perspective, species differences and regulation of peptide transporters must be clarified. In addition, considering overall drug absorption, efforts must be directed toward the identification and characterization of the basolateral-type peptide transporter. Lastly, the characterization of the expression of peptide transporters in the brain, lung, liver, and spleen may open new avenues for the delivery of peptide-like drugs in peripheral tissues.

REFERENCES

Adibi, S. A., 1997, The oligopeptide transporter (Pept-1) in human intestine: Biology and function, *Gastroenterology* **113**:332–340.

Boll, M., Herget, M., Wagener, M., Weber, W. M., Markovich, D., Biber, J., Clauss, W., Murer, H., and Daniel, H., 1996, Expression cloning and functional characterization of the kidney cortex high-affinity proton-coupled peptide transporter, *Proc. Natl. Acad. Sci. USA* **93**:284–289.

Daniel, H., and Adibi, S. A., 1993, Transport of β-lactam antibiotics in kidney brush border membrane. Determinants of their affinity for the oligopeptide/H^+ symporter, *J. Clin. Invest.* **92**:2215–2223.

Daniel, H., and Adibi, S. A., 1994, Functional separation of dipeptide transport and hydrolysis in kidney brush border membrane vesicles, *FASEB J.* **8**:753–759.

Daniel, H., and Herget, M., 1997, Cellular and molecular mechanisms of renal peptide transport, *Am. J. Physiol.* **273**:F1–F8.

Daniel, H., Morse, E. L., and Adibi, S. A., 1991, The high and low affinity transport systems for dipeptides in kidney brush border membrane respond differently to alterations in pH gradient and membrane potential, *J. Biol. Chem.* **266**:19917–19924.

Dantzig, A. H., and Bergin, L., 1990, Uptake of the cephalosporin, cephalexin, by a dipeptide transport carrier in the human intestinal cell line, Caco-2, *Biochim. Biophys. Acta* **1027**:211–217.

Dixon, C., and Mizen, L. W., 1977, Absorption of amino penicillins from everted rat intestine, *J. Physiol. (Lond.)* **269**:549–559.

Döring, F., Dorn, D., Bachfischer, U., Amasheh, S., Herget, M., and Daniel, H., 1996, Functional analysis of a chimeric mammalian peptide transporter derived from the intestinal and renal isoforms, *J. Physiol. (Lond.)* **497**:773–779.

Dyer, J., Beechey, R. B., Gorvel, J.-P., Smith, R. T., Wootton, R., and Shirazi-Beechey, S. P., 1990, Glycyl-L-proline transport in rabbit enterocyte basolateral-membrane vesicles, *Biochem. J.* **269**:565–571.

Fei, Y.-J., Kanai, Y., Nussberger, S., Ganapathy, V., Leibach, F. H., Romero, M. F., Singh, S. K., Boron, W. F., and Hediger, M. A., 1994, Expression cloning of a mammalian proton-coupled oligopeptide transporter, *Nature* **368**:563–566.

Fei, Y.-J., Liu, W., Prasad, P. D., Kekuda, R., Oblak, T. G., Ganapathy, V., and Leibach, F. H., 1997, Identification of the histidyl residue obligatory for the catalytic activity of the human H^+/peptide cotransporters PEPT1 and PEPT2, *Biochemistry* **36**:452–460.

Ganapathy, M. E., Brandsch, M., Prasad, P. D., Ganapathy, V., and Leibach, F. H., 1995, Differential recognition of β-lactam antibiotics by intestinal and renal peptide transporters, PEPT1 and PEPT2, *J. Biol. Chem.* **270**:25672–25677.

Ganapathy, V., and Leibach, F. H., 1983, Role of pH gradient and membrane potential in dipeptide transport in intestinal and renal brush-border membrane vesicles from the rabbit. Studies with L-carnosine and glycyl-L-proline, *J. Biol. Chem.* **258**:14189–14192.

Han, H., and Amidon, G. L., 1996, Intestinal absorption of valacyclovir, a novel prodrug of acyclovir, in the rat jejunum, *Pharmaceut. Res.* **13**:S-246.

Hidalgo, I. J., Bhatnagar, P., Lee, C.-P., Miller, J., Cucullino, G., and Smith, P. L., 1995, Structural requirements for interaction with the oligopeptide transporter in Caco-2 cells, *Pharmaceut. Res.* **12**:317–319.

Hori, R., Okano, T., Kato, M., Maegawa, H., and Inui, K., 1988, Intestinal absorption of cephalosporin antibiotics: Correlation between intestinal absorption and brush-border membrane transport, *J. Pharm. Pharmacol.* **40**:646–647.

Hori, R., Tomita, Y., Katsura, T., Yasuhara, M., Inui, K., and Takano, M., 1993, Transport

of bestatin in rat renal brush-border membrane vesicles, *Biochem. Pharmacol.* **45**:1763–1768.
Hu, M., Subramanian, P., Mosberg, H. I., and Amidon, G. L., 1989, Use of the peptide carrier system to improve the intestinal absorption of L-α-methyldopa: Carrier kinetics, intestinal permeabilities, and *in vitro* hydrolysis of dipeptidyl derivatives of L-α-methyldopa, *Pharmaceut. Res.* **6**:66–70.
Inui, K., Okano, T., Takano, M., Kitazawa, S., and Hori, R., 1983, Carrier-mediated transport of amino-cephalosporins by brush border membrane vesicles isolated from rat kidney cortex, *Biochem. Pharmacol.* **32**:621–626.
Inui, K., Okano, T., Takano, M., Saito, H., and Hori, R., 1984, Carrier-mediated transport of cephalexin via the dipeptide transport system in rat renal brush-border membrane vesicles, *Biochim. Biophys. Acta* **769**:449–454.
Inui, K., Okano, T., Maegawa, H., Kato, M., Takano, M., and Hori, R., 1988, H^+ coupled transport of p.o. cephalosporins via dipeptide carriers in rabbit intestinal brush-border membranes: Difference of transport characteristics between cefixime and cephradine, *J. Pharmacol. Exp. Ther.* **247**:235–241.
Inui, K., Yamamoto, M., and Saito, H., 1992, Transepithelial transport of oral cephalosporins by monolayers of intestinal epithelial cell line Caco-2: Specific transport systems in apical and basolateral membranes, *J. Pharmacol. Exp. Ther.* **261**:195–201.
Kato, M., Maegawa, H., Okano, T., Inui, K., and Hori, R., 1989, Effect of various chemical modifiers on H^+ coupled transport of cephradine via dipeptide carriers in rabbit intestinal brush-border membranes: Role of histidine residues, *J. Pharmacol. Exp. Ther.* **251**:745–749.
Kimura, T., Endo, H., Yoshikawa, M., Muranishi, S., and Sezaki, H., 1978, Carrier-mediated transport systems for aminopenicillins in rat small intestine, *J. Pharmacobio-Dyn.* **1**:262–267.
Kimura, T., Yamamoto, T., Mizuno, M., Suga, Y., Kitade, S., and Sezaki, H., 1983, Characterization of aminocephalosporin transport across rat small intestine, *J. Pharmacobio-Dyn.* **6**:246–253.
Kitagawa, S., Takeda, J., Kaseda, Y., and Sato, S., 1997, Inhibitory effects of angiotensin-converting enzyme inhibitors on cefroxadine uptake by rabbit small intestinal brush border membrane vesicles, *Biol. Pharmaceut. Bull.* **20**:449–451.
Kramer, W., Girbig, F., Gutjahr, U., Kleemann, H.-W., Leipe, I., Urbach, H., and Wagner, A., 1990, Interaction of renin inhibitors with the intestinal uptake system for oligopeptides and β-lactam antibiotics, *Biochim. Biophys. Acta* **1027**:25–30.
Leibach, F. H., and Ganapathy, V., 1996, Peptide transporters in the intestine and the kidney, *Annu. Rev. Nutr.* **16**:99–119.
Liang, R., Fei, Y.-J., Prasad, P. D., Ramamoorthy, S., Han, H., Yang-Feng, T. L., Hediger, M. A., Ganapathy, V., and Leibach, F. H., 1995, Human intestinal H^+/peptide cotransporter. Cloning, functional expression, and chromosomal localization, *J. Biol. Chem.* **270**:6456–6463.
Liu, W., Liang, R., Ramamoorthy, S., Fei, Y.-J., Ganapathy, M. E., Hediger, M. A., Ganapathy, V., and Leibach, F. H., 1995, Molecular cloning of PEPT2, a new member of the H^+/peptide cotransporter family, from human kidney, *Biochim. Biophys. Acta* **1235**:461–466.
Matsumoto, S., Saito, H., and Inui, K., 1994, Transcellular transport of oral cephalosporins

in human intestinal epithelial cells, Caco-2: Interaction with dipeptide transport systems in apical and basolateral membranes, *J. Pharmacol. Exp. Ther.* **270**:498–504.

Matsumoto, S., Saito, H., and Inui, K., 1995, Transport characteristics of ceftibuten, a new cephalosporin antibiotic, via the apical H^+/dipeptide cotransport system in human intestinal cell line Caco-2: Regulation by cell growth, *Pharmaceut. Res.* **12**:1483–1487.

Matthews, D. M., 1975, Intestinal absorption of peptides, *Physiol. Rev.* **55**:537–608.

Meredith, D., and Boyd, C. A. R., 1995, Dipeptide transport characteristics of the apical membrane of rat lung type II pneumocytes, *Am. J. Physiol.* **269**:L137–L143.

Miyamoto, K., Shiraga, T., Morita, K., Yamamoto, H., Haga, H., Taketani, Y., Tamai, I., Sai, Y., Tsuji, A., and Takeda, E., 1996, Sequence, tissue distribution and developmental changes in rat intestinal oligopeptide transporter, *Biochim. Biophys. Acta* **1305**:34–38.

Miyamoto, Y., Ganapathy, V., and Leibach, F. H., 1986, Identification of histidyl and thiol groups at the active site of rabbit renal dipeptide transporter, *J. Biol. Chem.* **261**:16133–16140.

Miyamoto, Y., Coone, J. L., Ganapathy, V., and Leibach, F. H., 1988, Distribution and properties of the glycylsarcosine-transport system in rabbit renal proximal tubule. Studies with isolated brush-border-membrane vesicles, *Biochem. J.* **249**:247–253.

Muranushi, N., Yoshikawa, T., Yoshida, M., Oguma, T., Hirano, K., and Yamada, H., 1989, Transport characteristics of ceftibuten, a new oral cephem, in rat intestinal brush-border membrane vesicles: Relationship to oligopeptide and amino β-lactam transport, *Pharmaceut. Res.* **6**:308–312.

Naasani, I., Sato, K., Iseki, K., Sugawara, M., Kobayashi, M., and Miyazaki, K., 1995, Comparison of the transport characteristics of ceftibuten in rat renal and intestinal brush-border membranes, *Biochim. Biophys. Acta* **1231**:163–168.

Nakanishi, T., Tamai, I., Sai, Y., Sasaki, T., and Tsuji, A., 1997, Carrier-mediated transport of oligopeptides in the human fibrosarcoma cell line HT1080, *Cancer Res.* **57**:4118–4122.

Nakashima, E., and Tsuji, A., 1985, Mutual effects of amino-β-lactam antibiotics and glycylglycine on the transmural potential difference in the small intestinal epithelium of rats, *J. Pharmacobio-Dyn.* **8**:623–632.

Nakashima, E., Tsuji, A., Mizuo, H., and Yamana, T., 1984, Kinetics and mechanism of *in vitro* uptake of amino-β-lactam antibiotics by rat small intestine and relation to the intact-peptide transport system, *Biochem. Pharmacol.* **33**:3345–3352.

Ogihara, H., Saito, H., Shin, B.-C., Terada, T., Takenoshita, S., Nagamachi, Y., Inui, K., and Takata, K., 1996, Immuno-localization of H^+/peptide cotransporter in rat digestive tract, *Biochem. Biophys. Res. Commun.* **220**:848–852.

Okano, T., Inui, K., Maegawa, H., Takano, M., and Hori, R., 1986a, H^+ coupled uphill transport of aminocephalosporins via the dipeptide transport system in rabbit intestinal brush-border membranes, *J. Biol. Chem.* **261**:14130–14134.

Okano, T., Inui, K., Takano, M., and Hori, R., 1986b, H^+ gradient-dependent transport of aminocephalosporins in rat intestinal brush-border membrane vesicles. Role of dipeptide transport system, *Biochem. Pharmacol.* **35**:1781–1786.

Ramamoorthy, S., Liu, W., Ma, Y.-Y., Yang-Feng, T. L., Ganapathy, V., and Leibach, F. H., 1995, Proton/peptide cotransporter (PEPT2) from human kidney: Functional characterization and chromosomal localization, *Biochim. Biophys. Acta* **1240**:1–4.

Ries, M., Wenzel, U., and Daniel, H., 1994, Transport of cefadroxil in rat kidney brush-border membranes is mediated by two electrogenic H^+-coupled systems, *J. Pharmacol. Exp. Ther.* **271**:1327–1333.

Saito, H., and Inui, K., 1993, Dipeptide transporters in apical and basolateral membranes of the human intestinal cell line Caco-2, *Am. J. Physiol.* **265**:G289–G294.

Saito, H., Okuda, M., Terada, T., Sasaki, S., and Inui, K., 1995, Cloning and characterization of a rat H^+/peptide cotransporter mediating absorption of β-lactam antibiotics in the intestine and kidney, *J. Pharmacol. Exp. Ther.* **275**:1631–1637.

Saito, H., Terada, T., Okuda, M., Sasaki, S., and Inui, K., 1996, Molecular cloning and tissue distribution of rat peptide transporter PEPT2, *Biochim. Biophys. Acta* **1280**:173–177.

Saito, H., Motohashi, H., Mukai, M., and Inui, K., 1997, Cloning and characterization of a pH-sensing regulatory factor that modulates transport activity of the human H^+/peptide cotransporter, PEPT1, *Biochem. Biophys. Res. Commun.* **237**:577–582.

Silbernagl, S., Ganapathy, V., and Leibach, F. H., 1987, H^+ gradient-driven dipeptide reabsorption in proximal tubule of rat kidney. Studies *in vivo* and *in vitro, Am. J. Physiol.* **253**:F448–F457.

Takahashi, K., Nakamura, N., Terada, T., Okano, T., Futami, T., Saito, H., and Inui, K., 1998, Interaction of β-lactam antibiotics with H^+/peptide cotransporters in rat renal brush-border membranes, *J. Pharmacol. Exp. Ther.* **286**:1037–1042.

Takuwa, N., Shimada, T., Matsumoto, H., and Hoshi, T., 1985, Proton-coupled transport of glycylglycine in rabbit renal brush-border membrane vesicles, *Biochim. Biophys. Acta* **814**:186–190.

Tamai, I., Nakanishi, T., Hayashi, K., Terao, T., Sai, Y., Shiraga, T., Miyamoto, K., Takeda, E., Higashida, H., and Tsuji, A., 1997, The predominant contribution of oligopeptide transporter PepT1 to intestinal absorption of β-lactam antibiotics in the rat small intestine, *J. Pharm. Pharmacol.* **49**:796–801.

Temple, C. S., Stewart, A. K., Meredith, D., Lister, N. A., Morgan, K. M., Collier, I. D., Vaughan-Jones, R. D., Boyd, C. A. R., Bailey, P. D., and Bronk, J. R., 1998, Peptide mimics as substrates for the intestinal peptide transporter, *J. Biol. Chem.* **273**:20–22.

Terada, T., Saito, H., Mukai, M., and Inui, K., 1996, Identification of the histidine residues involved in substrate recognition by a rat H^+/peptide cotransporter, PEPT1, *FEBS Lett.* **394**:196–200.

Terada, T., Saito, H., Mukai, M., and Inui, K., 1997a, Characterization of stably transfected kidney epithelial cell line expressing rat H^+/peptide cotransporter PEPT1: Localization of PEPT1 and transport of β-lactam antibiotics, *J. Pharmacol. Exp. Ther.* **281**:1415–1421.

Terada, T., Saito, H., Mukai, M., and Inui, K., 1997b, Recognition of β-lactam antibiotics by rat peptide transporters, PEPT1 and PEPT2, in LLC-PK$_1$ cells, *Am. J. Physiol.* **273**:F706–F711.

Terada, T., Saito, H., and Inui, K., 1998, Interaction of β-lactam antibiotics with histidine residue of rat H^+/peptide cotranporters, PEPT1 and PEPT2, *J. Biol. Chem.* **273**:5582–5585.

Thwaites, D. T., Brown, C. D. A., Hirst, B. H., and Simmons, N. L., 1993, Transepithelial glycylsarcosine transport in intestinal Caco-2 cells mediated by expression of H^+-coupled carriers at both apical and basal membranes, *J. Biol. Chem.* **268**:7640–7642.

Tomita, Y., Katsura, T., Okano, T., Inui, K., and Hori, R., 1990, Transport mechanisms of bestatin in rabbit intestinal brush-border membranes: Role of H^+/dipeptide cotransport system, *J. Pharmacol. Exp. Ther.* **252**:859–862.

Tsuji, A., Nakashima, E., Kagami, I., and Yamana, T., 1981, Intestinal absorption mechanism of amphoteric β-lactam antibiotics I: Comparative absorption and evidence for saturable transport of amino-β-lactam antibiotics by *in situ* rat small intestine, *J. Pharmaceut. Sci.* **70**:768–772.

Tsuji, A., Terasaki, T., Tamai, I., and Hirooka, H., 1987, H^+ gradient-dependent and carrier-mediated transport of cefixime, a new cephalosporin antibiotic, across brush-border membrane vesicles from rat small intestine, *J. Pharmacol. Exp. Ther.* **241**:594–601.

Tsuji, A., Tamai, I., Nakanishi, M., and Amidon, G. L., 1990, Mechanism of absorption of the dipeptide α-methyldopa-phe in intestinal brush-border membrane vesicles, *Pharmaceut. Res.* **7**:308–309.

Wenzel, U., Gebert, I., Weintraut, H., Weber, W.-M., Clauss, W., and Daniel, H., 1996, Transport characteristics of differently charged cephalosporin antibiotics in oocytes expressing the cloned intestinal peptide transporter PepT1 and in human intestinal Caco-2 cells, *J. Pharmacol. Exp. Ther.* **277**:831–839.

Zingerle, M., Silbernagl, S., and Gekle, M., 1997, Reabsorption of the nephrotoxin ochratoxin A along the rat nephron *in vivo, J. Pharmacol. Exp. Ther.* **280**:220–224.

11

Antigenic Peptide Transporter

Vashti G. Lacaille and Matthew J. Androlewicz

1. INTRODUCTION

The transporter associated with antigen processing (TAP) plays a critical role in the major histocompatibility complex (MHC) class I antigen processing pathway by transporting antigenic peptides from the cytosol to the lumen of the endoplasmic reticulum (ER), where the loading of newly synthesized class I molecules takes place. TAP is a member of the large superfamily of ATP-binding cassette (ABC) transporters which possess a characteristic domain structure and transport a wide variety of substrates across membranes in an ATP-dependent manner (for review see Higgins, 1992). Other well-known transporters from this group include P-glycoprotein, which is associated with multidrug resistance (MDR) (Gros *et al.,* 1986), the cystic fibrosis transductance regulator (Riordan *et al.,* 1989) and the sterile 6 transporter in yeast (STE6), which transports the a-type mating factor across the plasma membrane (Kuchler *et al.,* 1989). Since the discovery of the TAP genes in 1990, much progress has been made toward an understanding of the structure and function of this important transporter. It is not the goal of the present work to provide an exhaustive review of TAP developments to date. An excellent review by Tim Elliott (1997) already exists on this subject. Rather, we would like to present an overview of the molecular biology of TAP in the context of TAP as a drug

---

Vashti G. Lacaille and Matthew J. Androlewicz • Immunology Program, H. Lee Moffitt Cancer Center and Research Institute, and Department of Biochemistry and Molecular Biology, University of South Florida College of Medicine, Tampa, Florida 33612.
Membrane Transporters as Drug Targets, edited by Amidon and Sadée. Kluwer Academic/Plenum Publishers, New York, 1999.

target. There is a wealth of information on the substrate specificity of TAP and on the inhibition of TAP by viral proteins. In addition, studies have been initiated using a rational approach toward the design of specific TAP inhibitors based on knowledge of the TAP–peptide interaction. Taken together, the data indicate that TAP has the potential to be a drug target, and in certain instances the inhibition of TAP function may be of therapeutic value, such as in the case of tissue transplantation and autoimmune disease. Peptidomimetics appears to be a promising approach toward the development of specific TAP inhibitors.

2. TAP AND THE MHC CLASS I ANTIGEN PROCESSING PATHWAY

2.1. Introduction

Major histocompatibility complex (MHC) class I molecules present antigenic peptides to CD8+ cytotoxic T lymphocytes (CTL). Recognition of the target cell by CTL initiates a chain of events which ultimately leads to the lysis and death of the target cell. In this way, cells which harbor foreign antigens, i.e., viruses, bacteria, or tumor antigens, can be destroyed by the immune system. The MHC class I antigen processing and presentation pathway therefore plays a major role in the immune response to disease. MHC class I antigens (HLA-A, B, and C in humans, H-2-K, D, and L in mice) are highly polymorphic and each locus contains several alleles, each with its own unique peptide-binding specificity (for review see Bjorkman and Parham, 1990). The class I alleles of a given organism are capable of presenting a wide variety of antigenic peptides derived from endogenous foreign proteins. Class I molecules are expressed as a membrane-bound heavy chain ($\sim$45 kDa) associated with a soluble light chain ($\sim$12 kDa) called β_2-microglobulin (β_2m). X-ray crystal structure analysis of the class I heterodimer revealed the existence of a groove in the α-1 and α-2 domains of the heavy chain which contained bound peptide (Bjorkman *et al.,* 1987). Indeed, proper folding of the heterodimer occurs only in the presence of peptide. The optimal size of the peptide for binding to the groove is 8–10 amino acids. The primary function of TAP is to supply these peptides to the class I molecules at the site of assembly, i.e., the ER.

The vast majority of the antigenic peptides presented by class I molecules are generated in the cell cytosol. Degradation of the antigens is mediated through the action of the proteasome, a large multicatalytic protease (Goldberg and Rock, 1992; Tanaka *et al.,* 1997), which is the primary mode of protein degradation in the cytosol. Antigens are degraded in a ubiquitin- and ATP-dependent manner. Because the peptides are generated in the cytosol and the site of class I assembly is in the ER, there must be a mechanism for the delivery of the peptides to the ER. The existence of a peptide transporter for this purpose was hypothesized prior to the actual discovery of the transporter genes (Townsend *et al.,* 1985). In certain antigen processing mutant cell lines, class I molecules were found to be unstable and devoid of peptide (DeMars *et al.,* 1985; Salter and Cresswell, 1986), and it was speculated that the cells lacked, or possessed a mutated, gene or genes that coded for the transporter. In 1990, genes which possessed sequence homology to the ABC family of transporter proteins were discovered in human, mouse, and rat (Spies *et al.,* 1990; Trowsdale *et al.,* 1990; Monaco *et al.,* 1990; Deverson *et al.,* 1990). The location of the genes, termed TAP.1 and TAP.2, corresponded to the regions of the genome which were missing in the mutant cells, and therefore were clear candidates for coding for the putative peptide transporter. Subsequent experiments showed that antigen presentation was restored in the mutant cells upon reintroduction of the TAP genes (Spies and Demars, 1991; Attaya *et al.,* 1992), and *in vitro* functional analyses showed that TAP did transport peptides in an ATP-dependent manner (Androlewicz *et al.,* 1993; Shepherd *et al.,* 1993; Neefjes *et al.,* 1993). It was therefore clear that TAP was the antigenic peptide transporter, a crucial element in the class I antigen processing pathway.

2.2. Current Model of MHC Class I Antigen Processing

Since the discovery of TAP, many of the molecular details of class I assembly and loading have come to light. Shown in Fig. 1 is the current model of MHC class I assembly in the ER. Antigens, targeted for processing and presentation, are ubiquinated and degraded by the proteasome into smaller peptide fragments which become substrates for TAP. Many of the processing events that take place in the cytosol remain in question, such as, is there a physical association between the proteasome and TAP, and are molecular chaperones involved in the shuttling of peptides from the proteasome to TAP? In any case, once generated, the peptides are transported across the ER membrane in an ATP-dependent manner by TAP. Inside the ER, the loading of the class I molecules with peptide involves a complex series of interactions which includes roles for other accessory molecules. Multiple components physically associate with TAP in the ER lumen. Among these are the class I molecules themselves (Ortmann *et al.,* 1994; Suh *et al.,* 1994), a novel 48-kDa glycoprotein called tapasin (Sadasivan *et al.,* 1996), and the molecular chaperone calreticulin (Sadasivan *et al.,* 1996). All of these molecules come together in a way which is thought to facilitate the loading of the empty class I molecule with peptide. Once the class I molecule is loaded, the multimeric complex dissociates and the stable class I molecule consisting of heavy chain, β_2m, and peptide leaves the ER en route to the cell surface. The precise role of each of the components of the TAP multimeric complex in peptide loading is not clear, although it is speculated that tapasin mediates TAP association with class I because class I molecules in mu-

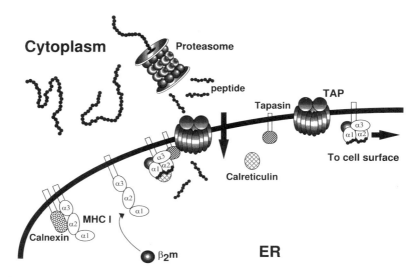

Figure 1. Model of MHC class I antigen processing. Antigens are degraded in the cytoplasm into smaller peptides through the action of the proteasome. The peptides are transported across the ER membrane by the transporter associated with antigen processing (TAP). Newly synthesized class I heavy chains associate with calnexin early in the assembly pathway. Upon the association of β_2-microglobulin (β_2m) with the heavy chain, calnexin dissociates, and may be replaced by calreticulin. The class I heterodimer then associates with TAP to form a large multimeric complex of class I heavy chain, β_2m, calreticulin, tapasin, and TAP. The class I molecule, associated with TAP, is loaded with peptide and then the multimeric complex dissociates into individual components. The loaded class I molecule is then released from the ER and is transported to the cell surface. See text for further details.

tant cells that lack tapasin fail to associate with TAP (Grandea et al., 1995). Furthermore, the formation of the complex is not absolutely required for peptide loading of class I molecules, as class I molecules in TAP-deficient cells can be successfully loaded with peptides that are derived from leader sequences (Wei and Cresswell, 1992), or are targeted to the ER by leader sequences (Anderson et al., 1991). In addition to calreticulin, the molecular chaperone calnexin is also involved in class I assembly (Sugita and Brenner, 1994); however, calnexin associates with the class I heavy chain early in the biosynthetic pathway, and does not appear to be part of the multimeric complex with TAP (at least in humans) (Sadasivan et al., 1996). Finally, while the class I molecule appears to be the terminal acceptor of the antigenic peptide, other molecules have been shown to bind to peptides in the ER, such as protein disulfide isomerase (Lammert et al., 1997; Spee and Neefjes, 1997), gp96 (Spee and Neefjes, 1997; Arnold et al., 1997), Bip (Levy et al., 1991), calreticulin (Spee and Neefjes, 1997), and tapasin (Li et al., 1997). These molecules may play an intermediate role in the loading of the class I molecule, or may bind only those peptides which do not bind to the class I molecule.

3. TAP STRUCTURE AND FUNCTION

The ABC transporter superfamily is highly conserved throughout evolution, and plays an important role in transport from bacteria to humans (Higgins, 1992). In addition to TAP, other well-known examples include the hemolysin B (Hly B) (Felmlee *et al.*, 1985) and oligopeptide (Opp ABCDE) (Hiles *et al.*, 1987) transporters in bacteria, the yeast sterile 6 gene (STE6), which transports the a mating factor (Kuchler *et al.*, 1989), the mammalian P-glycoprotein (Gros *et al.*, 1986) associated with multidrug resistance (MDR), and the cystic fibrosis transmembrane conductance regulator (CFTR) (Riordan *et al.*, 1989). The hallmark of all these transporters is the possession of highly conserved ATP-binding domains which contain the Walker A and Walker B motifs (Walker *et al.*, 1982). In addition, they possess a series of hydrophobic transmembrane segments, usually 12 in number, which contain the ligand binding and transport regions. The domain structure of TAP is shown in Fig. 2. TAP is composed of two separate polypeptides, TAP.1 and TAP.2, which come together to form a single functional transporter. ABC transporter proteins can be expressed as the product of one to four separate genes, however, regardless of the number of polypeptides expressed, the classic domain struc-

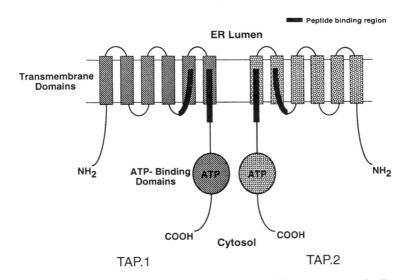

Figure 2. Domain organization of TAP. TAP is a member of the ABC transporter superfamily, and is expressed as a heterodimer of two proteins, TAP.1 and TAP.2. Each subunit is comprised of approximately six transmembrane domains and one ATP-binding domain. The expression of both TAP.1 and TAP.2 is required to generate a functional transporter. The putative peptide-binding regions, as determined by photo-cross-linking, are shown by dark bars. Adapted from Nijenhuis and Hammerling (1996).

ture of the ABC transporter is maintained. TAP.1 and TAP.2 are encoded within the class II region of the MHC, and possess 60–70% homology at the protein level with the highest homology occurring within the ATP-binding domains. Hydropathy analysis has revealed that TAP.1 and TAP.2 each contain between 6 and 10 transmembrane segments (Nijenhuis and Hammerling, 1996; Momburg et al., 1996). The actual transmembrane topology is in the process of being worked out experimentally (Gileadi and Higgins, 1997).

3.1. TAP Substrate Specificity

The substrate specificity of TAP has been analyzed in some detail over the past few years. TAP specificity is quite species-dependent. Human TAP appears to be the least restrictive with regard to peptide specificity, while mouse TAP is the most restrictive, and rat TAP selectivity lies somewhere between human and mouse. The substrate specificity studies have been performed utilizing two major assay systems. One measures the competition for the TAP-dependent transport of a reporter peptide which possesses a glycosylation acceptor site (Neefjes et al., 1993; Androlewicz and Cresswell, 1994). Transport of the reporter peptide into the lumen of the ER results in the glycosylation of the peptide by the natural glycosylation machinery of the cell, and the degree of peptide transport is directly related to the degree of peptide glycosylation. The second assay measures competition for peptide binding to TAP (van Endert et al., 1994; Meyer et al., 1994). This assay does not measure peptide transport, but peptide binding and transport are closely coupled, and the TAP peptide binding assay is a valid way to measure substrate specificity. In fact, because the transport assay involves competition at the level of peptide binding, peptide transport, as well as peptide glycosylation, the peptide binding assay is a more direct way to measure substrate specificity.

The combined results of many experiments reveal that, across the species tested, TAP possesses a similar peptide length specificity. All the data are in agreement that the minimum peptide length required for peptide binding and transport is 8–9 amino acids (Androlewicz and Cresswell, 1994; van Endert et al., 1994; Momburg et al., 1994a; Schumacher et al., 1994a). The maximum peptide length is more loosely restricted. Peptides of up to 40 amino acids have been reported to be transported by TAP (Koopmann et al., 1996). In general, the transport of peptides in the 8- to 16-amino acid range is quite efficient, while the transport efficiency of longer peptides goes down as the peptides increase in length. The length specificity of TAP essentially correlates with the length specificity of MHC class I molecules, i.e., 8–10 amino acids. However, the TAP-dependent transport of longer peptides may play a critical role in getting peptide epitopes into the ER, which, in their shorter versions, are not good substrates for TAP. In support of this, there is strong evidence to suggest that peptide trimming can occur in the ER (Snyder et al., 1994; Elliott et al., 1995), which can result in the generation of the shorter, optimal class I binding peptides from the longer peptides.

In contrast to peptide length specificity, TAP peptide sequence specificity varies considerably among the species tested. Early competition studies focused on the C-terminal amino acid of the peptide because the C-terminal residue was important as a secondary anchor residue for binding to certain class I molecules (Rammensee et al., 1993). In addition, the proteasome subunits that are encoded within the MHC region, i.e., LMP2 and LMP7, were thought to enhance the generation of peptides that possessed hydrophobic or basic residues at the C-terminus (Driscoll et al., 1993; Gaczynska et al., 1993). What was found was that human TAP possessed just a minor degree of specificity at the C-terminus of the peptide (Androlewicz and Cresswell, 1994; Momburg et al., 1994b), while mouse TAP possessed a significant preference to transport peptides that contained a hydrophobic residue (aliphatic and aromatic) at the C-terminus (Schumacher et al., 1994a; Momburg et al., 1994b). The specificity of the mouse transporter clearly has an important functional consequence, as the majority of mouse MHC class I peptide-binding motifs possess a hydrophobic residue at the C-terminus (Rammensee et al., 1993). Rat TAP was shown to be an interesting combination of the mouse and human transporters. Rat TAP exists in two forms which have major functional consequences for class I expression in the rat. The rat TAP.2 gene comes in two allelic versions, TAP.2^a and TAP.2^u, and each one provides for a unique specificity when complexed with rat TAP.1. TAP.1/TAP.2^a possesses a specificity which is similar to human TAP, i.e., a slight preference for particular C-terminal residues, while the TAP.1/TAP.2^u possesses a specificity which is very similar to mouse TAP, i.e., a preference for hydrophobic C-terminal residues (Heemels and Ploegh, 1994; Momburg et al., 1994b). This TAP allelism in the rat results in a significant difference in the pools of peptides which the rat class I molecules can sample in the ER, and has profound effects on class I expression as shown by the rat cim phenomenon (Powis et al., 1991). Other alleles of TAP that exist in the human and mouse have not been shown to possess any such functional consequences (Daniel et al., 1997; Obst et al., 1995; Schumacher et al., 1994b).

Additional studies on the peptide specificity of TAP revealed that human TAP is more selective than originally thought. By measuring competition for peptide binding to TAP in microsomes of >250 peptides with defined amino acid substitutions, van Endert et al. (1995) were able to identify a peptide-binding motif for human TAP. The motif highlighted certain regions of the peptide which played an important role in peptide recognition by TAP. In contrast to the peptide binding motifs which have been described for MHC class I molecules (Rammensee et al., 1995), the motif for TAP did not possess primary or secondary anchor positions with a strict requirement for one or two amino acids. Instead, the motif for TAP was more subtle and less defined. The important regions of the peptide were the first three amino acids at the N-terminus and the single C-terminal amino acid. TAP prefers strongly hydrophobic residues in position 3 (P3) and hydrophobic or charged residues in P2. In addition, a proline residue in P1 or P2 or an aromatic or acidic residue in P1 has strong deleterious effects. At the C-terminus, hydrophobic and positively charged residues are favored, while negatively charged residues are not. An additional study by Uebel *et al.* (1997) further supports the idea of a peptide binding motif for human TAP. These investigators took advantage of combinatorial peptide technology to address peptide recognition by TAP without the bias of a defined sequence from which to start. This concept is based on the use of a randomized peptide library of defined length (nine amino acids) which possesses one defined amino acid at a particular position. In this way, the effects of an individual residue at a given position can be determined independently of sequence context. The peptide libraries are screened by competition for the binding of a reporter peptide to TAP in microsomes in a manner analogous to that by van Endert *et al.* (1995) described above. Using this combinatorial approach, the investigators were able to more or less confirm the peptide-binding motif for human TAP. They go on to suggest a model of peptide interaction with TAP, where the points of contact between peptide and TAP are at the N- and C-termini of the peptide, leaving the central portion of the peptide unassociated with TAP. In this way, longer peptides could be accommodated by TAP due to the flexibility and lack of contact of the central portion of the peptide with TAP. This would be analogous to the way longer peptides can bind to class I molecules by bulging out from the center of the peptide binding groove (Guo *et al.,* 1992).

In the final analysis, TAP does preselect peptides that are favorable for binding to class I molecules. However, the stringency by which this preselection takes place is not near the level of stringency required for peptide binding to class I. In the case of HLA, the peptide-binding motifs generally do contain basic or hydrophobic residues at the C-terminus, as well as anchor residues in positions 1–3 of the peptide that favor binding to TAP. In the case of H2, the peptide motifs also correspond to the specificity of mouse TAP, i.e., the motifs contain hydrophobic residues at the C-terminus. Additional studies on the peptide selectivity of mouse TAP will most certainly identify a peptide-binding motif for the mouse transporter. It has already been shown that a proline residue in position 3 has a strong deleterious effect on peptide transport by mouse TAP (Neisig *et al.,* 1995). A nice feature of the TAP system is the capacity of the transporter to transport longer peptides. In this way, class I-binding peptides which happen to be poor substrates for TAP, such as HLA-B7-binding peptides, which contain proline as an anchor residue at position 2, can be transported as longer precursors and then trimmed in the ER to the appropriate size for binding to class I molecules. In summary, TAP provides an abundant supply of peptides that are predisposed to bind to class I molecules, and it is ultimately the class I molecule itself which determines which epitopes will be presented to T cells.

3.2. Characterization of the TAP Peptide-Binding Site

Much effort has been directed at trying to identify the peptide-binding site on TAP. The data on peptide binding and transport by TAP indicate that there exists a single peptide-binding site. The development of photoreactive peptide analogues has greatly facilitated the study of the TAP peptide-binding site (Androlewicz and Cresswell, 1994; Androlewicz *et al.,* 1994; Nijenhuis *et al.,* 1996; Nijenhuis and Hammerling, 1996; Wang *et al.,* 1996; Ahn *et al.,* 1996b). In fact, the first description of a peptide binding site on TAP was through the use of a photoreactive peptide analogue derived from the natural HLA-A3-binding peptide Nef7B (QV-PLRPMTYK) (Androlewicz and Cresswell, 1994). In this study it was shown that peptide binding to TAP was not dependent on ATP, and this formed the basis for the two-step model of peptide binding and transport by TAP. In addition, it was shown that the Nef7B photopeptide was able to label both TAP.1 and TAP.2 subunits. Subsequent analysis by a variety of groups using photoreactive peptides revealed that in order for photolabeling to occur, both subunits of TAP had to be expressed (Androlewicz *et al.,* 1994; Nijenhuis *et al.,* 1996; Wang *et al.,* 1996; Ahn *et al.,* 1996b). In other words, the expression of single-chain TAP.1 or TAP.2 molecules was not enough to provide a peptide-binding site. This led to the conclusion that TAP.1 and TAP.2 must come together to form a functional binding site, and that the binding site is comprised of elements of both TAP.1 and TAP.2. The photo-cross-linking approach was taken a step further by the generation of antipeptide antisera to particular regions of TAP.1 and TAP.2 in order to map the peptide-binding region through the immunoprecipitation of tryptic fragments of photolabeled TAP with the different antisera (Nijenhuis and Hammerling, 1996). These investigators were able to identify two regions within TAP.1 and TAP.2 which contribute to the peptide-binding site. The regions are comprised of both membrane and cytosolic portions of the two transmembrane segments closest to the ATP-binding domains (see Fig. 2). The authors go on to propose a model in which the transmembrane segments of TAP.1 and TAP.2 come together to form a pore in the membrane, with the peptide recognition site existing at the cytosolic mouth of the pore in close proximity to the ATP-binding domains. This model takes into consideration the role of ATP hydrolysis in the transport of the peptide into the lumen of the ER. The close proximity of the peptide-binding site and the ATP-binding domains may facilitate the transduction of a conformational change upon ATP hydrolysis which results in release of the peptide and opening of the pore.

4. VIRAL INHIBITION OF TAP

Viruses have developed elaborate mechanisms to interfere with the MHC class I antigen processing pathway, including interference with TAP function. One of the first mechanisms for viral-induced class I downregulation was described for adenovirus (Burgert and Kvist, 1985; Andersson *et al.*, 1985). Adenovirus codes for the glycoprotein E3/19K, which binds to human class I molecules and retains them in the ER. This leads to reduced levels of class I on the cell surface and consequently reduced recognition of adenoviral epitopes by T cells. Viral inhibition of class I presentation can occur at multiple steps along the processing pathway. For example, the Epstein–Barr nuclear antigen (EBNA-1) can prevent its own antigenic processing by the existence of a long stretch of repeated glycine and alanine residues, which is thought to interfere with the ubiquitination and degradation of the antigen through the proteasome (Levitskaya *et al.*, 1995). As a result, CTLs against the EBNA-1 protein have never been detected. Human cytomegalovirus (HCMV) possesses several early genes which interfere with class I presentation. US3 is an immediate early gene that causes retention of class I molecules in the ER (Ahn *et al.*, 1996a), in a manner analogous to the adenoviral E3/19K protein. US2 and US11 are early genes which mediate the dislocation of class I chains from the ER to the cytosol where they are deglycosylated and degraded by the proteasome (Jones *et al.*, 1995; Wiertz *et al.*, 1996). US6 is another early gene which has recently been shown to inhibit TAP function (Hengel *et al.*, 1997; Ahn *et al.*, 1997; Lehner *et al.*, 1997). Therefore, in cells infected with HCMV, class I presentation is inhibited by a multipronged approach. Class I molecules are first retained in the ER by US3, and are then dislocated or translocated to the cytosol for degradation by US2 and US11. In addition, US6 inhibits the flow of peptides into the ER for assembly with class I. And, as if that were not enough, HCMV encodes another gene (UL18) which possesses homology to the class I heavy chain itself (Beck and Barrell, 1988). It is thought that the class I homologue serves as a decoy molecule on the cell surface to prevent lysis of the cell by natural killer (NK) cells, which recognize the lack of class I molecules on cells and destroy them accordingly (Ljunggren and Karre, 1990). In this way, the infected cell is able to avoid lysis by both cytotoxic T cells and NK cells.

4.1. Herpes Simplex Virus ICP47 Protein

Herpes simplex virus (HSV) encodes a small cytosolic protein (ICP47) which has the capacity to bind to TAP and strongly inhibit its peptide transport function (Hill *et al.*, 1995; Fruh *et al.*, 1995). The study of ICP47 action provides a glimpse of the possibilities of the specific inhibition of TAP by a small molecule. ICP47 is an 88-amino acid protein encoded by the immediate early gene US12 (Murchie and McGeoch, 1982). It effectively blocks antigen presentation to HSV-specific CD8+ T cells in infected human cells, but is not effective in blocking antigen presentation in mouse cells (York *et al.*, 1994). Analysis of the mechanism of action of ICP47 has revealed that it binds to TAP with a high affinity (K_D ~50 nM, or 10-fold higher than the highest affinity peptides) (Ahn et al., 1996b; Tomazin et al., 1996). It was shown that peptides can compete for the binding of ICP47 to TAP at 100- to 1000-fold molar excess, and that ICP47 completely blocks the binding of peptides to TAP. This indicates that ICP47 binds to the peptide recognition site of TAP, and effectively shuts down peptide transport by preventing other peptides from binding to TAP. The binding of ICP47 to TAP does not interfere with the binding of ATP to TAP, and, consistent with peptide binding to TAP, ICP47 does not require ATP to bind. Furthermore, ICP47 remains stably bound to TAP and is not transported (Ahn et al., 1996b; Tomazin et al., 1996). All the data so far indicate that ICP47 is a potent inhibitor of TAP function.

As previously discussed, TAP prefers to bind peptides in the 8- to 16-amino acid range, while longer peptides can be accommodated to a certain extent. The question arises as to whether full-length ICP47 or a fragment thereof is required for TAP inhibition. The studies up to this point were carried out using full-length recombinant ICP47 produced in bacteria. In order to delineate the minimal active region of ICP47, two studies were performed using synthetic truncated versions of ICP47 (Galocha et al., 1997; Neumann et al., 1997). The truncated versions of ICP47 were tested for their ability to inhibit peptide binding to TAP in microsomes (Neumann et al., 1997), or to inhibit peptide translocation by TAP in streptolysin O-permeabilized cells (Galocha et al., 1997). Both studies were in close agreement for the minimal core sequence required for the full inhibition of TAP. The region was made up of N-terminal residues 2–35 (Galocha et al., 1997) or 3–34 (Neumann et al., 1997). The remainder of the molecule was not required for TAP inhibition. It was also shown that a set of charged residues within this N-terminal fragment was critical for activity, and that these charged residues were conserved in ICP47 derived from both HSV-1 and HSV-2 (Neumann et al., 1997). Furthermore, in an earlier study it was shown that ICP47 (residues 1–53) was largely unstructured in solution, but in the presence of lipid membranes an α-helical structure was induced (Beinert et al., 1997). It was speculated that the α-helix served as a membrane anchor and could effectively increase the local concentration of ICP47 around TAP. This helical region was mapped by computer modeling to residues 3–13. Taken together, the data suggested a model in which ICP47 is composed of three functional domains (see Fig. 3). The first domain is the α-helical region (residues 3–13), which may serve to target the inhibitor to the membrane and stabilize binding to TAP. The second domain (residues 14–34) most likely interacts with the peptide-binding site of TAP through the set of conserved charged amino acids. The remainder of the molecule (residues 35–88), which is not required for TAP inhibitory activity, forms the third domain. This domain possesses the most highly conserved segment between ICP47 from HSV-1 and HSV-2 (residues 35–52), and it is interesting to speculate that the domain plays some, as yet unknown, functional role in ICP47 action.

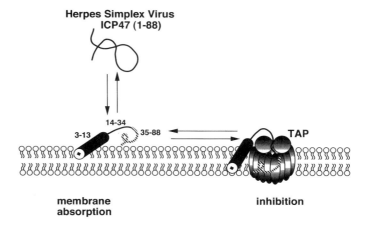

Figure 3. Model of the inhibition of TAP by ICP47. ICP47 inhibits TAP function through its tight association with the peptide-binding site. ICP47 contains three distinct regions. Residues 3–13 adopt an α-helical conformation when in contact with lipid membranes and is thought to anchor and stabilize the inhibitor. Residues 14–34 interact with the peptide-binding site of TAP and block the binding of other peptides, thus inhibiting TAP function. The minimum fragment of ICP47 that possesses activity is residues 3–34. Residues 35–88 are not required for ICP47 activity, but may possess an as-yet-unknown function. Adapted with permission from Beinert et al. (1997). Copyright 1997 American Chemical Society.

4.2. Human Cytomegalovirus US6 Protein

The only other viral gene product described thus far which is capable of directly inhibiting TAP function is the HCMV US6 protein (Hengel et al., 1997; Ahn et al., 1997; Lehner et al., 1997). The US6 gene encodes a 22-kDa transmembrane glycoprotein that is localized to the ER. The mechanism by which US6 inhibits TAP appears to be fundamentally different from that of ICP47. US6 does not interfere with peptide binding to TAP, and TAP inhibitory activity is a result of interactions on the ER side of the membrane as opposed to the cytosolic side. US6 association with TAP in the ER lumen results in the inhibition of peptide translocation by TAP and subsequent lack of stable class I assembly with peptide. Although US6 was shown to associate physically with the TAP multimeric complex, which consists of TAP, class I, tapasin, and calreticulin (Sadasivan et al., 1996; Ortmann et al., 1997), the presence of class I and tapasin was not required for the US6 block of peptide transport by TAP (Hengel et al., 1997). It was further suggested that US6 interacts directly with the TAP heterodimer, although this remains to be proven. In another study, the luminal domain, as opposed to the transmembrane or cytoplasmic domains, of US6 was shown to possess the TAP blocking activity, as revealed by experiments which utilized chimeric or truncated versions of US6 (Ahn *et al.,* 1997). Taken together, the data suggest that the luminal domain of US6 physically associates with TAP and inhibits peptide transport from the ER side of the transporter. One can speculate that US6 blocks the transport of the peptide at the point of exit from the putative translocation pore of TAP. This is in stark contrast to ICP47, which prevents peptide binding to TAP on the cytosolic side. Further elucidation of the viral mechanisms of TAP inhibition will provide insight for the development of specific inhibitors of TAP that could be used as pharmacological agents.

5. TAP AS A DRUG TARGET

TAP plays a critical role in delivering peptide to class I molecules. Without a source of peptides, class I molecules remain unstable and are expressed at low levels on the cell surface. This in turn leads to a reduced capacity to present class I-associated antigens and the possible failure of CTL to recognize the affected cell. In most cases it would be advantageous to stimulate or enhance TAP function in order to facilitate a vigorous immune response. However, in some instances, it would be advantageous to inhibit class I-associated antigen presentation, i.e., in the case of transplantation, where class I mismatches can cause tissue rejection, or in the case of autoimmune disease, where the presentation of self-antigens results in tissue destruction. Therefore, drugs which can specifically inhibit TAP could be of therapeutic value. In addition, specific inhibitors of TAP are valuable research tools to probe the mechanisms of antigen processing, as in the case of the viral inhibitor ICP47. The search for drugs which inhibit TAP is in its infancy, as revealed by the literature, where only a handful of reports address this issue. Nevertheless, TAP is an important target for drug intervention and it is only a matter of time before effective anti-TAP agents are developed.

5.1. Peptidomimetics

A current trend in the development of therapeutic drugs is the rational design of compounds based on what is known about the biology of the target protein. This type of approach has been successful in the development of anticancer agents such as the prenyltransferase inhibitors, which prevent the prenylation of the oncogene Ras, leading to inhibition of its oncogenic activity (Sebti and Hamilton, 1997). These compounds are peptidomimetics which mimic the carboxy-terminal tetrapeptide prenylation site (CAAX) on Ras and prevent its normal processing. In general, the goal of peptidomimetics is to generate nonpeptide compounds which are resistant to proteolytic degradation, and which have enhanced cellular uptake through hydrophobic modifications. The information that has accrued with regard to TAP substrate specificity has the potential to be utilized in the rational design of TAP inhibitors, and the specific inhibition of TAP lends itself to the peptidomimetic approach. First, the substrate for TAP is a peptide, and second, TAP is capable of binding and transporting peptides which possess large, bulky hydrophobic side chains (see below). However, the minimal length of the peptide substrate for TAP (8–9 amino acids) is prohibitive because successful peptidomimetic compounds are generally based on a length of 5 amino acids or less (Sebti and Hamilton, 1997). Nevertheless, the rational approach toward the development of anti-TAP drugs using peptidomimetics is very promising.

An obvious requirement for a peptide inhibitor of TAP is the high-affinity binding of the inhibitor to the peptide-binding site of TAP, but without its transport across the membrane. A perfect example of this is ICP47. ICP47 binds tightly to the peptide-binding site of TAP, and is not transported. It therefore inhibits TAP by preventing other peptides from binding and being transported. One problem with ICP47 is that it is too long to be an effective peptidomimetic (the minimum fragment which retains activity is 32 amino acids). Therefore, other peptide candidates which are shorter, ideally 8–9 amino acids, should be sought which bind tightly to TAP, but which are not transported. Finding such a peptide will most likely be difficult, because peptide binding and transport by TAP are closely linked mechanistically, and no such peptide has been described.

5.2. Synthesis of a TAP Inhibitor

In one study the investigators attempted to create an inhibitor of TAP based on peptide modification (Gromme et al., 1997). Here, peptides were synthesized that possessed an extended polylysine side chain attached to the ε-amino group of the lysyl residue at position 2 or 7 of the peptides AKDNATRDY or ARDNATKDY, respectively. The polylysine side chains were of varying lengths (from 1 to 20 residues, K_n) and were coupled end to end so that the side chain itself would not act as a normal peptide. The modified peptides were tested for competition of reporter peptide transport, and by direct translocation into the ER as measured by glycosylation. Surprisingly, all the peptides, including the ones with K_{20} side chains, were able to compete equally well for reporter peptide transport. However, the direct translocation of peptides which possessed K_n side chains of 10 or more was dramatically reduced. As a control, it was shown that the polylysine side chains themselves were not substrates for TAP. The data indicated that peptides which possessed large side chains (up to 166 Å) could bind efficiently to TAP and prevent the transport of a reporter peptide, but that the longer chain derivatives, i.e. K_{15} or K_{20}, were not transported across the membrane, or were transported with very low efficiency, and hence could be considered TAP inhibitors. Two points should be made from this analysis: (1) TAP possesses an extraordinarily large side-chain exclusion limit for peptide binding (which was hinted at by an earlier study, see below), and (2) even though the K_{15} or K_{20} peptide derivatives could technically be considered TAP inhibitors, from a practical standpoint, they would be difficult to work with *in vivo* and do not lend themselves to the peptidomimetic approach.

5.3. Steric Requirements for Peptide Binding to TAP

To ascertain the capacity of TAP to handle peptide substrates which possessed large, bulky side groups, such as photoaffinity or fluorescent probes, Uebel *et al.* (1995) synthesized peptides which contained the bulky, hydrophobic amino acids β-[1-naphthyl]-alanine or ε-dansyl-lysine at various positions of the peptide RRYNASTEL. The peptide variants were tested by competition for peptide binding to TAP in microsomes. The result of this analysis was that, in almost all cases, the introduction of the bulky side chains resulted in peptides with an equal or higher affinity for TAP than the original peptide. This was the first report which systematically looked at the effect of bulky, hydrophobic side chains on peptide binding to TAP, and to show that TAP could readily accommodate these peptide modifications. This result was confirmed by the study of Gromme *et al.* (1997) in which large, bulky pivaloyl, benzoyl, or dansyl groups were incorporated into a model peptide, and it was observed that the modified peptides competed better for peptide transport than the unmodified peptide. Also, it was shown that the modified peptides were transported into the ER. Interestingly, these large, bulky groups could be inserted at any position along the peptide and not have a deleterious effect on peptide binding or transport, indicating that they could be in positions which are thought to have direct contact with TAP, i.e., N-terminal positions 1–3 and the C-terminal residue. These results helped to explain why the photoreactive peptide analogues which had been used so successfully in earlier studies worked so well for labeling TAP. In addition, they suggested that any peptide modifications made to study the structure and function of TAP, i.e., photolabels, fluorescent labels, or biotin, could be carried out successfully, and that TAP peptide substrates would be amenable to peptidomimetics.

5.4. Incorporation of Reduced Peptide Bonds and D-Amino Acids into TAP Substrates

An aspect of the peptidomimetic approach is the design of compounds which are less peptide-like than the original substrate. One way to do this is through the incorporation of reduced peptide bonds into the backbone, i.e., a CH_2-NH bond instead of the CO—NH peptide bond. These pseudopeptide analogues are highly resistant to proteolysis, and therefore offer an advantage over normal peptide substrates in vaccine and drug development. This approach was used successfully to generate pseudopeptide analogues which can efficiently mimic the binding of an antigenic peptide to the mouse class I molecule K^d (Guichard *et al.,* 1995). Another way to make peptides less peptide-like is through the incorporation of D-amino acids instead of L-amino acids. This results in a change in the orientation of the side chain of the modified amino acid and allows one to determine the stereospecificity of a particular peptide interaction. In addition, retro–inverso peptide analogues contain all D-amino acids, in which the original sequence is reversed, resulting in the exchange of the natural CO—NH peptide bond with NH—CO bonds. Retro–inverso peptides are able to mimic their L-amino acid counterparts because the side chains are in the same relative orientation in both peptides; however, one problem is that the charges at the ends of the peptide are reversed. It has been shown that retro–inverso peptides can take the place of their natural counterparts with regard to antibody recognition (Benkirane *et al.,* 1995). Furthermore, as with peptides that contain reduced peptide bonds, peptides that contain D-amino acids are much more resistant to proteolysis.

The binding and transport of peptides that contain reduced peptide bonds and D-amino acids by TAP has been analyzed to a certain extent. Gromme *et al.* (1997) looked at the transport of peptides which contained one or two isosteric double bonds in which the CO—NH peptide bond was replaced with a CH=CH bond. This is another form of peptide bond reduction which does not allow hydrogen bonds to form with the peptide backbone. The double bonds were placed between positions 7 and 8 or positions 2 and 3 and positions 7 and 8 of the parent peptide DFGNKTFGY (DFGNKTF=GY or DF=GNKTF=GY). Surprisingly, the double-bond isosters were shown to compete for peptide transport and be transported into the ER more efficiently than the parent peptide. These results suggest that peptide backbone interactions with TAP may not be crucial for peptide transport, and that double-bond isosters, which are resistant to proteases, may be useful in the development of TAP peptidomimetics. Gromme *et al.* (1997) and Uebel *et al.* (1997) looked at the peptide binding and transport of peptides which contained D-amino acids including retro–inverso peptides. The upshot of these studies was that the incorporation of D-amino acids into TAP peptide substrates had a dramatic deleterious effect on peptide binding and transport. The regions of the peptide that were the most sensitive to D-amino acid inclusion were the N-terminal positions 1–3 and the C-terminal position 9, consistent with the most sensitive regions as defined by the peptide-binding motif for human TAP (van Endert *et al.,* 1995). Again, the central portion of the peptide could easily tolerate D-amino acids, further indicating that this region probably has little or no interaction with TAP. Peptides which contained all D-amino acids in the original order and retro–inverso peptides were extremely poor substrates for TAP. The data suggest that stereochemistry, charge distribution, and orientation of the peptide backbone may all play a role in TAP substrate specificity. In addition, the retro–inverso peptide approach may not be too useful in the development of TAP inhibitors. However, peptides that contain a combination of L and D-amino acids in the right positions, i.e., D-amino acids in the central region and L-amino acids toward the ends of the peptide, may prove of some utility in the development of inhibitors.

5.5. Allosteric Modulation of TAP Activity

It should not be overlooked that TAP function may be modulated by compounds which interact with TAP at a site other than the peptide binding site. P-Glycoprotein, for instance, has the capacity to transport a wide variety of chemotherapeutic and cytotoxic agents out of the cell, such as vinblastine, doxorubicin, etoposide, and colchicine (Gottesman and Pastan, 1993). In addition to these anticancer and cytotoxic drugs, there is a group of compounds including verapamil, azidopine, and cyclosporin A that are not cytotoxic themselves but can reverse multidrug resistance in cells. These drugs, known as reversing agents, are thought to act as competitive inhibitors of P-glycoprotein. Yet, recent evidence also suggests that some of these reversing agents may act as noncompetitive inhibitors of P-glycoprotein (Dey *et al.,* 1997). This implies that the agent binds to a site on the molecule which is distinct from the substrate-binding site. Given this information for P-glycoprotein, it is intriguing to speculate that such allosteric modulators of TAP may also exist. In a recent study which looked at a possible role for TAP in the multidrug-resistance phenomenon, it was shown that peptide transport by TAP could be inhibited by etoposide and vincristine, albeit at relatively high concentrations (Izquierdo *et al.,* 1996). Due to the requirements of the TAP peptide-binding site, it is doubtful that these drugs are competitive inhibitors of peptide transport, and the results suggest that they may act through an allosteric mechanism. Further investigation into the interaction of TAP with drugs such as etoposide may lead to the identification of a distinct class of TAP inhibitors.

6. CONCLUDING REMARKS

TAP is a promising drug target for the modulation of MHC class I antigen processing and presentation. In certain cases it may be advantageous to inhibit TAP function and therefore class I expression, i.e., in tissue transplantation and autoimmune disease. To be effective, TAP-specific drugs must be able to cross the plasma membrane and therefore they should be lipophilic in nature. The most promising approach for the development of anti-TAP drugs is through the rational design of drugs using peptidomimetics. The most effective inhibitors will be those that bind to the peptide-binding region of TAP, but are not transported. The viral inhibitor ICP47 shows how effective this strategy can be. The search for compounds that inhibit TAP will be facilitated by the use of combinatorial peptide and chemical libraries and the development of a rapid screening assay for TAP inhibition. The capacity for TAP to handle substrates which possess large, bulky hydrophobic groups should be an advantage in the development of anti-TAP drugs, as long as the appropriate groups are found which allow for binding, but not transport of the drug. The recent study of Gromme *et al.* (1997) marks the beginning of the search for anti-TAP drugs, and underscores how far we have to go to produce effective, practical compounds. In addition, further studies on the precise substrate specificity of TAP, such as the one by Uebel *et al.* (1997), will provide valuable information with regard to drug development and the molecular understanding of peptide transport by TAP. Ultimately, X-ray structural analysis of TAP/peptide complexes will be required for a full understanding of TAP interaction with peptide.

ACKNOWLEDGMENTS

We would like to thank Said Sebti, William Dalton, and Robert Tampe for helpful discussions. We acknowledge the Molecular Imaging Core Facility at the Moffitt Cancer Center.

REFERENCES

Ahn, K., Angulo, A., Ghazal, P., Yang, Y., and Fruh, K., 1996a, Human cytomegalovirus inhibits antigen presentation by a sequential multistep process, *Proc. Natl. Acad. Sci. USA* **93**:10990–10995.

Ahn, K., Meyer, T. H., Uebel, S., Sempe, P., Djaballah, H., Yang, Y., Peterson, P. A., Fruh, K., and Tampe, R., 1996b, Molecular mechanism and species specificity of TAP inhibition by herpes simplex virus protein ICP47, *EMBO J.* **15**:3247–3255.

Ahn, K., Gruhler, A., Galocha, B., Jones, T. R., Wiertz, E. J. H. J., Ploegh, H. L., Peterson, P. A., Yang, Y., and Fruh, K., 1997, The ER-luminal domain of the HCMV glycoprotein US6 inhibits peptide translocation by TAP, *Immunity* **6**:613–621.

Anderson, K., Cresswell, P., Gammon, M., Hermes, J., Williamson, A., and Zweerink, H., 1991, Endogenously synthesized peptide with an endoplasmic reticulum signal sequence sensitizes antigen processing mutant cells to class I-restricted cell mediated lysis, *J. Exp. Med.* **174**:489–492.

Andersson, M., Paabo, S., Nilsson, T., and Peterson, P. A., 1985, Impaired intracellular transport of class I MHC antigens as a possible means for adenoviruses to evade immune surveillance, *Cell* **43**:215–222.

Androlewicz, M. J., and Cresswell, P., 1994, Human transporters associated with antigen processing possess a promiscuous peptide-binding site, *Immunity* **1**:7–14.

Androlewicz, M. J., Anderson, K. S., and Cresswell, P., 1993, Evidence that transporters associated with antigen processing translocate a major histocompatibility complex class I-binding peptide into the endoplasmic reticulum in an ATP-dependent manner, *Proc. Natl. Acad. Sci. USA* **90**:9130–9134.

Androlewicz, M. J., Ortman, B., van Endert, P. M., Spies, T., and Cresswell, P., 1994, Characteristics of peptide and major histocompatibility complex class I/B$_2$-microglobulin binding to the transporters associated with antigen processing (TAP1 and TAP2), *Proc. Natl. Acad. Sci. USA* **91**:12716–12720.

Arnold, D., Wahl, C., Faath, S., Rammensee, H.-G., and Schild, H., 1997, Influences of transporter associated with antigen processing (TAP) on the repertoire of peptides associated with the endoplasmic reticulum-resident stress protein gp96, *J. Exp. Med.* **186**:461–466.

Attaya, M., Jameson, S., Martinez, C. K, Hermel, E., Aldrich, C., Forman, J., Fischer Lindahl, K., Bevan, M. J., and Monaco J. J., 1992, Ham-2 corrects the class I antigen processing defect in RMA-S cells, *Nature* **355**:647–649.

Beck, S., and Barrell, B. G., 1988, Human cytomegalovirus encodes a glycoprotein homologous to MHC class-I antigens, *Nature* **331**:269–272.

Beinert, D., Neumann, L., Uebel, S., and Tampe, R., 1997, Structure of the viral TAP-inhibitor ICP47 induced by membrane association, *Biochemistry* **36**:4694–4700.

Benkirane, N., Guichard, G., Regenmortel, M. H. V. V., Briand, J.-P., and Muller, S., 1995, Cross-reactivity of antibodies to retro-inverso peptidomimetics with the parent protein histone H3 and chromatin core particle, *J. Biol. Chem.* **270**:11921–11926.

Bjorkman, P. J., and Parham, P., 1990, Structure, function, and diversity of class I major histocompatibility complex molecules, *Annu. Rev. Biochem.* **59**:253–288.

Bjorkman, P. J., Saper, M. A., Samraoui, B., Bennett, W. S., Strominger, J. L., and Wiley, D. C., 1987, Structure of the human class I histocompatibility antigen, HLA-A2, *Nature* **329**:506–512.

Burgert, H. G., and Kvist, S., 1985, An adenovirus type 2 glycoprotein blocks cell surface expression of human histocompatibility class I antigens, *Cell* **41**:987–997.

Daniel, S., Caillat-Zucman, S., Hammer, J., Bach, J.-F., and Endert, P. M. v., 1997, Absence of functional relevance of human transporter associated with antigen processing polymorphism for peptide selection, *J. Immunol.* **159**:2350–2357.

DeMars, R., Rudersdorf, R., Chang, C., Pertersen, J., Strandtmann, J., Korn, N., Sidwell, B., and Orr, H. T., 1985, Mutations that impair a posttranscriptional step in expression of HLA-A and -B antigens, *Proc. Natl. Acad. Sci. USA* **82**:8183–8187.

Deverson, E. V., Gow, I. R., Coadwell, W. J., Monaco, J. J., Butcher, G. W., and Howard, J. C., 1990, MHC class II region encoding proteins related to the multidrug resistance family of transmembrane transporters, *Nature* **348**:738–741.

Dey, S., Ramachandra, M., Pastan, I., Gottesman, M. M., and Ambudkar, S. V., 1997, Evidence for two nonidentical drug-interaction sites in the human P-glycoprotein, *Proc. Natl. Acad. Sci. USA* **94**:10594–10599.

Driscoll, J., Brown, M. G., Finley, D., and Monaco, J. J., 1993, MHC-linked LMP gene products specifically alter peptidase activities of the proteasome, *Nature* **365**:262–264.

Elliott, T., 1997, Transporter associated with antigen processing, *Adv. Immunol.* **65**:47–109.

Elliott, T., Willis, A., Cerundolo, V., and Townsend, A., 1995, Processing of major histcompatibility class I-restricted antigens in the endoplasmic reticulum, *J. Exp. Med.* **181**:1481–1491.

Felmlee, T., Pellett, S., and Welch, R. A., 1985, Nucleotide sequence of an *Escherichia coli* chromosomal hemolysin, *J. Bacteriol.* **163**:94–105.

Fruh, K., Ahn, K., Djaballah, H., Sempe, P., Endert, P. M. v., Tampe, R., Peterson, P. A., and Yang, Y., 1995, A viral inhibitor of peptide transporters for antigen presentation, *Nature* **375**:415–418.

Gaczynska, M., Rock, K. L., and Goldberg, A. L., 1993, γ-Interferon and expression of MHC genes regulate peptide hydrolysis by proteasomes, *Nature* **365**:264–267.

Galocha, B., Hill, A., Barnett, B. C., Dolan, A., Raimondi, A., Cook, R. F., Brunner, J., McGeoch, D. J., and Ploegh, H. L., 1997, The active site of ICP47, a herpes simplex virus-encoded inhibitor of the major histocompatibility complex (MHC)-encoded peptide transporter associated with antigen processing (TAP), maps to the NH_2-terminal 35 residues, *J. Exp. Med.* **185**:1565–1572.

Gileadi, U., and Higgins, C. F., 1997, Membrane topology of the ATP-binding cassette transporter associated with antigen presentation (TAP1) expressed in *Escherichia coli*, *J. Biol. Chem.* **272**:11103–11108.

Goldberg, A. L., and Rock, K. L., 1992, Proteolysis, proteasomes and antigen presentation, *Nature* **357**:375–379.

Gottesman, M. M., and Pastan, I., 1993, Biochemistry of multidrug resistance mediated by the multidrug transporter, *Annu. Rev. Biochem.* **62**:385–427.

Grandea, A. G., Androlewicz, M. J., Athwal, R. S., Geraghty, D. E., and Spies, T., 1995, Dependence of peptide binding by MHC class I molecules on their interaction with TAP, *Science* **270**:105–108.

Gromme, M., Valk, R. v. d., Sliedregt, K., Vernie, L., Liskamp, R., Hammerling, G., Koopmann, J.-O., Momburg, F., and Neefjes, J., 1997, The rational design of TAP inhibitors using peptide substrate modifications and peptidomimetics, *Eur. J. Immunol.* **27**:898–904.

Gros, P., Croop, J., and Housman, D., 1986, Multidrug resistance gene: Complete cDNA sequence indicates strong homology to bacterial transport proteins, *Cell* **47**:371–380.

Guichard, G., Calbo, S., Muller, S., Kourilsky, P., Briand, J.-P., and Abastado, J.-P., 1995, Efficient binding of reduced peptide bond pseudopeptides to major histocompatibility complex class I molecule, *J. Biol. Chem.* **270**:26057–26059.

Guo, H.-C., Jardetsky, T. S., Garrett, T. P. J., Lane, W. S., Strominger, J. L., and Wiley, D. C., 1992, Different length peptides bind to HLA-Aw68 similarly at their ends but bulge out in the middle, *Nature* **360**:364–366.

Heemels, M.-T., and Ploegh, H. L., 1994, Substrate specificity of allelic variants of the TAP peptide transporter, *Immunity* **1**:775–784.

Hengel, H., Koopman, J.-O., Flohr, T., Muranyi, W., Goulmy, E., Hammerling, G. J., Koszinowski, U. H., and Momburg, F., 1997, A viral ER-resident glycoprotein inactivates the MHC-encoded peptide transporter, *Immunity* **6**:623–632.

Higgins, C. F., 1992, ABC transporters: From microorganisms to man, *Annu. Rev. Cell. Biol.* **8**:67–113.

Hiles, I. D., Gallagher, M. P., Jamieson, D. J., and Higgins, C. F., 1987, Molecular characterization of the oligopeptide permease of *Salmonella typhimurium*, *J. Mol. Biol.* **195**:125–142.

Hill, A., Jugovic, P., York, I., Russ, G., Bennink, J., Yewdell, J., Ploegh, H., and Johnson, D., 1995, Herpes simplex virus turns off the TAP to evade host immunity, *Nature* **375**:411–415.

Izquierdo, M. A., Neefjes, J. J., Mathari, A. E. L., Flens, M. J., Scheffer, G. L., and Scheper, R. J., 1996, Overexpression of the ABC transporter TAP in multidrug-resistant human cancer cell lines, *Br. J. Cancer* **74**:1961–1967.

Jones, T. R., Hanson, L. K., Sun, L., Slater, J. S., Stenberg, R. M., and Campbell, A. E., 1995, Multiple independent loci within the human cytomegalovirus unique short region down-regulate expression of major histocompatibility complex class I heavy chains, *J. Virol.* **69**:4830–4841.

Koopmann, J.-O., Post, M., Neefjes, J. J., Hammerling, G. J., and Momburg, F., 1996, Translocation of long peptides by transporters associated with antigen processing (TAP), *Eur. J. Immunol.* **26**:1720–1728.

Kuchler, K., Sterne, R. E., and Thorner, J., 1989, *Saccharomyces cerevisiae* STE6 gene product: A novel pathway for protein export in eukaryotic cells, *EMBO J.* **8**:3973–3984.

Lammert, E., Stefanovic, S., Brunner, J., Rammensee, H.-G., and Schild, H., 1997, Protein disulfide isomerase is the dominant acceptor for peptides translocated into the endoplasmic reticulum, *Eur. J. Immunol.* **27**:1685–1690.

Lehner, P. J., Karttunen, J. T., Wilkinson, G. W. G., and Cresswell, P., 1997, The human cytomegalovirus US6 glycoprotein inhibits transporter associated with antigen processing-dependent peptide translocation, *Proc. Natl. Acad. Sci. USA* **94**:6904–6909.

Levitskaya, J., Coram, M., Levitsky, V., Imreh, S., Steigerwald-Mullen, P. M., Klein, G., Kurilla, M. G., and Masucci, M. G., 1995, Inhibition of antigen processing by the internal repeat region of the Epstein–Barr virus nuclear antigen-1, *Nature* **375**:685–688.

Levy, F., Gabathuler, R., Larsson, R., and Kvist, S., 1991, ATP is required for *in vitro* assembly of MHC class I antigens but not for transfer of peptides across the ER membrane, *Cell* **67**:265–274.

Li, S., Sjogren, H.-O., Hellman, U., Pettersson, R. F., and Wang, P., 1997, Cloning and functional characterization of a subunit of the transporter associated with antigen processing, *Proc. Natl. Acad. Sci. USA* **94**:8708–8713.

Ljunggren, H. G., and Karre, K., 1990, In search of the missing self: MHC molecules and NK cell recognition, *Immunol. Today* **11**:237–244.

Meyer, T. H., van Endert, P. M., Uebel, S., Ehring, B., and Tampe, R., 1994, Functional expression and the purification of the ABC transporter complex associated with antigen processing (TAP) in insect cells, *FEBS Lett.* **351**:443–447.

Momburg, F., Roelse, J., Hammerling, G. J., and Neefjes, J. J., 1994a, Peptide size selection by the major histocompatibility complex-encoded peptide transporter, *J. Exp. Med.* **179**:1613–1623.

Momburg, F., Roelse, J., Howard, J. C., Butcher, G. W., Hammerling, G. J., and Neefjes, J. J., 1994b, Selectivity of MHC-encoded peptide transporters from human, mouse and rat, *Nature* **367**:648–651.

Momburg, F., Armandola, E. A., Post, M., and Hammerling G. J., 1996, Residues in TAP2 peptide transporters controlling substrate specificity, *J. Immunol.* **156**:1756–1763.

Monaco, J. J., Cho, S., and Attaya, M., 1990, Transport protein genes in the murine MHC: Possible implications for antigen processing, *Science* **253**:1723–1726.

Murchie, M.-J., and McGeoch, D. J., 1982, DNA sequence analysis of an immediate-early gene region of the herpes simplex virus type I genome, *J. Gen. Virol.* **62**:1–15.

Neefjes, J. J., Momburg, F., and Hammerling, G. J., 1993, Selective and ATP-dependent translocation of peptides by the MHC-encoded transporter, *Science* **261**:769–771.

Neisig, A., Roelse, J., Sijts, A. J. A. M., Ossendorp, F., Feltkamp, M. C. W., Kast, W. M., Melief, C. J. M., and Neefjes, J. J., 1995, Major differences in transporter associated with antigen presentation (TAP)-dependent translocation of MHC class I-presentable peptides and the effect of flanking sequences, *J. Immunol.* **154**:1273–1279.

Neumann, L., Kraas, W., Uebel, S., Jung, G., and Tampe, R., 1997, The active domain of the herpes simplex virus protein ICP47: A potent inhibitor of the transporter associated with antigen processing (TAP), *J. Mol. Biol.* **272**:484–492.

Nijenhuis, M., and Hammerling, G. J., 1996, Multiple regions of the transporter associated with antigen processing (TAP) contribute to its peptide binding site, *J. Immunol.* **157**:5467–5477.

Nijenhuis, M., Schmitt, S., Armandola, E. A., Obst, R., Brunner, J., and Hammerling, G. J., 1996, Identification of a contact region for peptide on the TAP1 chain of the transporter associated with antigen processing, *J. Immunol.* **156**:2186–2195.

Obst, R., Armandola, E. A., Nijenhuis, M., Momburg, F., and Hammerling, G. J., 1995, TAP polymorphism does not influence transport of peptide variants in mice and humans, *Eur. J. Immunol.* **25**:2170–2176.

Ortmann, B., Androlewicz, M. J., and Cresswell, P., 1994, MHC class I/β_2-microglobulin complexes associate with TAP transporters before peptide binding, *Nature* **368**:864–867.

Ortmann, B., Copeman, J., Lehner, P. J., Sadasivan, B., Herberg, J. A., Grandea, A. J., Riddell, S. R., Tampe, R., Spies, T., Trowsdale, J., and Cresswell, P., 1997, A critical role for tapasin in the assembly and function of multimeric MHC class I-TAP complexes, *Science* **277**:1306–1309.

Powis, S. J., Howard, J. C., and Butcher, G. W., 1991, The major histocompatibility class II-linked *cim* locus controls the kinetics of intracellular transport of a classical class I molecule, *J. Exp. Med.* **173**:913–921.

Rammensee, H.-G., Falk, K., and Rotzschke, O., 1993, Peptides naturally presented by MHC class I molecules, *Annu. Rev. Immunol.* **11**:213–244.

Rammensee, H.-G., Friede, T., and Stefanovic, S., 1995, MHC ligands and peptide motifs: First listing, *Immunogenetics* **41**:178–228.

Riordan, J. R., Rommens, J. M., Kerem, B.-S., Alon, N., Rozmahel, R., Grzelczak, Z., Zielenski, J., Lok, S., Plavsic, N., Chou, J.-L., Drumm, M. L., Iannuzzi, M. C., Collins, F. S., and Tsui, L.-C., 1989, Identification of the cystic fibrosis gene: Cloning and characterization of complementary DNA, *Science* **245**:1066–1073.

Sadasivan, B., Lehner, P. J., Ortmann, B., Spies, T., and Cresswell, P., 1996, Roles for calreticulin and a novel glycoprotein, tapasin, in the interaction of MHC class I molecules with TAP, *Immunity* **5**:103–114.

Salter, R. D., and Cresswell, P., 1986, Impaired assembly and transport of HLA-A and -B antigens in a mutant TxB cell hybrid, *EMBO J.* **142**:943–949.

Schumacher, T. N. M., Kantesaria, D. V., Heemels, M.-T., Ashton-Rickardt, P. G., Shepherd, J. C., Fruh, K., Yang, Y., Peterson, P. A., Tonegawa, S., and Ploegh, H. L., 1994a, Peptide length and sequence specificity of the mouse TAP1/TAP2 translocator, *J. Exp. Med.* **179**:533–540.

Schumacher, T. N. M., Kantesaria, D., Serreze, D. V., Roopenian, D. C., and Ploegh, H. L., 1994b, Transporters from H-2^b, h-2^d, H-2^s, H-2^k, and H-2^{g7} (NOD/Lt) haplotype translocate similar peptides, *Proc. Natl. Acad. Sci. USA* **91**:13004–13008.

Sebti, S. M., and Hamilton, A. D., 1997, Inhibition of ras prenylation: A novel approach to cancer chemotherapy, *Pharmacol. Ther.* **74**:103–114.

Shepherd, J. C., Schumacher, T. N. M., Ashton-Rickard, P. G., Imaeda, S., Ploegh, H. L., Janeway, C. A. J., and Tonegawa, S., 1993, TAP1-dependent peptide translocation *in vitro* is ATP dependent and peptide selective, *Cell* **74**:577–584.

Snyder, H. L., Yewdell. J. W., and Bennink, J. R., 1994, Trimming of antigenic peptides in an early secretory compartment, *J. Exp. Med.* **180**:2389–2394.

Spee, P., and Neefjes, J., 1997, TAP-translocated peptides specifically bind proteins in the endoplasmic reticulum, including gp 96, protein disulfide isomerase and caltreticulin, *Eur. J. Immunol.* **27**:2441–2449.

Spies, T., and Demars, R., 1991, Restored expression of major histocompatibility class I molecules by gene transfer of a putative peptide transporter, *Nature* **351**:323–324.

Spies, T., Bresnahan, M., Bahram, S., Arnold, D., Blanck, G., Mellins, E., Pious, D., and DeMars, R., 1990, A gene in the human major histocompatibility complex class II region controlling the class I antigen presentation pathway, *Nature* **348**:744–747.

Sugita, M., and Brenner, M. B., 1994, An unstable b$_2$-microglobulin: major histocompatibility complex class I heavy chain intermediate dissociates from calnexin and then is stabilized by binding peptide, *J. Exp. Med.* **180**:2163–2171.

Suh, W.-K., Chen-Doyle, M. F., Fruh, K., Wang, K., Peterson, P. A., and Williams D. B., 1994, Interaction of MHC class I molecules with the transporter associated with antigen processing, *Science* **264**:1322–1326.

Tanaka, K., Tanahashi, N., Tsurumi, C., Yokota, K.-Y., and Shimbara, N., 1997, Proteasomes and antigen processing, *Adv. Immunol.* **64**:1–38.

Tomazin, R., Hill, A. B., Jugovic, P., York, I., Endert, P. v., Ploegh, H. L., Andrews, D. W., and Johnson, D. C., 1996, Stable binding of the herpes simplex virus ICP47 protein to the peptide binding site of TAP, *EMBO J.* **15**:3256–3266.

Townsend, A. R., Gotch, F. M., and Davey, J., 1985, Cytotoxic T cells recognize fragments of the influenza nucleoprotein, *Cell* **42**:457–467.

Trowsdale, J., Hanson, I., Mockridge, I., Beck, S., Townsend, A., and Kelly, A., 1990, Sequences encoded in the class II region of the MHC related to the 'ABC' superfamily of transporters, *Nature* **348**:741–744.

Uebel, S., Meyer, T. H., Kraas, W., Kienle, S., Jung, G., Wiesmuller, K.-H., and Tampe, R., 1995, Requirements for peptide binding to the human transporter associated with antigen processing revealed by peptide scans and complex peptide libraries, *J. Biol. Chem.* **270**:18512–18516.

Uebel, S., Kraas, W., Kienle, S., Wiesmuller, K.-H., Jung, G., and Tampe, R., 1997, Recognition principle of the TAP transporter disclosed by combinatorial peptide libraries, *Proc. Natl. Acad. Sci. USA* **94**:8976–8981.

van Endert, P. M., Tampe, R., Meyer, T. H., Tisch, R., Bach, J.-F., and McDevitt, H. O.,

1994, A sequential model for peptide binding and transport by the transporters associated with antigen processing, *Immunity* **1**:491–500.

van Endert, P. M., Riganelli, D., Greco, G., Fleischhauer, K., Sidney, J., Sette, A., and Bach, J.-F., 1995, The peptide-binding motif for the human transporter associated with antigen processing, *J. Exp. Med.* **182**:1883–1895.

Walker, J. E., Saraste, M., Runswick, M. J., and Gay, N. J., 1982, Distantly related sequences in the a- and b-subunits of ATP synthase, myosin, kinases and other ATP requiring enzymes and a common nuceotide binding fold, *EMBO J.* **1**:945–951.

Wang, P., Gyllner, G., and Kvist, S., 1996, Selection and binding of peptides to human transporters associated with antigen processing and rat cim-a and -b, *J. Immunol.* **157**:213–220.

Wei, M. J., and Cresswell, P., 1992, HLA-A2 molecules in an antigen-processing mutant cell contain signal-derived peptides, *Nature* **356**:443–446.

Wiertz, E. J. H. J., Jones, T. R., Sun, L., Bogyo, M., Geuze, H. J., and Ploegh, H. L., 1996, The human cytomegalovirus US11 gene product dislocates MHC class I heavy chains from the endoplasmic reticulum to the cytosol, *Cell* **84**:769–779.

York, I. A., Roop, C., Andrews, D. W., Riddell, S. R., Graham, F. L., and Johnson, D. C., 1994, A cytosolic herpes simplex virus protein inhibits antigen presentation to CD8+T lymphocytes, *Cell* **77**:525–535.

12

Nucleoside Transporters of Mammalian Cells

*Carol E. Cass, James D. Young,
Stephen A. Baldwin, Miguel A. Cabrita,
Kathryn A. Graham, Mark Griffiths,
Lori L. Jennings, John R. Mackey,
Amy M. L. Ng, Mabel W. L. Ritzel,
Mark F. Vickers, and Sylvia Y. M. Yao*

Carol E. Cass • Molecular Biology of Membranes Group, Membrane Transport Research Group, and Departments of Biochemistry and Oncology, University of Alberta, and Department of Experimental Oncology, Cross Cancer Institute, Edmonton, Alberta, Canada.   *James D. Young, Amy M. L. Ng, Mabel W. L. Ritzel, and Sylvia Y. M. Yao* • Membrane Transport Research Group and Department of Physiology, University of Alberta, Edmonton, Alberta, Canada.   *Stephen A. Baldwin and Mark Griffiths* • School of Biochemistry and Molecular Biology, University of Leeds, Leeds, United Kingdom.   *Miguel A. Cabrita and Mark F. Vickers* • Molecular Biology of Membranes Group, Membrane Transport Research Group, and Department of Biochemistry, University of Alberta, and Department of Experimental Oncology, Cross Cancer Institute, Edmonton, Alberta, Canada.   *Kathryn A. Graham and Lori L. Jennings* • Molecular Biology of Membranes Group, Membrane Transport Research Group, and Department of Oncology, University of Alberta, and Department of Experimental Oncology, Cross Cancer Institute, Edmonton, Alberta, Canada.   *John R. Mackey* • Department of Oncology, University of Alberta, and Departments of Experimental Oncology and Medicine, Cross Cancer Institute, Edmonton, Alberta, Canada.
Membrane Transporters as Drug Targets, edited by Amidon and Sadée. Kluwer Academic/Plenum Publishers, New York, 1999.

1. INTRODUCTION

1.1. Overview

Most mammalian cells possess nucleoside transporters that mediate the cellular uptake of physiologic purine and pyrimidine nucleosides for incorporation into cellular nucleotides and nucleic acids. A number of general reviews which summarize the characteristics of plasma membrane nucleoside transport processes have been published during the past decade (Jarvis, 1988; Plagemann *et al.,* 1988; Cabantchik, 1989; Gati and Paterson, 1989; Geiger and Fyda, 1991; Belt *et al.,* 1993; Cass, 1995; Griffith and Jarvis, 1996; Thorn and Jarvis, 1996). While the focus of most work has been on inwardly directed transport processes, nucleosides are also released from cells via mechanisms that function in the outward direction. Because the activity of the various nucleoside transporters determines levels of free nucleosides in the circulation, nucleoside transport processes, functioning in both directions, are believed to play a major role in the regulation of the physiological effects of nucleosides such as adenosine (Geiger and Fyda, 1991; Pelleg, 1993; Van Belle, 1993a,b; Lasley and Mentzer, 1996). Most nucleoside analogue drugs, including those with antineoplastic or antiviral activity, are relatively hydrophilic molecules and, because their pharmacologic targets are intracellular, transportability across plasma membranes is a critical determinant of pharmacologic action (Cass, 1995). There is a growing body of evidence that nucleoside transport processes of intracellular membranes may also play a critical role in the intracellular distribution of nucleosides and nucleoside analogue drugs (Boumah *et al.,* 1992; Mani *et al.,* 1998).

Although the individual nucleoside transport processes of mammalian cells differ in their permeant selectivities, the physiologic ribosyl and 2'-deoxyribosyl purine and pyrimidine nucleosides enter cells by mediated, nucleoside-specific mechanisms. Among the physiologic nucleosides, the purine nucleosides adenosine and guanosine and the pyrimidine nucleosides uridine and thymidine have been most commonly used as diagnostic permeants in assays of isotopic fluxes (Fig. 1A). Uridine appears to be a "universal" permeant in that it is transported by all of the processes identified thus far, whereas adenosine, which was once also considered to be a universal permeant, appears to exhibit a dual function, in that it is a high-affinity, high-activity permeant for some transporters and a high-affinity, low-activity (or inhibitory) permeant for others. Nucleoside drugs currently in clinical use in cancer therapy include the purine analogues cladribine and fludarabine and the pyrimidine analogues gemcitabine and cytarabine (Fig. 1B). The antiviral nucleoside drugs of current importance in clinical therapy of AIDS patients include the purine analogue didanosine and the pyrimidine analogues lamivudine, zalcitabine, and zidovudine (Fig. 1C).

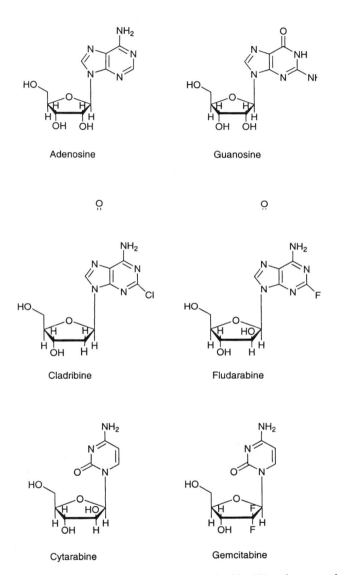

Figure 1. Chemical structures of selected physiologic nucleosides (A), anticancer nucleosides (B), and antiviral nucleosides (C).

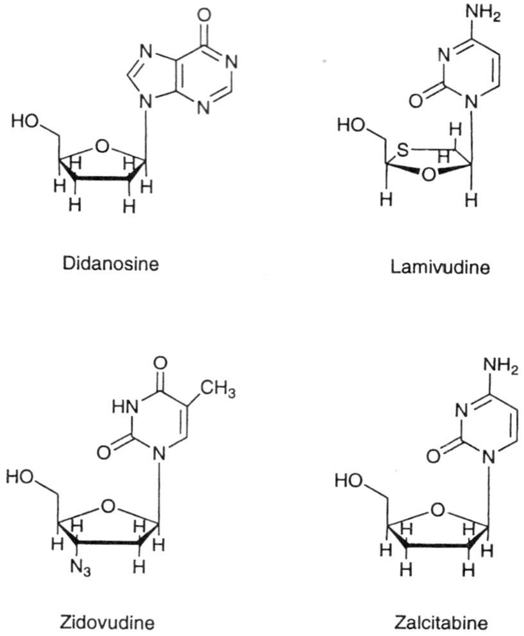

Didanosine   Lamivudine

Zidovudine   Zalcitabine

Figure 1. (*continued*)

1.2. The Heterogeneity of Nucleoside Transport Processes

Nucleoside transport processes in mammalian cells have been categorized into two groups, based on transport mechanisms: the equilibrative transporters, which are "facilitators" that are driven solely by the concentration gradient of nucleoside permeants, and the concentrative transporters, which are "cotransporters" that are driven by transmembrane sodium gradients. The equilibrative transporters have broad permeant selectivities and are found in most cell types, whereas the concentrative transporters are limited to specialized cell types and have relatively narrow permeant selectivities. The functional heterogeneity of nucleoside transport processes in various cell types and tissues has long been considered to be a reflection of the existence of a complex family of membrane proteins with diverse physiologic functions (Gati and Paterson, 1989; Ullman, 1989b; Geiger and Fyda, 1991; Belt *et al.*, 1993; Cass, 1995; Griffith and Jarvis, 1996). This prediction has been confirmed by analysis of predicted protein structures from molecularly cloned cDNAs, which, when introduced into transient expression systems, impart nucleoside transport activity to the recipient cells (Pajor and Wright, 1992; Craig *et al.*, 1994; Huang *et al.*, 1994; Che *et al.*, 1995; Hogue *et al.*, 1996; Yao *et al.*,

1996a,b, 1997; Griffiths *et al.*, 1997a,b; Ritzel *et al.*, 1997, 1998; Wang *et al.*, 1997b; Crawford *et al.*, 1998).

Nitrobenzylmercaptopurine ribonucleoside (NBMPR),* a tight-binding and selective inhibitor of nucleoside transport (Fig. 2), has been extraordinarily useful in the characterization of nucleoside transport processes in mammalian cells (for recent reviews, see Jarvis and Young, 1987; Paterson *et al.*, 1987; Cass, 1995; Griffith and Jarvis, 1996; Buolamwini, 1997). The equilibrative processes have been categorized on the basis of sensitivity to NBMPR (Belt *et al.*, 1993): one subtype (*es*) is exquisitely sensitive to inhibition by NBMPR ($IC_{50} < 5$ nM), whereas the other subtype (*ei*) is either resistant to NBMPR or is inhibited only at relatively high concentrations ($IC_{50} > 1$ μM). NBMPR binds with high affinity ($K_d < 5$ nM) to *es* nucleoside transporters at or near the permeant-binding site. Both *es*- and *ei*-mediated processes exhibit relatively broad permeant selectivities and accept purine and pyrimidine ribo- and deoxyribonucleosides as permeants. There is a large literature, dating back to the early 1970s, on the characteristics of *es*-mediated processes because of the availability of easily studied model systems, e.g., human erythrocytes, which possess only *es* transport activity. Much less is known about *ei*-mediated processes, which are typically found in cells and tissues together with other nucleoside transport processes. However, the recent molecular cloning and functional expression of rat and human cDNAs encoding membrane proteins with *ei* activity have provided new model systems in which *ei*-mediated transport processes can be studied in isolation (Griffiths *et al.*, 1997b; Yao *et al.*, 1997; Crawford *et al.*, 1998). Information about the characteristics of *ei* transporters is expected to increase rapidly.

The concentrative nucleoside transporters comprise several functional subtypes that differ in their substrate selectivities as well as in their tissue distributions and, with a few exceptions, are unaffected by high concentrations (>10 μM) of NBMPR, the prototypic tight-binding inhibitor of *es*- mediated processes (Belt *et al.*, 1993; Cass, 1995; Griffith and Jarvis, 1996). The concentrative nucleoside transporters are inwardly directed sodium/nucleoside symporters, which can translocate nucleosides against their concentration gradients through coupled movement of sodium down its electrochemical gradient. In human cells and tissues, as many as six concentrative nucleoside transport processes have been described on the basis of functional characteristics (Cass, 1995; Griffith and Jarvis,

---

*Abbreviations used: Cytarabine, 1-β-D-arabinofuranosyl cytosine; ara-CTP, cytarabine 5′-triphosphate; zidovudine, 3′-azido-3′-deoxythymidine; cladribine, 2-chlorodeoxyadenosine; *cif,* concentrative, insensitive to NBMPR, and accepts formycin B as a permeant; *cit,* concentrative, insensitive to NBMPR, and accepts thymidine as a permeant: *cs,* concentrative and sensitive to NBMPR; *csg,* concentrative, sensitive to NBMPR, and accepts guanosine as a permeant; CNT, concentrative nucleoside transporter; zalcitabine, 2′,3′-dideoxycytidine; didanosine, 2′,3′-dideoxyinosine; gemcitabine, 2′,2′-difluoro-2′-deoxycytidine; ENT, equilibrative nucleoside transporter; *ei,* equilibrative and insensitive to NBMPR; *es,* equilibrative and sensitive to NBMPR; h, human; NBMPR, nitrobenzylmercaptopurine riboside, 6-[(4-nitrobenzyl)thio]-9-β-D-ribofuranosyl purine; NT, nucleoside transporter; r, rat; RT-PCR, reverse transcriptase polymerase chain reaction.

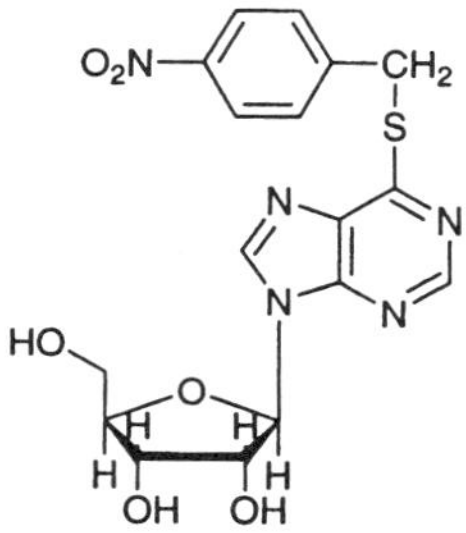

Nitrobenzylmercaptopurine
ribonucleoside
(Nitrobenzylthioinosine)

Figure 2. Chemical structure of nitrobenzylmercaptopurine ribonucleoside (also known as nitrobenzylthioinosine).

1996; Flanagan and Meckling-Gill, 1997; Wang and Giacomini, 1997). The terminologies currently in use for the concentrative transporters have followed either one of two conventions: (i) trivial names based on permeant selectivities (*cit, cif, cib, csg,* and *cs*), and (ii) numerical designations that signify the order of discovery (N1–N5). The correspondence between the terminologies and, where known, with the transporter proteins identified by molecular cloning of cDNAs (see Section 1.3 below) is summarized for nucleoside transport processes of human cells in Table I. The *cit, cif,* and *cib* transport processes are insensitive to NBMPR and differ in their permeant selectivities: *cit* transport processes exhibit preference for pyrimidine nucleosides and also accept adenosine (subtypes N2, N4) and guanosine (subtype N4) as permeants; *cif* transport processes exhibit preference for purine nucleosides and also accept uridine as a permeant; *cib* transport processes accept both pyrimidine and purine nucleosides as permeants. The *csg* and *cs* transport processes are inhibited by NBMPR, and, while their permeant selectivities have not been well defined, the *csg* process exhibits preference for guanosine and the *cs* process accepts adenosine analogues as permeants. The transporter proteins responsible for the *cit* (subtype N2) and *cif* processes have been identified by molecular cloning and functional expression of cDNAs from rat and human tissues (Huang *et al.,* 1994; Che *et al.,* 1995; Yao *et al.,* 1996b; Ritzel *et al.,* 1997, 1998; Wang *et al.,* 1997b), and understanding of relationships between their structure and function is increasing rapidly. The identity of the proteins responsible for the *cit* (subtype N4), *cib, csg, and cs* processes remains to be determined.

1.3. Recent Advances in the Molecular Biology of Nucleoside Transporters

Understanding of relationships among nucleoside transporters has been greatly advanced by the recent isolation and functional expression from rat (r) and

Table I.
Nucleoside Transport Processes of Human Cells[a]

NT process (acronym)	NT process (numerical)	Permeant selectivity	NT protein	Reference
Equilibrative transporters:				
es	na	Pur & pyr nuc	hENT1	Griffiths et al. (1997a)
ei	na	Pur & pyr nuc	hENT2	Griffiths et al. (1997b), Crawford et al. (1998)
Concentrative Transporters:				
cif	N1	Pur nuc, Urd	hCNT2[b]	Wang et al. (1997b), Ritzel et al. (1998)
cit	N2	Pyr nuc, Ado	hCNT1	Ritzel et al. (1997)
cit	N4	Pyr nuc, Ado, Guo	nd	Gutierrez et al. (1992), Gutierrez and Giacomini (1993)
cib	N3	Pur & pyr nuc	nd	Belt et al. (1993)
cs	N5	Ado analogues	nd	Paterson et al. (1993)
csg	N6[c]	Guo	nd	Flanagan and Meckling-Gill (1997)

[a]na, Not applicable; nd, not determined; pur, purine; pyr, pyrimidine; nuc, nucleoside; Urd, uridine, Ado, adenosine; Guo, guanosine.
[b]Also known as hSPNT.
[c]Terminology applied herein.

human (h) cells of cDNAs encoding nucleoside transporter (NT) proteins. These NT proteins comprise two structurally unrelated protein families that have been designated ENT and CNT, depending on whether they mediate, respectively, equilibrative (E) or concentrative (C) nucleoside transport processes (Huang et al., 1994; Yao et al., 1996b, 1997; Griffiths et al., 1997a,b; Ritzel et al., 1997, 1998). Members of the ENT family exhibit functional characteristics that are typical of *es*-mediated (e.g., hENT1, rENT1) or *ei*-mediated (e.g., hENT2, rENT2) transport processes. Members of the CNT family are Na^+–nucleoside symporters that catalyze inward transport of nucleosides with preference for pyrimidine nucleosides (e.g., hCNT1, rCNT1) or purine nucleosides (e.g., hCNT2, rCNT2), although both subtypes accept uridine and adenosine as permeants. The relationships of the nucleoside transporter proteins identified by molecular cloning to the nucleoside transport processes previously defined by functional studies are as follows: ENT1 mediates *es* transport activity; ENT2 mediates *ei* transport activity; CNT1 mediates *cit* (subtype N2) transport activity; and CNT2, which has also been termed SPNT (sodium purine nucleoside transporter), mediates *cif* transport activity. In the nomenclature for NT proteins proposed by Baldwin, Cass, and Young (Yao et al., 1996b; 1997; Griffiths et al., 1997a,b; Ritzel et al., 1997, 1998) and used in this review, functionally distinct members of the two families (ENT, CNT) are designated sequentially in order of discovery, e.g., ENT1 versus ENT2, and a lowercase prefix is used to designate the species or origin, e.g., rCNT1 for rat CNT1.

Comparisons of predicted protein sequences registered in the public data bases with those of the rat and human CNT and ENT proteins have established that the concentrative and equilibrative nucleoside transporters are members of two previously unrecognized protein superfamilies, with representatives, respectively, from bacteria and lower eukaryotes (see Fig. 4). These relationships were first noted when cDNAs for rCNT1 (Huang *et al.*, 1994) and hENT1 (Griffiths *et al.*, 1997a) were isolated and functionally characterized by expression in oocytes of *Xenopus laevis*. The striking correspondence between structural relationships and functional characteristics, as additional mammalian nucleoside transporter cDNAs have been molecularly cloned and expressed, has substantiated the division of the nucleoside transporters into the CNT (Che *et al.*, 1995; Yao *et al.*, 1996b; Ritzel *et al.*, 1997; Wang *et al.*, 1997b) and ENT families (Griffiths *et al.*, 1997b; Yao *et al.*, 1997; Crawford *et al.*, 1998). Members of the mammalian (human, rat, pig and mouse) CNT protein family have homologues in the nematode *Caenorhabditis elegans* and in several bacterial species, one of which, the NupC protein of *Escherichia coli*, has been shown to be a proton/nucleoside symporter (Craig *et al.*, 1994). NupC has close structural homologues of as-yet-unproven function in *Haemophilus influenzae* (Casari *et al.*, 1995), *Bacillus subtilis*, and *Helicobacter pylori*. Five ENT homologues of unknown function have been identified in *Caenorhabditis elegans*, one in *Saccharomyces cerevisae*, and one in *Arabidopsis thaliana;* the last-named is the first putative nucleoside transporter identified in plants.

Understanding of relationships among the mammalian nucleoside transporters has been greatly advanced by the isolation and functional expression of four rat and four human cDNAs encoding membrane proteins with nucleoside transport activity (Huang *et al.*, 1994; Che *et al.*, 1995; Yao *et al.*, 1996a,b, 1997; Griffiths *et al.*, 1997a,b; Ritzel *et al.*, 1997, 1998; Wang *et al.*, 1997b; Crawford *et al.*, 1998). These transporters comprise two structurally unrelated protein families (ENT, CNT) with quite different architectural designs. The mammalian ENT proteins identified thus far are predicted to have either 456 or 457 amino acid residues and 11 transmembrane domains that are connected by relatively short hydrophilic loops, except for the large hydrophilic loops connecting the first and second and the sixth and seventh transmembrane domains, which are thought to be located, respectively, at the extracellular and intracellular faces of the plasma membrane (Griffiths *et al.*, 1997a; Yao *et al.*, 1997). The mammalian CNT proteins identified thus far are predicted to have 648–658 amino acid residues and 14 transmembrane domains (Huang *et al.*, 1994), which, if the 14-domain prediction is correct, places the long N- and C-termini together with several large loops with charged residues at the same membrane face. The C-terminus of rCNT1 is glycosylated (Hamilton *et al.*, 1997), identifying this region of the protein as exofacial. The striking architectural differences between the two protein families suggests different evolutionary origins. Among the bacteria whose genomic sequences have been completely characterized e.g., *Escherichia coli, Haemophilus influenzae, He-*

licobactor pylori), there are open reading frames encoding CNT-like proteins, but none encoding ENT-like proteins. A CNT-related protein of *Escherichia coli* (termed NupC) has been shown to be a H^+/nucleoside symporter that exhibits selectivity for pyrimidine nucleosides (Craig *et al.*, 1994).

2. THE ENT FAMILY OF NUCLEOSIDE TRANSPORTERS

2.1. Molecular and Functional Characteristics of the ENT Family

2.1.1. THE ENT1 (*es*) SUBFAMILY

A cDNA encoding the human equilibrative nucleoside transporter (hENT1) was isolated from a placental cDNA library using amino acid sequence information obtained from the N-terminus of the purified human erythrocyte *es* transporter (Griffiths *et al.*, 1997a). The predicted protein is comprised of 456 amino acids (50 kDa) with 11 transmembrane domains and three potential N-linked glycosylation sites, of which one (Asn 48) lies in the large hydrophilic loop (Loop 1) between the first and second transmembrane domains and two others (Asn 277, 288) lie in the large hydrophilic loop between the sixth and seventh transmembrane domains. Earlier studies had established that the *es* transporter of human erythrocytes, which is readily identified by site-specific photolabeling of plasma membranes with ^{3}H-NBMPR (Jarvis and Young, 1987), is glycosylated and migrates in sodium dodecyl sulfate–polyacrylamide gel electrophoresis in the "band 4.5" region with the erythrocyte glucose transporter (Wu *et al.*, 1983; Kwong *et al.*, 1992). Because the glycosylation site(s) of the erythrocyte *es* transporter are known to be near one of its ends (Kwong *et al.*, 1993), current topology models are based on the assumption that Asn 48 is the site of N-linked glycosylation, thereby placing Loop 1 at the extracellular membrane face (Griffiths *et al.*, 1997a). Glycosylation of Loop 1 and its consequent external orientation has recently been confirmed in studies in which the glycosylation site (Asn 48) of hENT1 was eliminated by site-directed mutagenesis (Sundaram, M., Yao, S. Y. M., Chomey, E., Baldwin, S. A., Cass, C. E., and Young, J. D., unpublished results). The functional significance of N-linked glycosylation of hENT1 is unknown.

The hENT1 gene is localized to chromosome 6p21.1–p21.2 (Coe *et al.*, 1997) and its genomic sequence has recently been determined (Graham, K. A., Coe, I. R., Carpenter, P., Baldwin, S. A., Young, J. D., and Cass, C. E., unpublished results). The hENT1 transporter appears to be broadly distributed among cells and tissues, based on evidence of expression of hENT1 mRNA. cDNAs with >90% identity at the nucleic acid level to placental hENT1 cDNA have been observed in many normal human tissues, including fetal brain, liver, and spleen, adult adipose tissue, aortic endothelial cells, brain, breast, colon, heart, lung, ovary, placenta, prostate, and uterus (see TIGR Human Gene Index). Malignant cells also express the hENT1 gene: hENT1 mRNA, detected by hybridization analysis, is found in a variety of human cancer cell lines (Boleti *et al.*, 1997) and hENT1-like sequences are registered in the GenBank EST data base for several human cancers. These observations are consistent with the apparently ubiquitous cell and tissue distribution of *es* transport activity (Belt *et al.*, 1993; Cass, 1995; Griffith and Jarvis, 1996).

The rat homologue of hENT1 was isolated from a rat jejunal cDNA library (Yao *et al.*, 1997) and is 78% identical and 88% similar to hENT1 in amino acid sequence. The predicted rENT1 protein has one additional amino acid residue, a cysteine found just before transmembrane domain 8. The architectural features of rENT1 are similar to those of hENT1, although it has five potential glycosylation sites, three of which are located in the predicted extracellular loop between transmembrane domains 1 and 2.

Recombinant hENT1 and rENT1 are readily produced in oocytes of *Xenopus laevis* by microinjection of RNA transcripts synthesized *in vitro* from the hENT1 and rENT1 cDNAs (Griffiths *et al.*, 1997a; Yao *et al.*, 1997). These recombinant proteins exhibit saturable transport of uridine with characteristics typical of an *es*-type transporter. Transport is inhibited by both purine and pyrimidine nucleosides (e.g., guanosine, thymidine), suggesting the broad permeant selectivity characteristic of an *es*-type transporter. Both of the recombinant proteins, when produced in *Xenopus* oocytes, exhibit high sensitivities to inhibition by NBMPR, with IC_{50} values of 3.6 nM (hENT1) and 4.6 nM (rENT1). Transport mediated by hENT1 is also potently inhibited by coronary vasodilator drugs (dipyridamole, dilazep, draflazine; see Fig. 3), whereas rENT1-mediated transport is unaffected (Griffiths *et al.*, 1997a,b; Yao *et al.*, 1997). These results agree with earlier studies with native *es*- and *ei*-mediated nucleoside transport processes (Cass, 1995; Griffith and Jarvis, 1996). Previous studies with rat erythrocytes and cultured cell lines had established that rat cells possess a unique *es*-type nucleoside transport process, with the hallmark high sensitivity to NBMPR, but with resistance to inhibition by the vasodilator drugs (Griffith and Jarvis, 1996).

The permeant selectivities of recombinant hENT1 and rENT1 have been examined in *Xenopus* oocytes, using uridine as the diagnostic permeant for flux analysis (Griffiths *et al.*, 1997a; Yao *et al.*, 1997). The apparent K_m values for uridine transport by recombinant hENT1 and rENT1 are 0.24 and 0.15 mM, respectively, and hENT1- and rENT1-mediated uridine transport processes are inhibited by physiologic ribo- and 2-deoxyribonucleosides, but not by uracil, a nucleobase. Uridine transport in hENT1-producing oocytes is also inhibited by several anticancer nucleoside drugs, including cladribine, cytarabine, fludarabine, and gem-citabine (for structures, see Fig. 1B). These results are consistent with

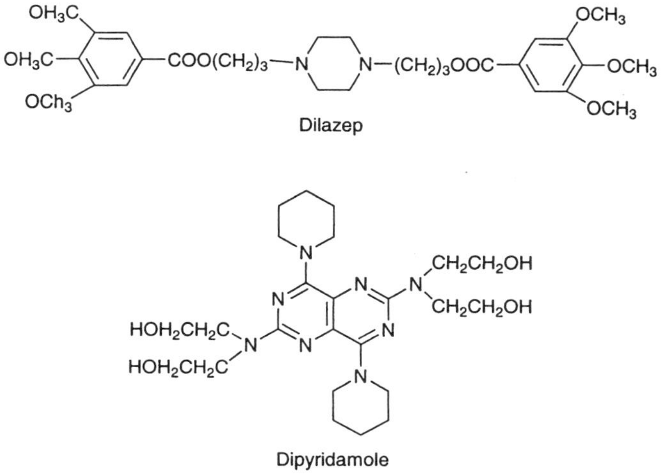

Figure 3. Chemical structures of dilazep and dipyridamole.

well-described characteristics of native *es*-type processes in both human and rat cells.

The high degree of sequence similarity between hENT1 and rENT1, coupled with the distinct functional differences in sensitivity to the coronary vasodilators (dipyridamole, dilazep, draflazine), with hENT1 being much more sensitive to these compounds than rENT1, provided an experimental approach to identify regions of the hENT1 involved in transport inhibition (Sundaram et al., 1998). Analysis of the transport characteristics of a series of recombinant reciprocal chimeras between hENT1 and rENT1 in *Xenopus* oocytes identified residues 100–231, encompassing transmembrane domains 3–6 of hENT1, as the major site of vasodilator action. These studies, which assessed the effects of dipyridamole on uridine transport by the recombinant chimeras, also eliminated involvement of (i) the large cytoplasmic loop between transmembrane domains 6 and 7 and (ii) the amino terminus up to the end of transmembrane 2 and including the large extracellular loop between transmembrane domains 1 and 2 in the inhibitory action of dipyridamole. Since dipyridamole is a competitive inhibitor of human *es* transport activity and NBMPR binding (Paterson et al., 1980; Jarvis et al., 1982; Jarvis, 1986; Deckert et al., 1994), it is likely that transmembrane domains 3–6 form part of the nucleoside- and NBMPR-binding site (Sundaram et al., 1998). The amino acid sequences of hENT1 and rENT1 in this region are 83% identical and 95% similar, thereby providing candidate residues for involvement in vasodilator binding.

2.1.2. THE ENT2 (*ei*) SUBFAMILY

The *es* transporter was identified as a band 4.5 protein in the early 1980s by photolabeling with ^{3}H-NBMPR (Wu *et al.*, 1983) and subsequently purified to homogeneity (Kwong *et al.*, 1988). However, the identity of membrane proteins responsible for *ei*-mediated transport remained a mystery until the cDNAs encoding rat and human proteins with *ei*-type activity were isolated and functionally expressed in *Xenopus* oocytes (Griffiths *et al.*, 1997b; Yao *et al.*, 1997). The key was the discovery in the GenBank database of a 36-kDa protein of murine and human origin, HNP36, that exhibited 44% similarity with the carboxyl-terminal two-thirds of hENT1, suggesting that HNP36 might represent a truncated ENT-like isoform (Griffiths *et al.*, 1997a). The HNP36 proteins are "delayed-early proliferative response" gene products of unknown function that appear to be located exclusively in the nucleolus of murine fibroblasts and rat intestinal epithelial cells (Williams and Lanahan, 1995). Two observations led to the prediction that the HNP36 proteins represented a good candidate for the elusive *ei*-type nucleoside transporters (Yao *et al.*, 1997): (i) mouse and human HNP36 are predicted to be integral membrane proteins with eight transmembrane domains that correspond to transmembrane domains 4–11 of hENT1, and (ii) the nucleotide sequence immediately upstream of the assigned start codon of the human HNP36 cDNA includes an open reading frame with 49% identity in predicted amino acid sequence to the amino-terminal region of hENT1. Using polymerase chain reaction (PCR) technology and primers based on the conserved regions of mouse and human HNP36, the first member of the ENT subfamily with *ei*-type transport activity (rENT2) was identified by isolation and functional characterization in *Xenopus* oocytes of a cDNA isolated from a rat jejunal library (Yao *et al.*, 1997). Using a similar PCR-based strategy, a cDNA encoding the human homologue (hENT2) was isolated shortly thereafter from a placental library (Griffiths *et al.*, 1997b). A cDNA encoding an identical predicted protein (also termed hENT2) was independently cloned by phenotypic selection by expression of a cDNA library from cultured HeLa cells in a nucleoside transport-deficient human leukemia cell line (Crawford *et al.*, 1998).

The hENT2 and rENT2 sequences are 93% similar and 88% identical at the amino acid level, and, as noted above (see Fig. 4), are about 50% identical in sequence to hENT1 and rENT1. Hydropathy analysis of the hENT2 and rENT2 protein sequences predicts the same membrane topology as was proposed for the ENT1 proteins, with the greatest sequence similarities between the putative transmembrane helices and the least between the two large hydrophilic loops linking transmembrane domains 1 with 2 and 6 with 7. hENT2 possesses three potential N-linked glycosylation sites, of which two (Asn 48, Asn 57) are found in the predicted extracellular loop between transmembrane domains 1 and 2 and one (Asn 225) is found in the predicted intracellular loop between transmembrane domains 6 and 7. The sequence of rENT2 possesses two potential N-linked glycosylation sites (Asn 47, Asn 56), which, allowing for the presence of a one-residue insertion before this point in the other ENT sequences, are conserved, respectively, among the ENT1/ENT2 proteins and the ENT2 proteins. It is not yet known which, if any, of the predicted exofacial glycosylation sites in hENT2 and rENT2 carry carbohydrate residues.

Functional studies of (i) recombinant hENT2 and rENT2 in *Xenopus* oocytes (Griffiths *et al.*, 1997b; Yao *et al.*, 1997) and (ii) recombinant hENT2 in stable transfectants established that the ENT2 proteins exhibit the hallmark features of *ei*-mediated nucleoside transport processes. Uridine transport is saturable, and the IC_{50} values for inhibition by NBMPR are >1 μM. The K_m values (hENT2, 0.20 mM; rENT2, 0.30 mM) for uridine transport in *Xenopus* oocytes (Griffiths *et al.*, 1997b; Yao *et al.*, 1997) are in agreement with those presented previously (Griffith and Jarvis, 1996) for rat erythrocytes (0.16 mM) and Ehrlich ascites cells (0.28 mM), whereas *ei*-mediated transport of uridine in cultured human K562 erythroleukemia cells exhibits a higher K_m value (1.1 mM), although it should be noted that the K562 studies (Boleti *et al.*, 1997) were conducted at 37°C, whereas the other studies, including those with recombinant hENT2 and rENT2, were conducted at 20–23°C. Uridine transport by recombinant hENT2 and rENT2 in *Xenopus* oocytes (Griffiths *et al.*, 1997b; Yao *et al.*, 1997) is relatively insensitive to inhibition by dilazep and dipyridamole (IC_{50} values >1 μM). Inhibition studies with physiologic nucleosides tested against ^{3}H-uridine suggest that hENT2 and rENT2 accept both purine and pyrimidine nucleosides as permeants. However, uridine transport is only partially inhibited by cytidine when recombinant hENT2 is produced in either *Xenopus* oocytes or cultured cells (Yao *et al.*, 1997; Crawford *et al.*, 1998). Of considerable interest is the observation that hypoxanthine inhibits uridine transport in a stable hENT2-producing cultured cell line (Crawford *et al.*, 1998), a result that supports the conclusion, from earlier studies with cultured human endothelial cells, that hypoxanthine is a permeant of the *ei* transport process (Osses *et al.*, 1996).

Studies of *ei*-mediated nucleoside fluxes with cultured cell lines have suggested a broad tissue distribution, although *ei*-processes are normally found as a minor transport component, almost always in the presence of *es*-mediated processes, and frequently in the presence of concentrative nucleoside transport processes. The tissue distribution of hENT2, as determined by hybridization analysis of RNA from a variety of human tissues, has shown that expression of hENT2 mRNA occurs in many tissues and at variable levels (Crawford *et al.*, 1998), with the greatest quantities in skeletal muscle.

The relationship between the ENT2 and HNP36 proteins is uncertain. HNP36 was localized in nucleoli with polyclonal antibodies raised against an epitope shared by HNP36 and hENT2, and the immunoreactive material observed in the HNP36 studies (Williams and Lanahan, 1995) may have been HNP36, as concluded, or hENT2, or a combination thereof. There is growing evidence that func-

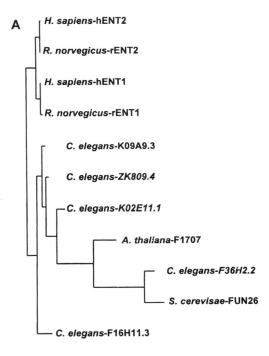
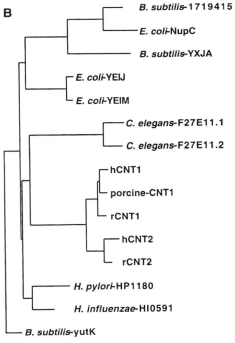

tional ENT proteins, with *es* or *ei* activities, are present in intracellular membranes. For example, nuclear envelopes have recently been shown to contain proteins (presumably hENT1 and hENT2) that mediate *es*- and *ei*-type transport, as indicated by functional reconstitution of transport activity in proteoliposomes prepared from detergent-solubilized membranes from cultured human choriocarcinoma cells (Mani *et al.*, 1998). Attempts to demonstrate directly that the HNP36 proteins have transport activity have thus far been unsuccessful. For example, human HNP36 is nearly identical to transmembrane domains 4–11 of hENT2, and the corresponding N-terminally truncated form of hENT2 (amino acid residues 130–456) does not mediate uridine transport in transiently transfected cultured cells (Crawford *et al.*, 1998). It is not known if HNP36 is produced *in vivo* as a functional protein.

3. THE CNT FAMILY OF NUCLEOSIDE TRANSPORTERS

3.1. Molecular and Functional Characteristics of the CNT Family

The functional characteristics of the concentrative transport processes of mammalian cells have been reviewed (Belt *et al.*, 1993; Cass, 1995; Griffith and Jarvis, 1996; Thorn and Jarvis, 1996). The Na^+-dependent nucleoside transporters are minor membrane proteins found only in certain tissue types, including intestinal and renal epithelia and liver. Nothing was known about the transporter proteins until cDNAs encoding the major concentrative nucleoside transporters (CNTs) of rat epithelial tissues were cloned by functional expression selection in *Xenopus* oocytes (Huang *et al.*, 1993, 1994; Che *et al.*, 1995). The identification of a structurally related protein (NupC) in bacteria with the characteristics of a H^+/nucleoside symporter (Craig *et al.*, 1994), together with the subsequent cloning and

---

←

Figure 4. Phylogenetic trees showing protein-sequence relationships between nucleoside transporters. GCG (Wisconsin) programs were used to create the dendograms for the equilibrative (A) and concentrative (B) nucleoside transporter families using sequences from the public databases. **Dendogram A accession numbers:** *H. sapiens* hENT1, #1845345; *R. norvegicus* rENT1, #2656137; *H. sapiens* (placenta) hENT2, #2754821; *H. sapiens* HeLa cells hENT2, #3811137; *R. norvegicus* rENT2, #2656139; *C. elegans* K09A9.3, #1515147; *C. elegans* ZK809.4, #1130665; *C. elegans* K02E11.1, #1469109; *C. elegans* F36H2.2, #1627922; *S. cerevisiae* FUN26, #486963; *C. elegans* F16H11.3, #1280131; *A. thaliana* F1707, #3176684. **Dendogram B accession numbers:** *E. coli* NupC, #141111; *B. subtilis* 1719415, #P39141; *B. subtilis* YXJA, #P42312; *H. sapiens* hCNT1, #2072782; *R. norvegicus* rCNT1, #628013; *S. scrofus* (porcine) CNT1, #3169734; *H. sapiens* hCNT2, #2665908 (see also hSPNT1, #2731439); *R. norvegicus* rCNT2, #2731439 (see also SPNT, #862387); *H. influenzae* HI0519, #1175856; *H. pylori* HP1180, #2314337; *B. subtilis* yutK, #e1184297; *E. coli* YEIM, #P33024; *E. coli* YEIJ, #P33021; *C. elegans* F27E11.1, AF016413.1, *C. elegans* F27E11.2, AF016413.2.

functional expression of human cDNAs encoding Na^+-dependent nucleoside transport proteins (Ritzel *et al.,* 1997, 1998; Wang *et al.,* 1997), established the CNT proteins as members of a previously unrecognized family of related transporter proteins.

3.1.1. THE CNT1 (*cit*) SUBFAMILY

The first mammalian member of the CNT subfamily of proteins to be recognized was cNT1 (Huang *et al.,* 1994), which has subsequently been designated as rCNT1 (Yao *et al.,* 1996b) with the adoption of a standard terminology for the CNT and ENT proteins. The identification of rCNT1 was achieved by functional expression screening of a rat intestinal library in *Xenopus* oocytes, following a lead that was provided earlier by the stimulation of Na^+-dependent uridine transport in oocytes when injected with rat jejunum poly(A)$^+$ RNA (Huang *et al.,* 1993). The sequence of rCNT1 predicts a protein of 648 amino acids (71 kDa) with a topology (14 transmembrane-spanning domains) that has served as the basis of the as-yet-unproven architectural model for other mammalian members of the CNT family (Che *et al.,* 1995; Yao *et al.,* 1996b; Ritzel *et al.,* 1997; Wang *et al.,* 1997b). The deduced amino acid sequence of rCNT1 contains numerous consensus sites for posttranslational modifications, including N-linked glycosylation sites and protein kinase C consensus phosphorylation sites. The functional significance of phosphorylation consensus sites has not yet been established. Northern analyses of rat tissues demonstrated that rCNT1 mRNA is produced in jejunum and kidney (Huang *et al.,* 1994), and amplication of mRNA from various regions of rat intestine using reverse transcriptase and the polymerase chain reaction (RT-PCR) with rCNT1-specific primers indicated that mRNA expression is greater in jejunum than in duodenum and ileum and there is no expression in colon (Wang *et al.,* 1997a). RT-PCR amplification of rat brain tissues with rCNT1-specific primers indicated that rCNT1 mRNA is produced in many regions, including cerebral cortex, cerebellum, hippocampus, striatuum, brain stem, superior colliculus, posterior hypothalamus, and choroid plexus (Anderson *et al.,* 1996). mRNA encoding a protein identical to rCNT1 has also been detected in rat liver by RT-PCR homology cloning, indicating that the pyrimidine-nucleoside transport activity of rat hepatocytes is mediated by rCNT1 (Felipe *et al.,* 1998).

The transport characteristics of rCNT1 have been studied by functional production of the recombinant protein by microinjection of RNA transcripts produced *in vitro* into *Xenopus* oocytes (Huang *et al.,* 1994; Yao *et al.,* 1996a,b), by transient transfection of cultured monkey kidney (COS) cells (Fang *et al.,* 1996) and human cervical carcinoma (HeLa) cells (Wang *et al.,* 1997a) and by stable transfection of a nucleoside-transport-defective clone of cultured murine leukemia (L1210) cells (Crawford *et al.,* 1998). Kinetic studies in the transient expression systems (Huang et al., 1994; Fang et al., 1996; Yao et al., 1996b; Wang et al., 1997a) yielded K_m values for uridine (37 and 21 μM), thymidine (12.5 and 7.4 μM), and adenosine (26 and 15 μM) that are comparable to those observed for native cit-type transporters. Earlier studies, which demonstrated that adenosine is a potent, apparently competitive inhibitor of the cit-type transporter of murine intestinal epithelium (Vijayalakshmi and Belt, 1988), led to the conclusion (Cass, 1995) that adenosine, like uridine, is a "universal" permeant for the CNT (and ENT) transporters. However, the observed V_{max} values for adenosine transport by recombinant rCNT1 are very low, suggesting that adenosine is a relatively poor permeant, possibly even an inhibitor (Fang et al., 1996; Yao et al., 1996b). This conclusion was confirmed in studies that assessed the relative abilities of uridine and adenosine to accelerate rCNT1-dependent efflux of uridine from preloaded oocytes (Yao et al., 1996b). rCNT1-mediated efflux of uridine from preloaded oocytes is accelerated by extracellular uridine and inhibited by extracellular adenosine, consistent with bidirectional operation of rCNT1 and suggesting that the rate of conversion of rCNT1 from the outward-facing conformation to the inward-facing conformation is greater when uridine is bound than when adenosine is bound. Thus, although rCNT1 binds adenosine and uridine with similar affinities, it kinetically favors transport of uridine and it is unlikely that it plays a physiologically significant role in cellular uptake of adenosine.

Studies with recombinant rCNT1 produced in *Xenopus* oocytes provided the first direct demonstration of mediated transport of the antiviral nucleosides zidovudine and zalcitabine (Huang et al., 1994; Yao et al., 1996a). The recombinant protein mediates Na^+-dependent influx of zidovudine and zalcitabine with K_m values of 0.5 mM and relative V_{max}/K_m ratios for zidovudine, zalcitabine, and uridine of 0.084, 0.068, and 1.00, respectively. Extracellular zidovudine and zalcitabine also accelerate rCNT1-mediated efflux of uridine from preloaded oocytes. The anticancer nucleosides cladribine and cytarabine are transported by recombinant rCNT1 in *Xenopus* oocytes (Wang et al., 1997a) and inhibit rCNT1-mediated transport of thymidine with IC_{50} values, respectively, of 61 μM and 1.88 mM (Wang et al., 1997a).

The relative abilities of several nucleoside drugs to inhibit rCNT1-mediated transport were assessed in transiently transfected COS cells (Fang et al., 1996), giving the following rank order of potency: 5-fluoro-2'-deoxyuridine > 5-iodo-2'-deoxyuridine > zidovudine > zalcitabine > cytarabine > gemcitabine. When the capacity of rCNT1 to transport anticancer nucleosides was further examined by comparing cytotoxicity in a transport-defective mouse leukemia cell line and a clonal derivative into which rCNT1 was stably transfected (Crawford et al., 1998; Mackey et al., 1998b), it was concluded that gemcitabine, 5-fluoro-2'-deoxyuridine, and 5-fluorouridine are relatively good permeants and cytarabine is a poor permeant of recombinant rCNT1.

Two related cDNAs encoding transporter proteins with cit-type activities (hCNT1) were isolated from human kidney by hybridization screening and RT-PCR amplification strategies (Ritzel et al., 1997). The predicted hCNT1a and hCNT1b proteins exhibit >99% identity of amino acid sequence (650 versus 648 residues, 71 kDa), with minor differences attributed to genetic polymorphisms and/or RT-PCR-induced errors. hCNT1 is 83% identical to rCNT1 and has a similar predicted topology, with 14 transmembrane domains. The tissue distribution of hCNT1 has not been determined. The hCNT1 gene is located on chromosome 15q25–26.

The demonstration that recombinant hCNT1 mediates Na^+-dependent transport of uridine in *Xenopus* oocytes provided the first evidence of the existence of the Na^+-dependent pyrimidine nucleoside-selective transporter in humans (Ritzel et al., 1997). Recombinant hCNT1 mediates inward transport of uridine with high affinity (K_m 42 μM) and is inhibited by adenosine, deoxyadenosine, thymidine, and cytidine, but not by guanosine or inosine. The apparent K_i values for adenosine and deoxyadenosine, which were calculated assuming competitive inhibition of uridine transport, are 50 and 46 μM, respectively, indicative of a high affinity of hENT1 for the adenine nucleosides. However, a direct comparison of transport of radiolabeled adenosine and uridine at the same extracellular concentration (10 μM) established that uridine transport is favored, with a uridine-to-adenosine flux ratio of 75:1. Adenosine-to-deoxyadenosine flux ratios were 4:1 for hCNT1 and 7:1 for rCNT1 (Ritzel et al., 1997). hCNT1, like rCNT1, also transports zidovudine and is inhibited by zalcitabine. Recombinant hCNT1 has been produced in cultured HeLa cells by transient transfection (Mackey et al., 1998b) and shown to transport gemcitabine with relatively high affinity (K_m 32 μM).

3.1.2. THE CNT2 (*cif*) SUBFAMILY

A Na^+-dependent adenosine transport activity of rat liver (Che et al., 1992) is mediated by a member of the CNT2 subfamily, initially demonstrated by isolation and functional expression in *Xenopus* oocytes of a cDNA from rat liver that encodes a transporter protein designated SPNT (Che et al., 1995). A cDNA encoding an almost identical protein, designated rCNT2, was isolated from rat intestine (Yao et al., 1996b). The SPNT and rCNT2 proteins, which differ in conservative substitutions of glycine for alanine and valine for isoleucine, respectively, at residues 419 and 522 of rCNT2, are both predicted to have 659 amino acids (72 kDa) with 14 transmembrane-spanning regions. There are five potential N-linked glycosylation sites and several consensus sites for protein kinase A and C phophosphorylation, suggesting that rCNT2 may be regulated by protein kinase-dependent mechanisms. rCNT2 is 64% identical to rCNT1 and the regions of greatest divergence are at the N- and C-termini. rCNT2 is 64% identical to rCNT1 and the regions of greatest divergence are at the N- and C-termini. rCNT2 appears to exhibit a broader tissue distribution that rCNT1, with multiple species of mRNA found in liver, jejunum, spleen, heart, and skeletal muscle by Northern analysis (Che *et al.*, 1995). RT-PCR amplification of rat brain mRNA with rCNT2-specific primers indicated that the distribution of rCNT2 mRNA in brain is similar to that of rCNT1 mRNA (Anderson *et al.*, 1996).

The functional characteristics of rCNT2/rSPNT are typical of a *cif*-type transporter, with permeant selectivity for uridine, adenosine, and other purine nucleosides (Che *et al.*, 1995; Yao *et al.*, 1996b; Schaner *et al.*, 1997; Wang *et al.*, 1997a). Kinetic studies, which were conducted in the *Xenopus* oocyte (Che *et al.*, 1995) and HeLa cell (Schaner *et al.*, 1997) expression systems, yielded K_m values for adenosine, uridine, and inosine, respectively, of 6, 21, and 28 µM. Although didanosine and cladribine inhibit rCNT2/rSPNT-mediated transport of inosine with IC_{50} values, respectively, of 13 and 46 µM, direct measurements of isotopic fluxes suggest that only cladribine is actually transported (Schaner *et al.*, 1997). As described below, there is evidence that didanosine is a permeant for hCNT2 (Ritzel *et al.*, 1998).

The molecular determinants of permeant selectivity in the CNT proteins of rat have been examined by a molecular cloning strategy in which various portions of the rCNT1 and rCNT2/rSPNT cDNAs were interchanged and the resulting constructs functionally expressed in *Xenopus* oocytes (Wang and Giacomini, 1997). When transmembrane domains 8 and 9 of rCNT2 are transplanted into rCNT1, the recombinant chimeric protein is purine nucleoside-selective, suggesting that the critical residues for distinguishing between the purine and pyrimidine moieties of nucleoside permeants are located in transmembrane domains 8 and 9 of the CNT proteins. The chimeric protein in which only domain 8 of rCNT2 is transplanted into rCNT1 transports both inosine and thymidine, a characteristic of the broadly selective and as yet molecularly unidentified *cib*-type transporter. The reciprocal rCNT2-to-rCNT1 transition is not functional. Further studies, in which individual amino acids are changed by site-directed mutagenesis, are needed to identify the particular residues that form the permeant-binding site of the CNT proteins.

The human *cif*-type transporter has been identified by molecular cloning of cDNAs from kidney (Wang *et al.*, 1997b) and intestine (Ritzel *et al.*, 1998) using RT-PCR homology cloning strategies and functional expression in *Xenopus* oocytes. The predicted proteins, termed hSPNT1 (Wang *et al.*, 1997b) and hCNT2 (Ritzel *et al.*, 1998), have 658 amino acids and are identical except for a polymorphism at residue 75 (arginine substituted by serine in hCNT2). hCNT2/hSPNT1, like other members of the CNT family, is predicted to possess 14 transmembrane-spanning domains. Its identity with rCNT2 and hCNT1 is 81% and 72%, respectively, with the greatest divergence in the N-terminal regions of the two proteins. Multiple mRNA species have been observed in a variety of human tissues, including heart, liver, skeletal muscle, kidney, intestine, pancreas, placenta, brain, and lung (Wang *et al.*, 1997b). Our investigations, however, failed to detect hCNT2 transcripts in liver (Ritzel et al., 1998). There are six consensus sites each for N-linked glycosylation and phosphorylation by protein kinase. The functional significance of these sites is not known. The presence of an *Alu* repetitive element of 282 base pairs in the 3' untranslated region of hCNT2 mRNA suggests that hCNT2 may be regulated by *Alu*-dependent pathways. The hCNT2 gene is localized to chromosome 15 at either 15q13–14 (Wang et al., 1997b) or 15q15 (Ritzel et al., 1998).

Kinetic studies conducted with recombinant hCNT2/hSPNT1 in *Xenopus* oocytes yielded respective K_m values for transport of inosine and adenosine of 4 and 8 μM (Wang et al., 1997b; Ritzel et al., 1998) and uridine of 40 μM (Ritzel et al., 1998) and 80 μM (Wang et al., 1997b). hCNT2 also mediates transport of guanosine, deoxyadenosine, and didanosine (Ritzel et al., 1998). Inosine transport is inhibited by adenosine, 2'-deoxyadenosine, guanosine, and uridine, with calculated K_i values for adenosine and 2'-deoxyadenosine, respectively, of 6 and 30 μM (Wang et al., 1997b; Ritzel et al., 1998). hCNT2-mediated transport of inosine is not inhibited by thymidine, cytidine, zidovudine, or zalcitabine (Ritzel et al., 1998).

A cDNA encoding a murine *cif*-type transporter, mCNT2, has also been described in a preliminary report (Patel et al., 1997), although the sequence has not been registered in the public databases. mCNT2 exhibits 93% and 61% identity to rCNT2 and rCNT1, respectively. Cultured COS cells transiently transfected with mCNT2 cDNA exhibit mCNT2-mediated Na^+-dependent uridine uptake.

4. "ORPHAN" NUCLEOSIDE TRANSPORTERS

Several minor Na^+-dependent nucleoside transport processes with functional characteristics that are distinct from those of the well-defined CNT1- and CNT2-mediated processes have been described in studies with intact cells or membrane vesicles (Belt et al., 1993; Gutierrez and Giacomini, 1993; Paterson et al., 1993; Gutierrez and Giacomini, 1994; Flanagan and Meckling-Gill, 1997). While the proteins responsible for these processes have not yet been identified, it seems likely that they will be members of the CNT or ENT protein families. Several of these rarely described activities are similar to *cit*- and *cif*-type processes, but exhibit key differences in their permeant selectivities and presumably in the underlying genetic sequences that encode the substrate-binding pockets. A putative nucleoside transporter of rabbit kidney, identified by its structural relationship to the intestinal Na^+/glucose cotransporter of mammalian cells, exhibits low levels of nucleoside transport activity when produced in *Xenopus* oocytes (Pajor and Wright, 1992) and its role as a bona fide nucleoside transporter is uncertain.

While most studies of nucleoside transport processes in mammalian cells have focused on transporters of the plasma membrane, the membranes of cellular organelles are also nucleoside permeability barriers. The compartmentation of enzymes of nucleoside metabolism within mitochondria and lysosomes suggests that organellar nucleoside transporters play a significant role in the intracellular distribution of nucleosides and nucleoside drugs. A nucleoside transport process of unknown origin has been kinetically characterized in isolated human lysosomes (Pisoni and Thoene, 1989).

4.1. Na$^+$-Dependent Transporters

4.1.1. UNIDENTIFIED TRANSPORTERS

A broadly selective process that mediates Na$^+$-dependent uptake of both purine and pyrimidine nucleosides has been described in freshly isolated human leukemic blasts (Belt *et al.*, 1992) and human colon cancer Caco-2 cells (Belt *et al.*, 1993). Na$^+$-dependent transport of thymidine (the diagnostic nucleoside for *cit* activity) is inhibited by both cytidine and formycin B, and Na$^+$-dependent transport of formycin B (the diagnostic nucleoside for *cif* activity) is inhibited by both inosine and thymidine. These results point to an additional transport activity with broad permeant selectivity. A similar activity has also been measured in the rabbit choroid plexus (Wu *et al.*, 1992) and in RNA derived from rat intestine (Huang *et al.*, 1993). Belt has termed the process *cib*, because it is concentrative, insensitive to inhibition by nM concentrations of NBMPR, and appears to possess broad permeant selectivity in that inward fluxes of uridine are inhibited by both purine and pyrimidine nucleosides. Characterization of the *cib* transporter has been difficult because of low levels of activity in the cell types in which it has been observed and because it evidently coexists with other nucleoside transporters. It is possible that *cib* activity could be mimicked by the coexistence in a single cell type of both *cit* and *cif* activities. Because of the technical difficulties associated with analysis of *cib*-mediated processes, its existence as a distinct entity will be questioned until the identification of a transporter protein, e.g., by molecular cloning of its cDNA, that mediates a transport process with *cib*-type activity. That the *cib* transporter may be a member of the CNT family is suggested by the broad permeant selectivity observed for a chimeric protein produced by transplanting the eighth transmembrane domain of rCNT2 (purine-nucleoside-selective) into rCNT1 (pyrimidine-nucleoside-selective) to produce a recombinant protein whose transport characteristics were examined in *Xenopus* oocytes (Wang and Giacomini, 1997); the chimera, which consisted of amino acid residues 1–300 of rCNT1, 297–300 of rCNT2, and 335–648 of rCNT1, transported both inosine and thymidine.

A Na^+/nucleoside cotransport process found in human renal brush-border membrane vesicles exhibits typical *cit*-type characteristics, with saturable inward transport of uridine with a Na^+–nucleoside coupling stoichiometry of 1:1. This *cit*-type nucleoside transport activity differs from the CNT1-mediated processes described in Section 3.1.1 in that it appears to also transport guanosine, based on this nucleoside's ability to inhibit inward fluxes of ^{3}H-thymidine. Inosine and formycin B, which are considered diagnostic permeants for *cif*-type transporters, are not substrates, whereas adenosine, 2'-deoxyadenosine, and 2-chloroadenosine (2-ClAdo), like guanosine, appear to also be permeants. It is highly likely that the protein responsible for this *cit*-type activity, which has been given the numerical designation N4 (Wang *et al.*, 1997a), is a member of the CNT family.

Two Na^+-dependent nucleoside transport processes with sensitivity to inhibition by nanomolar concentrations of NBMPR have been described in freshly isolated and cultured leukemic cells (Paterson *et al.*, 1993; Flanagan and Meckling-Gill, 1997). The first report (Paterson *et al.*, 1993) described a novel activity that was both concentrative and NBMPR-sensitive (*cs*) and thus highly unexpected, given the many examples in different cell types of the insensitivity of concentrative transport processes to NBMPR, as well as the transport inhibitors with vasodilator activity (e.g., dipyridamole, dilazep). The *cs* process was repeatedly observed, by analysis of time courses of uptake at 37°C of nucleoside drugs (cladribine, fludarabine) in the presence and absence of sodium, NBMPR, and dipyridamole, in freshly isolated leukemic cells from a small group of patients undergoing treatment with nucleoside analogues (Paterson *et al.*, 1993). Over several years, *cs* activity was found in leukemic blasts in 2 of 5 patients with acute myelogenous leukemia and in leukemic cells of 7 of 10 patients with chronic lymphocytic leukemia. A second NBMPR-sensitive concentrative nucleoside transport process is found in cultured human NB4 acute promyelocytic leukemia cells and murine L1210 leukemia cells (Flanagan and Meckling-Gill, 1997). This process is termed *csg,* because it is concentrative, sensitive to inhibition by nanomolar concentrations of NBMPR, and selective for guanosine. The *cs* and *csg* processes are not well defined because in cells in which they have been reported they exhibit low levels of activity and coexist with other nucleoside transport processes. The proteins responsible for mediating *cs* and *csg* activity have not been identified, but because of the high sensitivity to NBMPR may share structural features with the ENT proteins.

4.1.2. THE SNST1 PROTEIN

The first putative nucleoside transport protein to be identified by molecular cloning (Pajor and Wright, 1992) was a protein, designated SNST1, of rabbit kidney that is related to the rabbit intestinal Na^+-dependent glucose transporter SGLT1. SNST1 encodes a predicted protein of 672 amino acids that is 61% identical and 80% similar in sequence to rabbit SGLT1. There is no sequence similarity between SNST1 and either the ENT or the CNT protein families. The SNST1 cDNA was isolated by homology screening of a rabbit kidney library with a SGLT1-related hybridization probe and its identity was established by functional expression of a chimeric cDNA construct comprised of nucleotides 243–2150 of SNST1 together with the homologous 5' end and 3'-untranslated region of rabbit SGLT1. Although the recombinant SGLT1/SNST1 chimera, when produced in *Xenopus* oocytes, stimulates low levels of Na^+-dependent uptake of uridine that is inhibited by purine and pyrimidine nucleosides (i.e., a *cib*-type pattern), the function of SNST1 remains unclear. The rate of Na^+-dependent uridine transport observed when SNST1 cDNA is expressed in oocytes is only twofold above endogenous (background) levels, whereas a $>20,000$-fold stimulation is observed when rCNT1 cDNA is expressed in oocytes (Huang *et al.*, 1994). A *cib*-type transport activity has not been observed in the tissues (kidney, heart) in which the SNST1 message was reported (Conant and Jarvis, 1994; Griffith and Jarvis, 1996). One interpretation is that the physiologic substrate of SNST1 has not yet been identified and is likely to be some other low-molecular-weight metabolite for which there is overlapping permeant recognition with nucleosides.

4.2. Organellar Transporters

Two organelles are known to have nucleoside transport activities: lysosomes and mitochondria. Lysosomes have a full complement of transport systems (for review, see Pisoni and Thoene, 1991), including a transport system for nucleosides, first described in lysosomes isolated from human fibroblasts (Pisoni and Thoene, 1989). The lysosomal transport process exhibits broad permeant selectivity and is sensitive to NBMPR and dipyridamole. It is thought to function primarily in the export of nucleosides originating from degraded nucleic acids from lysosomes to the cytoplasm. Mitochondria isolated from rat liver possess the capacity for inward transport of thymidine and 2'-deoxyguanosine (Mitra and Bernstein, 1970; Watkins and Lewis, 1987) and exhibit high-affinity binding of NBMPR (King, K. M., Mani, R. S., and Cass, C. E., unpublished results); mitochondria from rat testis have also been shown to bind NBMPR (Camins *et al.*, 1996). Mitochondria contain many enzymes of nucleoside and nucleotide metabolism, suggesting, since the inner mitochondrial membrane is a stringent permeability barrier, that they also possess nucleoside-selective transporters that function in the movement of nucleosides between the cytosol and the inner mitochondrial matrix. A hydrophobic protein (ND4) encoded by the human mitochondrial genome, when produced in a thymidine-transport-deficient strain of yeast, enables salvage of extracellular thymidine, suggesting that ND4 may possess nucleoside transport capability (Hogue *et al.*, 1997).

A novel putative nucleoside transporter of organellar membranes was recently identified using a yeast functional complementation assay that was developed for selection of cDNAs encoding thymidine-selective transporter proteins (Hogue *et al.*, 1996). A partial cDNA from a murine cDNA library complemented the thymidine transport deficiency of recipient yeast by allowing growth in the presence of limiting concentrations of thymidine. The full cDNA, which failed to complement the thymidine transport deficiency, encodes a 26-kDa protein, termed MTP (for mouse transporter protein), comprised of 233 amino acids with four predicted transmembrane domains. Production in *Xenopus* oocytes of a truncated form of MTP lacking the carboxy-terminal 36 residue, altered membrane permeability, resulting in increased uptake of adenosine, uridine, and thymidine, whereas the complete protein failed to complement, evidently because it is targeted to intracellular, rather than plasma, membranes. MTP appears to be a protein of late endosomes and lysosomes (Cabrita, M. A., Hobman, T. C., Hogue, D. L., and Cass, C. E., unpublished results).

5. NUCLEOSIDE TRANSPORTERS AS DRUG TARGETS

5.1. Purinergic Receptors and Nucleoside Transporters

Adenosine and ATP are known to interact with specific purinergic receptors that are widely distributed throughout mammalian tissues to mediate a diversity of biological functions. The activation of purinergic receptors by nucleosides and nucleotides is influenced by nucleoside transport processes. Therefore, nucleoside transporters serve as pharmacologic targets that can modulate a multitude of physiologic effects elicited by purinergic receptors.

The first classification of purinergic receptors (Burnstock, 1978) was proposed on the basis that P1 and P2 purinoceptors are preferentially activated by adenosine and ATP, respectively. Further pharmacologic and biochemical studies identified several apparently distinct subtypes of purinergic receptors. Recently, molecular cloning technologies have isolated many of these entities. The current classification and nomenclature of purinergic receptors [IUPHAR Receptor Nomenclature Committee (Fredholm, 1994)] has recently been reviewed (Olah and Stiles, 1995; Windscheif, 1996) and is summarized below.

5.1.1. ADENOSINE/P1 PURINOCEPTOR SUBTYPES

Adenosine acts via P1 purinoceptors, which are also referred to as adenosine receptors. On the basis of their biochemical and pharmacologic properties, adenosine receptors are divided into three main groups: A1, A2, and A3. Four P1 purinoceptor subtypes have been identified by molecular cloning of cDNAs: A1,

A2a, A2b, and A3. Adenosine receptors are members of the rhodopsin-like G-protein-coupled receptor family. The activation of A1 and A3 receptors results in an inhibition of adenylate cyclase, while A2 receptor activation stimulates adenylate cyclase. The signal transduction mechanisms utilized by adenosine receptor subtypes are complex and not clearly defined.

5.1.2. ATP/P2 PURINOCEPTOR SUBTYPES

The ATP/P2 purinoceptors are classified into the two major groups, P2X and P2Y, on the basis of their transduction mechanisms. The P2X purinoceptor subtypes are ligand-gated ion channels; direct activation by ATP results in a depolarizing cationic current via an intrinsic ion channel. The P2Y purinoceptor subtypes are members of the G-protein-coupled receptor family. The activation of P2Y purinoceptors leads to modulation of membrane phosphoinositide metabolism and, hence, inositol-1,4,5-trisphosphate (IP3) and diacyl-glycerol (DAG) generation. However, multiple pathways are known to mediate the responses of P2 purinoceptor activation and further analysis is required to elucidate fully the intricate transduction mechanisms involved.

5.1.3. THERAPEUTIC IMPLICATIONS

There is considerable potential for modulation of purinergic mechanisms in the treatment of heart and vascular disease, stroke, neurologic disorders, and various other pathophysiologic processes. However, the ubiquitous nature of adenosine and ATP must be considered in strategies for the rational design of pharmacologic agents that will selectively target dysfunctions related to purinoceptor-mediated processes. Some selected examples of therapeutic implications for purinergic therapy are discussed below. For general reviews, see Jacobson *et al.* (1992), Olah and Stiles (1995), Guieu *et al.* (1996), Windscheif (1996), and Buolamwini (1997).

Purinergic mechanisms have been implicated in many neurologic disorders, including the cellular response to ischemic stroke, cerebral trauma, and the pathophysiology of epilepsy and alcoholism. Adenosine is a powerful neuromodulator of transmissions at both central and peripheral synapses. Reuptake of adenosine by nucleoside transporters is believed to be the primary mechanism by which purinoceptor-mediated processes are terminated. Inhibition of nucleoside transport activity can therefore have profound effects on purinergic neuromodulation. The existence of multiple nucleoside transporter subtypes in the central nervous system offers the potential to design selective agents for targeting certain regions of the brain. Propentofylline, a nucleoside transport inhibitor, has been shown to have promise as a neuroprotectant in the prevention of ischemia-induced brain damage during stroke (for review, see Parkinson *et al.*, 1994).

The therapeutic potential of adenosine in heart disease relies on its ability to protect against ischemia and reperfusion injury. The potentiation of the cardioprotective effects of adenosine by inhibition of nucleoside transport activity has been studied extensively. Since the release of adenosine in ischemia is transient and localized, the use of nucleoside transport inhibitors to block reuptake and thereby enhance the signaling effects of adenosine offers the potential to selectively target tissues that are under ischemic stress. Such an approach would circumvent systemic effects via widespread activation of purinergic receptors. Adenosine is effective as an antiarrhythmic drug in the treatment of paroxysmal supraventricular tachycardia and as an adjunct in aneurism surgery for the control of hypertension. Nucleoside uptake blockers, such as dipyridamole, are potent coronary vasodilators and inhibitors of platelet aggregation.

In addition to protecting against damage caused by acute cardiac and cerebral ischemic episodes, adenosine has been shown to precondition tissues, conferring resistance to future ischemic injury. Inhibitors of nucleoside transport are beneficial as agents for both myocardial and cerebral infarction. Nucleoside uptake blockers also prolong cardiac storage capabilities for heart transplantation, and protect heart tissue during cardiac surgery.

With new insights into processes mediated by purinergic mechanisms in multiple organic systems, one can propose a diverse array of therapeutic strategies related to modulation of adenosine levels in tissue fluids. Adenosine is implicated in events related to both inflammation and the immune response and has effects on the pituitary–adrenocortical axis, increasing the release of a number of hormones. Adenosine also plays a role in asthma by stimulating histamine release and leukotriene secretions from mast cells, while nucleotides such as ATP and UTP have shown clinical value in the treatment of cystic fibrosis. Purinergic functions in the renal system may offer a target for novel diuretics and agents that protect against acute renal failure. In the urogenital system, purine nucleotides affect bladder tone and offer a possible therapy for stress incontinence. In addition, the gastrointestinal tract has a wide distribution of purinoceptors which, if selectively targeted, would be useful in treatment of gut motility dysfunction. In some of these instances, the selective enhancement of nucleoside transporter activity may present an innovative strategy to diminish the physiologic actions mediated by purinergic receptors in some disease states and to augment their beneficial effects in others.

5.2. Anticancer Drugs and Nucleoside Transporters

Nucleoside transporters are a particularly promising target in cancer therapy, since pharmacologic manipulation of nucleoside transport activity has the potential to overcome nucleoside anticancer drug resistance and to improve the relationship of anticancer effect to normal tissue toxicity (therapeutic index) of the anticancer nucleoside drugs. Several comprehensive reviews of nucleoside drug transport are available (Paterson and Cass, 1986; Gati and Paterson, 1989; Belt et al., 1993; Cass, 1995; Griffith and Jarvis, 1996).

5.2.1. THE CLINICAL ANTICANCER NUCLEOSIDES

The anticancer nucleosides in routine clinical use are cytarabine (ara-C), which is effective in acute leukemia and some forms of lymphoma (Bolwell et al., 1988; Hiddemannn, 1991; Ross, 1991; Miller et al., 1992; Preti and Kantarjian, 1994; Cortes et al., 1996), clabridine (CdA), which is effective in chronic lymphocytic leukemia, low-grade lymphomas, and hairy cell leukemia (Hoffman, 1996; Piro, 1996; Saven, 1996; Tallman and Hakimian, 1996), and fludarabine (F-ara-A), which is effective in chronic lymphocytic leukemias and low-grade lymphoma (Keating et al., 1994; Plunkett and Gandhi, 1994; Keating et al., 1996). Gemcitabine (dFdC) is the newest nucleoside anticancer agent in routine clinical use, and has a broad-range activity in common solid tumors, including non-small-cell lung, breast, ovarian, pancreatic, and bladder cancers (Carmichael et al., 1996; Carmichael and Walling, 1996; Cronauer et al., 1996; Eisenhauer and Vermorken, 1996; Fagbemi and Stadler, 1998). Deoxycoformycin has largely been supplanted by the arrival of the newer purine nucleoside analogues (Brogden and Sorkin, 1993; Jaiyesimi et al., 1993; Cheson et al., 1994; Polliack, 1997). Although capecitabine (N4-pentyloxycarbonyl-5'-deoxy-5-fluorocytidine), an orally administered nucleoside prodrug of 5-fluorouracil (Budman et al., 1998), has recently been approved in the United States for use in patients with taxane-resistant metastatic breast cancer, its transportability has not yet been studied. The structures of several of these drugs are shown in Fig. 1B.

In the following discussion of nucleoside drug transport, functional studies have been interpreted in the context of current knowledge of the molecular biology of NT proteins. As described earlier in this review, the molecular identities of the NT proteins responsible for mediating the *es, ei, cit,* and *cif* processes in various cell types have proven thus far to be members of the ENT or CNT protein families. Where the molecular identity of the NT process has not yet been defined, the process is designated by its acronym (e.g., *es* or *ei*) linked with the name of the presumptive NT protein (e.g., hENT1 or rCNT1). Thus, the NBMPR-sensitive equilibrative process of human leukemic cells is termed "*es*/hENT1" and is catalyzed by a predicted protein whose existence has not yet been determined.

5.2.2. THE ROLE OF NUCLEOSIDE TRANSPORTERS IN DRUG RESISTANCE

The pharmacologic targets of the anticancer nucleosides are intracellular, and permeation through the plasma membrane is an obligatory first step in cytotoxicity (Cass, 1995). However, because these agents are hydrophilic molecules, their rates of passive diffusion through plasma membranes are low, and transporter-mediated uptake is the major route of drug influx. Thus, inefficient cellular uptake is a potential mechanism of resistance to anticancer nucleoside drugs. Nucleoside transport deficiency is a well-established mechanism of resistance to the anticancer nucleosides in cultured cell lines (Cohen *et al.*, 1979; Cass *et al.*, 1981; Aronow *et al.*, 1985; Belt and Noel, 1988; Ullman *et al.*, 1988; Ullman, 1989b; Hoffman, 1991; Aran and Plagemann, 1992). Results from clinical studies with cytarabine suggests that nucleoside transport deficiency may also be a mechanism of resistance *in vivo*. A comprehensive review of the role of nucleoside transporters in resistance to nucleoside drugs is available (Mackey *et al.*, 1998a).

One of the earliest examples of drug resistance caused by a deficiency in nucleoside transport was the AE1 clone, which was derived from S49 murine T-cell lymphoma cells by chemical mutagenization and subsequent selection in media containing toxic concentrations of adenosine (Cohen *et al.*, 1979; Cass *et al.*, 1981). AE1 cells exhibit greatly reduced uptake of physiologic nucleosides and high-level resistance to cytotoxic nucleosides, including cytarabine, 5-fluoro-2′-deoxyuridine, 5-fluorouridine (Cohen *et al.*, 1979), and gemcitabine (Mackey *et al.*, 1998b). AE1 cells lack *es*/mENT1-mediated transport activity and NBMPR-binding sites, suggesting a mutational loss of functional transporter protein. A similar transport-related resistant human cell line, ARAC-8C, was obtained from human T-lymphoblast CCRF-CEM cells that were selected for resistance to cytarabine (Ullman *et al.*, 1988; Ullman, 1989a). The parental CCRF-CEM cells exhibit primarily *es*/hENT1-mediated transport activity. ARAC-8C cells, like AE1 cells, appear to be resistant by virtue of an absence of functional hENT1 protein since they completely lack NBMPR-sensitive nucleoside transport activity and high-affinity NBMPR-binding sites, and both cell lines exhibit cross-resistance to a broad spectrum of nucleoside drugs. The latter characteristic is consistent with the broad permeant selectivity of the *es*/hENT1-mediated process. A cultured line of murine erythroleukemia cells selected for resistance to a cytotoxic adenosine analogue also exhibit markedly reduced levels of *es*/mENT1-mediated activity (Hoffman, 1991). High-level resistance to 5-fluoro-2′-deoxyuridine in human HCT-3 colon cancer cells is accompanied by an absence of drug uptake and NBMPR binding to resistant cells, with only minimal differences in activating or target enzymatic activities between the sensitive and resistant cells (Sobrero *et al.*, 1985, 1990). A deficiency in nucleoside transport, produced either pharmacologically or genetically, provides up to 1800-fold protection from gemcitabine cytotoxicity in cultured cell lines (Mackey *et al.*, 1998b). Although the molecular mechanisms have not been established, it is likely that mutations that have impaired ENT1 production and/or function are responsible for the transport deficiencies observed in many of the genetically resistant cell lines.

The role of nucleoside transport deficiency in clinical drug resistance is less clear, in large part because of the difficulties of performing transport studies on malignant cells derived from clinical specimens and of quantifying transporter abundance in malignant clones admixed with normal cells. Relationships between the cytotoxic effects of cytarabine and levels of *es*/hENT1-mediated transport processes have been shown in studies conducted with leukemic cells *in vitro*. The abundance of the *es*/hENT1 protein on the surface of freshly isolated human leukemic cells was estimated by flow cytometry with an impermeant fluorescent ENT inhibitor [5′-*S*-(2-aminoethyl)-N^6-(4-nitrobenzyl)-5′-thioadenosine] and was found to be positively correlated with sensitivity to cytarabine (Gati and Paterson, 1997). The rate of accumulation of ara-CTP is also positively correlated with total numbers of cell-associated NBMPR-binding sites in human leukemic blasts, and patients with the lowest rates of cytarabine uptake and NBMPR-binding site numbers were among those who failed therapy (Wiley *et al.*, 1985; Young *et al.*, 1985). At low cytarabine concentrations (<1 μM), *es*/hENT1-mediated transport is the rate-limiting step in ara-CTP incorporation into freshly isolated human leukemic blasts (White *et al.*, 1987). In a single patient with T-cell lymphoblastic leukemia, comparison of leukemic blasts prior to cytarabine therapy and after subsequent relapse revealed a 75% reduction in accumulation of ara-CTP accompanied by a reduction in NBMPR binding (Wiley *et al.*, 1987), suggesting that cytarabine chemotherapy had selected variant cells with reduced *es*/hENT1 activity. The role of nucleoside transport deficiency in clinical resistance to fludarabine, cladribine, and gemcitabine has not been explored.

5.2.3. STRATEGIES TO ENHANCE THERAPEUTIC INDEX

There are several strategies based on manipulation of nucleoside transport activity, by either increasing uptake by target malignant cells or decreasing uptake by dose-limiting normal cells, that could potentially increase the therapeutic index of anticancer nucleosides. One approach is to specifically enhance nucleoside transport capability in malignant cells by upregulating gene expression and/or by stimulating transport activity. There is some evidence that the level of *es*/hENT1-mediated transport varies with proliferation rate and in response to cytotoxic and mitogenic stimuli. High proliferation rates have been associated with increased NBMPR-binding site numbers in human thymocytes and leukemic blasts (Smith *et al.*, 1989; Wiley *et al.*, 1989). Although stimulation of freshly isolated myloid leukemic blasts with granulocyte-macrophage colony-stimulating factor increases *es*/hENT1 abundance and proliferation rates in some patients (Wiley *et al.*, 1994), "priming" of patients with acute myeloid leukemia with recombinant granulocyte-macrophage colony-stimulating factor does not enhance sensitivity to cytarabine (Lowenberg *et al.*, 1997). However, more information about mechanisms of regulation of the human ENT and CNT proteins is needed to design clinical strategies that will increase sensitivity of malignant cells to nucleoside drugs.

Results from studies with model systems suggest that inhibitors of the ENT-mediated processes (e.g., NBMPR, dipyridamole) can be used to protect cells selectively in dose-limiting normal tissues from toxicity by preventing cellular uptake of nucleoside drugs. It has long been known that inhibitors of equilibrative transport protect cultured cancer cells from the cytotoxic effects of a variety of cytotoxic nucleosides (Warnick *et al.*, 1972; Paterson *et al.*, 1979), and normal hematopoietic progenitor cells derived from human bone marrow are protected from the toxicity of *in vitro* exposures to tubercidin (7-deazaadenosine) by co-administration of NBMPR, dilazep, or dipyridamole (Cass *et al.*, 1992). If hematopoietic stem cells possess predominantly *es*/hENT1 activity (Allay *et al.*, 1997), greater therapeutic selectivity could possibly be achieved by protecting hematopoietic cells from particular nucleoside drugs with ENT inhibitors if the target malignant cells possess other nucleoside transporters for which the drug in question is a permeant.

An alternate strategy based on the use of inhibitors of the ENT-mediated processes is the enhancement of toxicity against malignant cells by selectively trapping nucleoside drugs inside cells by inhibition of drug efflux (Dagnino and Paterson, 1990). Dipyridamole enhances the cytotoxicity of vidarabine (9-β-D-arabinofuranosyladenine) in L1210/C2 cells, which have *es*/mENT1, *ei*/mENT2, and *cif*/mCNT2 transport activities, by causing intracellular accumulation of drug above steady-state levels because of (i) the inhibition of drug efflux via the bidirectional equilibrative transporters, and (ii) continuation of influx via the concentrative transporter. Thus, it may be possible to sensitize malignant cells that have the appropriate concentrative transporter to particular anticancer nucleosides by sequential administration of anticancer nucleoside, followed by ENT inhibitor, to selectively trap drug inside the malignant cells.

ENT inhibitors sometimes enhance the cytotoxicity of non-nucleoside anticancer drugs by blocking nucleoside salvage in malignant cells. Because most cells have *de novo* synthesis pathways for nucleotides, ENT inhibitors are not inherently cytotoxic, except in circumstances where cells have been made dependent upon nucleoside salvage. The antifolates methotrexate and trimetrexate inhibit dihydrofolate reductase and lead to impaired synthesis of purines and pyrimidine nucleotides, with dose-limiting toxicities of myelosuppression and mucositis. Antifolate cytotoxicity is enhanced by inhibition of ENT-mediated nucleoside salvage in cultured cells; for example, human colon cancer cells are rescued from methotrexate toxicity by thymidine and hypoxanthine in the absence, but not the presence, of dipyridamole, which potentiates the growth-inhibitory effects of methotrexate by blocking cellular uptake of thymidine (Van Mouwerik *et al.*, 1987). However, when patients are treated with graded dosages of methotrexate and high-dose dipyridamole administered by infusion, moderate to severe myelosuppression and/or mucositis occurs more frequently with the combination than with methotrexate alone (Willson *et al.,* 1989). The myelosuppression induced by antifolates is largely due to toxicity to cells in late stages of hematopoietic development, while early myeloid progenitor cells are relatively resistant. Although murine hematopoietic progenitor cells are spared from trimetrexate toxicity, if given access to thymidine and hypoxanthine, exposure to ENT inhibitors either *in vitro* or *in vivo* prevents thymidine rescue from trimetrexate toxicity (Allay *et al.,* 1997). These results suggest that myelotoxicity may limit the potential use of ENT inhibitors in combination with antifolates.

6. SUMMARY

In this review, we have summarized recent advances in our understanding of the biology of nucleoside transport arising from new insights provided by the isolation and functional expression of cDNAs encoding the major nucleoside transporters of mammalian cells. Nucleoside transporters are required for permeation of nucleosides across biological membranes and are present in the plasma membranes of most cell types. There is growing evidence that functional nucleoside transporters are required for translocation of nucleosides between intracellular compartments and thus are also present in organellar membranes. Functional studies during the 1980s established that nucleoside transport in mammalian cells occurs by two mechanistically distinct processes, facilitated diffusion and Na^+-nucleoside cotransport. The determination of the primary amino acid sequences of the equilibrative and concentrative transporters of human and rat cells has provided a structural basis for the functional differences among the different transporter subtypes. Although nucleoside transporter proteins were first purified from human erythrocytes a decade ago, the low abundance of nucleoside transporter proteins in membranes of mammalian cells has hindered analysis of relationships between transporter structure and function. The molecular cloning of cDNAs encoding nucleoside transporters and the development of heterologous expression systems for production of recombinant nucleoside transporters, when combined with recombinant DNA technologies, provide powerful tools for characterization of functional domains within transporter proteins that are involved in nucleoside recognition and translocation. As relationships between molecular structure and function are determined, it should be possible to develop new approaches for optimizing the transportability of nucleoside drugs into diseased tissues, for development of new transport inhibitors, including reagents that are targeted to the concentrative transporters, and, eventually, for manipulation of transporter function through an understanding of the regulation of transport activity.

REFERENCES

Allay, J. A., Spencer, H. T., Wilkinson, S. L., Belt, J. A., Blakley, R. L., and Sorrentino, B. P., 1997, Sensitization of hematopoietic stem and progenitor cells to trimetrexate using nucleoside transport inhibitors, *Blood* **90**:3546–3554.

Anderson, C. M., Xiong, W., Young, J. D., Cass, C. E., and Parkinson, F. E., 1996, Demonstration of the existence of mRNAs encoding N1/cif and N2/cit sodium/nucleoside cotransporters in rat brain, *Brain Res. Mol. Brain Res.* **42**:358–361.

Aran, J. M., and Plagemann, P. G., 1992, Nucleoside transport-deficient mutants of PK-15 pig kidney cell line, *Biochim. Biophys. Acta* **1110**:51–58.

Aronow, B., Allen, K., Patrick, J., and Ullman, B., 1985, Altered nucleoside transporters in mammalian cells selected for resistance to the physiological effects of inhibitors of nucleoside transport, *J. Biol. Chem.* **260**:6226–6233.

Belt, J. A., and Noel, L. D., 1988, Isolation and characterization of a mutant of L1210 murine leukemia deficient in nitrobenzylthioinosine-insensitive nucleoside transport, *J. Biol. Chem.* **263**:13819–13822.

Belt, J. A., Harper, E. H., Byl, J. A., and Noel, L. D., 1992, Sodium-dependent nucleoside transport in human myeloid leukemic cell lines and freshly isolated myeloblasts, *Proc. Am. Assoc. Cancer Res.* **33**:20.

Belt, J. A., Marina, N. M., Phelps, D. A., and Crawford, C. R., 1993, Nucleoside transport in normal and neoplastic cells, *Adv. Enzyme Regul.* **33**:235–252.

Boleti, H., Coe, I. R., Baldwin, S. A., Young, J. D., and Cass, C. E., 1997, Molecular identification of the equilibrative NBMPR-sensitive (es) nucleoside transporter and demonstration of an equilibrative NBMPR-insensitive (ei) transport activity in human erythroleukemia (K562) cells, *Neuropharmacology* **36**:1167–1179.

Bolwell, B. J., Cassileth, P. A., and Gale, R. P., 1988, High dose cytarabine: A review, *Leukemia* **2**:253–260.

Boumah, C. E., Hogue, D. L., and Cass, C. E., 1992, Expression of high levels of nitrobenzylthioinosine-sensitive nucleoside transporrt in cultured human choriocarcinoma (BeWo) cells, *Biochem. J.* **288**:987–996.

Brogden, R. N., and Sorkin, E. M., 1993, Pentostatin. A review of its pharmacodynamic and pharmacokinetic properties, and therapeutic potential in lymphoproliferative disorders, *Drugs* **46**:652–677.

Budman, D. R., Meropol, N. J., Reigner, B., Creaven, P. J., Lichtman, S. M., Berghorn, E., Behr, J., Gordon, R. J., Osterwalder, B., and Griffin, T., 1998, Preliminary studies of a novel oral fluoropyrimidine carbamate: Capecitabine, *J. Clin. Oncol.* **16**:1795–1802.

Buolamwini, J. K., 1997, Nucleoside transport inhibitors: Structure–activity relationships and potential therapeutic applications, *Curr. Med. Chem.* **4**:35–66.

Burnstock, G., 1978, A basis for distinguishing two types of purinergic receptor, in: *Cell Membrane Receptors of Drugs and Hormones: A Multidisciplinary Approach* (R. W. Straug and L. Bolis, eds.), Raven Press, New York, pp. 107–118.

Cabantchik, Z. I., 1989, Nucleoside transport across red cell membranes, *Meth. Enzymol.* **173**:250–263.

Camins, A., Jimenez, A., Sureda, F. X., Pallas, M., Escubedo, E., and Camarasa, J., 1996,

Characterization of nitrobenzylthioinosine binding sites in the mitochondrial fraction of rat testis, *Life Sci.* **58:**753–759.

Carmichael, J., and Walling, J., 1996, Phase II activity of gemcitabine in advanced breast cancer, *Semin. Oncol.* **23:**77–81.

Carmichael, J., Fink, U., Russell, R. S., Spittle, M. F., Harris, A. L., Spiessi, G., and Blatter, J., 1996, Phase II study of gemcitabine in patients with advanced pancreatic cancer, *Br. J. Cancer* **73:**101–105.

Casari, G., Andrade, M. A., Bork, P., Boyle, J., Daruvar, A., Ouzounis, C., Schneider, R., Tamames, J., Valencia, A., and Sander, C., 1995, Challenging times for bioinformatics, *Nature* **376:**647–648.

Cass, C. E., 1995, Nucleoside transport, in: *Drug Transport in Antimicrobial Therapy and Anticancer Therapy* (N. H. Georgopapadakou, ed.), Marcel Dekker, New York, pp. 403–451.

Cass, C. E., Kolassa, N., Uehara, Y., Dahlig-Harley, E., Harley, E. R., and Paterson, A. R., 1981, Absence of binding sites for the transport inhibitor nitrobenzylthioinosine on nucleoside transport-deficient mouse lymphoma cells, *Biochim. Biophys. Acta* **649:**769–777.

Cass, C. E., King, K. M., Montano, J. T., and Janowska-Wieczorek, A., 1992, A comparison of the abilities of nitrobenzylthioinosine, dilazep, and dipyridamole to protect human hematopoietic cells from 7-deazaadenosine (tubercidin), *Cancer Res.* **52:**5879–5886.

Che, M., Nishida, T., Gatmaitan, Z., and Arias, I. M., 1992, A nucleoside transporter is functionally linked to ectonucleotidases in rat liver canalicular membrane, *J. Biol. Chem.* **267:**9684–9688.

Che, M., Ortiz, D. F., and Arias, I. M., 1995, Primary structure and functional expression of a cDNA encoding the bile canalicular, purine-specific Na^+-nucleoside cotransporter, *J. Biol. Chem.* **270:**13596–13599.

Cheson, B. D., Vena, D. A., Foss, F. M., and Sorensen, J. M., 1994, Neurotoxicity of purine analogs: A review, *J. Clin. Oncol.* **12:**2216–2228.

Coe, I. R., Griffiths, M., Young, J. D., Baldwin, S. A., and Cass, C. E., 1997, Assignment of the human equilibrative nucleoside transporter (hENT1) to 6p21.1–p21.2, *Genomics* **45:**459–460.

Cohen, A., Ullman, B., and Martin, D. W., Jr., 1979, Characterization of a mutant mouse lymphoma cell with deficient transport of purine and pyrimidine nucleosides, *J. Biol. Chem.* **254:**112–116.

Conant, A. R., and Jarvis, S. M., 1994, Nucleoside influx and efflux in guinea-pig ventricular myocytes. Inhibition by analogues of lidoflazine, *Biochem. Pharmacol.* **48:**873–880.

Cortes, J. E., Talpaz, M., and Kantarjian, H., 1996, Chronic myelogenous leukemia: A review, *Am. J. Med.* **100:**555–570.

Craig, J. E., Zhang, Y., and Gallagher, M. P., 1994, Cloning of the *nupC* gene of *Escherichia coli* encoding a nucleoside transport system, and identification of an adjacent insertion element, IS 186, *Mol. Microbiol.* **11:**1159–1168.

Crawford, C. R., Patel, D. H., Naeve, C., and Belt, J. A., 1998, Cloning of the human equilibrative, nitrobenzylmercaptopurine riboside (NBMPR)-insensitive nucleoside transporter ei by functional expression in a transport-deficient cell line, *J. Biol. Chem.* **273:**5288–5293.

Crawford, C. R., Cass, C. E., Young, J. D., and Belt, J. A., n.d., Stable expression of a recombinant sodium-dependent, pyrimidine-selective nucleoside transporter (rCNT1) in a transport-deficient mouse leukemia cell line, *Biochem. Cell Biol.,* **76**:843–851.

Cronauer, M. V., Klocker, H., Talasz, H., Geisen, F. H., Hobisch, A., Radmayr, C., Bock, G., Culig, Z., Schirmer, M., Reissigl, A., Bartsch, G., and Konwalinka, G., 1996, Inhibitory effects of the nucleoside analogue gemcitabine on prostatic carcinoma cells, *Prostate* **28**:172–181.

Dagnino, L., and Paterson, A. R., 1990, Sodium-dependent and equilibrative nucleoside transport systems in L1210 mouse leukemia cells: Effect of inhibitors of equilibrative systems on the content and retention of nucleosides, *Cancer Res.* **50**:6549–6553.

Deckert, J., Hennemann, A., Bereznai, B., Fritze, J., Vock, R., Marangos, P. J., and Riederer, P., 1994, (3H)dipyridamole and (3H)nitrobenzylthioinosine binding sites at the human parietal cortex and erythrocyte adenosine transporter: A comparison, *Life Sci.* **55**:1675–1682.

Eisenhauer, E. A., and Vermorken, J. B., 1996, New drugs in gynecologic oncology, *Curr. Opin. Oncol.* **8**:408–414.

Fagbemi, S. O., and Stadler, W. M., 1998, New chemotherapy regimens for advanced bladder cancer, *Semin. Urol. Oncol.* **16**:23–29.

Fang, X., Parkinson, F. E., Mowles, D. A., Young, J. D., and Cass, C. E., 1996, Functional characterization of a recombinant sodium-dependent nucleoside transporter with selectivity for pyrimidine nucleosides (cNT1rat) by transient expression in cultured mammalian cells, *Biochem. J.* **317**:457–465.

Felipe, A., Valdes, R., Santo, B., Lloberas, J., Casado, J., and Pastor-Anglada, M., 1998, Na^+-dependent nucleoside transport in liver: Two different isoforms from the same gene family are expressed in liver cells, *Biochem. J.* **330**:997–1001.

Flanagan, S. A., and Meckling-Gill, K. A., 1997, Characterization of a novel Na^+-dependent, guanosine specific, nitrobenzylthioinosine sensitive transporter in acute promyelocytic leukemia cells, *J. Biol. Chem.* **272**: 18026–18032.

Fredholm, B. B., Abbracchio, M. P., Burnstock, G., Daly, J. W., Harden, T. K., Jacobson, K. A., Leff, P., and Williams, M., 1994, Nomenclature and classification of purinoceptors. *Pharmacol. Rev.* **46**:143–156.

Gati, W. P., and Paterson, A. R. P., 1989, Nucleoside transport, in *Red Blood Cell Membranes* (P. Agre and J. C. Parker, eds.), Marcel Dekker, New York, pp. 635–661.

Gati, W. P., and Paterson, A. R. P., 1997, Measurement of nitrobenzylthioinosine in plasma and erythrocytes: A pharmacokinetic study in mice, *Cancer Chemother. Pharmacol.* **40**:342–346.

Geiger, D. D., and Fyda, D. M., 1991, Adenosine Transport in Nervous System Tissues, in: *Adenosine in the Nervous System* (T. Stone, ed.), London, pp. 1–23.

Griffith, D. A., and Jarvis, S. M., 1996, Nucleoside and nucleobase transport systems of mammalian cells, *Biochim. Biophys. Acta Rev. Biomembr.* **1286**:153–181.

Griffiths, M., Beaumont, N., Yao, S. Y., Sundaram, M., Boumah, C. E., Davies, A., Kwong, F. Y., Coe, I., Cass, C. E., Young, J. D., and Baldwin, S. A., 1997a, Cloning of a human nucleoside transporter implicated in the cellular uptake of adenosine and chemotherapeutic drugs, *Nature Med.* **3**:89–93.

Griffiths, M., Yao, S.Y. M., Abidi, F., Phillips, S. E. V., Cass, C. E., Young, J. D., and Baldwin, S. A., 1997b, Molecular cloning and characterization of a nitrobenzylthioinosine-insensitive (ei) equilibrative nucleoside transporter from human placenta, *Biochem. J.* **328:**739–743.

Guieu, R., Couraud, F., Pouget, J., Sampieri, F., Bechis, G., and Rochat, H., 1996, Adenosine and the nervous system: Clinical implications, *Clin. Neuropharmacol.* **19:**459–474.

Gutierrez, M. M., and Giacomini, K. M., 1993, Substrate selectivity, potential sensitivity and stoichiometry of Na(+)-nucleoside transport in brush border membrane vesicles from human kidney, *Biochim. Biophys. Acta* **1149:**202–208.

Gutierrez, M. M., and Giacomini, K. M., 1994, Expression of a human renal sodium nucleoside cotransporter in *Xenopus laevis* oocytes, *Biochem. Pharmacol.* **48:**2251–2253.

Gutierrez, M. M., Brett, C. M., Ott, R. J., Hui, A. C., and Giacomini, K. M., 1992, Nucleoside transport in brush border membrane vesicles from human kidney, *Biochim. Biophys. Acta* **1105:**1–9.

Hamilton, S. R., Yao, S. Y. M., Ingram, J., Henderson, P. J. F., Gallagher, M. P., Cass, C. E., Young, J. D., and Baldwin, S. A., 1997, Anti-peptide antibodies as probes of the structure and subcellular distribution of the Na$^+$-dependent nucleoside transporter rCNT1, *J. Physiol.* (Lond.) **499P:**P50–P51.

Hiddemann, W., 1991, cytosine arabinoside in the treatment of acute myeloid leukemia: The role and place of high-dose regimens, *Ann. Hematol.* **62:**119–128.

Hoffman, J., 1991, Murine erythroleukemia cells resistant to periodate-oxidized adenosine have lowered levels of nucleoside transporter, *Adv. Exp. Med. Biol.* **309A:**443–446.

Hoffman, M. A., 1996, Cladribine for the treatment of indolent non-Hodgkin's lymphomas, *Semin. Hematol.* **33:**40–44.

Hogue, D. L., Ellison, M. J., Young, J. D., and Cass, C. E., 1996, Identification of a novel membrane transporter associated with intracellular membranes by phenotypic complementation in the yeast *Saccharomyces cerevisiae*, *J. Biol. Chem.* **271:**9801–9808.

Hogue, D. L., Ellison, M. J., Vickers, M., and Cass, C. E., 1997, Functional complementation of a membrane transport deficiency in *Saccharomyces cerevisiae* by recombinant ND4 fusion protein, *Biochem. Biophys. Res. Commun.* **238:**811–816.

Huang, Q. Q., Harvey, C. M., Paterson, A. R., Cass, C. E., and Young, J. D., 1993, Functional expression of Na(+)-dependent nucleoside transport systems of rat intestine in isolated oocytes of *Xenopus laevis*. Demonstration that rat jejunum expresses the purine-selective system N1 (cif) and a second, novel system N3 having broad specificity for purine and pyrimidine nucleosides, *J. Biol. Chem.* **268:**20613–20619.

Huang, Q. Q., Yao, S. Y., Ritzel, M. W., Paterson, A. R., Cass, C. E., and Young, J. D., 1994, Cloning and functional expression of a complementary DNA encoding a mammalian nucleoside transport protein, *J. Biol. Chem.* **269:**17757–17760.

Jacobson, K. A., van Galen, P. J., and Williams, M., 1992, Adenosine receptors: Pharmacology, structure–activity relationships, and therapeutic potential, *J. Med. Chem.* **35:**407–422.

Jaiyesimi, I. A., Kantarjian, H. M., and Estey, E. H., 1993, Advances in therapy for hairy cell leukemia. A review, *Cancer* **72:**5–16.

Jarvis, S. M., 1986, Nitrobenzylthioinosine-sensitive nucleoside transport system: Mechanism of inhibition by dipyridamole, *Mol. Pharmacol.* **30:**659–665.

Jarvis, S. M., 1988, Adenosine transporters, in: *Adenosine Receptors* (D. M. F. Cooper and C. Londos, eds.), Alan R. Liss, New York, pp. 113–123.

Jarvis, S. M., and Young, J. D., 1987, Photoaffinity labelling of nucleoside transport peptides, *Pharmacol. Therapeut.* **32**:339–359.

Jarvis, S. M., McBride, D., and Young, J. D., 1982, Erythrocyte nucleoside transport: Asymmetrical binding of nitrobenzylthioinosine to nucleoside permeation sites, *J. Physiol.* **324**:31–46.

Keating, M. J., O'Brien, S., Plunkett, W., Robertson, L. E., Gandhi, V., Estey, E., Dimopoulos, M., Cabanillas, F., Kemena, A., and Kantarjian, H., 1994, Fludarabine phosphate: A new active agent in hematologic malignancies, *Semin. Hematol.* **31**:28–39.

Keating, M. J., O'Brien, S., McLaughlin, P., Dimopoulos, M., Gandhi, V., Plunkett, W., Lerner, S., Kantarjian, H., and Estey, E., 1996, Clinical experience with fludarabine in hemato-oncology, *Hematol. Cell. Ther.* **38**(Suppl. 2):S83–S91.

Kwong, F. Y., Davies, A., Tse, C. M., Young, J. D., Henderson, P. J., and Baldwin, S. A., 1988, Purification of the human erythrocyte nucleoside transporter by immunoaffinity chromatography, *Biochem. J.* **255**:243–249.

Kwong, F. Y., Fincham, H. E., Davies, A., Beaumont, N., Henderson, P. J., Young, J. D., and Baldwin, S. A., 1992, Mammalian nitrobenzylthioinosine-sensitive nucleoside transport proteins. Immunological evidence that transporters differing in size and inhibitor specificity share sequence homology, *J. Biol. Chem.* **267**:21954–21960.

Kwong, F. Y., Wu, J. S., Shi, M. M., Fincham, H. E., Davies, A., Henderson, P. J., Baldwin, S. A., and Young, J. D., 1993, Enzymic cleavage as a probe of the molecular structures of mammalian equilibrative nucleoside transporters, *J. Biol. Chem.* **268**:22127–22134.

Lasley, R. D., and Mentzer, R. M., Jr., 1996, Myocardial protection: The adenosine story, *Drug Dev. Res.* **39**:314–318.

Lowenberg, B., Suciu, S., Archimbaud, E., Ossenkoppele, G., Verhoef, G. E., Vellenga, E., Wijermans, P., Berneman, Z., Dekker, A. W., Stryckmans, P., Schouten, H., Jehn, U., Muus, P., Sonneveld, P., Dardenne, M., and Zittoun, R., 1997, Use of recombinant GM-CSF during and after remission induction chemotherapy in patients aged 61 years and older with acute myeloid leukemia: Final report of AML-11, a phase III randomized study of the Leukemia Cooperative Group of European Organisation for the Research and Treatment of Cancer and the Dutch Belgian Hemato-Oncology Cooperative Group, *Blood* **90**:2952–2961.

Mackey, J. R., Baldwin, S. A., Young, J. D., and Cass, C. E., 1998a, Nucleoside transport and its significance for anticancer drug resistance, *Drug Resistance Updates* **1**:310–324.

Mackey, J. R., Mani, R. S., Selner, M., Mowles, D., Young, J. D., Belt, J. A., Crawford, C. R., and Cass, C. E., 1998b, Functional nucleoside transporters are required for gemcitabine and cytotoxicity in cancer cell lines, *Cancer Res.* **58**:4349–4357.

Mani, R. S., Hammond, J. R., Marjan, J. M. J., Graham, K. M., Young, J. D., Baldwin, S. A., and Cass, C. E., 1998, Demonstration of equilibrative nucleoside transporters (hENT1, hENT2) in nuclear envelopes of cultured human choriocarcinoma (BeWo) cells by functional reconstitution in proteoliposomes. *J. Biol. Chem.* **273**:30818–30825.

Miller, K. B., Kim, K., Morrison, F. S., Winter, J. N., Bennett, J. M., Neiman, R. S., Head, D. R., Cassileth, P. A., O'Connell, M. J., and Kyungmann, K., 1992, The evaluation of low-dose cytarabine in the treatment of myelodysplastic syndromes: A phase-III intergroup study, *Ann. Hematol.* **65:**162–168; Erratum, *Ann. Hematol.* **66:**164 (1993).

Mitra, R. S., and Bernstein, I. A., 1970, Thymidine incorporation into deoxyribonucleic acid by isolated rat liver mitochondria, *J. Biol. Chem.* **245:**1255–1260.

Olah, M. E., and Stiles, G. L., 1995, Adenosine receptor subtypes: Characterization and therapeutic regulation, *Annu. Rev. Pharmacol. Toxicol.* **35:**581–606.

Osses, N., Pearson, J. D., Yudilevich, D. L., and Jarvis, S. M., 1996, Hypoxanthine enters human vascular endothelial cells (ECV 304) via the nitrobenzylthioinosine-insensitive equilibrative nucleoside transporter, *Biochem. J.* **317:**843–848.

Pajor, A. M., and Wright, E. M., 1992, Cloning and functional expression of a mammalian Na^+-nucleoside cotransporter. A member of the SGLT family, *J. Biol. Chem.* **267:**3557–3560.

Parkinson, F. E., Rudolphi, K. A., and Fredholm, B. B., 1994, Propentofylline: A nucleoside transport inhibitor with neuroprotective effects in cerebral ischemia, *Gen. Pharmacol.* **25:**1053–1058.

Patel, D., Crawford, C., Naeve, C., and Belt, J., 1997, Molecular cloning and characterization of a mouse purine-selective concentrative nucleoside transporter, *Proc. Am. Assoc. Cancer Res.* **38:**A405 (Abstract).

Paterson, A. R. P., and Cass, C. E. (1986). Transport of nucleoside drugs in animal cells, in: *Membrane Transport of Antineoplastic Agents* (I. D. Goldman, ed.), Pergamon Press, Oxford, pp. 309–329.

Paterson, A. R., Yang, S. E., Lau, E. Y., and Cass, C. E., 1979, Low specificity of the nucleoside transport mechanism of RPMI 6410 cells, *Mol. Pharmacol.* **16:**900–908.

Paterson, A. R., Lau, E. Y., Dahlig, E., and Cass, C. E., 1980, A common basis for inhibition of nucleoside transport by dipyridamole and nitrobenzylthioinosine? *Mol. Pharmacol.* **18:**40–44.

Paterson, A. R., Gati, W. P., Vijayalakshmi, D., Cass, C. E., Mant, M. J., Young, J. D., and Belch, A. R., 1993, Inhibitor-sensitive, Na(+)-linked transport of nucleoside analogs in leukemia cells from patients, *Proc. Am. Assoc. Cancer Res.* **34:**A84 (Abstract).

Piro, L. D., 1996, Cladribine in the treatment of low-grade non-Hodgkin's lymphoma, *Semin. Hematol.* **33:**34–39.

Pisoni, R. L., and Thoene, J. G., 1989, Detection and characterization of a nucleoside transport system in human fibroblast lysosomes, *J. Biol. Chem.* **264:**4850–4856.

Pisoni, R. L., and Thoene, J. G., 1991, The transport systems of mammalian lysosomes, *Biochim. Biophys. Acta* **1071:**351–373.

Plagemann, P. G., Wohlhueter, R. M., and Woffendin, C., 1988, Nucleoside and nucleobase transport in animal cells, *Biochim. Biophys. Acta* **947:**405–443.

Plunkett, W., and Gandhi, V., 1994, Evolution of the arabinosides and the pharmacology of fludarabine, *Drugs* **47:**30–38.

Polliack, A., 1997, Hairy cell leukemia and allied chronic lymphoid leukemias: Current knowledge and new therapeutic options, *Leuk. Lymphoma* **26**(Suppl. 1):41–51.

Preti, A., and Kantarjian, H. M., 1994, Management of adult acute lymphocytic leukemia: Present issues and key challenges, *J. Clin. Oncol.* **12:**1312–1322.

Ritzel, M. W. L., Yao, S. Y. M., Huang, M. Y., Elliott, J. F., Cass, C. E., and Young, J. D.,

1997, Molecular cloning and functional expression of cDNAs encoding a human Na$^+$-nucleoside cotransporter (hCNT1), *Am. J. Physiol.* **272**:C707–C714.

Ritzel, M. W. L., Yao, S. Y. M., Ng, A. M. L., Mackey, J. R., Cass, C. E., and Young, J. D., 1998, Molecular cloning, functional expression and chromosomal localization of a cDNA encoding a human Na$^+$ nucleoside cotransporter (hCNT2) selective for purine nucleosides and uridine, *Mol. Membr. Biol.* **15**:203–211.

Ross, D. D., 1991, Cellular and pharmacologic aspects of drug resistance in acute myeloid leukemia, *Curr. Opin. Oncol.* **3**:21–29.

Saven, A., 1996, The Scripps Clinic experience with cladribine (2-CdA) in the treatment of chronic lymphocytic leukemia, *Semin. Hematol.* **33**:28–33.

Schaner, M. E., Wang, J., Zevin, S., Gerstin, K. M., and Giacomini, K. M., 1997, Transient expression of a purine-selective nucleoside transporter (SPNTint) in a human cell line (HeLa), *Pharmaceut. Res.* **14**:1316–1321.

Smith, C. L., Pilarski, L. M., Egerton, M. L., and Wiley, J. S., 1989, Nucleoside transport and proliferative rate in human thymocytes and lymphocytes, *Blood* **74**:2038–2042.

Sobrero, A. R., Moir, R. D., Bertino, J. R., and Handschumacher, R. E., 1985, Defective facilitated diffusion of nucleosides, a primary mechanism of resistance to 5-fluoro-2'-deoxyuridine in the HCT-8 human carcinoma line, *Cancer Res.* **45**:3155–3160.

Sobrero, A., Aschele, C., Guglielmi, A., Nobile, M. T., and Rosso, R., 1990, Resistance to 5-fluorouracil and 5-fluoro-2'-deoxyuridine mechanisms and clinical implications, *J. Chemother.* **2**:12–16.

Sundaram, M., Yao, S. Y. M., Cass, C. E., Baldwin, S. A., and Young, J. D., 1998, Chimaeric constructs between human and rat equilibrative nucleoside transporters (hENT1 and rENT1) reveal hENT1 structural domains interacting with coronary vasoactive drugs, *J. Biol. Chem.* **273**:21519–21525.

Tallman, M. S., and Hakimian, D., 1996, Current results and prospective trials of cladribine in chronic lymphocytic leukemia, *Semmin. Hematol.* **33**:23–27.

Thorn, J. A., and Jarvis, S. M., 1996, Adenosine transporters, *Gen. Pharmacol.* **27**:613–620.

Ullman, B., 1989a, Dideoxycytidine metabolism in wild type and mutant CEM cells deficient in nucleoside transport or deoxycytidine kinase, *Adv. Exp. Med. Biol.* **253B**:415–420.

Ullman, B., 1989b, Mutational analysis of nucleoside and nucleobase transport, in: *Resistance to Antineoplastic Drugs,* CRC Press, Boca Raton, Florida, pp. 293–315.

Ullman, B., Coons, T., Rockwell, S., and McCartan, K., 1988, Genetic analysis of 2',3'-dideoxycytidine incorporation into cultured human T lymphoblasts, *J. Biol. Chem.* **263**:12391–12396.

Van Belle, H., 1993a, Nucleoside transport inhibition for cardioprotection: New perspectives for the clinic, *Drug Dev. Res.* **28**:344–348.

Van Belle, H., 1993b, Nucleoside transport inhibition: A therapeutic approach to cardioprotection via adenosine? *Cardiovasc. Res.* **27**:68–76.

Van Mouwerik, T. J., Pangallo, C. A., Willson, J. K., and Fischer, P. H., 1987, Augmentation of methotrexate cytotoxicity in human colon cancer cells achieved through inhibition of thymidine salvage by dipyridamole, *Biochem. Pharmacol.* **36**:809–814.

Vijayalakshmi, D., and Belt, J. A., 1988, Sodium-dependent nucleoside transport in mouse

intestinal epithelial cells. Two transport systems with differing substrate specificities, *J. Biol. Chem.* **263**:19419–19423.
Wang, J., and Giacomini, K. M., 1997, Molecular determinants of substrate selectivity in Na$^+$-dependent nucleoside transporters, *J. Biol. Chem.* **272**:28845–28848.
Wang, J., Schaner, M. E., Thomassen, S., Su, S. F., Piquette-Miller, M., and Giacomini, K. M., 1997a, Functional and molecular characteristics of Na(+)-dependent nucleoside transporters, *Pharmaceut. Res.* **14**:1524–1532.
Wang, J., Su, S.-F., Dresser, M. J., Schnaner, M. E., Washington, C. B., and Giacomini, K. M., 1997b, Na$^+$-dependent purine nucleoside transporter from human kidney: Cloning and functional characterization, *Am. J. Physiol.* **273**:F1058–F1065.
Warnick, C. T., Muzik, H., and Paterson, A. R., 1972, Interference with nucleoside transport in mouse lymphoma cells proliferating in culture, *Cancer Res.* **32**:2017–2022.
Watkins, L. F., and Lewis, R. A., 1987, The metabolism of deoxyguanosine in mitochondria. Characterization of the uptake process, *Mol. Cell. Biochem.* **77**:71–77.
White, J. C., Rathmell, J. P., and Capizzi, R. L., 1987, Membrane transport influences the rate of accumulation of cytosine arabinoside in human leukemia cells, *J. Clin. Invest.* **79**:380–387.
Wiley, J. S., Taupin, J., Jamieson, G. P., Snook, M., Sawyer, W. H., and Finch, L. R., 1985, Cytosine arabinoside transport and metabolism in acute leukemias and T cell lymphoblastic lymphoma, *J. Clin. Invest.* **75**:632–642.
Wiley, J. S., Woodruff, R. K., Jamieson, G. P., Firkin, F. C., and Sawyer, W. H., 1987, Cytosine arabinoside in the treatment of T-cell acute lymphoblastic leukemia, *Aust. N. Z. J. Med.* **17**:379–386.
Wiley, J. S., Snook, M. B., and Jamieson, G. P., 1989, Nucleoside transport in acute leukaemia and lymphoma: Close relation to proliferative rate, *Br. J. Haematol.* **71**:203–207.
Wiley, J. S., Cebon, J. S., Jamieson, G. P., Szer, J., Gibson, J., Woodruff, R. K., McKendrick, J. J., Sheridan, W. P., Biggs, J. C., Snook, M. B., Brocklebank, A. M., Rallings, M. C., and Paterson, A. R. P., 1994, Assessment of proliferative responses to granulocyte-macrophage colony-stimulating factor (GM-CSF) in acute myeloid leukaemia using a fluorescent ligand for the nucleoside transporter, *Leukemia* **8**:181–185.
Williams, J. B., and Lanahan, A. A., 1995, A mammalian delayed-early response gene encodes HNP36, a novel, conserved nucleolar protein, *Biochem. Biophys. Res. Commun.* **213**:325–333.
Willson, J. K. V., Fisher, P. H., Remick, S. C., Tutsch, K. D., Grem, J. L., Nieting, L., Alberti, D., Bruggink, J., and Trump, D. L., 1989, Methotrexate and dipyridamole combination chemotherapy based upon inhibition of nucleoside salvage in humans, *Cancer Res.* **49**:1866–1870.
Windscheif, U., 1996, Purinoceptors: From history to recent progress. A review, *J. Pharm. Pharmacol.* **48**:993–1011.
Wu, J. S., Kwong, F. Y., Jarvis, S. M., and Young, J. D., 1983, Identification of the erythrocyte nucleoside transporter as a band 4.5 polypeptide. Photoaffinity labeling studies using nitrobenzylthioinosine, *J. Biol. Chem.* **258**:13745–13751.
Wu, X., Yuan, G., Brett, C. M., Hui, A. C., and Giacomini, K. M., 1992, Sodium-dependent nucleoside transport in choroid plexus from rabbit. Evidence for a single transporter for purine and pyrimidine nucleosides, *J. Biol. Chem.* **267**:8813–8818.

Yao, S. Y., Cass, C. E., and Young, J. D., 1996a. Transport of the antiviral nucleoside analogs 3'-azido-3'-deoxythymidine and 2,',3'-dideoxycytidine by a recombinant nucleoside transporter (rCNT) expressed in *Xenopus laevis* oocytes, *Mol. Pharmacol.* **50**:388–393.

Yao, S. Y., Ng, A. M., Ritzel, M. W., Gati, W. P., Cass, C. E., and Young, J. D., 1996b. Transport of adenosine by recombinant purine- and pyrimidine-selective sodium/nucleoside cotransporters from rat jejunum expressed in *Xenopus laevis* oocytes, *Mol. Pharmacol.* **50**:1529–1535.

Yao, S. Y., Ng, A. M., Muzyka, W. R., Griffiths, M., Cass, C. E., Baldwin, S. A., and Young, J. D., 1997, Molecular cloning and functional characterization of nitrobenzylthioinosine (NBMPR)-sensitive (es) and NBMPR-insensitive (ei) equilibrative nucleoside transporter proteins (rENT1 and rENT2) from rat tissues, *J. Biol. Chem.* **272**:28423–28430.

Young, I., Young, G. J., Wiley, J. S., and van der Weyden, M. B., 1985, Nucleoside transport and cytosine arabinoside (araC) metabolism in human T lymphoblasts resistant to araC, thymidine and 6-methylmercaptopurine riboside, *Eur. J. Cancer Clin. Oncol.* **21**:1077–1082.

13

Multidrug-Resistance Transporters

Jeffrey A. Silverman

1. INTRODUCTION

Approximately half of the one million new cancer cases annually in the United States will be treated with systemic chemotherapy. Unfortunately, this therapy will fail for the majority of these patients due to resistance of the tumors to the drugs. There are numerous mechanisms through which tumors become resistant to drugs; this chapter is limited to the action of a set of proteins known as the multidrug resistance (MDR) transporters. Following a brief discussion of the structure and function of the multidrug resistance transporters, this chapter focuses on the role of these proteins in the absorption and disposition of drugs and the potential consequences of modification of their function.

Several MDR transporters have been identified and characterized and the list of these proteins continues to grow as investigators seek new drug transporters. The first identified and best-characterized MDR transporter is the P-glycoprotein (P-gp) encoded by the MDR1 gene (Endicott and Ling, 1989; Borst *et al.*, 1993; Gottesman and Pastan, 1993; Germann, 1996). This P-gp is an active drug export pump which mediates the resistance to a large group of diverse compounds. P-gp is widely expressed in many organs, leading to the hypothesis that the physiological role of this protein is a protective mechanism against xenobiotics and endogenous metabolites. Recent investigations also suggest that P-gp-mediated transport of both parent drug and metabolites are important in absorption and disposition of drugs in tissues which express this protein such as the liver, kidney, and

---

Jeffrey A. Silverman • AvMax, Inc., South San Francisco, CA 94080.
Membrane Transporters as Drug Targets, edited by Amidon and Sadée. Kluwer Academic/Plenum Publishers, New York, 1999.

intestine (Cavet *et al.,* 1996; Mayer *et al.,* 1996; Su and Huang, 1996; Wacher *et al.,* 1996). Alternative multidrug-resistance transporters such as MRP1, the multidrug-resistance-related protein, have been identified in multidrug-resistant cells which do not express P-gp (Cole *et al.,* 1992). Additional MRP family members have recently been identified and are adding to our understanding of drug resistance, absorption, and disposition. The MRP1 and MRP2 (cMOAT) transporters secrete conjugated compounds and organic anions (Müller *et al.,* 1996a, b). Both the MDR and MRP proteins interact with numerous compounds, suggesting that these transporters may affect the pharmacokinetics of many drugs in addition to those used in cancer therapy.

2. MDR GENE FAMILY

The phenomenon of drug resistance has been investigated for nearly 30 years (Biedler and Riehm, 1970; Biedler, 1994). During this time, experiments in several laboratories have resulted in the identification, isolation, and characterization of P-glycoprotein, a major drug transporter, and the multidrug-resistance genes which encode it. Ling and co-workers observed that multidrug-resistant Chinese hamster ovary cells overexpress a cell surface protein which was subsequently named P-glycoprotein for its role in affecting the permeability of cytotoxic drugs (Juliano and Ling, 1976; Kartner *et al.,* 1983). Production of a monoclonal antibody, C219, against P-gp and subsequent screening of a cDNA expression library constructed from mRNA isolated from a multidrug-resistant cell line led to the isolation of a cDNA, pCHP1, which partially encodes a hamster P-gp (Kartner *et al.,* 1985; Riordan *et al.,* 1985). Use of this cDNA as a probe on Southern blots foretold the subsequent identification of the MDR multigene family. A molecular cloning approach employing the technique of in-gel renaturation and assignment of fragments to their homologous regions permitted Gros and co-workers to isolate a 120-kb domain from multidrug-resistant hamster cell lines and to demonstrate that the hamster *mdr* gene resides within that segment (Gros *et al.,* 1986c). Isolation of human carcinoma cell lines which were independently selected but cross-resistant to colchicine, vinblastine, and adriamycin led to the isolation of the human *MDR* gene and subsequently to the full-length cDNA (Roninson *et al.,* 1986; Shen *et al.,* 1986; Ueda *et al.,* 1987b). Screening of a mouse cDNA library constructed from a drug-sensitive cell line with the hamster *mdr* sequences led to the isolation of the first mouse *mdr* gene and demonstrated that expression of the wild-type protein, and not a mutant, is sufficient to acquire a drug-resistant phenotype (Gros *et al.,* 1986a).

Subsequently additional members of the *mdr* gene family have been isolated from humans, mice, hamsters, and rats (Table I) (Gros *et al.,* 1986b, 1988; Roninson *et al.,* 1986; van der Bliek *et al.,* 1987, 1988; Ng *et al.,* 1989; Silverman *et al.,* 1991; Deuchars *et al.,* 1992; Brown *et al.,* 1993). Combined, these investigations demonstrate that P-gp's are encoded by the small MDR gene family, comprised of two genes in humans, MDR1 and MDR2, and three genes in rodents, *mdr1a, mdr1b,* and *mdr2* (Table I). Other vertebrates have additional members of this gene family, but have not been fully investigated (Juranka *et al.,* 1989). Based on sequence identity and function the *mdr* genes are classified into two classes; the class I genes encode the drug transporter associated with multidrug resistance, whereas the class II genes encode a phospholipid transporter. In contrast to humans, rodents have two class I genes which are highly homologous, yet encode P-gp's which have distinct substrate specificities (Devault and Gros, 1990; Yang *et al.,* 1990).

3. P-GLYCOPROTEIN IS AN ABC FAMILY TRANSPORTER

During the initial characterization of the mouse *mdr1* cDNA a comparison with the Genbank database revealed a homology with the previously identified bacterial transporter, *hlyB*. This homology has since been extended to include a large family of transport proteins known as the ATP-binding cassette (ABC) transporters (Higgins, 1992). More than 100 ABC transporters have been identified in bacteria, plants, fungi, and animal cells. These transporters, for the most part, hydrolyze ATP to drive the flux of their substrate against a concentration gradient. The substrate range for these proteins is diverse and includes drugs, nutrients, amino acids, sugars, peptides, pigments, and metals. These transporters share a common organization and considerable amino acid homology around the nucleotide-binding domain, which suggests an evolutionary conservation. Examples of ABC transporters include *pfmdr*, which mediates resistance in the malarial parasite *Plasmodium falciparum*, the cystic fibrosis transmembrane regulator (CFTR), the cystic fibrosis gene product, and the antigenic peptide transporter (TAP) involved in MHC antigen presentation (reviewed in Higgins, 1992).

Table I.
Multidrug-Resistance Gene Family

	Class I	Class II
Human	*MDR1*	*MDR2/MDR3*
Hamster	*pgp1, pgp2*	*pgp3*
Mouse	*mdr1a/mdr3, mdr1b/mdr1*	*mdr2*
Rat	*mdr1a, mdr1b*	*mdr2*

4. STRUCTURE OF P-GLYCOPROTEIN

The physical properties of P-gp have been extensively investigated. P-gp is an integral membrane protein with a molecular weight of approximately 170 kDa. Comprised of 1280 amino acids, P-gp has bilateral symmetry with the amino and carboxyl halves each having six putative membrane-spanning domains and one intracellular ATP-binding site. This structural model has been largely substantiated by antibody mapping, site-directed mutagenesis, and biochemical analysis (Yoshimura et al., 1989; Zhang and Ling, 1991; Georges et al., 1993; Loo and Clarke, 1993, 1995). Epitope mapping with the MRK-16 monoclonal antibody demonstrated that the first and fourth predicted loops are extracellular (Georges et al., 1993). Similarly, antipeptide antibodies to Glu^{393}–Lys^{408} and Leu^{1206}–Thr^{1226} recognize their epitopes in permeabilized, but not intact cells, confirming their predicted intracellular location (Yoshimura et al., 1989). Mapping of the topology of cysteine residues into putative intra- or extracellular loops confirmed the 12-transmembrane-domain model for P-gp (Loo and Clarke, 1995). Recently, Rosenberg et al. used high-resolution electron microscopy to present a model for P-gp which is consistent with the available immunological and biochemical analysis (Rosenberg et al., 1997). At 2.5-nm resolution, P-gp appears to function as a monomer and have a 5-nm central pore which is closed on the cytoplasmic surface of the plasma membrane forming an aqueous compartment. Two 3-nm intracellular lobes were observed and are consistent with the predicted 200-amino acid nucleotide-binding domains. This also agrees with data suggesting that substrate binding and cross-linking agents interact at the cytoplasmic face of the membrane (Greenberger, 1993). Further structural characterization of P-gp awaits its crystallization and analysis by X-ray crystallography.

5. *MDR1* P-gp IS A PUMP OF CYTOTOXIC DRUGS

Increased expression of P-gp is associated with multidrug resistance, which is characterized by cross-resistance of cells to structurally and mechanistically distinct cytotoxic drugs (Table II). Exposure of cells *in vitro* to a single cytotoxic drug leads to increased expression of P-gp and a multidrug-resistant phenotype; however, the most persuasive evidence that P-gp is important in the resistant phenotype comes from gene transfer experiments. Transfection of the murine *mdr1* gene into drug-sensitive cells results in a 200-fold increase in resistance to adriamycin and cross-resistance to colchicine and vincristine (Gros et al., 1986a; Hammond et al., 1989). Transfection of the human *MDR1* cDNA into parental NIH 3T3 or KB cells also confers cross-resistance to colchicine, vinblastine, and doxorubicin (Ueda et al., 1987a). Similarly, retroviral transfer of the murine or human MDR

Table II.
Representative List of Compounds Which Interact with P-Glycoprotein

Anti-Cancer Agents		Other	
Daunorubicin	Mitomycin C	Digoxin	Gramicidin D
Doxorubicin	Paclitaxel	Progesterone	Diltiazem
Mitoxantrone	Docetaxel	Terfenidine	Erythromycin
Etoposide	Actinomycin D	Morphine	Ketoconozole
Teniposide	Colchicine	Loperimide	Nifedipine
Vinblastine	Topotecan	Celiprolol	Saquinavir
Vincristine		Rifampicin	Estrogen glucuronide

genes into cells leads to the development of drug-resistant cell lines without requiring drug selection (Guild et al., 1988; Pastan et al., 1988). The level of drug resistance in MDR-transfected cells correlates with the density of P-gp in the plasma membrane (Choi et al., 1991). Combined, these experiments emphasize that transfer of the single MDR gene and the subsequent expression of wild-type P-gp is sufficient to cause a drug-resistant phenotype and that mutation of this protein is unnecessary for this activity.

In addition to being resistant to the cytotoxic effect of anticancer drugs, cells which express P-gp are able to directionally transport substrates. Certain epithelial cell lines express P-gp on their apical surface, resulting in a particularly useful model in which to investigate substrate flux. When grown on semipermeable membranes separating two media chambers, these cells transport P-gp substrates from the basolateral compartment into the apical compartment. Cell lines which have polarized P-gp expression include the porcine kidney cell line LLC-PK1, the canine kidney cell line MDCK, and the human intestinal adenocarcinoma cell lines HCT-8 and Caco-2 (Hidalgo et al., 1989; Horio et al., 1990; Hunter et al., 1993; Zacherl et al., 1994; Cavet et al., 1996). Reflecting the variable levels of P-gp expression, LLC-PK1 cells transport ^{3}H-vinblastine more efficiently than MDCK cells (Horio et al., 1990). The expression of P-gp in LLC-PK1 cells has also been enhanced by stable transfection of MDR1 and the resulting cells were used to investigate the directional transport of drugs such as daunorubicin, digoxin, and cyclosporin A (CsA) (Tanigawara et al., 1992; Schinkel et al., 1995; Tanaka et al., 1997). HCT-8 and Caco-2 cells are derived from human intestinal tumors and have been used as in vitro models of intestinal drug absorption. These cells possess many characteristics of differentiated epithelial cells such as cellular polarity, formation of tight junctions, and directional transepithelial transport and are widely used to investigate the transepithelial flux of several P-gp substrates and the effect of potential reversal agents (Hidalgo et al., 1989; Artursson, 1990; Audus et al., 1990; Hunter et al., 1991; Zacherl et al., 1994).

6. P-GLYCOPROTEIN IS AN ATP-DEPENDENT PUMP

The amino acid sequence of P-gp suggests that there are two nucleotide-binding domains, one in each half of the molecule. Experiments with cells transiently transfected with the murine *mdr1b* into which point mutations were introduced demonstrated that the integrity of both of these sites is essential for P-gp activity (Azzaria *et al.,* 1989). Site-directed mutagenesis of either or both of these sites eliminates the ability of *mdr1b* to confer resistance, suggesting that the two halves of the molecule may function cooperatively to transport substrates. Georges *et al.* used the monoclonal antibodies C219 and C494 to show that both putative ATP-binding sites bind ATP and suggested that there is cooperativity between these sites (Georges *et al.,* 1991). Several laboratories have isolated plasma membrane vesicles or have partially purified P-gp from drug-resistant cells and expression systems. These investigations have demonstrated drug-stimulated ATPase activity and begun to characterize the drug and ATP-binding sites in this protein (Ambudkar *et al.,* 1992; Al-Shawi and Senior, 1993; Sharom *et al.,* 1993, 1995; Al-Shawi *et al.,* 1994). Reconstitution of partially purified P-gp into phospholipid bilayers results in ATP-dependent colchicine uptake against a concentration gradient, demonstrating functional drug transport. Liposomes reconstituted from P-gp-expressing cells have drug-dependent ATPase activity which is required for drug transport (Sharom *et al.,* 1993). Nonhydrolyzable ATP analogues are unable to support drug uptake in these plasma membrane vesicles, indicating that ATP hydrolysis is essential for transport (Doige and Sharom, 1992).

Association of ATPase activity with P-gp began with the recognition that the ATP analogue 8 azido-ATP binds to P-gp (Cornwell *et al.,* 1987). ATPase activity was then reported in immunoaffinity-purified P-gp (Hamada and Tsuruo, 1988). Subsequently, partially purified P-gp has been used to further characterize this ATPase activity; some drugs such as colchicine, nifedipine, amiodarone, and verapamil increase ATPase activity, whereas others such as N-ethylmaleimide inhibit this activity (Doige *et al.,* 1992; Shapiro and Ling, 1994). Daunomycin and vinblastine inhibit ATPase activity in some investigations, but increase ATPase activity in others, suggesting that modulation of ATPase activity is dependent on the experimental conditions and may not correlate with the suitability of a drug to be a transport substrate (Ambudkar *et al.,* 1992; Sharom *et al.,* 1993; Shapiro and Ling, 1994). Dependence on the lipid environment or the detergents used in protein isolation may result in variation in stimulation or inhibition of ATPase activity due to drugs. Phosphatidylcholine and phosphatidylserine were the most effective lipids at retaining ATPase activity of partially purified P-gp (Doige *et al.,* 1993).

7. *MDR2* P-gp IS A PHOSPHOLIPID TRANSPORTER

Despite a highly conserved sequence, the P-gp encoded by *mdr2* is unable to confer drug resistance. Transfection of the mouse *mdr2* or human MDR2 genes into drug-sensitive cells fails to convert these cells to a drug-resistant phenotype (Buschman and Gros, 1991; Schinkel *et al.*, 1991). Resolution of this conundrum came from Borst and co-workers upon their development of a mouse in which this gene was disrupted by homologous recombination (Smit *et al.*, 1993; Mauad *et al.*, 1994; Oude Elferink and Groen, 1995). These "knockout" mice develop severe liver pathology, nonsupportive cholangitis with portal inflamation and ductular proliferation, resulting from disruption of bile formation. The apparent cause of this pathology is the inability of the liver to secrete phospholipids, resulting in toxic injury to the liver due to secretion of free bile salts. Thus, it was proposed that the *mdr2*-encoded P-gp is indeed a transporter, not of drugs, but of phospholipids, in particular phosphatidylcholine. Additionally, its function is to translocate phosphatidylcholine from the inner to the outer membrane leaflet where, in the canalicular membrane, it forms buds in the presence of bile salts in the production of normal bile (Smit *et al.*, 1993). This hypothesis was supported by Ruetz and Gros, who demonstrated that *mdr2,* but not *mdr1*, expressed in yeast translocated phosphatidylcholine across a lipid bilayer (Ruetz and Gros, 1994, 1995). Smith and co-workers further demonstrated *mdr2* mediated transfer of phosphatidylcholine from the inner to the outer membrane leaflet in fibroblasts which were transfected with the human MDR2 gene (Smith *et al.*, 1994). Thus, the P-gp encoded by MDR2 is a transporter of phosphatidylcholine, but not of cytotoxic drugs.

8. *MDR1* TISSUE DISTRIBUTION

P-gp is widely expressed in normal tissues, many of which are associated with drug disposition or serve a barrier function. Its localization on the plasma membrane combined with the ability to export toxic compounds is consistent with the proposed protective role for this protein. One function of P-gp may have evolved to protect against dietary components and environmental toxins.

Fojo *et al.* examined the level of MDR1 mRNA expression in normal human tissues using slot blot analysis (Fojo *et al.*, 1987). High expression of MDR1 mRNA was observed in adrenal and kidney, and intermediate levels of expression were found in liver, lung, jejunum, colon, and rectum. Lower levels of MDR1 mRNA were found in brain, prostate, skin, muscle, spleen, bone marrow, stomach, ovary, and esophagus. These authors observed highly variable expression between samples and suggested that genetic or environmental factors affect the level of expression in different individuals. A similar pattern of expression was observed by Chin et al. (1989). P-gp was also observed in the endothelial cells in the brain and testes (Cordon-Cardo et al., 1989). Using a reverse transcriptase-polymerase chain reaction (RT-PCR) method, Bremer et al. found high levels (15–50 amole mRNA/ µg mRNA/RNA) of *MDR1* mRNA in intestine, kidney, liver, and placenta and low levels (0.2 amole/µg mRNA/RNA) in respiratory epithelium (Bremer et al., 1992). P-glycoprotein is also expressed at a moderate level in hematopoietic cells, notably lymphoid bone marrow (Chaudhary and Roninson, 1991; Chaudhary et al., 1992).

Immunohistochemistry with the MRK16 antibody revealed high levels of *MDR1*-encoded P-gp expression on the canalicular surface in liver, on the apical surface of the columnar epithelium in colon and jejunum, in the proximal tubules of the kidney, in the ductules in the pancreas, and in the cortex and medulla of the adrenal (Thiebaut et al., 1987). van Kalken et al. investigated the expression of P-gp in the human fetus using RNase protection assays and noted expression at the earliest time examined, 7 weeks (van Kalken et al., 1992). Immunohistochemical analysis showed staining in kidney by 11 weeks of gestation. By week 13 positive staining was observed in liver, adrenal, heart, and smooth muscle. Positive staining in the intestine, stomach, bile ducts, and brain were observed at week 28. The early expression and wide tissue distribution suggest an important role in transport of endogenous substrates or the importance of protecting those organs from xenobiotics during a critical development period.

9. EXPRESSION OF P-gp IN HUMAN CANCER

Increased expression of P-gp has been observed in many human tumors (reviewed in Goldstein, 1996). Goldstein and co-workers observed elevated *MDR1* mRNA levels in colon, kidney, adrenal, hepatic, and pancreatic cancers as well as some leukemias (Goldstein et al., 1989). Many tumors in tissues which normally express P-gp have elevated MDR1 expression; however, there is a considerable variation from tumor to tumor. Additionally, low or undetectable MDR1 expression was observed in many cancers, some of which were considered to be drug resistant, supporting the importance of alternate mechanisms of resistance. Tumors which are initially sensitive to chemotherapy may acquire a multidrug-resistant phenotype subsequent to chemotherapy. Such acquired resistance has been observed in breast tumors, adult and childhood acute lymphoblastic leukemia, adult acute myelocytic leukemia, sarcoma, and multiple myeloma (Goldstein et al., 1989; Goldstein, 1996). In an investigation of 42 renal cell carcinomas, Kanamuru et al. observed that the level of MDR1 expression was increased, but variable among tumors (Kanamaru et al., 1989). No such elevation of MDR1 levels was observed by Lai and co-workers in an examination of lung cancers and lung cancer cell lines (Lai et al., 1989). These and numerous additional investigations demonstrate that P-gp is expressed in many human tumors; however, it is not yet possible to predict the likely response of a tumor to chemotherapy based on the level of MDR1 expression. A more systematic approach with standardization of quantitative detection methods would facilitate informative correlations between studies and better assessment the impact of P-gp on clinical outcome.

10. MDR2 TISSUE DISTRIBUTION

P-gp encoded by the MDR2 gene is also widely expressed. Using RT-PCR, Chin et al. observed MDR2 expression in liver, kidney, adrenal gland, and spleen (Chin et al., 1989). Using RNase protection assays, Smit et al. (1994) found high levels of MDR2 in the liver and lower amounts in the adrenal gland, spleen, tonsil, striated muscle, and heart. MDR2 mRNA was undetectable in brain, stomach, lung, colon, pancreas, ovary, testis, gall bladder, bladder, salivary gland, mammary gland, placenta, duodenum, and thyroid. In contrast to previous investigations, in this study, no MDR2 expression was found in kidney. Immunohistochemical analysis of these tissues revealed that the canalicular membranes of hepatocytes contained MDR2 protein. No MDR2 P-gp was detected in colon or gall bladder with MDR2 antibodies, but staining was observed with C219, suggesting expression of MDR1. Despite expression of MDR2 mRNA in adrenal, muscle, heart, and spleen, no MDR2 P-gp was detected by immunohistochemistry.

11. MODULATION OF P-GLYCOPROTEIN

Soon after the identification of P-glycoprotein and its role in drug resistance in cancer, mechanisms to circumvent this resistance were sought (Ford and Hait, 1990). The first agents were adaptive uses of drugs which have alternate pharmacological activities, including verapamil, CsA, and trifluoperizine. Tsuruo and co-workers first showed that the calcium channel blocker verapamil also functions as an MDR reversal agent to increase the sensitivity of P-gp-expressing murine leukemia cells to vincristine (Tsuruo et al., 1981). Since that time numerous agents have been identified which reverse or sensitize multidrug-resistant cells and tumors using a variety of models and some of these drugs have progressed to human clinical trials (Ford and Hait, 1990; Lum et al., 1993; Ford, 1996). First-generation inhibitors include compounds from many categories of drugs which were developed for alternate therapeutic indications such as calcium channel blockers (verapamil), immunosuppressants (CsA, FK506), antiarrhythmic drugs (quinidine, amiodarone), and steroidal agents (progesterone, tamoxifen) (for an extensive review see Ford and Hait, 1990). These compounds increase the sensitivity of cells *in vitro* to cytotoxic substrate drugs, inhibit P-gp-mediated transport, modulate ATPase activity, and often block photoaffinity labeling of P-gp. Many of these agents have also been used in initial clinical trials (Ford and Hait, 1990; Lum *et al.*, 1993; Ford, 1996). The mechanisms by which these compounds inhibit P-gp is equally diverse, and often undetermined. In some cases, such as CsA, reversal agents are themselves substrates for P-gp and are actively transported (Gan *et al.*, 1996). Therefore their antagonism of drug resistance may in part be due to competitive inhibition. Many, but not all drugs which reverse multidrug resistance can inhibit photoaffinity labeling of P-glycoprotein with vinblastine, daunorubicin, or verapamil analogues (Akiyama *et al.*, 1988; Nogae *et al.*, 1989; Beck and Qian, 1992). Vinblastine and verapamil effectively compete for azidobenzoyl–daunorubicin binding more effectively than daunorubicin (Beck and Qian, 1992). Similarly quinidine and reserpine, strong P-gp inhibitors, block photoaffinity labeling by a vinblastine analogue, whereas weaker inhibitors such as chloroquine did not block binding (Akiyama *et al.*, 1988). This suggests that some reversal agents may share the same binding site or amino acids involved in drug binding.

The affinity of many of the first P-gp inhibitors is low and thus their clinical effectiveness is dose-limited due to their pharmacological activity. Dose-limiting toxicity of verapamil, cyclosporine and other early P-gp reversal agents has been observed (Lum *et al.*, 1993; Ferry *et al.*, 1996). For example, use of racemic verapamil is limited by its hypotensive effects; however, the stereoisomer R-verapamil is less cardiotoxic and is equipotent as a P-gp reversal agent. Thus this compound is replacing the racemic mixture in MDR-reversal trials. The use of the immunosuppressant CsA as a reversal agent is limited by hyperbilirubinemia, myelosuppression, headache, hypomagnesemia, and mild reversible hypertension (Lum *et al.*, 1993). Second-generation modulators with higher affinities and without undesired pharmacological activity have been and continue to be actively developed which specifically inhibit P-gp as their primary role. Unlike the earlier reversal agents, these new compounds are being developed specifically as potent inhibitors of P-gp.

Several newer P-gp reversal agents have been identified which lack the undesired pharmacological activities of first-generation inhibitors and which are effective at much lower concentrations. PSC 833 is a nonimmunosuppressive cyclosporine analogue which is 10 or more fold more potent than CsA as a P-gp inhibitor (Keller *et al.*, 1992; Twentyman, 1992; Ferry *et al.*, 1996). Early results with this compound suggest that it has a dramatic effect on the pharmacokinetics of etoposide and paclitaxel (Fisher et al., 1996). However, temporary ataxia is associated with PSC 833. This compound is currently being tested in numerous clinical trials for its ability to sensitize MDR tumors to chemotherapy. Additional promising reversal agents include GG918 (also known as GF120918), LY335979, and VX-710 among many others. At 0.05–0.01 µM, GG918 restored sensitivity of the multidrug-resistant MCF7/ADR cells to vincristine and competed with ^{3}H-azidopine labeling (Hyafil et al., 1993). Treatment of highly resistant leukemic cells with 0.1 µM GG918 cause a 40- and 57-fold sensitization to daunorubicin and mitoxantrone, respectively (Zhou et al., 1997). Similarly, 0.1 µM LY335979 fully restores sensitivity to vinblastine, doxorubicin etoposide, and taxol in multidrug-resistant CEM cells (Dantzig et al., 1996). This compound is a competitive inhibitor of vinblastine binding to P-gp with an apparent K_i of 0.06 µM, reduces photoaffinity labeling of P-gp by ^{3}H-azidopine, and increases the lifespan of mice bearing xenografts. Thus, it appears that this drug is a potent P-gp modulator. A new ligand of the FK 506-binding protein FKBP12, VX-710, has also recently been shown to effectively restore sensitivity of several drug-resistant cell lines to doxorubicin, vincristine, etoposide, and taxol (Germann et al., 1997b). Interestingly, this compound also reverses multidrug resistance to MRP expressing cells (see below) (Germann et al., 1997a). The above-mentioned compounds as well as other second-generation P-gp inhibitors are actively being tested in preclinical *in vitro* and *in vivo* models and clinical trials are underway to test their ability to modulate multidrug resistance in patients.

It is evident that many reversal agents enhance the cytotoxicity of anticancer drugs in *MDR1*-expressing cell lines *in vitro;* however, their effects are not straightforward in cancer patients. Many early trials of P-gp-reversal agents evaluated their effects in patients who had failed previous therapy; however, these studies often lacked appropriate crossover controls. Inter- and intrapatient variation in lean body mass, age, kidney, and liver function and drug disposition all contribute to obscure any increase in efficacy of the modifying agents (Lum et al., 1993; Relling, 1996). Current well-designed trials with newer resistance-modifying agents will allow optimization of anticancer drug dosage regimens; however, the recognition that P-gp has a significant role in both absorption and disposition of anticancer drugs suggests that a major function of resistance-modifying agents may be to alter the pharmacokinetics of the drugs rather than have a primary effect on tumor cells. There is no evidence to suggest that P-gp in tumors is different than in normal cells; thus changes in the pharmacokinetics of anticancer drugs by reversal agents likely affect both normal and neoplastic tissues. One possible effect of using reversal agents is to increase the efficacy of the anticancer agent in the cancer cells while simultaneously increasing the exposure of normal tissues to these drugs. The net effect of this may be to narrow the therapeutic range of the anticancer agent (Relling, 1996).

12. P-GLYCOPROTEIN IN DRUG ABSORPTION AND DISPOSITION

The role of P-gp in non-cancer drug disposition is not well established; however, evidence for the importance of this transporter in pharmacokinetics is expanding. The list of compounds which interact with P-gp is large and includes many drugs not used in cancer chemotherapy (Table III) (Gottesman and Pastan, 1993; Wacher *et al.*, 1995; Schuetz *et al.*, 1996a). Thus, P-gp potentially has a role in the pharmacokinetics of many drugs in addition to its role in cancer chemotherapy. As new data become available it is also increasingly evident that P-gp is important in the absorption of drugs as well as their excretion. For example, Su and Huang observed that inhibition of P-gp increased bioavailability of digoxin by increasing absorption as well as reducing excretion (Su and Huang, 1996). A similar phenomenon was observed with etoposide (Leu and Huang, 1995). It follows that if P-gp has a role in the pharmacokinetics of drugs, then modulation of P-gp through the use of reversal agents may alter the absorption and disposition of coadministered drugs. For example, in a phase I clinical trial, Lum *et al.* observed that high doses of CsA modulated the pharmacokinetics of etoposide (Lum *et al.*, 1992). Administration of the immunosuppressant increased the area under the curve (AUC) for etoposide by as much as 80%, decreased total clearance by 38%, and increased the steady-state volume of distribution by 46%. Both renal and nonrenal clearance of etoposide were significantly reduced by CsA.

The expression of P-gp in normal tissues suggests that inhibition of this transporter may result in altered pharmacokinetics with potentially toxic effects. Inhibition of P-gp in the endothelial cells at the blood–brain barrier may provide central nervous system access to many previously excluded drugs. Inhibition in the liver, kidney, and intestine may prolong drug clearance by lowering excretion. Another consequence of inhibition in those organs is altered bioavailability through increased drug absorption. Coadministration of a suboptimal dose of doxorubicin with CsA enhanced the effectiveness of the chemotherapy to levels equivalent to that of a 25% higher dose (van de Vrie *et al.*, 1993). These authors also later observed that addition of a chemosensitizer such as cyclosporine significantly enhanced the toxic side effects equivalent to increasing the dose of the anticancer

Table III.
Partial List of MDR-Reversal Agents

Verapamil	r-Verapamil	PSC 833
Cyclosporin A	Staurosporine	VX-710
FK506	Reserpine	LY335979
Rapamycin	Tamoxifen	BIBW22
Amiodarone	Quercetin	GG918 (GF120918)
Quinidine	Trifluoroperizine	

drug (van de Vrie et al., 1994). Cyclosporine did not alter the pattern of toxicity of doxorubicin in rats, which includes myelosuppression, leukopenia, thrombopenia, nephrotoxicity, and imbalances in cholesterol and triglyceride levels. No increase in histologically observable toxicity was found in organs which express high levels of P-gp, such as the liver and intestine. Similarly, coadministration of CsA with etoposide required a 50% reduction in dose as a result of decreased clearance (Lum et al., 1993). Consistent with inhibition of P-gp in the blood–brain barrier, liver, and hematopoietic stem cells, increased toxicities included nausea and vomiting, myelosuppression, and hyperbilirubinemia.

13. ROLE OF P-gp IN NORMAL TISSUES AND DRUG DISPOSITION

An informative new model of the physiological function of P-gp and its role in pharmacokinetics is the *mdr1a* knockout mouse. These mice were developed with a genetically disrupted *mdr1a* gene, the only MDR gene expressed in the intestine tissue. These mice are viable, have a normal lifespan, reproduce, and have no major pathology, but have altered pharmacokinetics of drugs such as ivermectin, digoxin, taxol, and vinblastine and are particularly useful models to investigate the function of P-gp in the blood–brain barrier and intestine (Schinkel et al., 1994, 1996; van Asperen et al., 1996; Sparreboom et al., 1997). Interpretation of drug elimination in other organs in these animals is potentially confounded by the expression of *mdr1b*, which yields a second murine drug-transporting P-gp. Double *mdr1a/mdr1b* knockout mice have recently been developed and will provide an additional model in which to examine further the role of these proteins in pharmacokinetics (Schinkel et al., 1997).

Examination of the pharmacokinetics of vinblastine in mice deficient in *mdr1a* P-gp confirms that this protein is important in the elimination of important cancer drugs (van Asperen et al., 1996). The terminal half-life of an intravenous bolus of vinblastine to mdr knockout mice was significantly longer than in the wild-type animals, fecal excretion was reduced, while urinary excretion and metabolism were largely unchanged. The increase in terminal half-life is therefore due to reduced elimination. Another consequence of this diminished clearance is the accumulation of vinblastine in tissues which normally express P-gp. This was particularly apparent in the brain (see below), heart, small intestine, skeletal muscle, and liver.

13.1. Intestine

Cell lines derived from the intestine have long been employed as models of drug absorption and intestinal permeability (Audus et al., 1990; Borchardt et al.,

1996). Cell lines such as Caco-2, T84, and HCT-8 express P-gp and have been utilized to investigate drug transport by this protein (Hidalgo et al., 1989; Hunter et al., 1993; Zacherl et al., 1994; Cavet et al., 1996). These cells form a polarized epithelial monolayer in culture and express high amounts of P-gp on the apical surface so that substrates for this pump are directionally transported from the basolateral to the apical surface. Vinblastine, for example, is vectorially transported in these cells and this flux is inhibited with P-gp inhibitors such as verapamil and cyclosporine (Hidalgo et al., 1989; Hunter et al., 1991). Similar studies have shown that these cell lines are suitable for investigation of P-gp-mediated transport of other drugs such as the β-adrenoreceptor blocker, celiprolol, an anticancer agent, etoposide, immunosuppressants, cyclosporine, and FK506 and peptides (Augustijns et al., 1993; Karlsson et al., 1993; Saeki et al., 1993; Leu and Huang, 1995; Nerurkar et al., 1996). These data are consistent with an excretory role for P-gp in the disposition of metabolites and xenobiotics. Caco-2 cells have also been used as a model for assessing intestinal drug absorption (Artursson and Karlsson, 1991). A good correlation between drug absorption rates *in vitro* and rat intestinal diffusion suggests that this is a suitable model for screening novel compounds for intestinal absorption.

Expression of P-gp in the intestine also suggests that inhibition of P-gp-mediated transport may be one mechanism leading to drug interactions with digoxin. Indeed, it has been observed that P-gp is involved in the absorption and elimination of digoxin in the intestine (Su and Huang, 1996). Digoxin transport in everted gut sacs is diminished with the P-gp inhibitor quinidine, C219 antibody, and a nonhydrolyzable ATP analogue. Using a single-pass *in situ* perfusion system, intestinal clearance of digoxin decreased from 29 to 11 ml/hr in the presence of the P-gp inhibitor quinidine. Furthermore, concomitant intravenous administration of digoxin with quinidine doubled plasma concentrations and decreased intestinal digoxin levels 40%. This effect is likely due to decreased clearance from organs, such as the liver and kidney, in addition to the intestine. Cavet *et al.* observed P-gp-mediated transport of digoxin in the Caco-2 intestinal epithelial cell line, suggesting that the intestine is important in digoxin elimination (Cavet *et al.*, 1996). Mayer *et al.* observed that the intestine contributes significantly to elimination and to reuptake of biliary excreted digoxin (Mayer *et al.*, 1996). In *mdr1a*-deficient mice a dramatic shift from fecal to urinary digoxin excretion occurred with both intravenous and orally administered drug. Intestinal excretion by P-gp accounted for approximately 16% of an intravenous dose within 90 min in the wild-type mouse, whereas in the *mdr1a* knockout mouse intestinal excretion was only 2%. Interestingly, biliary excretion of digoxin was unchanged in the *mdr1a*-deficient mice, suggesting alternate transporters are important for digoxin excretion in the liver.

Recent work on the development of fluorquinolone antimicrobial agents has led to the development of more potent agents. These compounds can be classified according to their major route of clearance; ofloxacin, temafloxacin, and lomefloxacin are renally excreted, whereas pefloxacin is hepatically excreted and norfloxacin, ciprofloxacin, and feroxacin are excreted through both pathways. Using an isolated rat intestinal model, Rabbaa *et al.* observed saturable and stereospecific intestinal elimination of ofloxacin (Rabbaa *et al.*, 1996). Verapamil and quinidine significantly reduced clearance of this drug, suggesting that P-gp is involved in the elimination of this drug. Norfloxacin and ciprofloxacin are actively and directionally transported across Caco-2 cell monolayers, further suggesting that these drugs are substrates for intestinal P-gp (Griffiths *et al.*, 1994).

Recent work with the *mdr1a* knockout mouse has clearly demonstrated the importance of P-gp as an intestinal barrier limiting the absorption of drugs. Oral administration of paclitaxel (Taxol™) to these mice resulted in a sixfold increase in the AUC for the plasma concentration versus time plot compared to wild-type mice (Sparreboom *et al.*, 1997). The oral bioavailability of paclitaxel in the *mdr1a*-deficient mice increased to 35% from 11% in wild-type animals. Also, following an intravenous dose the fecal excretion was significantly diminished in the knockout mice despite there being no difference in biliary excretion. This investigation suggests that wild-type P-gp is important in preventing the absorption of compounds from the intestinal lumen and in lowering enterohepatic recirculation of drugs. Inhibition of P-gp in wild-type mice by concomitant administration of PSC833 or cyclosporine with paclitaxel increases the oral bioavailability of the anticancer drug over 10-fold (van Asperen *et al.*, 1997). This increase is greater than observed in the *mdr1a*-deficient mice and is likely due to inhibition of CYP3A in the intestine and liver. These data provide direct evidence of the combined role of P-gp and CYP3A in limiting oral bioavailability of clinically significant drugs.

13.2. Liver

The liver serves many physiological roles including absorption and processing of nutrients and metabolism of endogenous compounds and xenobiotics. P-gp's are expressed in the liver and likely serve to transport these compounds to their correct compartments (Silverman and Schrenk, 1997). The *MDR1* gene is expressed at a low level in normal liver, but is inducible by chemotherapeutic drugs and xenobiotics (Kohno *et al.*, 1989; Chin *et al.*, 1990; Gant *et al.*, 1991; Morrow *et al.*, 1994; Hill *et al.*, 1996). Cholestasis or damage to the liver which causes a regenerative response also leads to increased MDR expression (Thorgeirsson *et al.*, 1987; Nakatsukasa *et al.*, 1993; Schrenk *et al.*, 1993). Transport of metabolites and xenobiotics out of the cell may be one role of P-gp in the liver. Recently, Gan *et al.* observed that P-gp transports cyclosporine metabolites in a Caco-2-polarized transport system (Gan *et al.*, 1996). These authors also suggested that P-gp increased exposure of cyclosporine to metabolic enzymes thereby effectively increasing the metabolism of the drug. P-gp was also demonstrated to directly impact CYP3A expression by regulating the level of the CYP3A inducer rifampicin (Schuetz et al., 1996b). Oral administration of rifampicin to *mdr1a*-deficient mice led to an 11-fold increase in hepatic drug levels compared to the wild-type animals. Increased rifampicin levels, in turn, caused an induction of CYP3A levels as detected by Western blot analysis and midazolam hydroxylation (Schuetz et al., 1996b). These results clearly demonstrate that P-gp can determine the expression of CYP3A, which can in turn determine the metabolism of orally administered drugs.

13.3. Kidney

Digoxin is a widely prescribed cardiac glycoside for the treatment of heart failure. Numerous pharmacologically significant drug interactions have been observed with digoxin which result in a decrease in renal clearance (Kelly and Smith, 1996). The molecular mechanism for this is largely unknown but may result from decreased P-gp-mediated excretion. Drugs which interact with digoxin include many P-gp inhibitors such as quinidine, verapamil, and cyclosporine. Localization of P-gp in the proximal tubules and an overlap of P-gp-reversal agents and drugs which adversely affect digoxin elimination led to a series of studies which demonstrate P-gp-mediated transport of this drug. de Lannoy and Silverman (1992) observed that inhibition of P-gp results in accumulation of ^{3}H-digoxin in MDR1-expressing Chinese hamster ovary cells. Additionally, MDR-1-transfected kidney epithelial cells, LLC-PK1, directionally transport ^{3}H-digoxin across a microporous membrane (Tanigawara et al., 1992; Schinkel et al., 1995). This flux was inhibited by P-gp-reversal agents such as verapamil and quinidine. Combined, these results suggest that P-gp may be involved in renal excretion of digoxin. However, the contribution of P-gp to renal excretion of digoxin is unclear. Despite the absence of *mdr1a*, renal clearance of this drug is increased in the *mdr1a*-knockout animals (Mayer et al., 1996). This may be due to increased expression of *mdr1b* in the kidney or due to expression of alternative transporters (Schinkel et al., 1994).

13.4. Blood–Brain Barrier

P-gp is localized in the endothelial cells of the capillaries of the cerebral cortex and cerebellum and was proposed as part of the protective barrier for the brain (Cordon-Cardo et al., 1989; Thiebaut et al., 1989). Indeed, P-gp expressed in brain capillaries is photoaffinity labeled with ^{125}I-arylazidoprazosin, a P-gp photoaffinity labeling reagent, and this labeling can be inhibited with PSC833, CsA, quinidine, and other drugs established to interact with P-gp (Jetté *et al.,* 1995). The *mdr1a* knockout mice provided the first functional evidence of a role for P-gp in the blood–brain barrier (Schinkel *et al.,* 1994). Administration of ivermectin, an acaracide, leads to significant neurotoxicity in the knockout mice due to accumulation of the drug in the central nervous system at doses which were harmless to wild-type and heterozygous mice. Thus, one apparent function of *mdr1a*-encoded P-gp is in the endothelial cells of the blood–brain barrier to prevent the entry of toxic drugs.

Recent investigation of other clinically important drugs also suggests that P-gp is a significant barrier to their use due to restriction of their penetration of the blood–brain barrier (Schinkel *et al.,* 1996). Domperidone, a dopamine antagonist, loperamide, an opioid, ondansetron, a seretonin antagonist, and, to a lesser extent, phenytoin were observed to be substrates for P-gp transport *in vitro*. Haloperidol, a dopamine antagonist, flunitrazepam, a benzodiazapam, and clozapine, a phenothiazine, were not transported by P-gp. In the *mdr1a* knockout mice ondansetron and loperamide accumulated to a higher level in brain than in wild-type mice. These data suggest that the failure of domperidone to cross the blood–brain barrier and hence not be useful as a dopamine antagonist may be due to its removal from the brain by P-gp. Haloperidol, in contrast, is not a P-gp substrate and does cross the blood–brain barrier. Lack of central opioid effectiveness by loperamide may also be due to the presence of P-gp, which prevents its accumulation in the brain. Intravenous administration of ^{3}H-digoxin, a P-gp substrate, similarly results in higher accumulation in the brain of *mdr1a* knockout mice versus wild-type animals (Mayer *et al.,* 1996).

Uptake of colchicine and vinblastine into the brain may also be restricted by P-gp. These drugs only poorly cross the blood–brain barrier in control animals; however, preinfusion of PSC833 or verapamil increased the uptake of these drugs and increased the volume of distribution in the brain (Drion *et al.,* 1996). Morphine and its metabolite, morphine-6β-glucuronide, have also been observed to be substrates for P-gp (Callaghan and Riordan, 1993; Schinkel *et al.,* 1995; Huwyler *et al.,* 1996). Uptake of these drugs into P-gp-expressing cells is increased in the presence of P-gp inhibitors. In addition, several synthetic opiates were able to compete for specific binding of ^{3}H-vinblastine to P-gp. Many opiate analgesics such as these do cross the blood–brain barrier, but are associated with tolerance; therefore, Callaghan and Riordan suggest that P-gp may in part regulate the amount of opioid reaching the brain and contribute to the tolerance phenomenon (Callaghan and Riordan, 1993). Following intravenous administration, the brains of *mdr1a*-deficient mice accumulated significantly more vinblastine than wild-type mice (van Asperen *et al.,* 1996). Additionally, the terminal half-life was prolonged due to reduced fecal excretion of the drug in the knockout animals. Altered accumulation was also observed in some other tissues; however, the presence of the *mdr1b* product with overlapping substrate affinities in these tissues may have diminished the effect of the loss of *mdr1a* in those organs.

14. THE MULTIDRUG RESISTANCE-ASSOCIATED PROTEIN

Over-expression of P-gp is detected in many multidrug-resistant tumors and cell lines; however, a subset of multidrug-resistant tumors and cell lines do not express this protein. This observation led to the isolation of the cDNA encoding another drug transporter, the multidrug-resistance-associated protein (MRP) (Cole *et al.*, 1992). Increased expression of this cDNA is associated with resistant cells isolated from a variety of tumors and is sometimes associated with amplification of its encoding gene. The MRP gene, located on chromosome 16 at band p13.1, encodes a 6.5- to 7-kb mRNA which is translated into a 1531-amino acid protein (Cole *et al.*, 1992; Grant *et al.*, 1994). Using synthetic peptides corresponding to the deduced amino acid sequence from the MRP cDNA, antisera were isolated which recognize a 190-kDa membrane protein which is overexpressed in HL60/ADR cells, a non-P-gp drug resistant cell line (Krishnamachary and Center, 1993).

Similar to P-gp, MRP is a membrane protein and a member of the ABC family of transporters. The highest degree of similarity to P-gp, as well as other ABC superfamily members, is in the nucleotide-binding domain. Cole and co-workers proposed a structural model which contains 11 or 12 membrane-spanning domains in the NH_2-terminal half of the protein and 5 or 6 transmembrane domains in the COOH-terminal half (Loe *et al.*, 1996b). Both halves are predicted to be glycosylated and the protein is phosphorylated, although the role of these modifications is undetermined (Krishnamachary and Center, 1993; Almquist *et al.*, 1995; Loe *et al.*, 1996b).

15. TISSUE DISTRIBUTION OF MRP EXPRESSION

MRP is expressed at different levels in many normal tissues (Zaman *et al.*, 1993; Flens *et al.*, 1996; Loe *et al.*, 1996b). Using Northern blot analysis, Cole *et al.* found expression in lung testes and peripheral blood monocytes, but not in placenta, brain, kidney, salivary gland, uterus, spleen, or liver (Cole *et al.*, 1992). Sensitive RNase protection assays detected significant levels of MRP mRNA in virtually all tissues examined, with the highest amounts in lung, adrenal, testes, bladder, thyroid, and spleen (Zaman *et al.*, 1993). Substantial expression was also found in some of the tissues previously reported to be negative, e.g., kidney, liver, and spleen. Western blot analysis revealed highest levels of MRP protein expression in the adrenal gland, lung, heart, and skeletal muscle. Lower amounts of protein were detected in liver, spleen, kidney, and erythrocytes (Flens et al., 1996). Immunohistochemical analysis with MRP-specific monoclonal antibodies identified the specific cell types which express this protein in many organs, including epithelial and muscle cells (Flens et al., 1996). For example, the respiratory columnar epithelial cells in lung, keratinocytes in skin, intestinal epithelial cells, but not brush border or goblet cells all stained positively for MRP. This pattern of expression suggests that, like P-gp, MRP may have a role in the protection of tissues against toxic xenobiotics.

16. MRP IS AN ATP-DEPENDENT DRUG TRANSPORTER

Several lines of investigation have established MRP as an ATP-dependent drug transporter with broad substrate specificity. Expression cloning using cDNA from the multidrug-resistant HL60R cell line resulted in the identification of MRP as the transferred gene associated with the emergence of a multidrug-resistant phenotype in previously drug-sensitive NIH 3T3 cells (Kruh et al., 1994). These cells have increased, but variable resistance to three unrelated cytotoxic drugs, adriamycin, vinblastine, and VP16. Similarly, transfection of the MRP cDNA into HeLa or SW-1573 cells also resulted in lines which are resistant to doxorubicin, daunorubicin, vincristine, and VP16, but not Taxol or cisplatin (Grant et al., 1994; Zaman et al., 1994). The intracellular accumulation was decreased and the drug efflux was increased in the transfectants. Furthermore, the increased levels of resistance are proportional to the amount of 190,000-kDa MRP protein expressed in the drug-resistant cells. Inside-out membrane vesicles prepared from MRP-transfected NIH 3T3 cells were used to establish that MRP pumps a wide variety of unaltered cytotoxic drugs (Paul et al., 1996). Saturable, ATP-dependent transport of daunorubicin, etoposide, and vincristine was observed with K_m values of 6.3, 4.4, and 4.2 µM, respectively. Spheroplasts isolated from yeast which have been transfected with MRP1 also provide evidence that this protein is a drug transporter. These vesicles mediate time-dependent transport of labeled adriamycin, demonstrating the functional transport of cytotoxic drugs by this protein (Ruetz et al., 1996).

Recent investigations suggest that the MRP1 transporter functions as the long-sought glutathione-S-conjugate (GS-X) pump. MRP mediates the transport of the glutathione conjugate leukotriene C_4 (LTC_4) in vesicles isolated from HL60/ADR cells (Jedlitschky et al., 1994). Similarly, increased transport of LTC_4 and S-dinitrophenylglutathione were also observed in the drug-resistant GLC_4/ADR cell line, which overexpresses MRP 150-fold relative to its parental cell line (Müller et al., 1994). One discrepancy which awaits resolution concerning the involvement of MRP in drug resistance is the observation that MRP confers resistance to drugs which are not thought to undergo modification. Furthermore, depletion of intracellular gluathione with buthathione sulfoximine (BSO) results in reversal of MRP-mediated resistance to doxorubicin, daunorubicin, vincristine, and VP16, supporting the hypothesis that MRP transports glutathione conjugates (Zaman *et al.*, 1995). However, BSO did not alter drug resistance to doxorubicin in H69AR cells, demonstrating that there is some variability of the effect of this agent in different model systems which are not yet explained (Loe *et al.*, 1996b). Drug resistance involving glutathione conjugates is a likely complex process involving glutathione synthesis, glutathione *S*-transferase, and the GS-X pump (Müller *et al.*, 1994, 1996a). In addition to its role in drug resistance, MRP has recently been suggested to transport endogenous conjugates. ATP-dependent transport of 17β-estradiol 3-β-glucuronide was observed in membrane vesicles isolated from MRP-transfected HeLa cells (Loe *et al.*, 1996a).

Hepatic excretion of phase II metabolites, conjugated with glutathione, glucuronide, or sulfate groups, is mediated by an ATP-dependent multispecific organic anion transport (cMOAT) system (Müller *et al.*, 1996a, b). Despite the data demonstrating an overlapping role of MRP1 with this function, low expression of this gene in the liver makes it unlikely that cMOAT functions are due to MRP1. The TR- and EHBR rats are defective in the excretion of glucuronide, glutathione, and sulfate-conjugated compounds and are used as models of the human Dubin–Johnson syndrome. A recent investigation suggests that a homologue of MRP1 encodes a transporter which functions as cMOAT and is defective in these rats (Paulusma *et al.*, 1996; Ito *et al.*, 1997). A rat MRP1 homologue was isolated and found to be expressed in normal rat liver, but not in TR- or EHBR rats due to mutations resulting in premature termination of transcription. This protein is expressed on the canalicular, but not the basolateral hepatic membranes. This MRP homologue is referred as MRP2 and is in part responsible for cMOAT activity and transport of organic anions in hepatocytes.

17. MODULATION OF MRP-MEDIATED RESISTANCE

Recognition that MRP is responsible for a distinct, non-P-gp-mediated, mechanism of drug resistance led to a search for ways to circumvent it. Initial investigations with compounds such as verapamil, nicardipine, cyclosporine, the cyclosporine analogue PSC833, amiodarone, genestein, among others, have reported increased drug accumulation, increased sensitivity, or altered drug distribution (Loe *et al.*, 1996b; Twentyman and Versantvoort, 1996). The ability of specific agents to reverse drug resistance varies among cell lines and among laboratories and likely reflects experimental conditions and the derivation of the cells (Twentyman and Versantvoort, 1996). Additionally, many of these compounds are only effective at concentrations at which they are toxic and would preclude their clinical use. Clearly, additional mechanistic information is required on how MRP causes drug resistance, in order to enable the design of more successful inhibitors. In one novel approach, recently a selective antisense oligonucleotide was used to inhibit MRP protein levels and was demonstrated to increase sensitivity of MRP-transfected cells to doxorubicin (Stewart *et al.*, 1996). The ability of these oligonucleotides to function *in vivo* awaits further investigation.

18. ADDITIONAL MULTIDRUG-RESISTANCE PROTEINS

18.1. Additional MRP Homologues

Kool and co-workers recently identified three additional MRP homologues through a search of human expressed sequence tag databases (Kool *et al.*, 1997). These genes were designated MRP3, MRP4, and MRP5 and are differentially expressed in human tissues. MRP5, like MRP1, is widely expressed in human tissues, whereas MRP4 is only expressed, at low levels, in lung, kidney, bladder, gall bladder, and tonsil. MRP3 is expressed highest in liver, like MRP2 (cMOAT), moderately in intestine and colon, and to a lesser extent in lung, kidney, bladder, and spleen. The physiological function of these new MRP proteins is undetermined.

18.2. Lung Resistance Protein

The lung resistance-related protein (LRP) was first identified in a non-small-cell lung cancer cell line which was P-gp-negative (Scheper *et al.*, 1993; Izquierdo *et al.*, 1996b). Monoclonal antibodies raised to these non-P-gp drug-resistant cells led to the identification of a 100-kDa protein which was overexpressed compared to drug-sensitive parental cells. Subsequent to cloning the cDNA encoding this protein, it was determined that this protein is the human major vault protein (Scheffer *et al.*, 1995). Vaults are recently described cellular organelles whose precise function is undetermined (Rome *et al.*, 1991). This protein is associated with the resistance of cells *in vitro* to several anticancer drugs such as doxorubicin, vincristine, and etoposide. Thus one potential function of vaults may be in resistance to toxic compounds. LRP is highly expressed in the bronchus, stomach, small and large intestine, adrenal cortex, and lymphocytes. Heterogeneous expression was observed in many other organs, e.g., kidney proximal tubules, pancreatic ductal cells, and hepatocytes (Izquierdo *et al.*, 1996a). This broad expression has led to the speculation that this protein may, like P-gp and MRP, be important in protection from toxins; however, more work is required to establish this hypothesis.

19. SUMMARY

P-glycoprotein was initially isolated due to its role in multidrug resistance to cancer chemotherapeutics. Recent work, however, makes it increasingly apparent that this transporter is also involved in the pharmacokinetics of many drugs. P-gp is strategically expressed in the luminal epithelial cells of organs often associated with drug absorption and disposition, for example, hepatocyte canalicular membrane, renal proximal tubules, and the intestinal mucosa. P-gp is also expressed in the endothelial cells comprising the blood–brain barrier. This localization clearly suggests the potential for this protein to serve as a protective mechanism against entry of toxic xenobiotics and also suggests that P-gp is well situated to participate in the removal of therapeutic agents. Numerous investigations with drugs such as digoxin, etoposide, cyclosporine, vinblastine, Taxol, loperamide, domperidone, and ondansteron demonstrate that P-gp has an important role in determining the pharmacokinetics of substrate drugs. Pharmacological modulation of P-gp function to increase drug bioavailability, both on a organismal and a cellular level, is one approach currently being explored to enhance therapeutic effectiveness. This approach is not without potential collateral consequences given the wide tissue distribution of P-gp. While animals deficient in P-gp are viable and without obvious abnormalities, the pharmacokinetics and toxic consequences of several compounds are significantly altered in these animals. Thus blockade of the protective P-gp barrier in humans may have adverse effects on substrate drugs. In particular, this situation may arise when several compounds which may be substrates compete for P-gp-mediated transport.

Additional multidrug transporters, notably MRP and family members, have been identified and may also determine the fate of pharmaceuticals. Further understanding the physiological role of each of the multidrug transporters is critical for determining their role in pharmacokinetics and for evaluating the consequences of modification of their activities. Such information is also important in the development of novel drugs which may be substrates for these transporters.

REFERENCES

Akiyama, S.-I., Cornwell, M. M., Kuwano, M., Pastan, I., and Gottesman, M. M., 1988, Most drugs that reverse multidrug resistance also inhibit photoaffinity labeling of P-glycoprotein by a vinblastine analog, *Mol. Pharmacol.* **33:**144–147.

Almquist, K. C., Loe, D. W., Hipfner, D. R., Mackie, J. E., Cole, S. P. C., and Deeley,

R. G., 1995, Characterization of the Mr 190,000 multidrug resistance protein (MRP) in drug-selected and transfected human tumor cells, *Cancer Res.* **55**:102–110.

Al-Shawi, M. K., and Senior, A. E., 1993, Characterization of the adenosine triphosphate activity of Chinese hamster P-glycoprotein, *J. Biol. Chem.* **268**:4197–4206.

Al-Shawi, M. K., Urbatsch, I. L., and Senior, A. E., 1994, Covalent inhibitors of P-glycoprotein in ATPase activity, *J. Biol. Chem.* **269**:8986–8992.

Ambudkar, S. V., LeLong, I. H., Zhang, J., Cardarelli, C. O., Gottesman, M. M., and Pastan, I., 1992, Partial purification and reconstitution of the human multidrug-resistance pump: Characterization of the drug-stimulatable ATP hydrolysis. *Proc. Natl. Acad. Sci. USA* **89**:8472–8476.

Artursson. P., 1990, Epithelial transport of drugs in cell culture. I: A model for studying the passive diffusion of drugs over intestinal absorbtive (Caco-2) cells, *J. Pharmaceut. Sci.* **79**:476–482.

Artursson, P., and Karlsson, J., 1991, Correlation between oral drug absorption in humans and apparent drug permeability coefficients in human intestinal epithelial (Caco-2) cells, *Biochem. Biophys. Res. Commun.* **175**:880–885.

Audus, K. L., Bartel, R. L., Hidalgo, I. J., and Borchardt, R. T., 1990, The use of cultured epithelial and endothelial cells for drug transport and metabolism studies, *Pharmaceut. Res.* **7**:435–451.

Augustijns, P. F., Bradshaw, T. P., Gan, L.-S. L., Hendren, R. W., and Thakker, D. R., 1993, Evidence for a polarized efflux system in Caco-2 cells capable of modulating cyclosporin A transport, *Biochem. Biophys. Res. Commun.* **197**:360–365.

Azzaria, M., Shurr, E., and Gros, P., 1989, Discrete mutations introduced in the predicted nucleotide-binding sites of the mdr1 gene abolish its ability to confer multidrug resistance, *Mol. Cell. Biol.* **9**:5289–5297.

Beck, W. T., and Qian X-dong, 1992, Photoaffinity substrates for P-glycoprotein, *Biochem. Pharmacol.* **43**:89–93.

Biedler, J. L., 1994, Drug resistance: Genotype *versus* phenotype, *Cancer Res.* **54**:666–678.

Biedler, J. L., and Riehm, H., 1970, Cellular resistance to actinomycin D in Chinese hamster cells *in vitro*; cross resistance, radioautographic and cytogenetic studies, *Cancer Res.* **30**:1174–1184.

Borchardt, R. T., Smith, P., and Wilson, G., 1996, General principles in the characterization and use of model systems for biopharmaceutical studies, in: *Models for Assessing Drug Absorption and Metabolism* (R. T. Borchardt, P. Smith, and G. Wilson, eds.), Plenum Press, New York, pp. 1–11.

Borst, P., Schinkel, A. H., Smit, J. J. M., Wagenaar, E., Van Deemter, L., Smith, A. J., Eijdems, W. H. M., Baas, F., and Zaman, G. J. R., 1993, Classical and novel forms of multidrug resistance and the physiological functions of P-glycoproteins in mammals, *Pharmacol. Ther.* **60**:289–299.

Bremer, S., Hoof, T., Wilke, M., Busche, R., Scholte, B., Riordan, J. R., Maass, G., and Tümmler, B., 1992, Quantitative expression patterns of multidrug-resistance P-glycoprotein (MDR1) and differentially spliced cystic-fibrosis transmembrane-conductance regulator mRNA transcripts in human epithelia, *Eur. J. Biochem.* **206**:137–149.

Brown, P. C., Thorgeirsson, S. S., and Silverman, J. A., 1993, Cloning and regulation of the rat *mdr2* gene, *Nucleic Acids Res.* **21**:3885–3891.

Buschman, E., and Gros, P., 1991, Functional analysis of chimeric genes obtained by exchanging homologous domains of the mouse *mdr1* and *mdr2* genes, *Mol. Cell. Biol.* **11**:595–603.

Callaghan, R., and Riordan, J. R., 1993, Synthetic and natural opiates interact with P-glycoprotein in multidrug-resistant cells, *J. Biol. Chem.* **268**:16059–16064.

Cavet, M. E., West, M., and Simmons, N. L., 1996, Transport and epithelial secretion of the cardiac glycoside, digoxin, by human intestinal epithelial (Caco-2) cells, *Br. J. Pharmacol.* **118**:1389–1396.

Chaudhary, P. M., and Roninson, I. B., 1991, Expression and activity of P-glycoprotein, a multidrug efflux pump, in human and hematopoietic stem cells, *Cell* **66**:85–94.

Chaudhary, P. M., Metchener, E. B., and Roninson, I. B., 1992, Expression and activity of the multidrug resistance P-glycoprotein in human peripheral blood lymphocytes, *Blood* **80**:2735–2739.

Chin, J. E., Soffir, R., Noonan, K. E., Choi, K., and Roninson, I. B., 1989, Structure and expression of the human *MDR* (P-glycoprotein) gene family, *Mol. Cell. Biol.* **9**:3808–3820.

Chin, K.-V., Tanaka, S., Darlington, G., Pastan, I., and Gottesman, M. M., 1990, Heat shock and arsenite increase expression of the multidrug resistance (MDR1) gene in human renal carcinoma cells, *J. Biol. Chem.* **265**:221–226.

Choi, K., Frommel, T. H. O., Stern, R. K., Perez, C. F., Kriegler, M., Tsuruo, T., and Roninson, I. B., 1991, Multidrug resistance after retroviral transfer of the human MDR1 gene correlates with P-glycoprotein density in the plasma membrane and is not affected by cytotoxic selection, *Proc. Natl. Acad. Sci. USA* **88**:7386–7390.

Cole, S. P. C., Bhardwaj, G., Gerlach, J. H., Mackie, J. E., Grant, C. E., Almquist, K. C., Stewart, A. J., Kurz, E. U., Duncan, A. M. V., and Deeley, R. G., 1992, Overexpression of a transporter gene in a multidrug-resistant human lung cancer cell line, *Science* **258**:1650–1654.

Cordon-Cardo, C., O'Brien, J. P., Casals, D., Rittman-Grauer, L., Biedler, J. L., Melamed, M. R., and Bertino, J. R., 1989, Multidrug-resistance gene (P-glycoprotein) is expressed by endothelial cells at blood–brain barrier sites, *Proc. Natl. Acad. Sci. USA* **86**:695–698.

Cornwell, M. M., Tsuruo, T., Gottesman, M. M., and Pastan, I., 1987, ATP-binding properties of P-glycoprotein from multidrug resistant KB cells, *FASEB J.* **1**:51–54.

Dantzig, A. H., Shepard, R. L., Cao, J., Law, K. L., Ehlhardt, W. J., Baughman, T. M., Bumol, T. F., and Starling, J. J., 1996, Reversal of P-glycoprotein-mediated multidrug resistance by a potent cyproyldibenzosuberane modulator, LY335979, *Cancer Res.* **56**:4171–4179.

de Lannoy, I. A. M., and Silverman, M., 1992, The MDR1 product, P-glycoprotein, mediates the transport of the cardiac glycoside, digoxin, *Biochem. Biophys. Res. Commun.* **189**:551–557.

Deuchars, K. L., Duthie, M., and Ling, V., 1992, Identification of distinct P-glycoprotein gene sequences in rat, *Biochim. Biophys. Acta* **1130**:157–165.

Devault, A., and Gros, P., 1990, Two members of the mouse *mdr* gene family confer multidrug resistance with overlapping but distinct drug specificities, *Mol. Cell. Biol.* **10**:1652–1663.

Doige, C. A., and Sharom, F. J., 1992, Transport properties of P-glycoprotein in plasma membrane vesicles from multidrug-resistant Chinese hamster ovary cells, *Biochim. Biophys. Acta* **1109**:161–171.

Doige, C. A., Yu, X., and Sharom, F. J., 1992, ATPase activity of partially purified P-glycoprotein from multidrug-resistant Chinese hamster ovary cells, *Biochim. Biophys. Acta* **1109**:149–160.

Doige, C. A., Yu, X., and Sharom, F. J., 1993, The effects of lipids and detergents on ATPase-active P-glycoprotein, *Biochim. Biophys. Acta* **1146**:65–72.

Drion, N., Lemaire, M., Lefauconnier, J.-M., and Scherrmann, J.-M., 1996, Role of P-glycoprotein in the blood–brain transport of colchicine and vinblastine, *J. Neurochem.* **67**:1688–1693.

Endicott, J. A., and Ling, V., 1989, The biochemistry of P-glycoprotein mediated multidrug resistance, *Annu. Rev. Biochem.* **58**:137–171.

Ferry, D. R., Traunecker, H., and Kerr, D. J., 1996, Clinical trials of P-glycoprotein reversal in solid tumours, *Eur. J. Cancer* **32A**:1070–1081.

Fisher, G. A., Lum, B. L., Hausdorff, J., and Sikic, B. I., 1996, Pharmacological considerations in the modulation of multidrug resistance, *Eur. J. Cancer* **32A**:1082–1088.

Flens, M. J., Zaman, G. J. R., van der Valk, P., Izquierdo, M. A., Schroeijers, A. B., Scheffer, G. L., van der Groep, P., de Haas, M., Meijer, C. J. L. M., and Scheper, R. J., 1996, Tissue distribution of the multidrug resistance protein, *Am. J. Pathol.* **148**:1237–1247.

Fojo, A. T., Ueda, K., Slamon, D. J., Poplack, D. G., Gottesman, M. M., and Pastan, I., 1987, Expression of a multidrug-resistance gene in human tumors and tissues, *Proc. Natl. Acad. Sci. USA* **84**:265–269.

Ford, J. M., 1996, Experimental reversal of P-glycoprotein-mediated multidrug resistance by pharmacological chemosensitizers, *Eur. J. Cancer* **32A**:991–1001.

Ford, J. M., and Hait, W. N., 1990, Pharmacology of drugs that alter multidrug resistance in cancer, *Pharmacol. Rev.* **42**:155–199.

Gan, L.-S. L., Moseley, A., Khosla, B., Augustijns, P. F., Bradshaw, T. P., Hendren, R. W., and Thakker, D. R., 1996, CYP3A-like cytochrome P450-mediated metabolism and polarized efflux of cyclosporin A in Caco-2 cells, *Drug Metab. Dispos.* **24**:344–349.

Gant, T. W., Silverman, J. A., Bisgaard, H. C., Burt, R. K., Marino, P. A., and Thorgeirsson, S. S., 1991, Regulation of 2-acetylaminofluorene and 3-methylcholanthrene mediated induction of mdr and cytochrome P450IA gene family expression in primary hepatocyte cultures and rat liver, *Mol. Carcin.* **4**:499–509.

Georges, E., Zhang, J.-T., and Ling, V., 1991, Modulation of ATP and drug binding by monoclonal antibodies against P-glycoprotein, *J. Cell. Physiol.* **148**:479–484.

Georges, E., Tsuruo, T., and Ling, V., 1993, Topology of P-glycoprotein as determined by epitope mapping of MRK-16 monoclonal antibody, *J. Biol. Chem.* **268**:1792–1798.

Germann, U. A., 1996, P-glycoprotein—A mediator of multidrug resistance in tumour cells, *Eur. J. Cancer* **32A**:927–944.

Germann, U. A., Ford, P. J., Shlaykhter, D., Mason, V. S., and Harding, M. W., 1997a, Chemosensitization and drug accumulation effects of VX-710, verapamil, cyclosporin A, MS-209 and GF120918 in multidrug resistant HL60/ADR cells expressing the multidrug resistance-associated protein MRP, *Anticancer Drugs* **8**:141–155.

Germann, U. A., Shlaykhter, D., Mason, V. S., Zelle, R. E., Duffy, J. P., Galullo, V., Armistead, D. M., Saunders, J. O., Boger, J., and Harding, M. W., 1997b, Cellular and bio-

chemical characterization of VX-710 as an chemosensitizer: Reversal of P-glycoprotein mediated multidrug resistance *in vitro, Anticancer Drugs* **8:**125–140.
Goldstein, L. J., 1996, *MDR1* gene expression in solid tumors, *Eur. J. Cancer* **32A:**1039–1050.
Goldstein, L. J., Galski, H., Fojo, A. T., Willingham, M., Lai, S.-L., Gazdar, A., Pirker, R., Green, A., Crist, W., Brodeur, G. M., Lieber, M., Cossman, J., Gottesman, M. M., and Pastan, I., 1989, Expression of a multidrug resistance gene in human cancers, *J. Natl. Cancer Inst.* **81:**116–124.
Gottesman, M. M., and Pastan, I., 1993, Biochemistry of multidrug resistance mediated by the multidrug transporter, *Annu. Rev. Biochem.* **62:**385–427.
Grant, C. E., Valdimarsson, G., Hipfner, D. R., Almquist, K. C., Cole, S. P. C., and Deeley, R. G., 1994, Overexpression of multidrug resistance-associated protein (MRP) increases resistance to natural product drugs, *Cancer Res.* **54:**357–361.
Greenberger, L. M., 1993, Major photoaffinity drug labeling sites for iodaryl azidoprazosin in P-glycoprotein are within or immediately C-terminal to transmembrane domains 6 and 12, *J. Biol. Chem.* **268:**11417–11425.
Griffiths, N. M., Hirst, B. H., and Simmons, N. L., 1994, Active intestinal secretion of the fluoroquinolone antibacterials ciprofloxacin, norfloxacin and pefloxacin; a common secretory pathway? *J. Pharmacol. Exp. Ther.* **269:**496–502.
Gros, P., Ben-Neriah, Y., Croop, J. M., and Housman, D. E., 1986a, Isolation and expression of a complementary DNA that confers multidrug resistance, *Nature* **323:**728–731.
Gros, P., Croop, J., and Housman, D., 1986b, Mammalian multidrug resistance gene: Complete cDNA sequence indicates strong homology to bacterial transport proteins, *Cell* **47:**371–380.
Gros, P., Croop, J., Roninson, I., Varshavsky, A., and Housman, D. E., 1986c, Isolation and characterization of DNA sequences amplified in multidrug-resistant hamster cells, *Proc. Natl. Acad. Sci. USA* **83:**337–341.
Gros, P., Raymond, M., Bell, J., and Housman, D., 1988, Cloning and characterization of a second member of the mouse *mdr* gene family, *Mol. Cell. Biol.* **8:**2770–2778.
Guild, B. C., Mulligan, R. C., Gros, P., and Housman, D. E., 1988, Retroviral transfer of a murine cDNA for multidrug resistance confers pleiotropic drug resistance to cells without prior drug selection, *Proc. Natl. Acad. Sci. USA* **85:**1595–1599.
Hamada, H., and Tsuruo, T., 1988, Purification of the 170–180 kilodalton membrane glycoprotein associated with multidrug resistance, *J. Biol. Chem.* **263:**1454–1458.
Hammond, J. R., Johnstone, R. M., and Gros, P., 1989, Enhanced efflux of 3-H-vinblastine from Chinese hamster ovary cells transfected with full length complementary DNA clone for the MDR1 gene, *Cancer Res.* **49:**3867–3871.
Hidalgo, I. J., Raub, T. J., and Borchardt, R. T., 1989, Characterization of the human colon carcinoma cell line (Caco-2) as a model system for intestinal epithelial permeability, *Gastroenterology* **96:**736–749.
Higgins, C. F., 1992, ABC transporters: From microorganisms to man, *Annu. Rev. Cell Biol.* **8:**67–113.
Hill, B. A., Brown, P. C., Preisegger, K.-H., and Silverman, J. A., 1996, Regulation of *mdr1b* gene expression in Fischer, Wistar and Sprague-Dawley rats *in vivo* and *in vitro, Carcinogenesis.* **17:**451–457.
Horio, M., Pastan, I., Gottesman, M. M., and Handler, J. S., 1990, Transepithelial transport

of vinblastine by kidney-derived cell lines. Application of a new kinetic model to estimate in situ K_m of the pump, *Biochim. Biophys. Acta* **1027**:116–122.

Hunter, J., Hirst, B. H., and Simmons, N. L., 1991, Epithelial secretion of vinblastine by human intestinal adenocarcinoma cell (HCT-8 and T84) layers expressing P-glycoprotein, *Br. J. Cancer* **64**:437–444.

Hunter, J., Hirst, B. H., and Simmons, N. L., 1993, Drug absorption limited by P-glycoprotein-mediated secretory drug transport in human intestinal epithelial Caco-2 cell layers, *Pharmaceut. Res.* **10**:743–749.

Huwyler, J., Drewe, J., Klusemann, C., and Fricker, G., 1996, Evidence for P-glycoprotein-modulated penetration of morphine-6-glucuronide into brain capillary endothelium, *Br. J. Pharmacol.* **118**:1879–1885.

Hyafil, F., Vergely, C., Du Vignaud, P., and Grand-Perret, T., 1993, In vitro and in vivo reversal of multidrug resistance by GF120918, an acridonecarboxamide derivative, *Cancer Res.* **53**:4595–4602.

Ito, K., Suzuki, H., Hirohashi, T., Kume, K., Shimizu, T., and Sugiyama, Y., 1997, Molecular cloning of canalicular multispecific organic anion transporter defective in EHBR, *Am. J. Physiol.* **272**:G16–G22.

Izquierdo, M. A., Scheffer, G. L., Flens, M. J., Giaccone, G., Broxterman, H. J., Meijer, C. J. L. M., van der Valk, P., and Scheper, R. J., 1996a, Broad distribution of the multidrug resistance related vault lung resistance protein in normal human tissues and tumors, *Am. J. Pathol.* **148**:877–887.

Izquierdo, M. A., Scheffer, G. L., Flens, M. J., Schroeijers, A. B., van der Valk, P., and Scheper, R. J., 1996b, Major vault protein LRP-related multidrug resistance, *Eur. J. Cancer.* **32A**:979–984.

Jedlitschky, G., Leier, I., Buchholz, U., Center, M., and Keppler, D., 1994, ATP-dependent transport of glutathione S-conjugates by the multidrug resistance-associated protein, *Cancer Res.* **54**:4833–4836.

Jetté, L., Murphy, G. F., Leclerc, J.-M., and Beliveau, R., 1995, Interaction of drugs with P-glycoprotein in brain capillaries, *Biochem. Pharmacol.* **50**:1701–1709.

Juliano, R. L., and Ling, V., 1976, A surface glycoprotein modulating drug permeability in Chinese hamster ovary cell mutants, *Biochim. Biophys. Acta* **455**:152–162.

Juranka, P. F., Zastawny, R. L., and Ling, V., 1989, P-glycoprotein: Multidrug-resistance and a superfamily of membrane-associated transport proteins, *FASEB J.* **3**:2583–2592.

Kanamaru, H., Kakehi, Y., Yoshida, O., Nakanishi, S., Pastan, I., and Gottesman, M. M., 1989, MDR1 RNA levels in human renal cell carcinomas: Correlation with grade and prediction of reversal of doxorubicin resistance by quinidine in tumor explants, *J. Natl. Cancer Inst.* **81**:844–849.

Karlsson, J., Kuo, S.-M., Ziemniak, J., and Artursson, P., 1993, Transport of celiprolol across human intestinal epithelial (Caco-2) cells: Mediation of secretion by multiple transporters including P-glycoprotein, *Br. J. Pharmacol.* **110**:1009–1016.

Kartner, N., Riordan, J. R., and Ling, V., 1983, Cell surface P-glycoprotein associated with multidrug resistance in mammalian cell lines, *Science* **221**:1285–1288.

Kartner, N., Evernden-Porelle, D., Bradley, G., and Ling, V., 1985, Detection of P-glycoprotein in multidrug resistant cell lines by monoclonal antibodies, *Nature* **316**:820–823.

Keller, R. P., Altermatt, H. J., Nooter, K., Poschmann, G., Laissue, J. A., Bollinger, P., and

Hiestand, P. C., 1992, SDZ PSC833, a non-immunosuppressive cyclosporine: Its potency in overcoming P-glycoprotein-mediated multidrug resistance of murine leukemia, *Int. J. Cancer* **50:**593–597.

Kelly, R., and Smith, T. W., 1996, Pharmacological treatment of heart failure, in: *Goodman and Gilman's The Pharmacological Basis of Therapeutics* (J. G. Hardman, L. E. Limbird, P. B. Molinoff, R. W. Ruddon, and A. G. Gilman eds.), McGraw-Hill, New York, pp. 809–838.

Kohno, K., Sato, S.-., Takano, H., Matsuo, K.-i., and Kuwano, M., 1989, The direct activation of human multidrug resistance gene by anticancer agents, *Biochem. Biophys. Res. Commun.* **164:**1415–1421.

Kool, M., de Haas, M., Scheffer, G. L., Scheper, R. J., van Eijk, M. J. T., Juijn, J. A., Baas. F., and Borst, P., 1997, Analysis of expression of *cMOAT* (MRP2), *MRP3, MRP4, MRP5* homologues of the multidrug resistance-associated protein gene (*MRP1*), in human cancer cell lines, *Cancer Res.* **57:**3537–3547.

Krishnamachary, N., and Center, M. S., 1993, The MRP gene associated with a non-P-glycoprotein multidrug resistance encodes a 190-kDa membrane bound glycoprotein, *Cancer Res.* **53:**3658–3661.

Kruh, G. D., Chan, A., Myers, K., Gaughan, K., Miki, T., and Aaronson, S. A., 1994, Expression complementary DNA library transfer establishes *mrp* as a multidrug resistance gene, *Cancer Res.* **54:**1649–1652.

Lai, S.-L., Goldstein, L. J., Gottesman, M. M., Pastan, I., Tsai, C.-M., Johnson, B. E., Muslshine, J. L., Ihde, D. C., Kayser, K., and Gazdar, A. F., 1989, MDR1 gene expression in lung cancer, *J. Natl. Cancer Inst.* **81:**1144–1150.

Leu, B.-L., and Huang, J.-d., 1995, Inhibition of intestinal P-glycoprotein and effects on etoposide absorption, *Cancer Chemother. Pharmacol.* **35:**432–436.

Loe, D. W., Almquist, K. C., Cole, S. P. C., and Deeley, R. G., 1996a, ATP-dependent 17β-estradiol 17-(β-D-glucuronide) transport by multidrug resistance protein (MRP), *J. Biol. Chem.* **271:**9683–9689.

Loe, D. W., Deeley, R. G., and Cole, S. P. C., 1996b, Biology of the multidrug resistance-associated protein, MRP, *Eur. J. Cancer* **32A:**945–957.

Loo, T. W., and Clarke, D. M., 1993, Functional consequences of phenylalanine mutations in the predicted transmembrane domain p-glycoprotein, *J. Biol. Chem.* **268:**19965–19972.

Loo, T. W., and Clarke, D. M., 1995, Membrane topology of a cysteine-less mutant of human P-glycoprotein, *J. Biol. Chem.* **270:**843–848.

Lum, B. L., Kaubisch, S., Yahanda, A. M., Adler, K. M., Jew, L., Ehsan, M. N., Brophy, N. A., Halsey, J., Gosland, M. P., and Sikic, B. I., 1992, Alteration of etoposide pharmacokinetics and pharmacodynamics by cyclosporine in a phase I trial to modulate multidrug resistance, *J. Clin. Oncol.* **10:**1635–1642.

Lum, B. L., Gisher, G. A., Brophy, N. A., Yahanda, A. M., Adler, K. M., Kaubisch, S., Halsey, J., and Sikic, B. I., 1993, Clincal trials of modulation of multidrug resistance, *Cancer* (Suppl.) **72:**3502–3514.

Mauad, T. H., van Nieuwkerk, C. M. J., Dingemans, K. P., Smit, J. J. M., Schinkel, A. H., Notenboom, R. G. E., van den Bergh Weerman, M. A., Verkruisen, R. P., Groen, A. K., Oude Elferink, R. P. J., Borst, P., and Offerhaus, G. J. A., 1994, Mice with homozygous disruption of the *mdr2* P-glycoprotein gene: A novel animal model for studies of

nonsuppurative inflammatory cholangitis and hepatocarcinogenesis, *Am. J. Pathol.* **145:**1237–1245.
Mayer, U., Wagenaar, E., Beijnen, J. H., Smit, J. W., Meijer, D. K. F., van Asperen, J., Borst, P., and Schinkel, A. H., 1996, Substantial excretion of digoxin via the intestinal mucosa and prevention of long-term digoxin accumulation in the brain by the mdr1a P-glycoprotein, *Br. J. Pharmacol.* **119:**1038–1044.
Morrow, C. S., Nakagawa, M., Goldsmith, M. E., Madden, M. J., and Cowan, K. H., 1994, Reversible transcription activation of *mdr1* by sodium butyrate treatment of human colon cancer cells, *J. Biol. Chem.* **269:**10739–10746.
Müller, M., Meijer, C., Zaman, G. J. R., Borst, P., Scheper, R. J., Mulder, N. H., de Vries, E. G. E., and Jansen, P. L. M., 1994, Overexpression of the gene encoding the multidrug resistance-associated protein results in increased ATP-dependent glutathione S-conjugate transport, *Proc. Natl. Acad. Sci. USA* **91:**13033–13037.
Müller, M., de Vries, E. G. E., and Jansen, P. L. M., 1996a, Role of the multidrug resistance protein (MRP) in glutathione S-conjugate transport in mammalian cells, *J. Hepatol.* **24:**100–108.
Müller, M., Roelofsen, H., and Jansen, P. L. M., 1996b, Secretion of organic anions by hepatocytes: Involvement of homologues of the multidrug resistance protein, *Sem. Liver Dis.* **16:**211–220.
Nakatsukasa, H., Silverman, J. A., Gant, T. W., Evarts, R. P., and Thorgeirsson, S. S., 1993, Expression of multidrug resistance genes in rat liver during regeneration and after carbon tetrachloride intoxication, *Hepatology* **18:**1202–1207.
Nerurkar, M. M., Burton, P. S., and Borchardt, R. T., 1996, The use of surfactants to enhance the permeability of peptides through Caco-2 cells by inhibition of an apically polarized efflux system, *Pharmaceut. Res.* **13:**528–534.
Ng, W. F., Sarangi, F., Zastawny, R. L., Veniot-Drebot, L., and Ling, V., 1989, Identification of members of the P-glycoprotein multigene family, *Mol. Cell. Biol.* **9:**1224–1232.
Nogae, I., Kohno, K., Kikuchi, J., Kuwano, M., Akiyama, S.-I., Kiue, A., Suzuki, K.-I., Yoshida, Y., Cornwell, M. M., Pastan, I., and Gottesman, M. M., 1989, Analysis of structural features of dihydropyridine analogs needed to reverse multidrug resistance and to inhibit photoaffinity labelling of P-glycoprotein, *Biochem. Pharmacol.* **38:**519–527.
Oude Elferink, R. P. J., and Groen, A. K., 1995, The role of mdr2 P-glycoprotein in biliary lipid secretion. Cross-talk between cancer research and biliary physiology, *J. Hepatol.* **23:**617–625.
Pastan, I., Gottesman, M. M., Ueda, K., Lovelace, E., Rutherford, A. V., and Willingham, M. C., 1988, A retrovirus carrying an MDR1 cDNA confers multidrug resistance and polarized expression of P-glycoprotein in MDCK cells, *Proc. Natl. Acad. Sci. USA* **85:**4486–4490.
Paul, S., Breuninger, L. M., Tew, K. D., Shen, H., and Kruh, G. D., 1996, ATP-dependent uptake of natural product cytotoxic drugs by membrane vesicles establishes MRP as a broad specificity transporter, *Proc. Natl. Acad. Sci. USA* **93:**6929–6934.
Paulusma, C. C., Bosma, P. J., Zaman, G. J. R., Bakker, C. T. M., Otter, M., Scheffer, G. L., Scheper, R. J., Borst, P., and Oude Elferink, R. P. J., 1996, Congenital jaundice in rats with a mutation in a multidrug resistance-associated protein gene, *Science* **271:**1126–1128.

Rabbaa, L., Dautrey, S., Colas-Linhart, N., Carbon, C., and Farinotti, R., 1996, Intestinal elimination of ofloxacin entantiomers in the rat: Evidence of a carrier-mediated process, *Antimicrob. Agents Chemother.* **40**:2126–2130.

Relling, M. V., 1996, Are the major effects of P-glycoprotein modulators due to altered pharmacokinetics of anticancer drugs? *Ther. Drug Monit.* **18**:350–356.

Riordan, J. R., Deuchars, K., Kartner, N., Alon, N., Trent, J., and Ling, V., 1985, Amplification of P-glycoprotein genes in multidrug-resistant cell mammalian cell lines, *Nature* **316**:817–819.

Rome, L. H., Kedersha, N. L., and Chungai, D. C., 1991, Unlocking vaults: Organelles in search of a function, *Trends Cell Biol.* **1**:47–50.

Roninson, I. B., Chin, J. E., Choi, K., Gros, P., Housman, D. E., Fojo, A., Shen, D.-W., Gottesman, M. M., and Pastan, I., 1986, Isolation of human *mdr* DNA sequences amplified in multidrug-resistant KB carcinoma cells, *Proc. Natl. Acad. Sci. USA* **83**:4538–4542.

Rosenberg, M. F., Callaghan, R., Ford, R. C., and Higgins, C. F., 1997, Structure of the multidrug resistance P-glycoprotein to 2.5 nm resolution determined by electron microscopy and image analysis, *J. Biol. Chem.* **272**:10685–10694.

Ruetz, S., and Gros, P., 1994, Phosphotidylcholine translocase: A physiological role for the *mdr2* gene, *Cell* **77**:1071–1081.

Ruetz, S., and Gros, P., 1995, Enhancement of Mdr2-mediated phosphotidylcholine translocation by the bile salt taurocholate, *J. Biol. Chem.* **270**:25388–25395.

Ruetz, S., Brault, M., Kast, C., Hemenway, C., Heitman, J., Grant, C. E., Cole, S. P. C., Deeley, R. G., and Gros, P., 1996, Functional expression of the multidrug resistance-associated protein in the yeast *Saccharomyces cerevisiae, J. Biol. Chem.* **271**:4154–4160.

Saeki, T., Ueda, K., Tanigawara, Y., Hori, R., and Komano, T., 1993, Human P-glycoprotein transports cyclosporin A and FK506, *J. Biol. Chem.* **268**:6077–6080.

Scheffer, G. L., Wijngaard, P. L. J., Flens, M. J., Izquierdo, M. A., Slovak, M. L., Pinedo, H. M., Meijer, C. J. L. M., Clevers, H. C., and Scheper, R. J., 1995, The drug resistance-related protein LRP is the human major vault protein, *Nature Med.* **1**:578–582.

Scheper, R. J., Broxterman, H. J., Scheffer, G. L., Kaaijk, P., Dalton, W. S., van Heijningen, T. H. M., van Kalken, C. K., Slovak, M. L., de Vries, E. G. E., van der Valk, P., Meijer, C. J. L. M., and Pindeo, H. M., 1993, Overexpression of a Mr 110,000 vesicular protein in non-P-glycoprotein-mediated multidrug resistance, *Cancer Res.* **53**:1475–1479.

Schinkel, A. H., Roelofs, M. E. M., and Borst, P., 1991, Characterization of the human *MDR3* P-glycoprotein and its recognition by P-glycoprotein-specific monoclonal antibodies, *Cancer Res.* **51**:2628–2635.

Schinkel, A. H., Smit, J. J. M., van Tellingen, O., Beijnen, J. H., Wagenaar, E., van Deemter, L., Mol, C. A. A. M., van der Valk, M. A., Robanus-Maandag, E. C., te Riele, H. P. J., Berns, A. J. M., and Borst, P., 1994, Disruption of the mouse *mdr1a* P-glycoprotein gene leads to a deficiency in the blood–brain barrier and to increased sensitivity to drugs, *Cell* **77**:491–502.

Schinkel, A. H., Wagenaar, E., van Deemter, L., Mol, C. A. A. M., and Borst, P., 1995, Absence of the mdr1a P-glycoprotein in mice affects tissue distribution and pharmacokinetics of dexamethasone, digoxin and cyclosporine A, *J. Clin. Invest.* **96**:1698–1705.

Schinkel, A. H., Wagenaar, E., Mol, C. A. A. M., and van Deemter, L., 1996, P-glycopro-

tein in the blood–brain barrier of mice influences the brain penetration and pharmacological activity of many drugs, *J. Clin. Invest.* **97**:2517–2524.
Schinkel, A. H., Mayer, U., Wagenaar, E., Mol, C. A. A. M., van Deemter, L., Smit, J. J. M., van der Valk, M. A., Voordouw, A. C., Spits, H., van Tellingen, O., Zijlmans, J. M. J. M., Fibbe, W. E., and Borst, P., 1997, Normal viability and altered pharmacokinetics in mice lacking mdr1-type (drug transporting) P-glycoproteins, *Proc. Natl. Acad. Sci. USA* **94**:4028–4033.
Schrenk, D., Gant, T. W., Preisegger, K.-H., Silverman, J. A., Marino, P., and Thorgeirsson, S. S., 1993, Induction of multidrug resistance gene expression during cholestasis in rats and nonhuman primates, *Hepatology* **17**:854–860.
Schuetz, E. G., Beck, W. T., and Schuetz, J. D., 1996a, Modulators and substrates of P-glycoprotein and cytochrome p4503A coordinately up-regulate these proteins in human colon carcinoma cells, *Mol. Pharmacol.* **49**:311–318.
Schuetz, E. G., Schinkel, A. H., Relling, M. V., and Schuetz, J. D., 1996b, P-glycoprotein: A major determinant of rifampicin-inducible expression of cytochrome P4503A in mice and humans, *Proc. Natl. Acad. Sci. USA* **93**:4001–4005.
Shapiro, A. B., and Ling, V., 1994, ATPase activity of purified and reconstituted P-glycoprotein from chinese hamster ovary cells, *J. Biol. Chem.* **269**:3745–3754.
Sharom, F. J., Yu, X., and Doige, C. A., 1993, Functional reconstitution of drug transport and ATPase activity in proteoliposomes containing partially purified P-glycoprotein, *J. Biol. Chem.* **268**:24197–24202.
Sharom, F. J., Yu, X., Chu, J. W. K., and Doige, C. A., 1995, Characterization of the ATPase activity of P-glycoprotein from multidrug-resistant Chinese hamster ovary cells, *Biochem. J.* **308**:381–390.
Shen, D.-W., Cardarelli, C., Hwang, J., Cornwell, M. M., Richert, N., Ishi, S., Pastan, I., and Gottesman, M. M., 1986, Multiple drug resistant human KB carcinoma cells independantly selected for high-level resistance to colchicine, adriamycin or vinblastine show changes in expression of specific proteins, *J. Biol. Chem.* **261**:7762–7770.
Silverman, J. A., and Schrenk, D., 1997, Expression of the multidrug resistance genes in the liver, *FASEB J.* **11**:308–313.
Silverman, J. A., Raunio, H., Gant, T. W., and Thorgeirsson, S. S., 1991, Cloning and characterization of a member of the rat multidrug resistance (*mdr*) gene family, *Gene* **106**:229–236.
Smit, J. J. M., Schinkel, A. H., Oude Elferink, R. P. J., Groen, A. K., Wagenaar, E., van Deemter, L., Mol, C. A. A. M., Ottenhoff, R., van der Lugt, N. M. T., van der Valk, M. A., van Roon, M. A., Offerhaus, G. J. A., Berns, A. J. M., and Borst, P., 1993, Homozygous disruption of the murine *mdr2* p-glycoprotein gene leads to a complete absence of phospholipid from bile and to liver disease, *Cell* **75**:451–462.
Smit, J. J. M., Schinkel, A. H., Mol, C. A. A. M., Majoor, D., Mooi, W. J., Jongsma, A. P. M., Lincke, C. R., and Borst, P., 1994, Tissue distribution of the human MDR3 P-glycoprotein, *Lab. Invest.* **71**:638–649.
Smith, A. J., Timmermans-Hereijgers, J. L. P. M., Roelofsen, B., Wirtz, K. W. A., van Blitterswijk, W. J., Smit, J. J. M., Schinkel, A. H., and Borst, P., 1994, The human MDR3 P-glycoprotein promotes translocation of phosphotidylcholine through the plasma membrane of fibroblasts from transgenic mice, *FEBS Lett.* **354**:263–266.
Sparreboom, A., van Asperen, J., Mayer, U., Schinkel, A. H., Smit, J. W., Meijer, D. K. F.,

Borst, P., Nooijen, W. J., Beijnen, J. H., and van Tellingen, O., 1997, Limited oral bioavailability and active epithelial excretion of paclitaxel (Taxol) caused by P-glycoprotein in the intestine, *Proc. Natl. Acad. Sci. USA* **94**:2031–2035.

Stewart, A. J., Canitrot, Y., Baracchini, E., Dean, N. M., and Deely, R. G., 1996, Reduction of expression of the multidrug resistance protein, MRP, in human tumor cells by antisense phophorothioate oligonuclotides, *Biochem. Pharmacol.* **51**:461–469.

Su, S.-F., and Huang, J.-D., 1996, Inhibition of the intestinal digoxin absorption and exsorption by quinidine, *Drug Metab. Dispos.* **24**:142–147.

Tanaka, K., Hirai, M., Tanigawara, Y., Ueda, K., Takano, M., Hori, R., and Inui, K.-I., 1997, Relationship between expression level of P-glycoprotein and daunorubicin transport in LLC-PK1 cells transfected with human *MDR1* gene, *Biochem. Pharmacol.* **53**:741–746.

Tanigawara, Y., Okamura, N., Hirai, M., Yasuhara, M., Ueda, K., Kioka, N., Komano, T., and Hori, R., 1992, Transport of digoxin by human P-glycoprotein expressed in a porcine kidney epithelial cell line (LLC-PK1), *J. Pharmacol. Exp. Ther.* **263**:840–845.

Thiebaut, F., Tsuruo, T., Hamada, H., Gottesman, M. M., Pastan, I., and Willingham, M. C., 1987, Cellular localization of the multidrug resistance gene product P-glycoprotein in normal human tissues, *Proc. Natl. Acad. Sci. USA* **84**:7735–7738.

Thiebaut, F., Tsuruo, T., Hamada, H., Gottesman, M. M., Pastan, I., and Willingham, M. C., 1989, Immunohistochemical localization in normal tissues of different epitopes in the multidrug transport protein P170: Evidence for localization in brain capillaries and crossreactivity on one antibody with a muscle protein, *J. Histochem. Cytochem.* **37**:159–164.

Thorgeirsson, S. S., Huber, B. E., Sorrell, S., Fojo, A. T., Pastan, I., and Gottesman, M. M., 1987, Expression of the multidrug-resistance gene in hepatocarcinogenesis and regenerating rat liver, *Science* **236**:1120–1122.

Tsuruo, T., Iida, H., Tsukagoshi, S., and Sakurai, Y., 1981, Overcoming of vincristine resistance in P388 leukemia *in vivo* and *in vitro* through enhanced cytotoxicity of vincristine and vinblastine by verapamil, *Cancer Res.* **41**:1967–1972.

Twentyman, P. R., 1992, Cyclosporins as drug resistance modifiers, *Biochem. Pharmacol.* **43**:109–117.

Twentyman, P. R., and Versantvoort, C. H. M., 1996, Experimental modulation of MRP (multidrug resistance-associated protein)-mediated resistance, *Eur. J. Cancer* **32A**:1002–1009.

Ueda, K., Cardarelli, C., Gottesman, M. M., and Pastan, I., 1987a, Expression of a full length cDNA for the human "MDR1" gene confers resistance to colchicine, doxorubicin and vinblastine, *Proc. Natl. Acad. Sci. USA* **84**:3004–3008.

Ueda, K., Clark, D. P., Chen, C.-J., Roninson, I. B., Gottesman, M. M., and Pastan, I., 1987b, The human multidrug resistance (*mdr1*) gene: cDNA cloning and transcription initiation, *J. Biol. Chem.* **262**:505–508.

van Asperen, J., Schinkel, A. H., Beijnen, J. H., Nooijen, W. J., Borst, P., and van Tellingen, O., 1996, Altered pharmacokinetics of vinblastine in mdr1a P-glycoprotein-deficient mice, *J. Natl. Cancer Inst.* **88**:994–999.

van Asperen, J., van Tellingen, O., Sparreboom. A., Schinkel, A. H., Borst, P., Nooijen, W. J., and Beijnen, J. H., 1997, Enhanced oral bioavailability of paclitaxel in mice

treated with the P-glycoprotein blockers SDZ PSC 833 or cyclosporin A, *Proc. Am. Assoc. Cancer Res.* **38**:5.
van der Bliek, A. M., Baas, F., Ten Houte de Lange, T., Kooiman, P. M., van der Velde-Koerts, T., and Borst, P., 1987, The human mdr3 gene encodes a novel P-glycoprotein homologue and gives rise to alternatively spliced mRNAs in liver, *EMBO J.* **6**:3325–3331.
van der Bliek, A. M., Kooiman, P. M., Schneider, C., and Borst, P., 1988, Sequence of *mdr3* cDNA encoding a human p-glycoprotein, *Gene* **71**:401–411.
van de Vrie, W., Gheuens, E. E., Durante, N. M., De Bruijn, E. A., Marquet, R. L., van Oosterom, A. T., and Eggermont, A. M., 1993, *In vitro* and *in vivo* chemosensitizing effect of cyclosporin A on an intrinsic multidrug-resistant rat colon tumour, *J. Cancer Res. Clin. Oncol.* **119**:609–614.
van de Vrie, W., Jonker, A. M., Marquet, R. L., and Eggermont, A. M. M., 1994, The chemosensitizer cyclosporin A enhances the toxic side-effects of doxorubicin in the rat, *J. Cancer Res. Clin. Oncol.* **120**:533–538.
van Kalken, C. K., Giaccone, G., van der Valk, P., Kuiper, C. M., Hadisaputro, M. M. N., Bosma, S. A. A., Scheper, R. J., Meijer, C. J. L. M., and Pinedo, H. M., 1992, Multidrug resistance gene (P-glycoprotein) expression in the human fetus, *Am. J. Pathol.* **141**:1063–1072.
Wacher, V. J., Wu, C.-Y., and Benet, L. Z., 1995, Overlapping substrate specificities and tissue distribution of cytochrome P450 3A and P-glycoprotein: Implications for drug delivery and activity in cancer chemotherapy, *Mol. Carcin.* **13**:129–134.
Wacher, V. J., Salphati, L., and Benet, L. Z., 1996, Active secretion and enterocytic drug metabolism barriers to drug absorption, *Adv. Drug Deliv. Rev.* **20**:99–112.
Yang, C.-P. H., Cohen, D., Greenberger, L. M., Hsu, S. I.-H., and Horwitz, S. B., 1990, Differential transport of two *mdr* gene products are distiguished by progesterone, *J. Biol. Chem.* **265**:10282–10288.
Yoshimura, A., Kuwazuru, Y., Sumizawa, T., Ichikawa, M., Ikeda, S.-I., Uda, T., and Akiyama, S.-I., 1989, Cytoplasmic orientation and two-domain structure of the multidrug transporter, P-glycoprotein, demonstrated with sequence-specific antibodies, *J. Biol. Chem.* **264**:16282–16291.
Zacherl, J., Hamilton, G., Thalhammer, T., Riegler, M., Cosentini, E. P., Ellinger, A., Bischof, G., Schweitzer, M., Teleky, B., Koperna, T., and Wenzl, E., 1994, Inhibition of P-glycoprotein-mediated vinblastine transport across HCT-8 intestinal carcinoma monolayers by verapamil, cyclosporine A and SDZ PSC 833 in dependence on extracellular pH, *Cancer Chemother. Pharmacol.* **34**:125–132.
Zaman, G. J. R., Versantvoort, C. H. M., Smit, J. J. M., Eijdems, E. W. H. M., de Haas, M., Smith, A. J., Broxterman, H. J., Mulder, N. H., de Vries, E. G. E., Baas, F., and Borst, P., 1993, Analysis of the expression of *MRP*, the gene for a new putative transmembrane drug transporter, in human multidrug resistant lung cancer cell lines, *Cancer Res.* **53**:1747–1750.
Zaman, G. J. R., Flens, M. J., van Leusden, M. R., de Haas, M., Mülder, H. S., Lankelma, J., Pindeo, H. M., Scheper, R. J., Baas, F., Broxterman, H. J., and Borst, P., 1994, The human multidrug resistance-associated protein MRP is a plasma membrane drug-efflux pump, *Proc. Natl. Acad. Sci. USA* **91**:8822–8826.
Zaman, G. J. R., Lankelma, J., van Tellingen, O., Beijnen, J., Dekker, H., Paulusma, C.,

Oude Elferink, R. P. J., Baas, F., and Borst, P., 1995, Role of glutathione in the export of compounds from cells by the multidrug-resistance-associated protein, *Proc. Natl. Acad. Sci. USA* **92:**7690–7694.

Zhang, J.-T., and Ling, V., 1991, Study of membrane orientation and glycosylated extracellular loops of mouse P-glycoprotein by *in vitro* translation, *J. Biol. Chem.* **266:** 18224–18232.

Zhou, D. C., Simonin, G., Faussat, A. M., Zittoun, R., and Marie, J. P., 1997, Effect of the multidrug inhibitor GG918 on drug sensitivity of human leukemic cells, *Leukemia* **11:**1516–1522.

14

Transporters for Bile Acids and Organic Anions

Hiroshi Suzuki and Yuichi Sugiyama

1. INTRODUCTION

The liver is one of the most important organs in the detoxification of xenobiotics. Compounds in the circulating blood are taken up by hepatocytes and then are metabolized and/or excreted into the bile. Many kinds of drugs and their metabolites are transported across the sinusoidal and bile canalicular membranes via carrier-mediated mechanisms. The molecular mechanisms for hepatic drug metabolism have been studied extensively and *in vivo* drug metabolism, along with drug–drug interactions, can be predicted from the *in vitro* data obtained with the cloned cDNA product, and interindividual differences in metabolic activity can be explained at the genetic level (Iwatsubo *et al.*, 1996, 1997; Ito *et al.*, 1998a, b). For example, the mechanism for polymorphism has been clarified and consequently its diagnosis using peripheral blood has become possible; in addition, the mechanism for enzyme induction has been clarified (Schmidt and Bradfield, 1996). However, there is currently little information on hepatic transport.

Recently, the cDNA cloning of several hepatic transporters has been performed, and its application to drug development is expected. Since some transporters are specifically expressed on hepatocyte plasma membranes, they can be used as a target for drug delivery to the liver. Moreover, the cloned transporter can be used for high-throughput screening of drug candidates to examine their trans-

---

Hiroshi Suzuki and Yuichi Sugiyama • Graduate School of Pharmaceutical Sciences, University of Tokyo, Hongo, Bunkyo-ku, Tokyo 113-0033, Japan.
Membrane Transporters as Drug Targets, edited by Amidon and Sadée. Kluwer Academic/Plenum Publishers, New York, 1999.

port activity, and since the affinity for these transporters has been shown to be a determinant factor for the hepatobiliary excretion of some series of derivatives, this application is particularly important. In this chapter, the nature of transporters for bile acids and non-bile acid organic anions in hepatocytes is summarized from a pharmacokinetic point of view. A review article on hepatic transport of anionic amino acids, fatty acids, and monocarboxylate is available (Petzinger, 1994). Transporters for prostaglandins have also been cloned recently from rats and humans (Kanai *et al.*, 1995; Lu *et al.*, 1996; Chan *et al.*, 1998).

2. TRANSPORT ACROSS THE SINUSOIDAL MEMBRANE

Transport across the sinusoidal membrane is the initial process of hepatic metabolism and/or biliary excretion. The transport properties across the sinusoidal membrane have been characterized in detail by examining uptake into the isolated sinusoidal membrane vesicles (SMVs) and isolated and/or primary cultured hepatocytes (reviewed by Petzinger, 1994; Oude Elferink *et al.*, 1995; Yamazaki *et al.*, 1996c). We have found a nice 1:1 correlation in the transport activity across the sinusoidal membrane between isolated hepatocytes and perfused liver, and shown that *in vivo* hepatic uptake under physiological conditions can be predicted from *in vitro* uptake data into isolated hepatocytes by considering the number of hepatocytes/g liver (Miyauchi *et al.*, 1993). The cDNA species for the transporters responsible for the hepatic uptake of these ligands have been cloned, and the transport properties of the cDNA products have been characterized (reviewed by Meier, 1995; Müller and Jansen, 1997; Meier *et al.*, 1997). Hepatic transporters for these organic anions are different from those for organic cations (such as tetraethylammonium and 1-methyl-4-phenylpyridinium), mediated, at least in part, by OCT1 (organic cation transporter) (reviewed by Zhang *et al.*, 1998; Koepsell, 1998). In this section, the sinusoidal transport of these anionic ligands is summarized.

2.1. Na^+-Dependent Transport

2.1.1. THE CONCEPT OF "MULTISPECIFIC BILE ACID TRANSPORTER"

Bile acids are efficiently taken up by hepatocytes from the blood circulation via a specific mechanism(s). The mechanism for the hepatic uptake of bile acids has been studied extensively, with particular focus on the uptake of taurocholic acid (TC) and cholic acid (CA) as compounds representing conjugated and unconjugated bile acids, respectively (reviewed by Suchy, 1993; Petzinger, 1994;

Meier, 1995; Hagenbuch and Meier, 1996; Meier *et al.,* 1997; Stieger and Meier, 1998). Hepatocellular uptake experiments revealed that both TC and CA are taken up via Na^+-dependent and -independent active transport mechanisms. According to cis-inhibition experiments in hepatocytes, the Na^+-independent uptake of both TC and CA is mediated by a common mechanism shared with other non-bile acid organic anions (reviewed by Suchy, 1993; Petzinger, 1994; Meier, 1995; Hagenbuch and Meier, 1996; Meier *et al.,* 1997). The Na^+-independent uptake of bile acids is described in Section 2.2. In this section, the Na^+-dependent uptake of bile acids is summarized briefly.

Hepatocellular uptake studies reveal that approximately 80% of TC uptake is mediated by an Na^+-dependent mechanism, whereas the contribution of Na^+-dependent uptake for CA is approximately 40% (Yamazaki *et al.,* 1993a). Kinetic analysis indicate that the K_m for the Na^+-dependent uptake of TC and CA is 13–15 μM (Anwer and Hegner, 1978; Nakamura *et al.,* 1996) and 58 μM (Anwer and Hegner, 1978), respectively. In addition, cis-inhibition experiments suggest that the Na^+-dependent uptake of TC and CA share a common transport system (Yamazaki *et al.,* 1993a). The stoichiometry for Na^+:TC has been suggested to be 2:1 (Yamazaki *et al.,* 1992b; Hagenbuch and Meier, 1996).

The substrate specificity of the Na^+–TC cotransport system was also examined by cis-inhibition experiments. Zimmerli *et al.* (1989) reported that the uptake into SMVs was competitively inhibited by other bile acids (such as taurochenodeoxycholate and chenodeoxycholate) and steroids (such as progesterone and 17-β-estradiol-3-sulfate), bumetanide, furosemide, verapamil, and phalloidin, and suggested a broad substrate specificity for this transport system. Based on these findings, the concept was proposed that Na^+–TC cotransport is mediated by a "multispecific bile acid transporter." As discussed in Section 2.1.3, however, recent molecular biological studies have revealed that the apparent broad specificity is mediated by several distinct transporters, although several bile acids can be substrates for the Na^+–TC cotransporter. The relationship between the chemical structure of a series of bile acid derivatives and their affinity for the Na^+-dependent bile acid transporter has been reviewed (Suchy, 1993; Petzinger, 1994).

2.1.2. NA^+-DEPENDENT UPTAKE OF DRUGS: INTERACTION WITH TAUROCHOLIC ACID

Using isolated hepatocytes, mutual cis-inhibition has been observed between TC and several compounds. These include bumetanide (Petzinger *et al.,* 1989; Blitzer *et al.,* 1982), ONO-1301,* a prostaglandin I_2 receptor agonist (Imawaka

---

*ONO-1301: 7,8-dihydro-5-[(E)-[[a-(3-pyridyl)-benzylidene]aminooxy]ethyl]-1-naphthyloxy]acetic acid.

and Sugiyama, 1998), and several kinds of small peptides (reviewed by Meier *et al.*, 1997) including octreotide (somatostatin analogue; Terasaki *et al.*, 1995; Yamada *et al.*, 1997a) and BQ-123 (Nakamura *et al.*, 1996) (Fig. 1). This section summarizes the transport properties of BQ-123, an anionic cyclic pentapeptide with the ability to antagonize endothelin (Nakamura *et al.*, 1996).

After intravenous administration to rats, BQ-123 is eliminated rapidly from plasma as revealed by its high total body clearance (50 ml/min/kg), comparable to the rate of hepatic blood flow (Nakamura *et al.*, 1996). Within 1 hr after injection, 86% of the dose is excreted in its intact form into the bile (Nakamura *et al.*, 1996). The mechanism for the hepatic uptake of BQ-123 has been examined using isolated rat hepatocytes (Nakamura *et al.*, 1996). BQ-123 was extensively taken up by isolated hepatocytes via Na$^+$-dependent (K_m = 6 μM and V_{max} = 140 pmole/min/10^6 cells) and Na$^+$-independent (K_m = 12 μM and V_{max} = 390 pmole/min/10^6 cells) mechanisms (Fig. 2). Both Na$^+$-dependent and -independent transport are mediated by active transport systems, since marked reduction in uptake was observed following incubation of the cells with metabolic inhibitors. TC competitively inhibited the Na$^+$-dependent and -independent transport of BQ-123 with K_i values of 9 and 10 μM, respectively (Fig. 2). These K_i values were comparable with the K_m values for the Na$^+$-dependent (13 μM) and Na$^+$-independent (25 μM) uptake of TC. The results of these cis-inhibition studies performed with hepatocytes are consistent with the hypothesis that the uptake of BQ-123 is mediated by the bile acid transporter (Nakamura *et al.*, 1996). As discussed in the next section, molecular biological studies reveal that this hypothesis is not true. The mechanism for the efficient transport of BQ-123 across the bile canalicular membrane is described in Section 3.2.

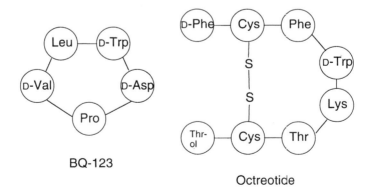

Figure 1. Chemical structure of BQ-123 and octreotide.

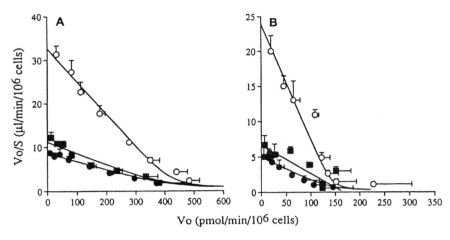

Figure 2. Effect of DBSP (10 μM; ●) and TC (20 μM; ■) on Na^+-dependent (A) and Na^+-independent (B) uptake of BQ-123 by isolated rat hepatocytes. Solid line represents the fitted line. Taken with permission from Nakamura *et al.* (1996). Copyright 1996 Williams & Wilkins.

2.1.3. MOLECULAR CLONING AND FUNCTIONAL ANALYSIS OF Ntcp

In order to clarify the detailed mechanism for hepatic uptake, it is essential to clone the cDNA encoding the transporter. Meier and his collaborators cloned the Na^+–TC cotransporting polypeptide (Ntcp) from rat hepatocytes based on expression cloning in *Xenopus laevis* oocytes (Hagenbuch *et al.*, 1991). The open reading frame of this transporter was found to consist of 1086 bp, encoding a polypeptide composed of 362 amino acids (Table I) (Hagenbuch *et al.*, 1991). Hydropathy plot analysis suggested that Ntcp has seven transmembrane domains (Meier *et al.*, 1997). Using antibodies against the polypeptide residues, it has been shown that Ntcp is selectively located on the basolateral membrane of rat hepatocytes (Stieger *et al.*, 1994; Ananthanarayanan *et al.*, 1994). Western blot analysis indicated that the molecular mass of Ntcp in isolated rat basolateral membrane (51 kDa) is shifted to 33.5 kDa by treatment with *N*-glycohydrolase F, suggesting glycosylation of Ntcp under physiological conditions (Stieger *et al.*, 1994). Using the radiation inactivation method, however, it has been suggested that the minimal functional molecular mass of the Na^+-dependent bile acid transporter is 170 kDa, consistent with the hypothesis that the functional transporter is composed of a complex including Ntcp molecule(s) (Elsner and Ziegler, 1989).

The transport properties of cloned Ntcp have been determined in cRNA-in-

Table I.
Transporters Responsible for the Sinusoidal Uptake of Bile Acids and Non-Bile Acid Organic Anions[a]

Transporter	Species	Number of amino acids	Observed molecular mass (kDa)	Chromosomal localization	Number of TM domains	Expression
oatp1	Rats	670	80	ND	12	Liver, kidney (apical membrane of S3 segment), choroid plexus (apical membrane)
oatp2	Rat	661	ND	ND	12	Liver, brain (cerebral cortex, cerebellum), kidney
oatp3	Rat	670	ND	ND	12	Kidney, liver
OATP	Human	670	ND	12q12	12	Liver, brain (cerebral cortex, cerebellum), kidney, testis
OAT2	Rat	535	ND	ND	12	Kidney, liver
OAT3	Rat	536	ND	ND	12	Liver, kidney, brain, eye
Ntcp	Rat	362	51	6q24	7	Liver
NTCP	Human	349	50	14q24.1–24.2	7	Liver

	Specific sequence in promoter region	References
oatp1	—	Kullak-Ublick et al. (1995), Jacquemin et al. (1994), Bergwerk et al. (1996), Meier et al. (1997), Angeletti et al. (1997)
oatp2	—	Noé et al. (1997)
oatp3	—	Abe et al. (1998)
OATP	AP-1 and 3, C/ERB, GRE, HNF-1, 3, 4, and 5, OCT, Pit-1	Kullak-Ublick et al. (1995, 1996a, 1997a)
OAT2		Sekine et al. (1998)
OAT3		Kusuhara et al. (1999)
Ntcp	AP1, BARE, HNF1 and 3, NF1, OCT, Sp1, SRE, S/TRE	Hagenbuch et al. (1991), Stieger et al. (1994), Karpen et al, (1996)
NTCP	—	Hagenbuch and Meier (1994), Clark et al. (1995)

[a]TM, Transmembrane; ND, not determined. The substrate specificity of each transporter is described in the text. Refer to Table 2 in Meier et al. (1997) for quantitative comparison of the transport activities of these cloned transporters.

jected oocytes and cDNA-transfected mammalian cells (reviewed by Meier, 1995; Meier et al., 1997). The K_m value for TC was determined to be 25 μM for cRNA-injected oocytes (Hagenbuch et al., 1991) and 17–30 μM (Boyer et al., 1994; Kouzuki et al., 1998a), 34 μM (Schroeder et al., 1998), and 10–15 μM (Torchia et al., 1996) for cDNA-transfected COS-7, CHO, and rat hepatoma McArdle RH-7777 cell lines, respectively. Western blot analysis indicated that the molecular mass of Ntcp in cRNA-injected oocytes and Ntcp-transfected COS-7 cells is approximately 41 kDa (Hagenbuch et al., 1991) and 33 kDa (Boyer et al., 1994; Kouzuki et al., 1998), respectively, indicating less extensive glycosylation. The comparability in K_m values between hepatocytes and oocytes and cultured mammalian cells expressing Ntcp suggests that glycosylation may not affect the transport properties of this transporter.

In Ntcp-transfected CHO cells, Na$^+$-dependent uptake of glycocholate (GCA; K_m = 27 μM; V_{max} = 241 pmole/min/mg protein), taurochenodeoxycholate (K_m = 5 μM; V_{max} = 787 pmole/min/mg protein), and tauroursodeoxycholate (K_m = 14 μM; V_{max} = 584 pmole/min/mg protein) was also observed, indicating that Ntcp can transport all bile salt derivatives that cis-inhibit Ntcp-mediated TC uptake (K_m = 34 μM; V_{max} = 744 pmole/min/mg protein) (Schroeder et al., 1998). In addition, Ntcp-mediated uptake of estrone-3-sulfate (K_m = 27 μM; V_{max} = 451 pmole/min/mg protein) has been reported (Schroeder et al., 1998). The V_{max}/K_m values for these ligands are in the order taurochenodeoxycholate > tauroursodeoxycholate > TC > estrone-3-sulfate > glycocholate, which is consistent with observations in isolated hepatocytes (Schroeder et al., 1998). Recently, it was demonstrated that thyroxine (T4), 3,5,3'-triiodo-L-tyronine (T3), and its metabolites are transported via Ntcp (Friesema et al., 1999).

The expression of Ntcp in COS-7 cells was characterized 48 hr after transfection with lipofectamine (Kouzuki et al., 1998). Although the Ntcp mRNA levels in COS-7 cells are 60- to 70-fold higher than in hepatocytes, Western blot analysis indicated that the expression of Ntcp in the crude membrane from Ntcp-transfected COS-7 cells was comparable with that from short-term (4 hr) cultured hepatocytes (Kouzuki et al., 1998). In addition, the kinetic parameters for the uptake of TC were comparable between the two kinds of cells (K_m = 17.4 μM and V_{max} = 1.45 nmole/min/mg protein for Ntcp-transfected COS-7 cells, and K_m = 17.7 μM and V_{max} = 1.63 nmole/min/mg protein for cultured hepatocytes) (Kouzuki et al., 1998). These results suggest that the Ntcp-transfected cells can be used to predict quantiatively the hepatocellular uptake of ligands by normalizing the expression level.

The contribution of Ntcp to the hepatocellular uptake of ligands may be determined by examining the uptake of ligands into hepatocytes and into Ntcp-transfected cells followed by normalization of the ligand uptake using TC as a reference compound; the contribution is described by the following equation:

contribution

$$= \frac{CL_{uptake} \text{ of ligands into Ntcp-transfected cells}/CL_{uptake} \text{ of TC into Ntcp-transfected cells}}{CL_{uptake} \text{ of ligands into hepatocytes}/CL_{uptake} \text{ of TC into hepatocytes}}$$

where CL_{uptake} represents the uptake clearance, which is determined by dividing the initial uptake velocity by the substrate concentration in the medium (Kouzuki et al., 1998). We examined the uptake of CA and GCA into Ntcp-transfected COS-7 cells; the contribution of Ntcp was 40% and 80%, respectively (Kouzuki et al., 1998). This method is valid if the Na^+-dependent uptake of TC by hepatocytes is predominantly mediated by Ntcp. This assumption has been justified by the finding that the simultaneous injection of an antisense oligonucleotide against Ntcp with total rat liver mRNA almost completely abolished the Na^+-dependent uptake of TC (Hagenbuch et al., 1996). Although von Dippe et al. (1996) reported that microsomal epoxide hydrolase may be partly responsible for the Na^+-dependent hepatic uptake of bile acids, the contribution of this transporter may be minor.

With this method, approximately 64% of the Na^+-dependent uptake of ONO-1301 was found to be mediated by Ntcp (Kouzuki et al., 1998). Although cis-inhibition studies in hepatocytes revealed an interaction between the transport of TC and bumetanide (Petzinger et al., 1989; Blitzer et al., 1982), the transport of bumetanide was found to be predominantly mediated by other transporter(s); using oocytes, it was found that Na^+-dependent transport of TC and bumetanide is coded by different mRNA fractions in rat liver (Honscha et al., 1993). In Ntcp-transfected CHO cells, the uptake of bumetanide was not significant, although the Ntcp-mediated transport of TC was inhibited by this ligand (Schroeder et al., 1998). In the same manner, the transport of BQ-123 in COS-7 cells was not stimulated by Ntcp transfection (Kouzuki et al., 1998). These results indicate the difficulty in identifying the responsible transporter from only cis-inhibition experiments.

2.1.4. HUMAN HOMOLOGUE OF Ntcp

Based on the homology with rat Ntcp, the human Na^+-dependent bile acid transporter (NTCP) was cloned (Table I). The human NTCP consists of 349 amino acids, showing 77% amino acid identity with rat Ntcp (Hagenbuch and Meier, 1994). In cRNA-injected oocytes, the K_m of NTCP for TC was determined to be 6 μM, which is lower than the K_m values determined in human hepatocytes and in isolated sinusoidal membrane (34–62 μM; Meier et al., 1997). The rank order for human NTCP-mediated transport under linear conditions is taurochenodeoxycholate > tauroursodeoxycholate > TC > estrone-3-sulfate > glycocholate, which is consistent with the results for rat Ntcp (Meier et al., 1997).

2.1.5. BILE ACID TRANSPORTER AS A TARGET FOR DELIVERY

Since Ntcp is exclusively expressed on the sinusoidal membrane of the liver, this transporter may be used as a target for drug delivery to the liver (Kramer and Wess, 1996). Kramer and his collaborators examined this hypothesis using the chlorambucil–TC complex, which was prepared by covalently linking this antitumor drug to the 3-hydroxy position of TC (Fig. 3) (Kramer et al., 1992). Using isolated rat hepatocytes, they demonstrated the interaction between this complex and the Na^+-dependent bile acid transporter (Kramer et al., 1992). Moreover, they demonstrated that this complex is excreted preferentially into bile, whereas chlorambucil per se is predominantly excreted into urine after injection into rats (Kramer et al., 1992).

Using an excised carcinoma specimen from patient liver, Kullak-Ublick et al. (1997b) examined the transport of this conjugated compound. Although NTCP mRNA levels in carcinoma were 56 ± 27% that in nonmalignant peritumor tissue, the extent of Na^+-dependent uptake of this compound into mRNA-injected oocyte was similar in carcinoma and nonmalignant tissues, suggesting the presence of other Na^+-dependent bile acid transporter(s) in the carcinoma (Kullak-Ublick et al., 1997b). In addition, they examined the uptake of this conjugate into NTCP cRNA-injected oocytes (Kullak-Ublick et al., 1997b). The initial velocity for the Na^+-dependent uptake of chlorambucil–TC complex was approximately 40% that of TC (Kullak-Ublick et al., 1997b). The kinetic parameters for the uptake were determined as follows: for TC and chlorambucil–TC complex, K_m was 6 and 11 μM and V_{max} was 30 and 11 fmole/oocyte/min, respectively (Kullak-Ublick et al., 1997b).

Marin et al. (1998) reported that the hepatic uptake of cisplatin is markedly enhanced by preparing its complex (Bamet-H2) with bis-cholyglycinate. More-

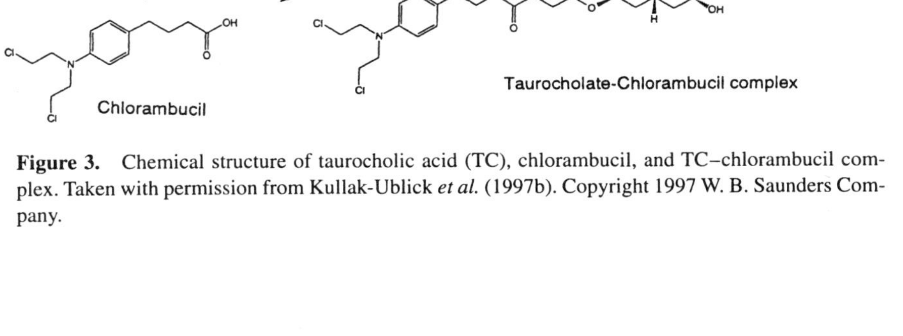

Figure 3. Chemical structure of taurocholic acid (TC), chlorambucil, and TC–chlorambucil complex. Taken with permission from Kullak-Ublick et al. (1997b). Copyright 1997 W. B. Saunders Company.

over, after intravenous administration to rats, approximately 85% of the dose was excreted into the bile within 2 hr. Coadministation of Bamet-H2 and cholyglycinate suggested the presence of partial cross-inhibition in both hepatic uptake and biliary excretion processes (Marin *et al.,* 1998). Although no tumor specificity was ascribed to these complexes, this method may be useful for drug targeting to the liver. Indeed, bile acids have been used in delivery of 3-hydroxy-3-methylglutaryl–coenzyme A (HMG-CoA) reductase inhibitors to the liver (Fig. 4) (reviewed by Kramer and Wess, 1996). Moreover, a desired low-density lipoprotein receptor upregulating activity of thyroid hormones in hepatocytes was attained by conjugating L-triiodothyronine to CA (Stephan *et al.,* 1992).

2.2. Na$^+$-Independent Transport

2.2.1. THE CONCEPT OF "MULTISPECIFIC ORGANIC ANION TRANSPORTER"

By examining the uptake of bile acids by isolated and cultured hepatocytes, it has been demonstrated that approximately 20% and 60% of the uptake of TC and CA, respectively, is mediated by an Na$^+$-independent mechanism with K_m values of 19, 25, and 57 μM (Yamazaki *et al.,* 1993b; Nakamura *et al.,* 1996; Anwer and Hegner, 1978) and 39 μM (Anwer and Hegner, 1978), respectively. Non-bile acid organic anions (such as bilirubin, bromosulfophthalein [BSP], dibromosulfophthalein [DBSP], and 1-anilino-8-naphthalenesulfonate), are also taken up by hepatocytes via Na$^+$-independent active transport (Yamazaki *et al.,* 1992a); in addition, mutual competitive inhibition between some of them and bile acids in terms of hepatocellular uptake has been reported (reviewed by Meier, 1995). Moreover, as summarized by Meier (1995), Na$^+$-independent uptake of bile acids is inhibit-

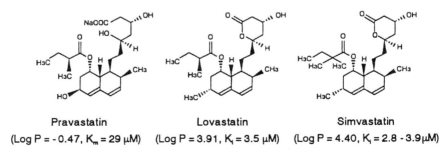

Figure 4. Chemical structure of HMG-CoA reductase inhibitors. Affinity for the hepatic uptake system and lipophilicity are also shown.

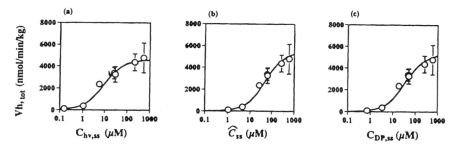

Figure 5. Dose dependence of hepatic elimination rate of pravastatin ($V_{h,\text{tot}}$) at steady state. The $V_{h,\text{tot}}$ of pravastatin, determined as the difference between the arterial and hepatic venous blood concentration, is plotted against (a) hepatic venous plasma concentration, (b) mean logarithmic concentration, and (c) the theoretical concentration for hepatic intrinsic clearance defined by the dispersion model. Solid line represents the fitted line. Taken with permission from Yamazaki *et al.* (1996a). Copyright 1996 Plenum Publishing Corp.

ed by cardiac glycosides and other neutral steroids, and some small peptides. Due to this broad substrate specificity, it has been proposed that the hepatic uptake of organic anions is mediated by a multispecific organic anion transporter (Meier, 1995). However, as discussed later in Section 2.2.4, recent molecular biological studies have revealed that the apparent broad specificity is mediated by several distinct transporters with partially overlapping substrate specificity.

In addition to the previously described organic anions, many clinically important drugs are taken up by hepatocytes predominately via Na^+-independent active transport. These include benzylpenicillin (Tsuji *et al.*, 1986), an HMG-CoA reductase inhibitor (pravastatin; Fig. 5) (Yamazaki *et al.*, 1993b), an angiotensin-converting enzyme (ACE) inhibitor (temocaprilat; Fig. 6) (Ishizuka *et al.*, 1998a), and a new quinolone antibiotic (grepafloxacin) (Sasabe *et al.*, 1997). In addition, approximately 70% of the uptake of ONO-1301 is mediated by the Na^+-independent mechanism (Imawaka and Sugiyama, 1998). The active transport of conjugated metabolites of E3040,* which is used in the treatment of colon ulcer, was also mediated by an Na^+-independent mechanism; although E3040 glucuronide and sulfate exhibited mutual competitive inhibition, the uptake of the sulfate was mediated by a high-affinity and high-capacity transport system compared with that of the glucuronide (Takenaka *et al.*, 1997). The following section summarizes the characteristics of hepatocellular uptake of pravastatin (Yamazaki *et al.*, 1993b). The transport of temocaprilat across the sinusoidal membrane is described in Sectin 3.2.1. Section 3.2.2 discusses the transport of the above-mentioned compounds across the bile canalicular membrane.

*E3040: 6-hydroxy-5,7-dimethyl-2-methylamino-4-(3-pyridylmethyl)benzothiazole.

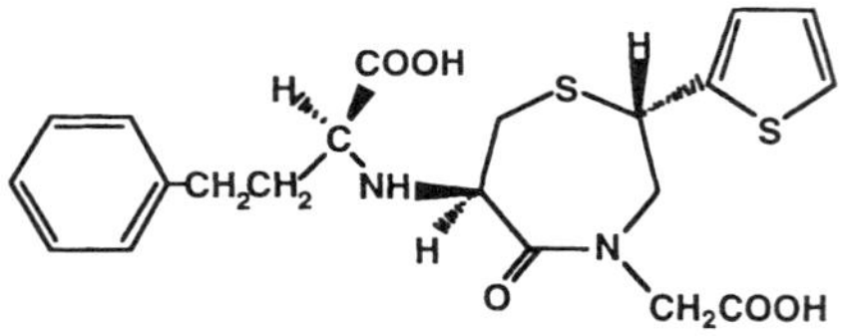

Figure 6. Chemical structure of temocaprilat.

2.2.2. THE ROLE OF MULTISPECIFIC ORGANIC ANION TRANSPORTER IN DRUG DISPOSITION: KINETIC ANALYSIS OF HEPATIC UPTAKE OF PRAVASTATIN

HMG-CoA reductase inhibitors are used clinically to decrease the plasma concentration of cholesterol due to their inhibitory effect on the biosynthesis of this compound in the liver. The inhibition of cholesterol synthesis in other tissues may lead to the appearance of side effects, and therefore the selectivity of the inhibitory effect on the liver is of importance. Pravastatin, a hydrophilic HMG-CoA reductase inhibitor, exhibits relatively selective inhibition of hepatic cholesterol synthesis compared with the more highly lipophilic inhibitors, such as lovastatin and simvastatin (Fig. 4) in rodents (Koga et al., 1990; Tsujita et al., 1986). In order to examine the hypothesis that pravastatin is efficiently taken up by hepatocytes via a specific transporter, the disposition of this compound was studied (Yamazaki et al., 1993b).

In in vivo experiments, the uptake of pravastatin into several tissues was examined at early time points after intravenous administration in order to determine the initial uptake of this compound (Yamazaki et al., 1996d). The analysis indicated that pravastatin is taken up by the liver and the kidney, but not by other tissues examined. These results suggest the presence of a specific transport system for this compound on the sinusoidal membrane of the hepatocytes (Yamazaki et al., 1996d). Pravastatin was intravenously infused to rats at different rates, and the concentration dependence of the rate of hepatic elimination was determined from the difference between the arterial and hepatic venous blood concentration (Fig. 5; Yamazaki et al., 1996a). Kinetic analysis with such pharmacokinetic models as the well-stirred, parallel-tube, and dispersion models yielded K_m values of 6.0, 34, and 23 μM, respectively (Fig. 5). The rate of elimination under linear conditions, calculated by V_{max}/K_m was 19, 4.9, and 5.7 ml/min/g liver for the respective kinetic models, which is comparable with those estimated from the initial rate of uptake by isolated hepatocytes ($K_m = 29$ μM, $V_{max}/K_m = 2.4$ ml/min/g liver). These results suggest that the hepatic intrinsic clearance of pravastatin at steady state is regulated by the uptake process, followed by rapid metabolism and/or biliary excretion with minimal efflux to the circulating blood. Moreover, it was indicated that the uptake experiments with isolated hepatocytes can be used to predict quantitatively the in vivo hepatic uptake (Yamazaki et al., 1996a).

Transporters for Bile Acids and Organic Anions

In order to clarify the mechanism for efficient uptake, the uptake of pravastatin by isolated hepatocytes was further characterized (Yamazaki et al., 1993b). Analysis indicated that the uptake of pravastatin is mediated by an Na^+-independent active transport system which is competitively inhibited by DBSP ($K_i = 6.3$ μM), CA ($K_i = 14$ μM), and TC ($K_i = 22$ μM). In addition, pravastatin inhibited the Na^+-independent uptake of TC with a K_i (19 μM) comparable to the K_m value of pravastatin. Competitive inhibition was observed by simvastatin, a more highly lipophilic HMG-CoA reductase inhibitor, with a K_i of 3.9 μM, an order of magnitude smaller than the K_m of pravastatin, indicating that simvastatin has a higher affinity for the transporter than pravastatin (Fig. 4). These findings suggest that the reason for the liver-specific distribtion of pravastatin *in vivo* compared with other lipophilic HMG-CoA reductase inhibitors is not that the liver has a higher affinity for pravastatin than for other inhibitors, but that the extrahepatic distribution of pravastatin is smaller because of its hydrophilic nature compared with other inhibitors. This suggestion is supported by the finding of Scott (1990) and Koga *et al.* (1990). Scott (1990) showed that lovastatin exhibited linear uptake by human skin fibroblasts to almost the same extent as in the liver, whereas pravastatin was taken up to a very small degree by these nonhepatic cells. In the same manner, Koga *et al.* (1990) found that lovastatin was efficiently taken up by nonhepatic cells (such as rat spleen cells and mouse L-cells) as well as by hepatocytes, whereas selective uptake of pravastatin was seen in hepatocytes.

Pravastatin molecules taken up by hepatocytes are efficiently excreted into bile, providing the pharmacokinetic basis for the presence of enterohepatic circulation (Yamazaki *et al.*, 1996b). The large pool size of pravastatin in the gastrohepatobiliary system is maintained by transporters on the sinusoidal and canalicular membranes, which leads to the large exposure of this drug to liver (Yamazaki *et al.*, 1996b). The transport of pravastatin across the bile canalicular membrane is discussed in Section 3.2.

2.2.3. MOLECULAR CLONING OF ORGANIC ANION-TRANSPORTING POLYPEPTIDE (oatp-1)

The mechanism for the Na^+-independent uptake of organic anions remains to be clarified. In order to get some insight into the mechanism, isolated rat hepatocytes were incubated for different time periods (1–30 min) in the presence and absence of FCCP (carbonylcyanide *p*-trifluoromethoxyphenyl hydrazone) (2 μM) or rotenone (0.2 or 30 μM) to decrease the cellular ATP content (Yamazaki *et al.*, 1993a). Under these conditions, the initial velocity of the uptake of TC, CA, and pravastatin was determined (Yamazaki *et al.*, 1993a). Within 5 min of exposure to these metabolic inhibitors, the cellular ATP content fell to less than one-fifth of the control, and the initial velocity of the total uptake of CA, Na^+-independent uptake of CA, TCA, and pravastatin decreased in parallel with the reduction in cellular

ATP. In contrast, Na^+-dependent uptake of TC remained virtually unchanged for a 5-min incubation with these metabolic inhibitors; a significant decrease was observed over longer incubation periods, suggesting that the inwardly directed Na^+ gradient is maintained to some extent at early time points after initiation of ATP depletion. The initial velocity of the total uptake of CA and pravastatin was shown to have a saturable relation to cellular ATP content, irrespective of different exposure times to these metabolic inhibitors. These results suggest that the Na^+-independent uptake of bile acids and organic anions may be driven either by ATP hydrolysis or by an as-yet-unidentified ion gradient that disappears more rapidly than the Na^+ gradient (Yamazaki et al., 1993a).

In order to clarify the detailed mechanism of hepatic uptake, Meier and collaborators cloned the organic anion transporting polypeptide (oatp-1) from rat hepatocytes based on expression cloning in *Xenopus laevis* oocytes (Table I) (Jacquemin et al., 1994). The open reading frame of this transporter consisted of 2010 bp encoding a polypeptide composed of 670 amino acids. Hydropathy plot analysis suggested that oatp-1 has 12 transmembrane domains (Meier et al., 1997). Using antibodies against the polypeptide residues, it has been shown that oatp-1 is selectively confined to the basolateral membrane in rat hepatocytes (Bergwerk et al., 1996). Western blot analysis indicated that the molecular mass of oatp-1 in isolated rat basolateral membrane (80 kDa) is shifted to 65 kDa after deglycosylation with *N*-glycanase (Bergwerk et al., 1996). However, results using the radiation inactivation method suggest that the minimal functional molecular mass of the Na^+-independent CA transporter is 107 kDa (Elsner and Ziegler, 1992). Northern blot analysis indicated expression of oatp-1 in liver and kidney and, to a lesser extent, in brain, lung, skeletal muscle, and proximal colon (Jacquemin et al., 1994; Bergwerk et al., 1996). Moreover, immunohistochemical studies revealed that the antibody against oatp-1 cross-reacted with the protein located on the apical membrane of the S3 segment of the rat renal proximal tubule (Bergwerk et al., 1996) and the protein located on the brush border membrane of the choroid plexus (Angeletti et al., 1997).

The transport properties of oatp-1 have been characterized using oocytes injected with cRNA and mammalian cells transfected with cDNA; it has been shown that oatp-1 mediates the Na^+-independent uptake of BSP [$K_m = 1.5$ μM in oocytes (Jacquemin et al., 1994) and 3.3 μM in HeLa cells (Shi et al., 1995)] and TC [$K_m = 50$ μM in oocytes (Kullak-Ublick et al., 1994) and 14 (Satlin et al., 1997) and 27 μM (Kanai et al., 1996) in HeLa cells]. Moreover, estradiol-17β-D-glucuronide ($E_2$17βG) is a substrate of oatp [$K_m = 3$ μM in oocytes (Bossuyt et al., 1996a) and HeLa cells (Kanai et al., 1996) and 20 μM in COS-7 cells (Kouzuki et al., 1998b)]. In addition, it has been recently shown that oatp-1 expressed on oocytes mediates the transport of steroid conjugates (estrone-3-sulfate; $K_m = 4.5$ μM), neutral steroids (ouabain, $K_m = 1.7$ mM; aldosterone, $K_m = 15$ nM; and cortisol, $K_m = 13$ μM), glutathione conjugates [2,4-dinitrophenyl-*S*-glutathione (DNP-SG;

K_m = 408 µM) and leukotriene C_4 (LTC$_4$; K_m = 0.27 µM)] (Li et al., 1998), and even some amphipathic organic cations such as N-(4,4-azo-n-pentyl)-21-deoxyajmaline (APD-ajmaline; Bossuyt et al., 1996a) and rocuronium (van Montfoort et al., 1998). Ochratoxin A (K_m = 17 µM in oocytes; Kontaxi et al., 1996) and CRC 200 (K_m = 30 µM in oocytes; Eckhardt et al., 1996), a peptide-based thrombin inhibitor, are also the substrates for oatp-1. Recently, it was demonstrated that T4, T3, and its metabolites are also transported via oatp-1 (Friesema et al., 1999). Meier et al. (1997) summarized the transport activity of oatp-1 expressed in oocytes and suggested that the initial velocity of the ligand uptake mediated by oatp-1 under linear conditions would be in the order estrone-3-sulfate > aldosterone > $E_2$17βG > tauroursodeoxycholate > taurochenodeoxycholate > BSP > CRC 200 > TC > GCA, ADP-ajmaline > CA > ouabain. Based on these transport characteristics, oatp-1 represents a polyspecific and multivalent transport system able to accept a variety of structurally unrelated and differently charged amphipathic organic substrates.

Although oatp-1 has been cloned, the mechanism for the uptake mediated by this transporter remains to be clarified. Satlin et al. (1997) examined the rate of TC-dependent HCO_3^- efflux from HCO_3^--loaded HeLa cells transfected with oatp-1 cDNA. By measuring the intracellular pH with 2′,7′-bis(carboxyethyl)-5(6)-carboxyfluorescein, they concluded that TC is taken up in exchange for the efflux of HCO_3^-. The oatp-1-mediated transport may be accounted for, at least in part, by this mechanism (Satlin et al., 1997).

More recently, using cRNA-injected oocytes, Li et al. (1998) examined the hypothesis that substrates are taken up into the cells in exchange for the efflux of glutathione (GSH) via oatp-1. They demonstrated that the oatp-1-mediated uptake of LTC$_4$, DNP-SG, and TC is inhibited by GSH (1 mM) and that the efflux of GSH is stimulated by adding TC (50 µM) or BSP (1 µM) to the medium (Li et al., 1998). The amount of TC-stimulated GSH efflux was compared with the amount of TC taken up into oocytes under the same experimental conditions. The analysis indicated a stoichiometry of 1:1 exchange (Li et al., 1998).

The contribution of oatp-1 to the Na^+-independent uptake of ligands into rat hepatocytes remains to be clarified. The contribution was determined based on the method described for Ntcp, using $E_2$17βG as a reference compound for oatp-1 (Kouzuki et al., 1999). Although the oatp-1 mRNA levels in COS-7 cells were approximately 10-fold higher than in the liver, the expression of oatp-1 in transfected COS-7 cells was approximately one-seventh that in liver. The K_m value for $E_2$17βG was determined to be 20 µM in transfected COS-7 cells, a value comparable with that determined in 4-hr cultured rat hepatocytes (12 µM). Moreover, the V_{max} value for this ligand in COS-7 cells was approximately one-seventh that in hepatocytes (1.30 vs. 0.175 nmole/min/mg protein). Although Western blot analysis indicated less extensive glycosylation in oatp-1 in cDNA-transfected HeLa and COS-7 cells (the molecular mass of oatp-1 is approximately 70 and 72 kDa, respectively), the comparability in the K_m value suggests that glycosylation may not affect the transport properties of oatp-1. The contribution of oatp-1 to the Na^+-independent uptake of TC and CA is approximately 60% and 50%, respectively, whereas the corresponding values for the sulfate conugates of E3040 and estrone-3-sulfate are 27% and 21%, respectively (Kouzuki et al., 1999).

The contribution of oatp-1 to the Na^+-independent uptake of anionic drugs was also examined (Ishizuka et al., 1999). Approximately 50% of the uptake of temocaprilat was mediated by oatp-1 (Ishizuka et al., 1998a) (see Section 3.2.1), although the contribution of this transporter to the uptake of pravastatin and BQ-123 was negligible (Kouzuki et al., 1999) as far as COS-7 cells were used. Moreover, approximately 20% of the uptake of E3040 sulfate was accounted for by oatp-1, whereas the contribution of this transporter was minimal for E3040 glucuronide (Kouzuki et al., 1999). These results suggest multiplicity in the Na^+-independent transport mechanism for organic anions across the sinusoidal membrane; it is necessary to assume the presence of multiple transporters to provide a molecular basis for the concept that the hepatic uptake of organic anions is mediated by a "multispecific organic anion transporter," which has been established from a kinetic analysis of the experimental data in hepatocytes and/or SMVs.

2.2.4. MOLECULAR CLONING OF oatp-2 AND 3

A homologue of oatp-1, oatp-2 has recently been cloned from rat brain (Table I) (Noé et al., 1997) and rat retina (Abe et al., 1998). The open reading frame of this transporter consists of 1983 bp, encoding a polypeptide composed of 661 amino acids. oatp-2 exhibited an amino acid sequence identity of 77% to oatp-1 and 73% to human OATP. Northern blot analysis indicated that the expression of this transporter is extensive in liver as well as in brain and kidney. In cRNA-injected oocytes, TC, CA, $E_2 17\beta G$, estrone-3-sulfate, ouabain, and digoxin were taken up in an Na^+-independent manner, with K_m values of 35, 46, 3, 11, 470, and 0.24 µM, respectively (Noé et al., 1997). Moreover, thyroid hormones (T4, K_m = 6.54 µM, and T3, K_m = 5.87 µM) were determined as substrates for oatp-2 (Abe et al., 1998). Although the transport properties of oatp-2 resemble those of oatp-1, high-affinity uptake of digoxin was observed for oatp-2, suggesting that this transporter is responsible for the hepatic uptake of cardiac glycosides. oatp-2 has also been shown to accept organic cations (such as APD-ajmaline and rocuronium) as substrates (van Montfoort et al., 1998). In addition, oatp-3 (670 amino acids, amino acid sequence identity of 80% to rat oatp-1) was shown to be expressed to some extent in the liver (Abe et al., 1998). oatp-3 expressed in oocytes also accepts TC, T4 (K_m = 4.93 µM), and T3 (K_m = 7.33 µM). Although OAT-K1 and K2 have been cloned as homologues of oatp, their expression level in the liver is minimal (Saito et al., 1996; Masuda et al., 1999).

Transporters for Bile Acids and Organic Anions

2.2.5. HUMAN HOMOLOGUE OF oatp-1

Based on the homology with rat oatp-1, the human Na^+-independent organic anion transporter (OATP) has been cloned from liver (Table I) (Kullak-Ublick et al., 1995). The human OATP consists of 670 amino acids, showing 65% amino acid identity with rat oatp-1. The substrate specificity was examined in cRNA-injected oocytes (Bossuyt et al., 1996b). Although the substrate specificity was very similar to that of oatp-1, the OATP-mediated uptake of bile acids, organic anions, and ouabain is much less extensive than that by oatp-1 following injection of the same amount of cRNA. In contrast, ADP-ajmaline, rocuronium, methylquinine ($K_m = 5$ μM), and methylquinidine ($K_m = 26$ μM) (van Montfoort et al., 1998), and CRC 220 were transported by OATP much more efficiently than by oatp-1. It has also been demonstrated that dehydroepiandrosterone sulfate, a neurosteroid, is also transported by OATP expressed on oocytes ($K_m = 6.6$ μM) (Kullak-Ublick et al., 1998). Northern blot analysis indicated extensive expression in brain (cerebral cortex and cerebellum), liver, lung, kidney, and testis (Kullak-Ublick et al., 1995). Based on the homology score of OATP to rat oatp, it is hypothesized that OATP may not be a homologue of oatp-1, 2, or 3.

2.2.6. MOLECULAR CLONING OF oat FAMILY

Recently, oat-1 was identified as a transporter for the renal uptake of organic anions in exchange for endogenous substrates (such as α-ketoglutarate) (Sekine et al., 1997; Sweet et al., 1997). As homologues of oat-1, oat-2 and 3 were cloned (Sekine et al., 1998; Kusuhara et al., 1999). oat-2, previously referred to as NLT (new liver-specific transport protein), can transport salicylate ($K_m = 88.8$ μM), acetylsalicylate, prostaglandin E_2, dicarboxylates, and p-aminohippurate in an Na^+-independent manner when expressed in oocytes (Sekine et al., 1997). In addition, oat-3 is also expressed in the liver, brain, kidney, and eye (Kusuhara et al., 1999). oat-3 was found to transport p-aminohippurate ($K_m = 65$ μM), ochratoxin A ($K_m = 0.74$ μM), and estrone-3-sulfate ($K_m = 2.3$ μM), along with cimetidine, an organic cation (Kusuhara et al., 1999). Although the contribution of oatp-2/3 and oat-2/3 to drug uptake by hepatocytes remains to be clarified, these transporters may account for the multiplicity of Na^+-independent uptake of organic anions across the sinusoidal membrane.

3. TRANSPORT ACROSS THE CANALICULAR MEMBRANE

Compounds taken up by hepatocytes are then excreted into bile in their intact form or after metabolic conversion. The mechanism for the transport across the bile canalicular membrane can be characterized using the isolated canalicular membrane vesicles (CMVs). Using CMVs, it has been shown that several kinds of primary active transporters are responsible for ligand efflux from hepatocytes into bile across this membrane. Although CMVs are composed of rightside-out and inside-out membrane vesicles, ATP-dependent uptake into CMVs is accounted for by uptake into inside-out vesicles, representing the efflux from hepatocytes into bile under physiological conditions. The ATP-dependent transporters located on the bile canalicular membrane include (1) the MDR1 product, which is responsible for the extrusion of amphipathic neutral and cationic compounds including certain antitumor drugs (Kamimoto et al., 1989; Watanabe et al., 1992, 1995; reviewed by Oude Elferink et al., 1995, Kusuhara et al., 1998), (2) the MDR3 product, responsible for the biliary excretion of phospholipids such as phosphatidylcholine (Smit et al., 1993; Schinkel et al., 1997; reviewed by Oude Elferink et al., 1995, 1997), (3) the bile salt export pump (BSEP), responsible for the biliary exretion of conjugated bile acids (Table II) (Gerloff et al., 1998; reviewed by Stieger and Meier, 1998), and (4) the canalicular multispecific organic anion transporter (cMOAT) responsible for the excretion of many kinds of organic anions including the conjugated metabolites formed within hepatocytes (Table II) (reviewed by Oude Elferink et al., 1995; Yamazaki et al., 1996c; Keppler and König, 1997; Suzuki and Sugiyama, 1998). These transporters possess two highly conserved ATP-binding cassette (ABC) regions and are therefore referred to as ABC transporters (Higgins, 1992). Hereditary defects in the expression of these ABC transporters have been identified as the pathogenesis of several kinds of diseases. Mutations in the genes for BSEP (Strautnieks et al., 1998) and MDR3 (Deleuze et al., 1996; de Vree et al., 1998) cause Type II and III progressive familial intrahepatic cholestasis, respectively (reviewed by Oude Elferink and van Berge Henegouwen, 1998; Arrese et al., 1998; Trauner et al., 1998). In addition, Dubin–Johnson syndrome associated with jaundice results from a defect in the expression of cMOAT (Kartenbeck et al., 1996; Paulusma et al., 1997; Wada et al., 1998; Toh et al., 1998; reviewed by Paulusma and Oude Elferink, 1997).

3.1. Transport of Bile Acids across the Bile Canalicular Membrane

The mechanism of transport of bile acids across the bile canalicular membrane has been clarified by examining uptake into CMVs. TC is taken up by CMVs in an ATP-dependent manner (Adachi et al., 1991; Nishida et al., 1991; Müller et al., 1991; Stieger et al., 1992). The transporter responsible for the ATP-dependent efflux of TC across the bile canalicular membrane is referred to as BSEP. The K_m value of TC for BSEP has been determined as 2, 7.5, 26, and 47 µM (Table III) (Stieger et al., 1992; Müller et al., 1991; Nishida et al., 1991; Adachi et al., 1991).

Table II.
Transporters Responsible for the Extrusion of Bile Acids and Non-Bile Acid Organic Anions[a]

Transporter	Species	Number of amino acids	Observed molecular mass (kDa)	Chromosomal localization	Number of TM domains	Expression	References
spgp	Rat	1321	160	2q36	12?	Canicular membrane, subcanalicular vesicles	Gerloff et al. (1998)
cMOAT (MRP2)	Rat	1541	190	ND	17?	Canalicular membrane, intestine, renal proximal tubule (apical membrane)	Paulusma et al. (1996), Büchler et al. (1996), Ito et al. (1997), Madon et al. (1997), Schaub et al. (1997)
cMOAT (MRP2)	Human	1545	ND	10q24	17?	Canalicular membrane	Taniguchi et al. (1996), Kartenbeck et al. (1996), Kool et al. (1997)
MLP2 (MRP3)	Rat	1523	ND	ND	17?	EHBR liver, intestine, lung	Hirohashi et al. (1998)
MLP2 (MRP3)	Human	1527	ND	17	17?	Basolateral membrane	Kool et al. (1997), Kiuchi et al. (1998), König et al. (1999)
MLP1 (MRP6)	Rat	1502	ND	ND	17?	Liver, duodenum, kidney	Hirohashi et al. (1998)

[a]TM, Transmembrane; ND, not determined.
[b]Hipfner et al. (1997).

A cis-inhibition experiment revealed that the BSEP-mediated uptake of TC is reduced by CA and other bile acid derivatives (Nishida et al., 1995). Although it has been reported that TC is taken up by CMV in a membrane-potential-dependent manner, studies with highly purified rat membrane fractions have revealed that the membrane-potential-dependent uptake mechanism of TC is located on the endoplasmic reticulum (Kast et al., 1994). ATP-dependent transport of TC into human CMVs has also been examined. The K_m for TC in human CMVs has been determined as 4.2 (Wolters et al., 1992) and 20.6 (Niinuma et al., 1998) μM, which is comparable to that in rat CMVs (Table III).

Cis-inhibition studies on the ATP-dependent uptake of TC into CMVs *in vitro* may be useful in predicting the inhibitory effect of ligands of interest on the biliary excretion of TC *in vivo*; PSC833, a potent inhibitor of P-glycoprotein, reduced the TC uptake into CMVs at higher concentration (IC_{50} of 0.2 μM; Böhme et al., 1994). *In vivo,* at higher doses, PSC833 also reduced the biliary excretion of TC, resulting in reduced bile acid-dependent bile flow rate (Böhme et al., 1994).

Molecular biological studies have been performed on BSEP (reviewed by Suchy et al., 1997, Steiger and Meier, 1998). Ortiz et al. (1997) focused on the fact that secretory vesicles and a vacuole-enriched fraction purified from *Saccharomyces cerevisiae* also exhibited ATP-dependent TC uptake and disrupted several kinds of genes encoding the yeast ABC transporters. Using this method, they identified a cDNA encoding a polypeptide of 1661 amino acid as the yeast BSEP (Tables II and III).

Another approach to clone BSEP has been reported by Brown et al. (1995).

Table III.
Kinetic Parameters for ATP-Dependent Uptake of TC

Reference	Source	K_m for TC (μm)	V_{max} for TC (nmol/min/mg protein)	K_m for ATP (mM)
Adachi et al. (1991)	Rat CMVs	4.7	0.81	0.064
Nishida et al. (1991)	Rat CMVs	26	0.15	0.67
Müller et al. (1991)	Rat CMVs	7.5		0.80
Stieger et al. (1992)	Rat CMVs	2.1	0.81	
Wolters et al. (1992)	Human CMVs	4.2 (4 subjects)	0.0065–0.188 (4 subjects)	0.25 (3 subjects)
Niinuma et al. (1999)	Rat CMVs	12.6	1.44	
Niinuma et al. (1999)	Human CMVs	20.6 (3 subjects)	0.625 (0.040–1.40) (3 subjects)	
Ortiz et al. (1997)	Vacular fraction from yeast	63		
Gerloff et al. (1998)	Membrane vesicles from Sf9 cells expressing rat spgp	4.3		

They established a cell line which is resistant against GCA by incubating hepatoma-derived HTC cells in the presence of its methyl ester. Since the resistant cells were associated with an increased rate of cellular extrusion of GCA, it has been suggested that BSEP was induced in this cell line. Indeed, the injection of poly A^+ RNA from these resistant cells stimulated the efflux of GCA from *Xenopus laevis* oocytes (Brown *et al.*, 1995), irrespective of an endogenous transporter for the extrusion of taurocholate in oocytes (Shneider and Moyer, 1993). Moreover, the ATP-dependent transport of TC into the membrane vesicles from the resistant cells was higher than that from the parental HTC cells (Luther *et al.*, 1997). By analyzing the poly A^+ RNA species from the resistant cells, BSEP cDNA can be cloned.

More recently, Gerloff *et al.* (1997) focused on a partial cDNA of sister of P-glycoprotein (spgp), a novel protein closely related to MDR1 (Table II) (Childs *et al.*, 1995). They cloned the full length of spgp cDNA from rat liver, which encodes a polypeptide consisting of 1320 amino acids. The sequence alignment indicated that this cDNA is similar to that of rat mdr1 (70.2%) and mdr2 (69.8%) rather than human multidrug-resistance-associated protein (MRP1) (51.2%; see Section 3.2) and rat cMOAT (49.8%; see Section 3.2). Efflux of TC from oocytes was stimulated by injection of its cRNA. Moreover, the membrane vesicles isolated from Sf9 cells infected with recombinant baculovirus containing spgp cDNA could take up TC in an ATP-dependent manner with a K_m of 4.3 µM. Sister of P-gp-mediated transport was in the order taurochenodeoxycholate > tauroursodeoxycholate > TC > GCA = CA (Gerloff *et al.*, 1998), which is the same as that observed in rat CMVs (Stieger *et al.*, 1992). Immunohistochemical studies indicated the presence of the cDNA product on the canalicular membrane and subcanalicular vesicles (Gerloff *et al.*, 1998). These results suggest that spgp encodes BSEP (Gerloff *et al.*, 1998). Mutation in human spgp may be the pathogenesis of progressive familial intrahepatic cholestasis (Strautnieks *et al.*, 1998). Many glucuronide or sulfate conjugates of bile acids are transported predominantly via cMOAT (Oude Elferink *et al.*, 1995; Suzuki and Sugiyama, 1998).

3.2. Transport of Organic Anions across the Bile Canalicular Membrane

By examining ATP-dependent uptake into CMVs, it has been established that a variety of organic anions can be substrates for cMOAT. These include glutathione conjugates [such as LTC_4 (Ishikawa *et al.*, 1990), oxidized glutathione (GSSG; Fernandez-Checa *et al.*, 1992), and DNP-SG (Kobayashi *et al.*, 1990)], glucuronide conjugates [such as bilirubin glucuronides (Nishida *et al.*, 1992a; Jedlitschky *et al.*, 1997)]. The substrate specificity of cMOAT resembles that of MRP1, a primary active transporter overexpressed on the plasma membrane of several tumor cells which have acquired multidrug resistance, in that these conjugated metabolites are also transported by MRP1 (Jedlitschky *et al.,* 1994, 1996; Leier *et al.,* 1994; Müller *et al.,* 1994; Loe *et al.,* 1996a, b; reviewed by Oude Elferink *et al.,* 1995; Loe *et al.,* 1996c; Lautier *et al.,* 1996; Keppler and Kartenbeck, 1996; Yamazaki *et al.,* 1996c; Kusuhara *et al.,* 1998); given the similarity in structure between MRP1 and cMOAT (Section 3.2.5), cMOAT has also been referred to as MRP2 (Keppler and Kartenbeck, 1996; Müller and Jansen, 1997, 1998). In addition, mutant rat strains whose cMOAT function is hereditarily defective have been established (Oude Elferink *et al.,* 1995; Yamazaki *et al.,* 1996c). These include Groningen Yellow (GY) and transport-deficient (TR⁻) rats derived from Wistar rats and the Eisai hyperbilirubinemic rat (EHBR) from Sprague-Dawley (SD) rats. Due to the impaired biliary excretion of bilirubin glucuronides, these mutant rats exhibit hyperbilirubinemia and therefore are used as animal models to study the pathogenesis of Dubin–Johnson syndrome found in humans. By examining the difference in the hepatobiliary transport of organic anions between normal and mutant rats, it has been found that clinically important anionic drugs can be substrates for cMOAT (Oude Elferink *et al.,* 1995; Yamazaki *et al.,* 1996c; Kusuhara *et al.,* 1998; Suzuki and Sugiyama, 1998). The substrates for cMOAT are summarized in Table IV. Molecular aspects of the cMOAT/MRP family have been summarized elsewhere (Deeley and Cole, 1997; Müller and Jansen, 1998; Suzuki and Sugiyama, 1998). In the next section, the role of cMOAT in drug disposition is described, focusing on the biliary excretion of temocaprilat (Fig. 6), an ACE inhibitor (Ishizuka *et al.,* 1997).

3.2.1. THE ROLE OF cMOAT IN DRUG DISPOSITION: KINETIC ANALYSIS OF THE BILIARY EXCRETION OF TEMOCAPRILAT

ACE inhibitors are used in the treatment of high blood pressure. Many active forms of ACE inhibitors, such as captoprilat, cilazaprilat, ramiprilat, and spiraprilat, are excreted predominantly into the urine (Ishizuka *et al.,* 1997). In contrast, the biliary excretion of temocaprilat is important (Ishizuka *et al.,* 1997); in rats, 85–90% of the administered dose is excreted into urine. In humans, 36–44% of the orally administered dose of temocapril (prodrug) is excreted into feces and 17–24% is excreted into urine, although absorption is almost complete. The presence of an excretion route other than the urinary one provides pharmacokinetic and subsequent pharmacodynamic advantage, particularly in the treatment of patients with renal failure (Ishizuka *et al.,* 1997). In patients with renal failure, the area under the curve (AUC) and C_{max} of captopril and enalapril are markedly increased, so it is necessary to reduce the dosage and/or to change the dosage interval because these ACE inhibitors are eliminated primarily via renal excretion (Lowenthal *et al.,* 1985; Sica, 1992). In contrast, alterations in these pharmacokinetic parameters

Table IV.
Substrate Specificity of cMOAT

Ligand	Methodology	Kinetic parameters in membrane vesicles[a]	Clearance for the uptake in membrane vesicles[b] (μl/min/mg protein)
Glutathione conjugates and cysteinyl leukotrienes			
glutathione disulfide	Hepatocytes (Oude Elferink et al., 1990)		
	In vivo (Oude Elferink et al., 1989b)		
	Rat CMVs (Fernandez-Checa et al., 1992)		
Leukotriene C_4	In vivo (Huber et al., 1987)		
	Rat CMVs (Ishikawa et al., 1990)	K_m 0.25	100 at 10 nM
	Rat CMVs (Niinuma et al., 1999)		268 at 0.1 μM
	Human CMVs (Niinuma et al., 1999)		25 at 0.1 μM
	Human cMOAT recombinant in MDCK II (König et al., 1998)	K_m 1.1	
Leukotriene D_4	In vivo (Huber et al., 1987)		
	Rat CMVs (Ishikawa et al., 1990)	K_m 1.5	47 at 10 nM
Leukotriene E_4	In vivo (Huber et al., 1987)		
	Rat CMVs (Ishikawa et al., 1990)	K_m >10	14 at 10 nM
N-acetyl leukotriene E_4	In vivo (Huber et al., 1987)		
	Rat CMVs (Ishikawa et al., 1990)	K_m 5.2	37 at 10 nM
2,4-Dinitrophenyl-S-glutathione	In vivo (Oude Elferink et al., 1989b)		
	Perfused liver (Oude Elferink et al., 1989b)		
	Hepatocytes (Oude Elferink et al., 1990)		
	Rat CMVs (Kobayashi et al., 1990)	K_m 4, V_{max} 0.18 K_m (ATP) 0.26 mM	45
	Rat CMVs (Akerboom et al., 1991)	K_m 71, V_{max} 0.34	4.8
	Rat CMVs (Niinuma et al., 1999)	K_m 21, V_{max} 1.22	58
	Rat cMOAT recombinant in NIH3T3 (Ito et al., 1998c)	K_m 0.175 μM	
	Human CMVs (Niinuma et al., 1999)	K_m 0.19 mM, V_{max} 0.70	3.7
Bromosulfophthalein glutathione	In vivo (Takikawa et al., 1991; Jansen et al., 1993)		

(continued)

Table IV. (*Continued*)
Substrate Specificity of cMOAT

Ligand	Methodology	Kinetic parameters in membrane vesicles[a]	Clearance for the uptake in membrane vesicles[b] (μl/min/mg protein)
	Rat CMVs (Kitamura et al., 1990)		
	Rat CMVs (Niinuma et al., 1999)		121 at 0.2 μM
	Human CMVs (Niinuma et al., 1999)		18 at 0.2 μM
Glutathionyl-bromoisovalerylurea	In vivo (Polhuijs et al., 1989)		
Glutathione-bimane	Hepatocytes (Oude Elferink et al., 1993)		
Glutathione conjugates of heavy metals			
Antimony	In vivo (Gyurasics et al., 1992)		
Arsenite	In vivo		
Bismuth	In vivo (Gyurasics et al., 1992)		
Cadmium	In vivo (Dijkstra et al., 1996)		
Copper	In vivo (Dijkstra et al., 1996)		
Silver	In vivo (Dijkstra et al., 1996)		
Zinc	In vivo (Dijkstra et al., 1996)		
Glucuronide conjugates			
Bilirubin monoglucuronide	In vivo (Jansen et al., 1993)		
	Perfused liver (Jansen et al., 1993)		
	Rat CMVs (Jedlitschky et al., 1997)	K_i 0.12 on leukotriene C_4	18 at 0.5 μM
	HepG2 (Jedlitschky et al., 1997)	K_i 0.28 on leukotriene C_4	17 at 0.5 μM
Bilirubin diglucuronide	In vivo (Jansen et al., 1993)		
	Perfused liver (Jansen et al., 1993)		
	Rat CMVs (Nisida et al., 1992a)	K_m 71 V_{max} 17 K_m (ATP) 0.37 mM	240
	Rat CMVs (Jedlitschky et al., 1997)	K_i 0.10 on leukotriene C_4	17 at 0.5 μM
	HepG2 (Jedlitschky et al., 1997)	K_i 0.40 on leukotriene C_4	9 at 0.5 μM
$E_2 17\beta G$	In vivo (Takikawa et al., 1991, 1996)		

Compound	Source (Reference)	Kinetics	Other
	Rat CMVs (Vore et al., 1996)	K_m 75 V_{max} 598 pmole/min/mg protein	8[c]
	Rat CMVs (Niinuma et al., 1999)		34 at 0.06 μM
	Human CMVs (Niinuma et al., 1999)		15 at 0.06 μM
	Human cMOAT recombinant in MDCK II (König et al., 1998)		
Triiodothyronine-glucuronide	In vivo (Oude Elferink et al., 1995)		3.0
p-Nitrophenyl glucuronide	Rat CMVs (Kobayashi et al., 1991)	K_m 7.2	
		K_m 20 V_{max} 0.060	
		K_m (ATP) 0.27 mM	
1-Naphthol-β-D-glucuronide	Perfused liver (de Vries et al., 1989)		
E3040 glucuronide	In vivo (Takenaka et al., 1995b)		
	Perfused liver (Takenaka et al., 1995a)		
	Rat CMVs (Niinuma et al., 1999)	K_m 3.8 V_{max} 0.36	95
	Human CMVs (Niinuma et al., 1999)	K_m 14 V_{max} 0.78	56
SN-38 glucuronide, carboxylate	In vivo (Chu et al., 1997a)		
	Rat CMVs (Chu et al., 1997b)	K_m 0.96 V_{max} 0.25	260
SN-38 glucuronide, lactone	In vivo (Chu et al., 1997a)		
	Rat CMVs (Chu et al., 1997b)	K_m 2.30 V_{max} 0.17	74
R-Grepafloxacin-glucuronide	In vivo (Sasabe et al., 1998b)		
	Rat CMVs (Sasabe et al., 1998a)	K_m 17 V_{max} 3.7	218
S-Grepafloxacin-glucuronide	In vivo (Sasabe et al., 1998b)		
	Rat CMVs (Sasabe et al., 1998a)	K_m 10 V_{max} 1.3	130
Grepafloxacin-glucuronide (racemic)	Human CMVs (Niinuma et al., 1999)		11
Glycyrrhizin	In vivo (Shimamura et al., 1996)		5.5
Liquiritigenin glucuronides	In vivo (Shimamura et al., 1994)		
Bile Salt conjugates			
Cholate 3-O-glucuronide	In vivo (Kuipers et al., 1989)		
Lithocholate 3-O-glucuronide	In vivo (Kuipers et al., 1989)		
Chenodeoxycholate 3-O-glucuronide	In vivo (Takikawa et al., 1991)		
Nordeoxycholate 3-O-glucuronide	In vivo (Oude Elferink et al., 1995)		

(*continued*)

Table IV. (Continued)
Substrate Specificity of cMOAT

Ligand	Methodology	Kinetic parameters in membrane vesicles[a]	Clearance for the uptake in membrane vesicles[b] (μl/min/mg protein)
Nordeoxycholate 3-sulfate	*In vivo* (Oude Elferink *et al.*, 1995)		
Lithocholate 3-sulfate	*In vivo* (Takikawa *et al.*, 1991)		
Taurolithocholate 3-sulfate	*In vivo* (Kuipers *et al.*, 1988)		
Glycolithocholate 3-sulfate	*In vivo* (Kuipers *et al.*, 1988)		
Taurochenodeoxycholate 3-sulfate	*In vivo* (Kuipers *et al.*, 1988)		
Nonconjugated organic anions			
Bromosulfophthalein	*In vivo* (Takikawa *et al.*, 1991)		
	Rat CMVs (Nishida *et al.*, 1992b)	K_m 31 V_{max} 33 K_m (ATP) 0.26 mM	1.06
Dibromosulfophthalein	*In vivo* (Jansen *et al.*, 1987; Sathirakul *et al.*, 1993 a, b)		
	Perfused liver (Jansen *et al.*, 1987)		
Carboxyfluorescein	Hepatocyte couplets (Kitamura *et al.*, 1990)		
Reduced folates			
• 5-Methyltetrahydrofolate	*In vivo* (Kusuhara *et al.*, 1988)		
	Rat CMVs (Kusuhara *et al.*, 1988)	K_m 126 V_{max} 272 pmole/min/mg protein	2.2
• Tetrahydrofolate	*In vivo* (Kusuhara *et al.*, 1998)		
	Rat CMVs (Kusuhara *et al.*, 1998)	K_i on DNP-SG 358	
• 5,10-Methylenetetrahydrofolate	*In vivo* (Kusuhara *et al.*, 1998)		
	Rat CMVs (Kusuhara *et al.*, 1998)	K_i on DNP-SG 269	
L-Methotrexate	*In vivo* (Masuda *et al.*, 1997)		
	Rat CMVs (Masuda *et al.*, 1997)	K_m 0.30 mM V_{max} 1.45	4.8
	Rat CMVs (Niinuma *et al.*, 1999)		9.9 at 0.2 μM

Compound	Source	K_m / K_i	Additional
D-Methotrexate	Human CMVs (Niinuma et al., 1999)	K_i on L-methotrexate 93	0.13 at 0.2 μM
CPT-11, carboxylate	Rat CMVs (Masuda et al., 1997)		
	In vivo (Chu et al., 1997a)		
	Rat CMVs (Chu et al., 1997b)	K_m 236 V_{max} 1.99	8.4
SN-38, carboxylate	In vivo (Chu et al., 1997a)		
	Rat CMVs (Chu et al., 1997b)	K_m 69 V_{max} 2.85	41
Ampicillin	In vivo (Verkade et al., 1990)		
Ceftriaxone	In vivo (Oude Elferink and Jansen, 1994)		
Cefodizime	In vivo (Sathirakul et al., 1993, 1994)		
R-Grepafloxacin	In vivo (Sasabe et al., 1997)		
	Rat CMVs (Sasabe et al., 1997)	K_i on DNP-SG 4.39 mM	
S-Grepafloxacin	In vivo (Sasabe et al., 1997)		
	Rat CMVs (Sasabe et al., 1997)	K_i on DNP-SG 3.70 mM	
Pravastatin	In vivo (Yamazaki et al., 1996b, 1997)		
	Rat CMVs (Yamazaki et al., 1996b, 1997, Niinuma et al., 1999)		6.1 at 0.2 μM
	Human CMVs (Niinuma et al., 1999)		1.9 at 0.2 μM
Temocaprilate	In vivo (Ishizuka et al., 1997)		
	Rat CMVs (Isizuka et al., 1997)	K_m 92.5 V_{max} 1.14	12
BQ-123	In vivo (Shin et al., 1997)		
	Rat CMVs (Shin et al., 1997)	K_m 29.4 V_{max} 1.47	50
	Rat CMVs (Niinuma et al., 1999)		17 at 0.1 μM
	Human CMVs (Niinuma et al., 1999)		0.82 at 0.1 μM

[a] K_m and K_i values in μM, unless otherwise noted; V_{max} values in nmole/min/mg protein, unless otherwise noted. K_m (ATP) represents K_m for ATP. DNP-SG, 2,4-Dinitrophenyl-S-glutathione.
[b] For substrates whose K_m and V_{max} values have been reported, clearance for the uptake was calculated as V_{max}/K_m. For other substrates, clearance for the uptake was calculated by dividing the initial velocity of the uptake by the substrate concentration in the medium.
[c] Contribution of P-glycoprotein has been postulated by the authors.

are minimal for temocaprilat because of the presence of the biliary excretion pathway (Oguchi et al., 1993). In order to examine the mechanism for the efficient biliary excretion of temocarpilat, the hepatocellular transport of this compound was examined.

To examine the transport process across the sinusoidal membrane, the uptake of temocaprilat into isolated rat hepatocytes was examined (Ishizuka et al., 1998a). The analysis indicated that the uptake of temocaprilat is mediated by an Na^+-independent active transport system (K_m = 21 μM) which was completely inhibited by DBSP with IC_{50} of 2–3 μM. In contrast, $E_2 17\beta G$ inhibited approximately 50% of temocaprilat uptake with an IC_{50} of ~1 μM, suggesting the presence of multiple transport systems for temocaprilat. The uptake of temocaprilat was examined in oatp-1-transfected COS-7 cells. Although temocaprilat was revealed to be a substrate for oatp-1, the contribution of oatp-1 to the hepatic uptake was approximately 50%, which is consistent with the inhibition study in isolated hepatocytes. The uptake of temocaprilat into hepatocytes was significantly inhibited by 100 μM enalaprilat, cilazaprilat, benazeprilat, and delaprilat, suggesting that a series of ACE inhibitors can be substrates for the transport system for temocaprilat. However, this hypothesis remains to be confirmed experimentally (Ishizuka et al., 1998a).

The transport of temocaprilat across the bile canalicular membrane was compared between SD rats and EHBR (Ishizuka et al., 1997). In EHBR, the amount of temocaprilat excreted into the bile up to 6 hr after intravenous administration of this compound (25%) was much lower than in SD rats (78%) (Fig. 7). Consequently, reduced plasma disappearance of temocaprilat was observed in EHBR; indeed, in EHBR, the plasma AUC up to 6 hr was more than six-fold highr than in SD rats (Fig. 7). The analysis indicated that the clearance for the biliary excretion of temocaprilat in EHBR (0.25 ml/min/kg) was much lower than in SD rats (5.00 ml/min/kg). These results indicate that temocaprilat can be a substrate for cMOAT (Ishizuka et al., 1997).

This conclusion was further confirmed using CMVs (Ishizuka et al., 1997). The uptake of temocaprilat into CMVs from SD rats, but not from EHBR, was markedly stimulated in the presence of ATP (Fig. 8). Kinetic analysis indicated that the ATP-dependent uptake of temocaprilat is mediated by a saturable transport system with a K_m of 92.5 μM. Other ACE inhibitors which are primarily excreted into urine (such as benazeprilat, cilazaprilat, delaprilat, enalaprilat, and imidaprilat) did not affect the ATP-dependent uptake of temocaprilat into the CMVs even at 200 μM. These results suggest that the affinity for cMOAT is the predominant factor in determining the biliary excretion of a series of ACE inhibitors. The ACE inhibitors examined in our study may be taken up by hepatocytes via a mechanism shared with temocaprilat. Although temocaprilat is excreted into bile with the aid of cMOAT, the biliary excretion of other ACE inhibitors is minimal due to the low affinity for cMOAT (Ishizuka et al., 1997). Consequently, these drugs are trans-

Transporters for Bile Acids and Organic Anions

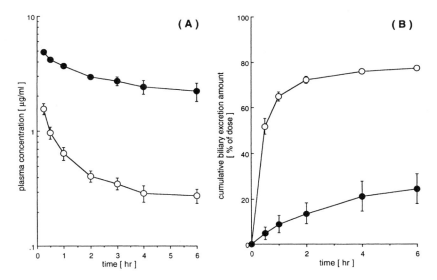

Figure 7. Time profiles for plasma concentration (A) and biliary excretion (B) of temocaprilat after intravenous administration to SD rats and EHBR. Taken with permission from Ishizuka *et al.* (1997). Copyright 1997 Williams & Wilkins.

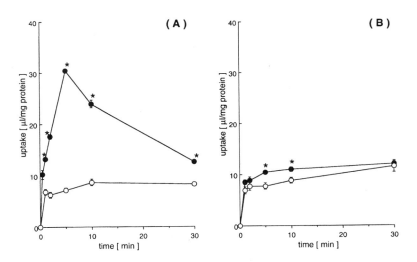

Figure 8. Uptake of temocaprilat into CMVs from SD rats (A) and EHBR (B) in the presence and absence of ATP. Taken with permission from Ishizuka *et al.* (1997). Copyright 1997 Williams & Wilkins.

ferred into the blood circulation across the sinusoidal membrane, and undergo subsequent urinary excretion mediated by glomerular filtration (Ishizuka et al., 1997).

3.2.2. THE ROLE OF cMOAT IN DRUG DISPOSITION: EFFICIENT BILIARY EXCRETION MEDIATED BY cMOAT

The efficient biliary excretion of substrates listed in Table IV is predominantly mediated by cMOAT. Accordingly, the affinity to cMOAT, along with that to the transporters for hepatic uptake, may be an important factor in determining the excretion route (biliary or urinary) of β-lactam and new quinolone antibiotics (Table IV). Moreover, efficient enterohepatic circulation of reduced folate and pravastatin (Section 2.2.2) is conferred by cMOAT (Table IV). The efficient biliary excretion of E3040 glucuronide after administration of E3040 (Takenaka et al., 1995b) provides a pharmacological advantage to this drug, since the biliary excreted glucuronide is hydrolyzed by the enzymes of the intestinal flora, resulting in the exposure of the target organ (large intestine) to the parental drug. In contrast to E3040 glucuronide, its sulfated conjugate is not a substrate for cMOAT, since the biliary excretion of E3040 sulfate was comparable between normal rats and EHBR (Takenaka et al., 1995a, b). Moreover, the uptake of E3040 sulfate into CMVs from both rat strains was comparable without a stimulatory effect of ATP (Takenaka et al., 1995a). It was also found that the biliary excretion of liquirithigenin di-sulfate is not predominantly mediated by cMOAT (Shimamura et al., 1994). These results suggest that the sulfated conjugates of xenobiotics may not be good substrates for cMOAT, which is in marked contrast to the efficient biliary excretion of sulfated bile acids mediated by cMOAT (Table IV).

It was also found that BQ-123 (Fig. 1; see Section 2.1.2), an anionic small peptide, can be a substrate for cMOAT. The efficient biliary excretion of BQ-123 is accounted for by its carrier-mediated uptake and by extrusion mediated by cMOAT. In contrast to BQ-123, octreotide (Fig. 1), a cationic small peptide, was also taken up by CMVs in an ATP-dependent manner from both normal and mutant rats, suggesting that primary active transporter(s) other than cMOAT mediates the efficient biliary excretion of this peptide (Yamada et al., 1997a). In the same manner, the rat P-glycoprotein (mdr1 product) can also extrude $E_2 17\beta G$ (Vore et al., 1996), although the canalicular transport of this conjugate is mediated predominantly by cMOAT (Takikawa et al., 1991, 1996).

It also has been demonstrated that plural transporters mediate the canalicular transport of CPT-11, a camptothecin analogue, and its metabolites (Chu et al., 1997a, b; Sugiyama et al., 1998). The high-affinity transport system of CPT-11 (carboxylate form; K_m = 3.39 μM, V_{max} = 0.12 nmole/min/mg protein) along with the low-affinity transport system of the carboxylate (K_m = 75 μM, V_{max} = 0.36 nmole/min/mg protein) and lactone forms of SN-38 glucuronide (K_m = 82 μM, V_{max} = 0.16 nmole/min/mg/protein) observed in normal rat CMVs were also detectable in CMVs from EHBR (Fig. 9 and Table IV) (Chu et al., 1997b). Based on the concentration-dependent inhibitory effect of verapamil, it has been suggested that mdr1 P-glycoprotein may be responsible for the uptake of these substrates. Multiplicity for the transport of organic anions across the bile canalicular membrane is described in Section 3.2.6.

It is difficult to determine the substrate specificity of cMOAT only from *in vivo* and/or *in situ* experiments. Although the biliary excretion of indocyanine green is reduced in EHBR, kinetic analysis suggested that the impaired intracellular transport of this compound may account for reduced excretion rather than impaired transport across the canalicular membrane (Sathirakul et al., 1993a, b).

3.2.3. PREDICTION OF *IN VIVO* cMOAT FUNCTION FROM CMV DATA

As discussed in the previous section, it is possible to predict qualitatively *in vivo* biliary excretion from *in vitro* data with CMV. Moreover, quantitative prediction should be possible. We compared the *in vivo* and *in vitro* (Yamazaki et al., 1997). *In vivo*, pravastatin was infused to rats to determine the transport clearance across the bile canalicular membrane, which is defined as the biliary excretion rate divided by hepatic unbound concentration of pravastatin at steady state. The K_m value for *in vivo* clearance was determined by analyzing its dose dependence. *In vitro*, the ATP-dependent uptake of pravastatin was kinetically analyzed to determine its K_m value. The K_m value determined *in vivo* (181 μM) was comparable with that determined *in vitro* (223 μM) (Yamazaki et al., 1997).

Such an *in vivo/in vitro* correlation was also examined for temocaprilat in mice, rats, rabbits, and dogs (Ishizuka et al., 1998b). The clearance for the efflux of temocaprilat across the bile canalicular membrane, defined as the biliary excretion rate divided by the hepatic unbound concentration, in these species was compared with the ATP-dependent uptake of DNP-SG, a typical substrate for cMOAT, into CMVs isolated from the respective species. A nice correlation was observed between *in vivo* efflux clearance and *in vitro* uptake clearance into CMVs. These results suggest that CMVs are useful for the quantitative prediction of biliary excretion *in vivo* (Ishizuka et al., 1998b).

3.2.4. STUDIES WITH HUMAN CMVs

The uptake of several ligands was examined in CMVs isolated from six human subjects (Niinuma et al., 1999). The ATP-dependent transport activity in human CMVs was comparable with that in rat CMV for glucuronide conjugates (such as $E_2 17\beta G$, E3040 glucuronide, and grepafloxacin glucuronide), but not for other ligands; the ATP-dependent uptake of the glutathione conjugates (such as LTC_4,

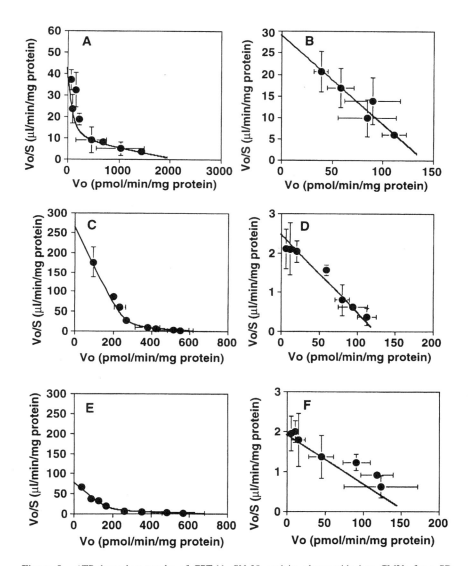

Figure 9. ATP-dependent uptake of CPT-11, SN-38, and its glucuronide into CMVs from SD rats and EHBR. Solid line represents the fitted line. Upper, uptake of carboxylate form of CPT-11 into CMVs from SD rats (A) and EHBR (B); middle, uptake of carboxylate form of SN-38 glucuronide into CMVs from SD rats (C) and EHBR (D); lower, uptake of lactone form of SN-38 glucuronide into CMVs from SD rats (E) and EHBR (F). Taken with permission from Chu et al. (1997b). Copyright 1997 American Association for Cancer Research, Inc.

DNP-SG, and glutathione conjugate of BSP), pravastatin, and BQ-123 in humans was approximately 5–10 times lower than in rats (Table IV). Kinetic analysis of the transport of DNP-SG revealed that the difference in the transport activity is due to the difference in K_m values (21 and 193 μM for rats and humans, respectively) rather than V_{max} values (1220 and 700 pmole/min/mg protein for rats and humans, respectively) (Table IV). These kinetic parameters were comparable for E3040 glucuronide between rats and humans (K_m = 4 and 14 μM and V_{max} = 360 and 780 pmole/min/mg protein for rats and humans, respectively) (Table IV; Niinuma *et al.*, 1998). The difference in the transport properties between the two species should be furter investigated by examining the transport function of the cloned rat and human cMOAT cDNA products.

3.2.5. MOLECULAR CLONING OF cMOAT

To elucidate the molecular mechanism for transport mediated by cMOAT, it is essential to isolate the cDNA for cMOAT. Since (1) the substrate specificity of cMOAT resembles that of MRP1 and (2) a series of primary active tansporters contain highly conserved ABC regions within their structure, we and others have cloned cMOAT cDNA from rat liver (Table II) (Paulusma *et al.*, 1996; Büchler *et al.*, 1996; Ito *et al.*, 1996, 1997; Madon *et al.*, 1997; reviewed by Müller and Jansen, 1997, 1998; Keppler and König, 1997; Paulusma and Oude Elferink, 1997). The open reading frame of cMOAT cDNA consists of 4623 bp, encoding a polypeptide composed of 1521 amino acids. Similarity in the profile of hydropathy plot (Fig. 10) betweeen MRP1 and cMOAT suggests that cMOAT has 17 transmembrane domains (Loe *et al.*, 1996c; Lautier *et al.*, 1996). Using antibodies against the polypeptide residues, it has been shown that cMOAT is selectively located on the canalicular membrane of rat hepatocytes (Paulusma *et al.*, 1996; Büchler *et al.*, 1996). In addition, the selective loss of this transporter was demonstrated in TR⁻ and EHBR liver. This reduction was accounted for by the decrease in the transcripts. Using reverse transcriptase-polymerase chain reaction (RT-PCR), the mechanism for the impaired expression of cMOAT in mutant rats was clarified. In TR⁻, one base deletion at amino acid 395 resulted in the introduction of a stop codon at amino acid 401 (Paulusma *et al.*, 1996), whereas in EHBR, one base replacement (T to C) at amino acid 855 was responsible for the introduction of a stop codon (Ito *et al.*, 1997).

The human cMOAT was initially cloned from a cisplatin-resistant tumor cell line (Taniguchi *et al.*, 1996). Paulusma *et al.* (1997) also cloned cMOAT from human liver based on the homology with rat cMOAT. The cDNA for cMOAT consists of 4635 bp, encoding a polypeptide composed of 1545 amino acids. Using immunohistochemical studies, the selective loss of this transporter from the bile canalicular membrane was demonstrated in patients suffering from Dubin–John-

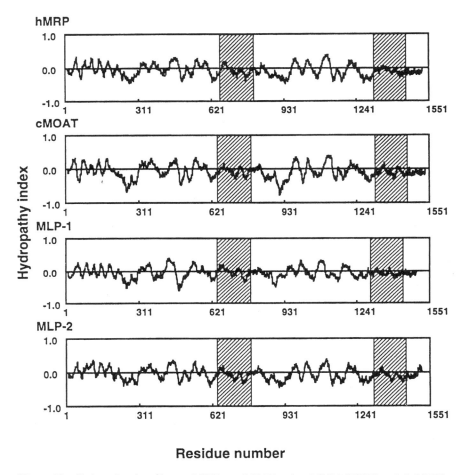

Figure 10. Hydropathy plot of human MRP1, rat cMOAT, and rat MLP-1 (MRP6) and -2 (MRP3). Taken with permission from Hirohashi *et al.* (1998). Copyright 1998 Williams & Wilkins.

son syndrome (Kartenbeck *et al.*, 1996). The mechanism for the defective expression of cMOAT in Dubin–Johnson syndrome has recently been clarified (Paulusma *et al.*, 1997; Wada *et al.*, 1998; Toh *et al.*, 1999). In addition, it is plausible that the tumor cells may acquire resistance against the antitumor drugs listed in Table I by overexpressing cMOAT. Indeed, the overexpression of cMOAT was observed in several kinds of cisplatin-resistant tumor cell lines (Taniguchi *et al.*, 1996; Kool *et al.*, 1997; Borst *et al.*, 1997). In addition, Koike *et al.* (1997) indicated that the transfection of antisense against cMOAT into resistant tumor cells results in increased sensitivity against certain antitumor drugs (also see Loe *et al.*, 1996c; Lautier *et al.*, 1996; Kusuhara *et al.*, 1998).

The transport properties of the cMOAT cDNA product have also been examined. Madon *et al.* (1997) injected cMOAT cRNA into oocytes and found accelerated export of DNP-SG. They also demonstrated that transfection of cMOAT cDNA into COS-7 cells results in an increase in the export of DNP-SG. To estimate quantitatively the function of cMOAT cDNA product, we transfected cMOAT cDNA to NIH/3T3 cells to obtain colonies stably expressing the cDNA product (Ito *et al.*, 1998c). Although endogenous activity of the ATP-dependent transport of glutathione conjugates was observed in vector-transfected NIH/3T3 cells, the transfection of cMOAT cDNA resulted in an increase in transport (Fig. 11). Kinetic analysis indicated that the K_m for the ATP-dependent transport of DNP-SG was approximately 175 nM (Fig. 11), which was much lower than that observed in rat CMVs (approximately 20 μM; Table IV) (Ito *et al.*, 1998c). The difference in the transport properties between transfected cells and CMVs could be accounted for by the difference in the lipid composition of the plasma membrane on which the transporter is expressed. Alternatively, it is possible that cMOAT acts as a modulator of unidentified transporter, since the K_m value of DNP-SG in the transfected NIH/3T3 cells was comparable with that mediated by a transporter which is endogenously expressed in the parental NIH/3T3 cells (Ito *et al.*, 1998c). These possibilities should be examined in order to establish a method to quantify the

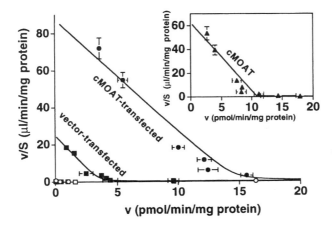

Figure 11. Uptake of DNP-SG into membrane vesicles from NIH/3T3 cells transfected with plasmid vector with (●, ○) or without (■, □) rat cMOAT cDNA. The uptake was examined in the presence (closed symbols) and absence (open symbols) of ATP. The inset represents the ATP-dependent uptake mediated by transfected rat cMOAT product, determined as the difference between cMOAT- and vector-transfected membrane vesicles. Taken with permission from Ito *et al.* (1998c). Copyright 1998 The American Society for Biochemistry and Molecular Biology, Inc.

cMOAT function. Recently, van Aubel et al. (1998) demonstrated that the rabbit protein initially cloned as epithelial basolateral chloride conductance regulator acts like cMOAT, by examining the ATP-dependent transport of LTC_4 and $E_217\beta G$ into membrane vesicles isolated from Sf9 cells infected by recombinant baculoviruses containing this gene. Their approach may be convincing, since a high expression of exogenous proteins is expected in this experimental system.

In order to examine the expression of rat cMOAT in polarized cells, cDNA was transfected to MDCK cells (Kinoshita et al., 1998). The cMOAT function was examined by measuring the efflux of glutathione-bimane (GS-B) into the apical and basal compartments after preloading the cells with precursor (monochlorobimane, MCB; Oude Elferink et al., 1993). Although symmetrical efflux of GS-B was observed in the vector-transfected MDCK cells, the apical efflux exceeded to the basal efflux in cMOAT-transfected cells, suggesting that cMOAT may be expressed on the apical membrane after transfection (Kinoshita et al., 1998). This result is consistent with the localization of cMOAT under physiological conditions in that cMOAT expression is confined to the bile canalicular (apical) membrane of hepatocytes. Evers et al. (1998) also examined the expression of human cMOAT in MDCKII cells. They demonstrated the apical localization of transfected cDNA product by immunohistochemistry, along with the apical efflux of DNP-SG after preloading its precursor (1-chloro-2,4-dinitrobenzene) (Evers et al., 1998). Moreover, Nies et al. (1998) demonstrated the apical localization of transfected rat cMOAT cDNA product in WIF-B, a polarized rat hepatoma/human fibroblast hybrid cell line. The cMOAT function was confirmed by demonstrating the apical vacuole accumulation of Fluo-3 after preloading its ester (Nies et al., 1998). Moreover, ATP-dependent transport of Fluo-3 (K_m = 3.7 μM) into the membrane vesicles isolated from cMOAT-transfected cells was demonstrated (Nies et al., 1998). If the transporters responsible for the hepatic uptake of organic anions can also be expressed on the basolateral membrane of cMOAT-transfected cells, vectorial transport from blood to bile across the hepatocytes can be predicted by examining the transcellular transport across such transfected cells (Kinoshita et al., 1998).

Characterization of the transport properties of human cMOAT was initially performed by using cultured human cells which constitutively express cMOAT, although the contribution of other transporter(s) should be taken into consideration. Jedlitschky et al. (1997) found the ATP-dependent uptake of bilirubin glucuronide into membrane vesicles isolated from HepG2 cells, which express cMOAT. By comparing the transport properties between parental HepG2 and HepG2 transfected with plasmid containing the antisense sequence of cMOAT cDNA (Koike et al., 1997), the transport properties of human cMOAT may be clarified. Direct examination of transport kinetics mediated by human cMOAT was recently performed by König et al. (1998). Using isolated membrane vesicles from human cMOAT-transfected HEK 283 cells, they found the K_m of LTC_4 and $E_217\beta G$ to be 1.1 and 7.2 μM, respectively (König et al., 1998).

3.2.6. MULTIPLICITY FOR THE TRANSPORT OF ORGANIC ANIONS ACROSS THE BILE CANALICULAR MEMBRANE

Although many studies have focused on cMOAT, several pieces of kinetic evidence suggest the presence of other transporter(s) responsible for the excretion of organic anions across the bile canalicular membrane. We found the presence of ATP-dependent uptake of E3040 glucuronide in CMVs isolated from EHBR, although it was approximately one-third or one-fourth that in CMVs from normal rats (Takenaka *et al.,* 1995a). In contrast, the uptake of DNP-SG in CMVs from EHBR was negligible. The ATP-dependent uptake of DNP-SG in CMVs from normal rats was completely inhibited by unlabeled E3040 glucuronide, whereas that of E3040 glucuronide was partially inhibited by unlabeled DNP-SG at a concentraton sufficient to saturate the uptake of DNP-SG (Niinuma *et al.,* 1997). In addition, the ATP-dependent uptake of E3040 glucuronide in CMVs from EHBR was not reduced in the presence of unlabeled DNP-SG (Niinuma *et al.,* 1997). Collectively, these data suggest the presence of other transporter(s) responsible for the excretion of E3040 glucuronide, along with cMOAT, which accepts both DNP-SG and E3040 glucuronide. Moreover, the presence of multiple pathways for the biliary excretion of anionic compounds has also been suggested from *in vivo* inhibition studies (Shimamura *et al.,* 1996; Sathirakul *et al.,* 1993a, b).

3.2.7. MOLECULAR CLONING OF MRP FAMILY

In order to examine whether members of the MRP family other than cMOAT is expressed in the liver, we performed RT-PCR using RNA from EHBR liver as a template. Using amplified cDNA fragments as probes, we were able to clone two kinds of transporters referred to as MRP-like protein (MLP)-1 and -2 (Table II) (Hirohashi *et al.,* 1998). Hydropathy plot analysis suggests that both MLP-1 and -2 also have 17 transmembrane domains (Fig. 10) (Loe *et al.,* 1996c; Lautier *et al.,* 1996). Sequence alignment suggested that MLP-1 and -2 are rat homologues of human MRP6 and 3, respectively (Kool *et al.,* 1997). Northern blot analysis suggests that the hepatic expression of MRP6 is comparable in normal rats and EHBR. In contrast, hepatic expression of MRP3 is observed in EHBR, but not in normal rats, suggesting that this transporter is inducible (Hirohashi *et al.,* 1998). It was also found that bilirubin and/or bilirubin glucuronide, along with phenobarbital, are inducers for the expression of MRP3 (Hirohashi *et al.,* 1997). Interestingly, expression of MRP3 has also been reported in human liver (Kool *et al.,* 1997; Kiuchi *et al.,* 1998). We examined the function of cloned rat MRP3 cDNA product by using the membrane vesicles isolated from MRP3-transfected HeLa and LLC-PK1 cells (Hirohashi *et al.,* 1999). It was demonstrated that glucuronide conjugates ($E_2 17\beta G$ and E3040 glucuronide) along with methotrexate are transported via MRP3, whereas LTC_4 and DNP-SG are poor substrates for this transporter (Hirohashi et al., 1999). Although Ortiz et al. (1999) demonstrated the canalicular localization of MRP3, König et al. (1999) and Kool et al. (1999) indicated the basolateral localization of this transporter. The physiological role of MRP3 remains to be clarified.

Recently, it was indicated that MRP6 is localized on the lateral membrane of hepatocytes, and transports BQ-123, but not TC, $E_2 17\beta G$, or LTC_4 (Madon et al., 1998).

4. CONCLUDING REMARKS

In this chapter, recent advances in the hepatobiliary transport of bile acids and non-bile acid organic anions have been summarized. In particular, the hepatocellular uptake of ligands can be quantitatively extrapolated from the data obtained with cloned cDNA products by considering the expression level. In addition, it is possible to extrapolate these *in vitro* data to the ligand uptake into hepatocytes *in vivo* (Miyauchi et al., 1993; Ishigami et al., 1995; Yamazaki et al., 1996a; Nakamura et al., 1996; Takenaka et al., 1997; Yamada et al., 1997b; Imawaka and Sugiyama, 1998; reviewed by Yamazaki et al., 1996c). Transport studies with transfected cells will be useful to predict quantitatively the ligand disposition in humans since in general a drug transporters are usually downregulated in human hepatic tumor cell lines and in cultured hepatocytes (von Dippe and Levy, 1990; Boyer et al., 1993; Liang et al., 1993; Kullak-Ublick et al., 1996b; reviewed by Müller and Jansen, 1997). To predict *in vivo* disposition, major transporters should be cloned and characterized.

Moreover, the mechanism for the regulation of the localization (Oude Elferink et al., 1993; Mukhopadhayay et al., 1997; Roelofsen et al., 1998) and expression of these transporters (Kauffmann et al., 1997, 1998; Kauffmann and Schrenk, 1998) is an important subject for further research (reviewed by Müller and Jansen, 1997, 1998). To understand the regulation of transcription, studies on the promoter region of transporters are essential (Karpen et al., 1996; Kullak-Ublick et al., 1997a). However, one must pay attention to the fact that the apparent hepatic intrinsic clearance is described as a function of influx and efflux processes across the sinusoidal membrane, intracellular metabolic process, and the excretion process across the bile canalicular membrane (reviewed by Yamazaki et al., 1996c).

REFERENCES

Abe, T., Kakyo, M., Sakagami, H., Tokui, T., Nishio, T., Tanemoto, M., Nomura, H., Hebert, S. C., Matsuno, S., Kondo, H., and Yawo, H., 1998, Molecular characterization and tissue distribution of a new organic anion transporter subtype (oatp3) that transports

thyroid hormones and taurocholate and comparison with oatp2, *J. Biol. Chem.* **273**:22395–22401.
Adachi, Y., Kobayashi, H., Kurumi, Y., Shouji, M., Kitano, M., and Yamamoto, T., 1991, ATP-dependent taurocholate transport by rat liver canalicular membrane vesicles, *Hepatology* **14**:655–659.
Akerboom, T. P. M., Narayanaswami, V., Kunst, M., and Sies, H., 1991, ATP-dependent S-(2,4-dinitrophenyl) glutathione transport in canalicular plasma membrane vesicles from rat liver, *J. Biol. Chem.* **266**:13145–13152.
Ananthanarayanan, M., Ng, O.-C., Boyer, J. L., and Suchy, F. J., 1994, Characterization of cloned rat liver Na^+–bile acid cotransporter using peptide and fusion protein antibodies, *Am. J. Physiol.* **267**:G637–G643.
Angeletti, R. H., Novikoff, P. M., Juvvadi, S. R., Fritschy, J. M., Meier, P. J., and Wolkoff, A. W., 1997, The choroid plexus epithelium is the site of the organic anion transport protein in the brain, *Proc. Natl. Acad. Sci. USA* **94**:283–286.
Anwer, M. S., and Hegner, D., 1978, Effect of Na^+ on bile acid uptake by isolated rat hepatocytes; evidence for a heterogenous system, *Hoppe-Seyler's Z. Physiol. Chem.* **359**:181–192.
Arrese, M., Ananthananarayanan, M., and Suchy, F., 1998, Hepatobiliary transport: Molecular mechanisms of development and cholestasis, *Pediat. Res.* **44**:141–147.
Bergwerk, A. J., Shi, X., Ford, A. C., Kanai, N., Jacquemin, E., Burk, R. D., Bai, S., Novikoff, P. M., Stieger, B., Meier, P. J., Schuster, V. L., and Wolkoff, A., 1996, Immunologic distribution of an organic anion transport protein in rat liver and kidney, *Am. J. Physiol.* **271**:G231–G238.
Blitzer, B. L., Ratoosh, S. L., Donovan, C. B., and Boyer, J. L., 1982, Effects of inhibitors of Na^+-coupled ion transport on bile acid uptake by isolated rat hepatocytes, *Am. J. Physiol.* **243**:G48–G53.
Böhme, M., Müller, M., Leier, I., Jedlitschky, G., and Keppler, D., 1994, Cholestasis caused by inhibition of the adenosine triphosphate-dependent bile salt transport in rat liver, *Gastroenterology,* **107**:255–265.
Borst, P., Kool, M., and Evers, R., 1997, Do cMOAT (MRP2), other MRP homologues, and LRP play a role in MDR? *Semin. Cancer Biol.* **8**:205–213.
Bossuyt, X., Müller, M., Hagenbuch, B., and Meier, P. J., 1996a, Polyspecific drug and steroid clearance by an organic anion transporter of mammalian liver, *J. Pharmacol. Exp. Ther.* **276**:891–896.
Bossuyt, X., Müller, M., and Meier, P. J., 1996b, Multispecific amphipathic substrate transport by an organic anion transporter of human liver, *J. Hepatol.* **25**:733–738.
Boyer, J. L., Hagenbuch, B., Ananthanarayanan, M., Suchy, F., Stieger, B., and Meier, P. J., 1993, Phylogenic and ontogenic expression of hepatocellular bile acid transport, *Proc. Natl. Acad. Sci. USA* **90**:435–438.
Boyer, J. L., Ng, O.-C., Ananthanarayanan, M., Hofmann, A. F., Schteingart, C. D., Hagenbuch, B., Stieger, B., and Meier, P. J., 1994, Expression and characterization of a functional rat liver Na^+ bile acid cotransport system in COS-7 cells, *Am. J. Physiol.* **266**:G382–G387.
Brown, R., Lomri, N., De Voss, J., Xie, M. H., Hua, T., Lidofsky, S. D., and Scharschmidt, B. F., 1995, Enhanced secretion of glycocholic acid in a specially adapted cell line as-

sociated with overexpression of apparently novel ATP-binding cassette proteins, *Proc. Natl. Acad. Sci. USA* **92**:5421–5425.

Büchler, M., König, J., Brom, M., Kartenbeck, J., Spring, H., Horie, T., and Keppler, D., 1996, cDNA cloning of the hepatocyte canalicular isoform of the multidrug resistance protein, cMrp, reveals a novel conjugate export pump deficient in hyperbilirubinemic mutant rats, *J. Biol. Chem.* **271**:15091–15098.

Chan, B. S., Satriano, J. A., Pucci, M., and Schuster, V. L., 1998, Mechanism of prostaglandin E_2 transport across the plasma membrane of HeLa cells and *Xenopus* oocytes expressing the prostaglandin transporter "PGI," *J. Biol. Chem.* **273**:6689–6697.

Childs, S., Yeh, R. L., Georges, E., and Ling, V., 1995, Identification of a sister gene to P-glycoprotein, *Cancer Res.* **55**:2029–2034.

Chu, X.-Y., Kato, Y., Niinuma, K., Sudo, K., Hakusui, H., and Sugiyama, Y., 1997a, Multispecific organic anion transporter is responsible for the biliary excretion of the camptothecin derivative irinotecan and its metabolites in rats, *J. Pharmacol. Exp. Ther.* **281**:304–314.

Chu, X.-Y., Kato, Y., and Sugiyama, Y., 1997b, Multiplicity of biliary excretion mechanisms for irinotecan, CPT-11, and its metabolites in rats, *Cancer Res.* **57**:1934–1938.

Clark, R. F., Cruts, M., Korenblat, K. M., He, C., Talbot, C., Van Broeckhoven, C., and Geate, A. M., 1995, A yeast artificial chromosome contiguity from human chromosome 14q24 spanning the Alzheimer's disease locus AD3, *Hum. Mol. Genet.* **8**:1347–1354.

Deeley, R. G., and Cole, S. P. C., 1997, Function, evolution and structure of multidrug resistance protein (MRP), *Semin. Cancer Biol.* **8**:193–204.

Deleuze, J.-F., Jacquemin, E., Dubuisson, C., Cresteil, D., Dumont, M., Erlinger, S., Bernard, O., and Hadchouel, M., 1996, Defect of multidrug-resistance 3 gene expression in a subtype of progressive familial intrahepatic cholestasis, *Hepatology* **23**:904–908.

de Vree, J. M., Jacquemin, E., Sturm, E., Cresteil, D., Bosma, P. J., Aten, J., Deleuze, J.-F., Desrochers, M., Burdelski, M., Bernard, O., Oude Elferink, R. P. J., and Hadchouel, M., 1998, Mutations in the MDR3 gene cause progressive familial intrahepatic cholestasis, *Proc. Natl. Acad. Sci. USA* **95**:282–287.

de Vries, M. H., Redegeld, F. A. M., Koster, A. Sj., Noordhoek, J., de Haan, J. G., Oude Elferink, R. P. J., and Jansen, P. L. M., 1989, Hepatic, intestinal and renal transport of 1-naphthol-β-D-glucuronide in mutant rats with hereditary-conjugated hyperbilirubinemia, *Naunyn-Schmiedeberg's Arch. Pharmacol.* **340**:588–592.

Dijkstra, M., Havinga, R., Vonk, R. J., and Kuipers, F., 1996, Bile secretion of cadmium, silver, zinc and copper in the rat: Involvement of various transport systems, *Life Sci.* **59**:1237–1246.

Eckhardt, U., Horz, J. A., Petzinger, E., Stüber, W., Reers, M., Dickneite, G., Daniel, H., Wagener, M., Hagenbuch, B., Stieger, B., and Meier, P. J., 1996, The peptide-based thrombin inhibitor CRC 220 is a new substrate of the basolateral rat liver organic anion-transporting polypeptide, *Hepatology* **24**:380–384.

Elsner, R., and Ziegler, K., 1989, Determination of the apparent functional molecular mass of the hepatocellular sodium-dependent taurocholate transporter by radiaton inactivation, *Biochim. Biophys. Acta* **983**:113–117.

Elsner, R., and Ziegler, K., 1992, Radiation inactivation of multispecific transport systems for bile acids and xenobiotics in basolateral rat liver plasma membrane vesicles, *J. Biol. Chem.* **267**:9788–9793.

Evers, R., Kool, M., van Deemter, L., Janssen, H., Calafat, J., Oomen, L. C. J. M., Paulusma, C. C., Oude Elferink, R. P. J., Baas, F., Schinkel, A. H., and Borst, P., 1998, Drug export activity of the human canalicular multispecific organic anion transporter in polarized kidney MDCK cells expressing cMOAT (MRP2) cDNA, *J. Clin. Invest.* **101**:1310–1319.

Fernandez-Checa, J. C., Takikawa, H., Horie, T., Ookhtens, M., and Kaplowitz, N., 1992, Canalicular transport of reduced glutathione in normal and mutant Eisai hyperbilirubinemic rats, *J. Biol. Chem.* **267**:1667–1673.

Friesema, E. C. H., Docter, R., Moerings, E. P. C. M., Stieger, B., Hagenbuch, B., Meier, P. J., Krenning, E. P., Hennemann, G., and Visser, T. J., 1999, Identification of thyroid hormone transporters, *Biochem. Biophys. Res. Commun.* **254**:497–501.

Gerloff, T., Stieger, B., Hagenbuch, B., Madon, J., Landmann, L., Roth, J., Hofmann, A. F., and Meier, P. J., 1998, The sister of P-glycoprotein represents the canalicular bile salt export pump of mammalian liver, *J. Biol. Chem.* **273**:10046–10050.

Gyurasics, A., Koszorüs, L., Varga, F., and Gregus, Z., 1992, Increased biliary excretion of glutathione is generated by the glutathione-dependent hepatobiliary transport of antimony and bismuth, *Biochem. Pharmacol.* **44**:1275–1281.

Hagenbuch, B., and Meier, P. J., 1994, Molecular cloning, chromosomal localization, and functional characterization of a human liver Na$^+$/bile acid cotransporter, *J. Clin. Invest.* **93**:1326–1331.

Hagenbuch, B., and Meier, P. J., 1996, Sinusoidal (basolateral) bile salt uptake systems of hepatocytes, *Semin. Liver Dis.* **16**:129–136.

Hagenbuch, B., Stieger, B., Foguet, M., Lübbert, H., and Meier, P. J., 1991, Functional expression cloning and characterization of the hepatocyte Na$^+$/bile acid cotransport system, *Proc. Natl. Acad. Sci. USA* **88**:10629–10633.

Hagenbuch, B., Scharschmidt, B. F., and Meier, P. J., 1996, Effect of antisense oligonucleotides on the expression of hepatocellular bile acid and organic anion uptake systems in *Xenopus laevis* oocytes, *Biochem. J.* **316**:901–904.

Higgins, C. F., 1992, ABC transporters: From microorganisms to man, *Annu. Rev. Cell Biol.* **8**:67–113.

Hipfner, D. R., Almquist, K. C., Leslie, E. M., Gerlach, J. H., Grant, C. E., Deeley, R. G., and Cole, S. P. C., 1997, Membrane topology of the multidrug resistance protein (MRP), *J. Biol. Chem.* **272**:23623–23630.

Hirohashi, T., Ogawa, K., Suzuki, H., Ito, K., Kume, K., Shimizu, T., and Sugiyama, Y., 1997, Molecular characterization of canalicular multispecific organic anion transporter and its related proteins expressed in rat liver, *Pharmaceut. Res.* **14**:S458–S459.

Hirohashi, T., Suzuki, H., Ito, K., Kume, K., Shimizu, T., and Sugiyama, Y., 1998, Hepatic expression of multidrug resistance-associated protein (MRP)-like proteins maintained in Eisai hyperbilirubinemic rats (EHBR), *Mol. Pharmacol.* **53**:1068–1075.

Hirohashi, T., Suzuki, H., and Sugiyama, Y., 1999, Characterization of transport properties of cloned rat MRP3, *J. Biol. Chem.*, **274**:15181–15185.

Honscha, W., Schulz, K., Müller, D., and Petzinger, E., 1993, Two different mRNAs from rat liver code for the transport of bumetanide and taurocholate in *Xenopus laevis* oocytes, *Eur. J. Pharmacol.* **246**:227–232.

Huber, M., Guhlmann, A., Jansen, P. L. M., and Kepper, D., 1987, Hereditary defect of hepatobiliary cysteinyl leukotriene elimination in mutant rats with defective hepatic anion excretion, *Hepatology* **7**:224–228.

Imawaka, H., and Sugiyama, Y., 1998, Kinetic study of the hepatobiliary transport of a new prostaglandin receptor agonist, *J. Pharmacol. Exp. Ther.* **284**:949–957.

Ishigami, M., Tokui, T., Komai, T., Tsukahara, K., Yamazaki, M., and Sugiyama, Y., 1995, Evaluation of the uptake of pravastatin by perfused rat liver and primary cultured rat hepatocytes, *Pharmaceut. Res.* **12**:1741–1745.

Ishikawa, T., Müller, M., Klünemann, C., Schaub, T., and Keppler, D., 1990, ATP-dependent primary active transport of cysteinyl leukotrienes across liver canalicular membrane: Role of the ATP-dependent transport system for glutathione S-conjugates, *J. Biol. Chem.* **265**:19279–19286.

Ishizuka, H., Konno, K., Naganuma, H., Sasahara, K., Kawahara, Y., Niinuma, K., Suzuki, H., and Sugiyama, Y., 1997, Temocaprilat, a novel angiotensin-converting enzyme inhibitor, is excreted in bile via an ATP-dependent active transporter (cMOAT) that is deficient in Eisai hyperbilirubinemic mutant rats (EHBR), *J. Pharmacol. Exp. Ther.* **280**:1304–1311.

Ishizuka, H., Konno, K., Naganuma, H., Nishimura, K., Kouzuki, H., Suzuki, H., Stieger, B., Meier, P. J., and Sugiyama, Y., 1998a, Transport of temocaprilat into rat hepatocytes: Role of organic anion transporting protein (oatp), *J. Pharmacol. Exp. Ther.* **287**:37–42.

Ishizuka, H., Suzuki, H., and Sugiyama, Y., 1998b, Species differences in the transport activity of cMOAT determined *in vivo* and *in vitro*, *Hepatology* **28**:424A.

Ito, K., Suzuki, H., Hirohashi, T., Kume, K., Shimizu, T., and Sugiyama, Y., 1996, Expression of a putative ATP-binding cassette region, homologous to that in multidrug resistance associated protein (MRP), is hereditarily defective in Eisai hyperbilirubinemic rats (EHBR), *Int. Hepatol. Commun.* **4**:292–299.

Ito, K., Suzuki, H., Hirohashi, T., Kume, K., Shimizu, T., and Sugiyama, Y., 1997, Molecular cloning of canalicular multispecific organic anion transporter defective in EHBR, *Am. J. Physiol.* **272**:G16–G22.

Ito, K., Iwatsubo, T., Kanamitsu, S., Nakajima, Y., and Sugiyama, Y., 1998a, Quantitative prediction of *in vivo* drug clearance and drug interactions from *in vitro* data on metabolism, together with binding and transport, *Annu. Rev. Pharmaol. Toxicol.* **38**:461–499.

Ito, K., Iwatsubo, T., Kanamitsu, S., Ueda, K., Suzuki, H., and Sugiyama, Y., 1998b, Prediction of pharmacokinetic alterations caused by drug–drug interactions; focusing on metabolic intreaction in the liver, *Pharmacol. Rev.* **50**:387–411.

Ito, K., Suzuki, H., Hirohashi, T., Kume, K., Shimizu, T., and Sugiyama, Y., 1998c, Functional analysis of a canalicular multispecific organic anion transporter cloned from rat liver, *J. Biol. Chem.* **273**:1684–1688.

Iwatsubo, T., Hirota, N., Ooie, T., Suzuki, H., and Sugiyama, Y., 1996, Prediction of *in vivo* drug disposition from *in vitro* data based on physiological pharmacokinetics, *Biopharmaceut. Drug Disposit.* **17**:273–310.

Iwatsubo, T., Horita, N., Ooie, T., Suzuki, H., Shimada, N., Chiba, K., Ishizaki, T., Green, C. E., Tyson, C. A., and Sugiyama, Y., 1997, Prediction of *in vivo* drug metabolism in the human liver from *in vitro* metabolism data, *Pharmacol. Ther.* **73**:147–171.

Jacquemin, E., Hagenbuch, B., Stieger, B., Wolkoff, A. W., and Meier, P. J., 1994, Expression cloning of a rat liver Na$^+$-independent organic anion transporter, *Proc. Natl. Acad. Sci. USA* **91**:133–137.

Jansen, P. L. M., Groothuis, G. M. M., Peters, W. H. M., and Meijer, D. F. M., 1987, Selective hepatobiliary transport defect for organic anions and neutral steroids in mutant rats with hereditary-conjugated hyperbilirubinemia, *Hepatology* **7**:71–76.

Jansen, P. L. M., van Klinken, J.-W., van Gelder, M., Ottenhoff, R., and Oude Elferink, R. P. J., 1993, Preserved organic anion transport in mutant TR$^-$ rats with a hepatobiliary secretion defect, *Am. J. Physiol.* **265**:G445–G452.

Jedlitschky, G., Leier, I., Buchholz, U., Center, M., and Keppler, D., 1994, ATP-dependent transport of glutathione-S-conjugates by the multidrug resistance-associated protein, *Cancer Res.* **54**:4833–4836.

Jedlitschky, G., Leier, I., Buchholz, U., Barnouin, K., Kurz, G., and Keppler, D., 1996, Transport of glutathione, glucuronate, and sulfate conjugates by the MRP gene-encoded conjugate export pump, *Cancer Res.* **56**:988–994.

Jedlitschky, G., Leier, I., Buchholz, U., Hummel-Eisenbeiss, J., Burchell, B., and Keppler, D., 1997, ATP-dependent transport of bilirubin glucuronides by the multidrug resistance protein MRP1 and its hepatocyte canalicular isoform MRP2, *Biochem. J.* **327**:305–310.

Kamimoto, Y., Gatmaitan, Z., Hsu, J., and Arias, I. M., 1989, The function of Gp170, the multidrug resistance gene product, in rat liver canalicular membrane vesicles, *J. Biol. Chem.* **264**:11693–11698.

Kanai, N., Lu, R., Satriano, J. A., Bao, Y., Wolkoff, A., and Schuster, V. L., 1995, Identification and characterization of a prostaglandin transporter, *Science* **268**:866–869.

Kanai, N., Lu, R., Bao, Y., Wolkoff, A. W., Vore, M., and Shuster, V. L., 1996, Estradiol 17β-D-glucuronide is a high-affinity substrate for oatp organic anion transporter, *Am. J. Physiol.* **270**:F326–F331.

Karpen, S. J., Sun, A.-Q., Kudish, B., Hagenbuch, B., Meier, P. J., Ananthanarayanan, M., and Suchy, F. J., 1996, Multiple factors regulate the rat liver basolateral sodium-dependent bile acid cotransporter gene promoter, *J. Biol. Chem.* **271**:15211–15221.

Kartenbeck, J., Leuschner, U., Mayer, R., and Keppler, D., 1996, Absence of the canalicular isoform of the MRP gene-encoded conjugate export pump from the hepatocytes in Dubin–Johnson syndrome, *Hepatology* **23**:1061–1066.

Kast, C., Stieger, B., Winterhalter, K. H., and Meier, P. J., 1994, Hepatocellular transport of bile acids: Evidence for distinct subcellular localizations of electrogenic and ATP-dependent taurocholate transport in rat hepatocytes, *J. Biol. Chem.* **269**:5179–5186.

Kauffmann, H. M., and Schrenk, D., 1998, Sequence analysis and functional characterization of the 5'-flanking region of the rat multidrug resistance protein 2 (MRP2) gene, *Biochem. Biophys. Res. Commun.* **245**:325–331.

Kauffmann, H. M., Keppler, D., Kartenbeck, J., and Schrenk, D., 1997, Induction of cMrp/cMoat gene expression by cisplatin, 2-acethylaminofluorene, or cycloheximide in rat hepatocytes, *Hepatology* **26**:980–985.

Kauffmann, H. M., Keppler, D., Grant, T. W., and Schrenk, D., 1998, Induction of hepatic mrp2 (cmrp/cmoat) gene expression in nonhuman primates treated with rifampicin or tamoxifen, *Arch. Toxicol.* **72**:763–768.

Keppler, D., and Kartenbeck, J., 1996, The canalicular conjugate export pump encoded by the cmrp/cmoat gene, *Prog. Liver Dis.* **14**:55–67.

Keppler, D., and König, J., 1997, Expression and localization of the conjugated export pump encoded by the MRP2 (cMRP/cMOAT) gene in liver, *FASEB J.* **11**:509–516.

Kinoshita, S., Suzuki, H., Ito, K., Kume, K., Shimizu, T., and Sugiyama, Y., 1998, Transfected rat cMOAT is functionally expressed on the apical membrane in Madin–Darby canine kidney (MDCK) cells, *Pharmaceut. Res.* **15**:1851–1856.

Kitamura, T., Jansen, P., Hardenbrook, C., Kamimoto, Y., Gatmaitan, Z., and Arias, I. M., 1990, Defective ATP-dependent bile canalicular transport of organic anions in mutant (TR$^-$) rats with conjugated hyperbilirubinemia, *Proc. Natl. Acad. Sci. USA* **87**:3557–3561.

Kiuchi, Y., Suzuki, H., Hirohashi, T., Tyson, C. A., and Sugiyama, Y., 1998, cDNA cloning and inducible expression of human multidrug resistance associated protein 3 (MRP3), *FEBS Lett.* **433**:149–152.

Kobayashi, K., Sogame, Y., Hara, H., and Hayashi, K., 1990, Mechanism of glutathione S-conjugate transport in canalicular and basolateral rat liver plasma membrane, *J. Biol. Chem.* **265**:7737–7741.

Kobayashi, K., Komatsu, S., Nishi, T., Hara, H., and Hayashi, K., 1991, ATP-dependent transpor for glucuronides in canalicular plasma membrane vesicles, *Biochem. Biophys. Res. Commun.* **176**:622–626.

Koepsell, H., 1998, Organic cation transporters in intestine, kidney, liver and brain, *Annu. Rev. Physiol.* **60**:243–266.

Koga, T., Shimada, Y., Kuroda, M., Tsujita, Y., Hasegawa, K., and Yamazaki, M., 1990, Tissue-selective inhibition of cholesterol synthesis *in vivo* by pravastatin sodium, a 3-hydroxymethylglutaryl coenzyme A reductase inhibitor, *Biochim. Biophys. Acta* **1045**:115–120.

Koike, K., Kawabe, T., Tanaka, T., Toh, S., Uchiumi, T., Wada, M., Akiyama, S., Ono, M., and Kuwano, M., 1997, A canalicular multispecific organic anion transporter (cMOAT) antisense cDNA enhances drug sensitivity in human hepatic cancer cells, *Cancer Res.* **57**:5475–5479.

König, J., Cui, Y., Leier, I., and Keppler, D., 1998, Localization and functional characterization of the apical multidrug resistance protein, MRP2, permanently expressed in transfected human and canine cells, *Hepatology,* **28**:427A.

König, J., Rost, D., Cui, Y., and Keppler, D., 1999, Characterization of the human multidrug resistance protein isoform MRP3 localized to the basolateral hepatocyte membrane, *Hepatology* **29**:1156–1163.

Kontaxi, M., Eckhardt, U., Hagenbuch, B., Stieger, B., Meier, P. J., and Petzinger, E., 1996, Uptake of the mycotoxin ochratoxin A in liver cells occurs via the cloned organic anion transporting polypeptide, *J. Pharmacol. Exp. Ther.* **279**:1507–1513.

Kool, M., de Haas, M., Scheffer, G. L., Scheper, R. J., van Eijk, M. J. T., Juijn, J. A., Baas, F., and Borst, P., 1997, Analysis of expression of cMOAT (MRP2), MRP3, MRP4, and MRP5, homologues of the multidrug resistance-associated protein gene (MRP1), in human cancer cell lines, *Cancer Res.* **57**:3537–3547.

Kool, M., van der Linden, M., de Haas, M., Scheffer, G. L., de Vree, J. M. L., Smith, A. J., Jansen, G., Peters, G. J., Ponne, N., Scheper, R. J., Oude Elferink, R. P. J., Baas, F., and Borst, P., 1999, MRP3, an organic anion transporter able to transport anti-cancer drugs, *Proc. Natl. Acad. Sci. (USA)* **96**:6914–6919.

Kouzuki, H., Suzuki, H., Ohashi, R., Ito, K., and Sugiyama, Y., 1998, Contribution of sodium taurocholate co-transporting polypeptide to the uptake of its possible substrates into rat hepatocytes, *J. Pharmacol. Exp. Ther.* **286**:1043–1050.

Kouzuki, H., Suzuki, H., Ohashi, R., Ito, K., and Sugiyama, Y., 1999, Contribution of organic anion transporting polypeptide to the uptake of ligands into rat hepatocytes, *J. Pharmacol. Exp. Ther.* **288**:627–634.

Kramer, W., and Wess, G., 1996, Bile acid transport systems as pharmaceutical targets, *Eur. J. Clin. Invest.* **26**:715–732.

Kramer, W., Wess, G., Schubert, G., Bickel, M., Girbig, F., Gutjahr, U., Kowalewski, S., Baringhaus, K.-H., Enhsen, A., Glombik, H., Müllner, S., Neckermann, G., Schulz, S., and Petzinger, E., 1992, Liver-specific drug targeting by coupling to bile acids, *J. Biol. Chem.* **267**:18598–18604.

Kuipers, F., Enserink, M., Havinga, R., van der Steen, A. B. M., Hardonk, M. J., Fevery, J., and Vonk, R. J., 1988, Separate transport systems for biliary secretion of sulfated and unsulfated bile acids in the rat, *J. Clin. Invest.* **81**:1593–1599.

Kuipers, F., Radominska, A., Zimniak, P., Little, J. M., Havinga, R., Vonk, R. J., and Lester, R., 1989, Defective biliary secretion of bile acid 3-O-glucuronides in rats with hereditary conjugated hyperbilirubinemia, *J. Lipid Res.* **30**:1835–1845.

Kullak-Ublick, G.-A., Hagenbuch, B., Stieger, B., Wolkoff, A. W., and Meier, P. J., 1994, Functional characterization of the basolateral rat liver organic anion transporting polypeptide, *Hepatology* **20**:411–416.

Kullak-Ublick, G.-A., Hagenbuch, B., Stieger, B., Schteingart, C. D., Hofmann, A. F., Wolkoff, A. W., and Meier, P. J., 1995, Molecular and functional characterization of an organic anion transporting polypeptide cloned from human liver, *Gastroenterology* **109**:1274–1282.

Kullak-Ublick, G.-A., Beuers, U., Meier, P. J., Domdey, H., and Paumgartner, G., 1996a, Assignment of the human organic anion transporting polypeptide (OATP) gene to chromosome 12p12 by fluorescence *in situ* hybridization, *J. Hepatol.* **25**:985–987.

Kullak-Ublick, G.-A., Beuers, U., and Paumgartner, G., 1996b, Molecular and functional characterization of bile acid transport in human hepatoblastoma JepG2 cells, *Hepatology* **23**:1053–1060.

Kullak-Ublick, G.-A., Beuers, U., Fahney, C., Hagenbuch, B., Meier, P. J., and Paumgartner, G., 1997a, Identification and functional characterization of the promoter region of the human organic anion transporting polypeptide gene, *Hepatology* **26**:991–997.

Kullak-Ublick, G.-A., Glasa, J., Böker, C., Oswald, M., Grützner, U., Hagenbuch, B., Stieger, B., Meier, P. J., Beuers, U., Kramer, W., Wess, G., and Paumgartner, G., 1997b, Chlorambucil-taurocholate is transported by bile acid carriers expressed in human hepatocellular carcinomas, *Gastroenterology* **113**:1295–1305.

Kullak-Ublick, G.-A., Fisch, T., Oswald, M., Hagenbuch, B., Meier, P. J., Beuers, U., and Paumgartner, G., 1998, Dehydroepiandrosterone sulfate (DHEAS): Identification of a carrier protein in human liver and brain, *FEBS Lett.* **424**:173–176.

Kusuhara, H., Suzuki, H., and Sugiyama, Y., 1998, The role of P-glycoprotein and canalicular multispecific organic anion transporter (cMOAT) in the hepatobiliary excretion of drugs, *J. Pharmaceut. Sci.* **87**:1025–1040.

Kusuhara, H., Sekine, T., Utsunomiya-Tate, N. Tsuda, M., Kojima, R., Cha, S. H., Sugiyama, Y., Kanai, Y., and Endou, H., 1999, Molecular cloning and characterization of a new multispecific organic anion transporter from rat brain, *J. Biol. Chem.* **274**:13675–13680.

Lautier, D., Canitrot, Y., Deeley, R. G., and Cole, S. P. C., 1996, Multidrug resistance mediated by the multidrug resistance protein (MRP) gene, *Biochem. Pharmacol.* **52**:967–977.

Leier, I., Jedlitschky, G., Buchholz, U., Cole, S. P. C., Deeley, R. G., and Keppler, D., 1994, The MRP gene encodes an ATP-dependent export pump for leukotriene C_4 and structurally related conjugates, *J. Biol. Chem.* **269**:27807–27810.

Li, L., Lee, T. K., Meier, P. J., and Ballatori, N., 1998, Identification of glutathione as a driving force and leukotriene C_4 as a substrate for oatp-1, the hepatic sinusoidal organic solute transporter, *J. Biol. Chem.* **273**:16184–16191.

Liang, D., Hagenbuch, B., Stieger, B., and Meier, P. J., 1993, Parallel decrease of Na^+-taurocholate cotransport and its encoding mRNA in primary cultures of rat hepatocytes, *Hepatology* **18**:1162–1166.

Loe, D. W., Almquist, K. C., Cole, S. P. C., and Deeley, R. G., 1996a, ATP-dependent 17β-estradiol 17-(β-D-glucuronide) transport by multidrug resistanc eprotein (MRP), *J. Biol. Chem.* **271**:9683–9689.

Loe, D. W., Almquist, K. C., Deeley, R. G., and Cole, S. P. C., 1996b, Multidrug resistance protein (MRP)-mediated transport of leukotriene C_4 and chemotherapeutic agents in membrane vesicles, *J. Biol. Chem.* **271**:9675–9682.

Loe, D. W., Deeley, R. G., and Cole, S. P. C., 1996c, Biology of the multidrug resistance-associated protein, MRP, *Eur. J. Cancer* **32A**:945–957.

Lowenthal, D. T., Irvin, J. D., Merrill, D., Saris, S., Ulm, E., Goldstein, S., Hichens, M., Klein, L., Till, A., and Harris, K., 1985, The effect of renal function on enalapril kinetics, *Clin. Pharmacol. Ther.* **38**:661–666.

Lu, R., Kanai, N., Bao, Y., and Schuster, V. L., 1996, Cloning, *in vitro* expression, and tissue distribution of a human prostaglandin transporter cDNA (hPGT), *J. Clin. Invest.* **98**:1142–1149.

Luther, T. T., Hammerman, P., Rahmaoui, C. M., Lee, P. P., Sela-Herman, S., Matula, G. S., Ananthanarayanan, M., Suchy, F. J., Cavalieri, R. R., Lomri, N., and Scharschmidt, B. F., 1997, Evidence for an ATP-dependent bile acid transport protein other than the canalicular liver ecto-ATPase in rats, *Gastroenterology* **113**:249–254.

Madon, J., Eckhardt, U., Gerloff, T., Stieger, B., and Meier, P. J., 1997, Functional expression of the rat liver canalicular isoform of the multidrug resistance-associated protein, *FEBS Lett.* **406**:75–78.

Madon, J., Hagenbuch, B., Gerloff, T., Landmann, L., Meier, P. J., and Stieger, B., 1998, Identificatino of a novel multidrug resistance-associated protein (mrp6) at the lateral plasma membrane of rat hepatocytes, *Hepatology* **28**:400A.

Marin, J. J. G., Herrera, M. C., Palomero, M. F., Macias, R. I. R., Monte, M. J., El-Mir, M. Y., and Villanueva, G. R., 1998, Rat liver transport and biotransformation of a cytostatic complex of bis-cholylglycinate and platinum (II), *J. Hepatol.* **28**:417–425.

Masuda, M., Iizuka, Y., Yamazaki, M., Nishigaki, R., Kato, Y., Niinuma, N., Suzuki, H., and Sugiyama, Y., 1997, Methotrexate is excreted into the bile by canalicular multispecific organic anion transporter (cMOAT) in rats, *Cancer Res.* **57**:3506–3510.

Masuda, S., Ibaramoto, K., Takeuchi, A., Satito, H., Hashimoto, Y., and Inui, K., 1999, Cloning and functional characterization of a new multispecific organic anion transporter, OAT-K2, in rat kidney, *Mol. Pharmacol.* **55**:743–752.

Meier, P. J., 1995, Molecular mechanisms of hepatic bile salt transport from sinusoidal blood into bile, *Am. J. Physiol.* **269**:G801–G812.

Meier, P. J., Eckhardt, U., Schroeder, A., Hagenbuch, B., and Stieger, B., 1997, Substrate specificity of sinusoidal bile acid and organic anion uptake systems in rat and human liver, *Hepatology* **26**:1667–1677.

Miyauchi, S., Sawada, Y., Iga, T., Hanano, M., and Sugiyama, Y., 1993, Comparison of the hepatic uptake clearances of fifteen drugs with a wide range of membrane permeabilities in isolated rat hepatocytes and perfused rat livers, *Pharmaceut. Res.* **10**:434–440.

Mukhopadhayay, S., Ananthanarayanan, M., Stieger, B., Meier, P. J., Suchy, F. J., and Anwer, M. S., 1997, cAMP increases liver Na^+-taurocholate cotransport by translocating transporter to plasma membranes, *Am. J. Physiol.* **273**:G842–G848.

Müller, M., and Jansen, P. L. M., 1997, Molecular aspects of hepatobiliary transport, *Am. J. Phyisol.* **272**:G1285–G1303.

Müller, J., and Jansen, P. L. M., 1998, The secretory function of the liver: New aspects of hepatobiliary transport, *J. Hepatol.* **28**:344–354.

Müller, M., Ishikawa, T., Berger, U., Klüneman, C., Lucka, L., Schreyer, A., Kannicht, C., Reutter, W., Kurz, G., and Kepper, D., 1991, ATP-dependent transport of taurocholate across the hepatocyte canalicular membrane mediated by a 110-kDa glycoprotein binding ATP and bile salt, *J. Biol. Chem.* **266**:18920–18926.

Müller, M., Meijer, C., Zaman, G. J. R., Borst, P., Scheper, R. J., Mulder, N. H., De Vries, E. G. E., and Jansen, P. L. M., 1994, Overexpression of the gene encoding the multidrug resistance-associated protein results in increased ATP-dependent glutathione S-conjugate transport, *Proc. Natl. Acad. Sci. USA* **91**:13033–13037.

Nakamura, T., Hisaka, A., Sawasaki, Y., Suzuki, Y., Fukami, T., Ishikawa, K., Yano, M., and Sugiyama, Y., 1996, Carrier-mediated active transport of BQ-123, a peptidic endothelin antagonist, into rat hepatocytes, *J. Pharmacol. Exp. Ther.* **278**:564–572.

Nies, A. T., Cartz, T., Brom, M., Leier, I., and Keppler, D., 1998, Expression of the apical conjugate export pump, Mrp2, in the polarized hepatoma cell line, WIF-B, *Hepatology* **28**:1332–1340.

Niinuma, K., Takenaka, O., Horie, T., Kobayashi, K., Kato, Y., Suzuki, H., and Sugiyama, Y., 1997, Kinetic analysis of the primary active transport of conjugated metabolites across the bile canalicular membrane: Comparative study between DNP-SG (S-(2,4-dinitrophenyl)-glutathione) and E3040-glucuronide, *J. Pharmacol. Exp. Ther.* **282**: 866–872.

Niinuma, K., Kato, Y., Suzuki, H., Tyson, C. A., Weizer, V., Dabbs, J. E., Froelich, R., Green, C. E., and Sugiyama, Y., 1999, Primary active transport of organic anions on bile canalicular membrane in humans, *Am. J. Physiol.* **276**:G1153–G1164.

Nishida, T., Gatmaitan, Z., Che, M., and Arias, I. M., 1991, Rat liver canalicular membrane vesicles contain an ATP-dependent bile acid transport system, *Proc. Natl. Acad. Sci. USA* **88**:6590–6594.

Nishida, T., Gatmaitan, Z., Roy-Chowdhury, J., and Arias, I. M., 1992a, Two distinct mechanisms for bilirubin glucuronide transport by rat bile canalicular membrane vesicles: Demonstration of defective ATP-dependent transport in rats (TR^-) with inherited conjugated hyperbilirubinemia, *J. Clin. Invest.* **90**:2130–2135.

Nishida, T., Hardenbrook, C., Gatmaitan, Z., and Arias, I. M., 1992b, ATP-dependent organic anion transport system in normal and TR^- rat liver canalicular membranes, *Am. J. Physiol.* **262**:G629–G635.

Nishida, T., Che, M., Gatmaitan, Z., and Arias, I. M., 1995, Structure-specific inhibition by

bile acids of adenosine triphosphate-dependent taurocholate transport in rat canalicular membrane vesicles, *Hepatology* **21**:1058–1062.
Noé, B., Hagenbuch, B., Stieger, B., and Meier, P. J., 1997, Isolation of a multispecific organic anion and cardiac glycoside transporter from rat brain, *Proc. Natl. Acad. Sci. USA* **94**:10346–10350.
Oguchi, H., Miyasaka, M., Koiwai, T., Tokunaga, S., Hara, K., Sato, K., Yoshie, T., Shioya, H., and Furuta, S., 1993, Pharmacokinetics of temocapril and enalapril in patients with arious degrees of renal insufficiency, *Clin. Pharmacokinet.* **24**:421–427.
Ortiz, D. F., St. Pierre, M. V., Abdulmessih, A., and Arias, I. M., 1997, A yeast ATP-binding cassette-type protein mediating ATP-dependent bile acid transport, *J. Biol. Chem.* **272**:15358–15365.
Ortiz, D. F., Li, S. H., Iyer, R., Zhang, X. M., Novikoff, P., and Arias, I. M., 1999, MRP3 a new ATP-binding cassette protein localized to the canalicular domain of the hepatocyte, *Am. J. Physiol.* **276**:G1493–G1500.
Oude Elferink, R. P. J., and Jansen, P. L. M., 1994, The role of the canalicular multispecific organic anion transporter in the disposal of endo- and xenobiotics, *Pharmacol. Ther.* **64**:77–97.
Oude Elferink, R. P. J., De Haan, J., Lambert, K. J., Hagey, L. R., Hofmann, A. F., and Jansen, P. L. M., 1989a, Selective hepatobiliary transport of nordeoxycholate side chain conjugates in mutant rats with a canalicular transport defect, *Hepatology* **9**:861–865.
Oude Elferink, R. P. J., Ottenhoff, R., Liefting, W., de Haan, J., and Jansen, P., 1989b, Hepatobiliary transport of glutathione and glutathione conjugate in rats with hereditary hyperbilirubinemia, *J. Clin. Invest.* **84**:476–483.
Oude Elferink, R. P. J., Ottenhoff, R., Liefting, W. G. M., Schoemaker, B., Groen, A. K., and Jansen, P. L. M., 1990, ATP-dependent efflux of GSSG and GS-conjugate from isolated rat hepatocytes, *Am. J. Physiol.* **258**:G699–G706.
Oude Elferink, R. P. J., Bakker, C. T. M., Roelofsen, H., Middelkoop, E., Ottenhoff, R., Heijn, M., and Jansen, P. L. M., 1993, Accumulation of organic anion in intracellular vesicles of cultured rat hepatocytes is mediated by the canalicular multispecific organic anion transporter, *Hepatology* **17**:434–444.
Oude Elferink, R. P. J., Meijer, D. K. F., Kuipers, F., Jansen, P. L. M., Groen, A. K., and Groothuis, G. M. M., 1995, Hepatobiliary secretion of organic compounds; molecular mechanisms of membrane transport, *Biochim. Biophys. Acta* **1241**:215–268.
Oude Elferink, R. P. J., Tytgat, G. N. J., and Groen, A. K., 1997, The role of mdr2 P-glycoprotein in hepatobiliary lipid transport, *FASEB J.* **11**:19–28.
Oude Elferink, R. P. J., and van Berge Henegouwen, G. P., 1998, Cracking the genetic code for benign recurrent and progressive familial intrahepatic cholestasis, *J. Hepatol.* **29**:317–320.
Paulusma, C. C., and Oude Elferink, R. P. J., 1997, The canalicular multispecific organic anion transporter and conjugated hyperbilirubinemia in rat and man, *J. Mol. Med.* **75**:420–428.
Paulusma, C. C., Bosma, P. J., Zaman, G. J., Bakker, C. T., Otter, M., Scheffer, G. L., Scheper, R. J., Borst, P., and Oude Elferink, R. P., 1996, Congenital jaundice in rats with a mutation in a multidrug resistance-associated protein gene, *Science* **271**:1126–1128.
Paulusma, C. C., Kool, M., Bosma, P. J., Scheffer, G. L., Ter Borg, F., Scheper, R. J., Tyt-

gat, G. N. J., Borst, P., Baas, F. and Oude Elferink, R. P. J., 1997, A mutation in the human canalicular multispecific organic anion transporter gene causes the Dubin-Johnson syndrome, *Hepatology,* **25:** 1539–1542.

Petzinger, E., 1994, Transport of organic anions in the liver: An update on bile acid, fatty acid, monocarboxylate, anionic amino acid, cholephilic organic anion, and anionic drug transport, *Rev. Physiol. Biochem. Pharmacol.* **123:**47–211.

Petzinger, E., Müller, N., Föllmann, W., Deutscher, J., and Kinne, R. K. H., 1989, Uptake of bumetanide into isolated rat hepatocytes and primary liver cell cultures, *Am. J. Physiol.* **256:**G78–G86.

Petzinger, E., Blumrichi, M., Brühl, B., Eckhardt, U., Föllmann, W., Honscha, W., Horz, J. A., Müller, N., Nickau, L., Ottallah-Kolac, M., Platte, H.-D., Schenk, A., Schuh, K., Schulz, K., and Schulz, S., 1996, What we have learned about bumetanide and the concept of multispecific bile acid/drug transporters from the liver, *J. Hepatol.* **24**(Suppl. 1)**:**42–49.

Polhuijs, M., Kuipers, F., Vonk, R. J., and Mulder, G. J., 1989, Stereoselectivity of glutathione conjugation: Blood elimination of alpha-bromoisovalerylurea enantiomers and biliary excretion of the conjugates in unanesthetized normal or congenitally jaundiced rats, *J. Pharmacol. Exp. Ther.* **249:**874–878.

Roelofsen, H., Soroka, C. J., Keppler, D., and Boyer, J. L., 1998, Cyclic AMP stimulates sorting of the canalicular organic anion transporter (Mrp2/cMOAT) to the apical domain in hepatocyte couplets, *J. Cell Sci.* **111:**1137–1145.

Saito, H., Masuda, S., and Inui, K., 1996, Cloning and functional characterization of a novel rat organic anion transporter mediating basolateral uptake of methotrexate in the kidney, *J. Biol. Chem.* **271:**20719–20725.

Sasabe, H., Kato, Y., Tsuji, A., and Sugiyama, Y., 1998a, Stereoselective hepatobiliary transport of the quinolone antibiotic grepafloxacin and its glucuronide in the rat, *J. Pharmacol. Exp. Ther.* **284:**661–668.

Sasabe, H., Tsuji, A., and Sugiyama, Y., 1998b, Carrier-mediated mechanism for the biliary excretion of the quinolone antibiotic grepafloxacin and its glucuronide in rats, *J. Pharmacol. Exp. Ther.* **284:**1033–1039.

Sasabe, H., Terasaki, T., Tsuji, A., and Sugiyama, Y., 1997, Carrier-mediated hepatic uptake of quinolone antibiotics in the rat, *J. Pharmacol. Exp. Ther.* **282:**162–171.

Sathirakul, K., Suzuki, H., Yamada, T., Hanano, M., and Sugiyama, Y., 1993a, Multiple transport systems for organic anions across the bile canalicular membrane, *J. Pharmacol. Exp. Ther.* **268:**65–73.

Sathirakul, K., Suzuki, H., Yasuda, K., Hanano, M., Tagaya, O., Horie, T., and Sugiyama, Y., 1993b, Kinetic analysis of hepatobiliary transport of organic anions in Eisai hyperbilirubinemic mutant rats, *J. Pharmacol. Exp. Ther.* **265:**1301–1312.

Satlin, L. M., Amin, V., and Wolkoff, A. W., 1997, Organic anion transporting polypeptide mediates organic anion/HCO_3^- exchange, *J. Biol. Chem.* **272:**26340–26345.

Schaub, T. P., Kartenbeck, J., König, J., Vogel, O., Witzgall, R., Kriz, W., and Keppler, D., 1997, Expression of the conjugate export pump encoded by the mrp2 in the apical membrane of kidney proximal tubules, *J. Am. Soc. Neprhol.* **8:**1213–1221.

Schinkel, A. H., Mayer, U., Wagenaar, E., Mol, C. A. A. M., van Deemter, L., Smit, J. J. M., van der Valk, M. A., Voordouw, A. C., Spits, H., van Tellingen, O., Zijlmans, J. M. J. M., Fibbe, W. E., and Borst, P., 1997, Normal viability and altered pharmacokinetics in

mice lacking mdr1-type (drug-transporting) P-glycoproteins, *Proc. Natl. Acad. Sci. USA* **94**:4028–4033.
Schmidt, J. V., and Bradfield, C. A., 1996, Ah receptor signaling pathways, *Annu. Rev. Cell Dev. Biol.* **12**:55–89.
Schroeder, A., Eckhardt, U., Stieger, B., Tynes, R., Schteingart, C. D., Hofmann, A. F., Meier, P. J., and Hagenbuch, B., 1998, Substrate specificity of the rat liver Na^+-bile salt cotransporter in *Xenopus laevis* oocytes and in CHO cells, *Am. J. Physiol.* **274**:G370–G375.
Scott, W. A., 1990, Hydrophilicity and the differential pharmacology of pravastatin, in: *Lipid Management: Pravastatin and the Differential Pharmacology of HMG-CoA Reductase Inhibitors* (C. Wood, ed.), Round Table Series No. 16, Royal Soc. Med. Service, London, pp. 17–25.
Sekine, T., Watanabe, N., Hosoyamada, M., Kanai, Y., and Endou, H., 1997, Expression cloning and characterization of a novel multispecific organic anion transporter, *J. Biol. Chem.* **272**:18526–18529.
Sekine, T., Cha, S. H., Tsuda, M., Apiwattanakul, N., Nakajima, N., Kanai, Y., and Endou, H., 1998, Identification of multispecific organic anion transporter 2 expressed predominantly in the liver, *FEBS Lett.* **429**:179–182.
Shi, X., Bai, S., Ford, A. C., Burk, R. D., Jacquemin, E., Hagenbuch, B., Meier, P. J., and Wolkoff, A. W., 1995, Stable inducible expression of a functional rat liver organic anion transport protein in HeLa cells, *J. Biol. Chem.* **270**:25591–25595.
Shimamura, H., Suzuki, H., Hanano, M., Suzuki, A., Tagaya, O., Horie, T., and Sugiyama, Y., 1994, Multiple systems for the biliary excretion of organic anions in rats: Liquiritigenin conjugates as model compounds, *J. Pharmacol. Exp. Ther.* **271**:370–378.
Shimamura, H., Suzuki, H., Tagaya, O., Horie, T., and Sugiyama, Y., 1996, Biliary excretion of glycyrrhizin in rats: Kinetic basis for the multiplicity in the bile canalicular transport of organic anions, *Pharmaceut. Res.* **13**:1833–1837.
Shin, H.-C., Kato, Y., Yamada, T., Niinuma, K., Hisaka, A., and Sugiyama, Y., 1997, Hepatobiliary transport mechanism for the cyclopentapeptide endothelin antagonist BQ-123, *Am. J. Physiol.* **272**:G979–G986.
Shneider, B., and Moyer, M. S., 1993, Characterization of endogenous carrier-mediated taurocholate efflux from *Xenopus laevis* oocytes, *J. Biol. Chem.* **268**:6985–6988.
Sica, D. A., 1992, Kinetics of angiotensin-converting enzyme inhibitor in renal failure, *J. Cardiovasc. Pharmacol.* **20**(Suppl. 10):S13–S20.
Smit, J. J., Schinkel, A. H., Oude Elferink, R. P., Groen, A. K., Wagenaar, E., van Deemter, L., Mol, C. A., Ottenhoff, R., van der Lugt, N. M., van Roon, M. A., van der Valk, M. A., Offerhaus, G. J. A., Berns, A. J. M., and Borst, P., 1993, Homozygous disruption of the murine mdr2 P-glycoprotein gene leads to a complete absence of phospholipid from bile and to liver diseases *Cell* **75**:451–462.
Stephan, Z. F., Yurachek, E. C., Sharif, R., Wasrary, J. M., Steele, R. E., and Howes, C., 1992, Reduction of cardiovascular and thyroxine-suppressing activities of $L-T_3$ by liver targeting with cholic acid, *Biochem. Pharmacol.* **43**:1969–1974.
Stieger, B., and Meier, P. J., 1998, Bile acid and xenobiotic transporters in liver, *Curr. Opin. Cell Biol.* **10**:462–467.
Stieger, B., O'Neill, B., and Meier, P. J., 1992, ATP-dependent bile-salt transport in canalicular rat plasma-membrane vesicles, *Biochem. J.* **284**:67–74.

Stieger, B., Hagenbuch, B., Landmann, L., Höchli, M., Schroeder, A., and Meier, P. J., 1994, In situ localization of the hepatocytic Na$^+$/taurocholate cotransporting polypeptide in rat liver, *Gastroenterology* **107**:1781–1787.
Strautnieks, S. S., Bull, L. N., Knisely, A. S., Kocoshis, S. A., Dahl, N., Arnell, H., Sokal, E., Dahan, K., Childs, S., Ling, V., Tanner, M. S., Kagalwalla, A. F., Nemeth, A., Pawlowska, J., Baker, A., Mieli-Vergani, G., Freimer, N. B., Gardiner, R. M., and Thompson, R. J., 1998, A gene encoding a liver-specific ABC transporter is mutated in progressive familial intrahepatic cholestasis, *Nature Genet.* **20**:233–238.
Suchy, F. J., 1993, Hepatocellular transport of bile acids, *Semin. Liver Dis.* **13**:235–247.
Suchy, F. J., Sippel, C. J., and Ananthanarayanan, M., 1997, Bile acid transport across the hepatocyte canalicular membrane, *FASEB J.* **11**:199–205.
Sugiyama, Y., Kato, Y., and Chu, X.-Y., 1998, Multiplicity of biliary excretion mechanisms for the camptothecin derivative irinotecan (CPT-11), its metabolite SN-38, and its glucuronide: Role of canalicular multispecific organic anion transporter and P-glycoprotein, *Cancer Chemother. Pharmacol.* **42**(Suppl):544–549.
Suzuki, H., and Sugiyama, Y., 1998, Excretion of GSSG and glutathione conjugates mediated by MRP1 and cMOAT/MRP2, *Semin. Liver Dis.* **18**:359–376.
Sweet, D. H., Wolff, N. A., and Pritchard, J. B., 1997, Expression cloning and characterization of ROAT1: The basolateral organic anion transporter in rat kidney, *J. Biol. Chem.* **272**:30088–30095.
Takenaka, O., Horie, T., Suzuki, H., Kobayashi, K., and Sugiyama, Y., 1995a, Kinetic analysis of hepatobiliary transport for conjugative metabolites in the perfused liver of mutant rats (EHBR) with hereditary conjugative hyperbilirubinemia, *Pharmaceut. Res.* **12**:1746–1755.
Takenaka, O., Horie, T., Suzuki, H., and Sugiyama, Y., 1995b, The biliary excretion systems for conjugative metabolites: Comparison of glucuronide and sulfate in Eisai hyperbilirubinemic rats (EHBR), *J. Pharmacol. Exp. Ther.* **274**:1362–1369.
Takenaka, O., Horie, T., Suzuki, H., and Sugiyama, Y., 1997, Carrier-mediated active transport of the glucuronide and sulfate of 6-hydroxy-5,7-dimethyl-2-methylamino-4-(3-pyridylmethyl)benzothiazole (E3040) into rat liver: Quantitative comparison of permeability in isolated hepatocytes, perfused liver and liver *in vivo, J. Pharmacol. Exp. Ther.* **280**:948–958.
Takikawa, H., Sano, N., Narita, T., Uchida, Y., Yamanaka, M., Horie, T., Mikami T., and Tagaya, O., 1991, Biliary excretion of bile acid conjugates in a hyperbilirubinemic mutant Sprague-Dawley rat, *Hepatology* **14**:352–360.
Takikawa, H., Yamazaki, R., Sano, N., and Yamanaka, M., 1996, Biliary excretion of estradiol-17β-glucuronide in the rat, *Hepatology* **23**:607–613.
Taniguchi, K., Wada, M., Kohno, K., Nakamura, T., Kawabe, T., Kawakami, M., Kagotani, K., Okumura, K., Akiyama, S., and Kuwano, M., 1996, A human canalicular multispecific organic anion transporter (cMOAT) gene is overexpressed in cisplatin-resistant human cancer cell lines with decreased drug accumulation, *Cancer Res.* **56**:4124–4129.
Terasaki, T., Mizuguchi, H., Itoho, C., Tamai, I., Lemaire, M., and Tsuji, A., 1995, Hepatic uptake of octreotide, a long-acting somatostatin analogue, via a bile acid transport system, *Pharmaceut. Res.* **12**:12–17.
Toh, S., Wada, M., Uhciumi, T., Inokuchi, A., Makino, Y., Horie, Y., Adachi, Y., Sakisaka,

S., and Kuwano, M., 1999, Genomic structure of the canalicular multispecific organic anion transporte r(MRP2/cMOAT) gene and mutations in the ATP-binding cassette region in Dubin–Johnson syndrome, *Am. J. Hum. Genet.* **64:**739–746.

Torchia, E. C., Shapiro, R. J., and Agellon, L. B., 1996, Reconstitution of bile acid transport in the rat hepatoma McArdle RH-7777 cell line, *Hepatology* **24:**206–211.

Trauner, M., Meier, P. J., and Boyer, J. L., 1998, Molecular pathogenesis of cholestasis, *N. Engl. J. Med.* **339:**1217–1227.

Tsuji, A., Terasaki, T., Takanosu, T., Tamai, I., and Nakashima, E., 1986, Uptake of benzylpenicillin, cefpiramide, and cefazolin by freshly prepared rat hepatocytes, *Biochem. Pharmacol.* **35:**1151–158.

Tsujita, Y., Juroda, M., Shimada, Y., Tanzawa, K., Arai, M., Kaneko, I., Tanaka, M., Masuda, H., Tarumi, C., Watanabe, Y., and Fujii, S., 1986, CS-514, a competitive inhibitor of 3-hydroxy-3-methylglutaryl coenzyme A reductase: Tissue-selective inhibition of sterol synthesis and hypolipodemic effect on various animal species, *Biochim. Biophys. Acta* **877:**50–60.

van Aubel, R. A. M. H., van Kuijck, M. A., Koenderink, J. B., Deen, P. M. T., van Os, C. H., and Russel, F. G. M., 1998, Adenosine triphosphate-dependent transport of anionic conjugates by the rabbit multidrug resistance-associated protein Mrp2 expressed in insect cells, *Mol. Pharmacol.* **53:**1062–1067.

van Montfoort, J. E., Hagenbuch, B., Muller, M., Meijer, D. K. F., and Meier, P. J., 1998, Transport of organic cations by the rat organic anion transporting polypeptides oatp1, oatp2 and the human OATP, *Hepatology* **28:**506A.

Verkade, H. J., Wolbers, M. J., Havinga, R. Uges, D. R. A., Vonk, R. J., and Kuipers, F., 1990, The uncoupling of biliary lipid from bile acid secretion by organic anions in the rat, *Gastroenterology* **99:**1485–1492.

von Dippe, P., and Levy, D., 1990, Expression of the bile acid transport protein during liver development and in hepatoma cells, *J. Biol. Chem.* **265:**5942–5945.

von Dippe, P., Amoui, M., Stellwagen, R. H., and Levy, D., 1996, The functional expression of sodium-dependent bile acid transport in Madin–Darby canine kidney cells transfected with the cDNA for microsomal epoxide hydrolase, *J. Biol. Chem.* **271:**18176–18180.

Vore, M., Hoffman, T., and Gosland, M., 1996, ATP-dependent transport of β-estradiol 17-(β-D-glucuronide) in rat canalicular membrane vesicles, *Am. J. Physiol.* **271:**G791–G798.

Wada, M., Toh, S., Taniguchi, K., Nakamura, T., Uchiumi, T., Kohno, K., Yoshida, I., Kimura, A., Sakisaka, S., Adachi, Y., and Kuwano, M., 1998, Mutations in the canalicular multispecific organic anion transporter (cMOAT) gene, a novel ABC transporter, in patients with hyperbilirubinemia II/Dubin–Johnson syndrome, *Human Mol. Genet.* **7:**203–207.

Watanabe, T., Miyauchi, S., Sawada, Y., Iga, T., Hanano, M., Inaba, M., and Sugiyama, Y., 1992, Kinetic analysis of hepatobiliary transport of vincristine in perfused rat liver: Possible roles of P-glycoprotein in biliary excretion of vincristine, *J. Hepatol.* **16:**77–88.

Watanabe, T., Suzuki, H., Sawada, Y., Naito, M., Tsuruo, T., Inaba, M., Hanano, M., and Sugiyama, Y., 1995, Induction of hepatic P-glycoprotein enhances biliary excretion of vincristine in rats, *J. Hepatol.* **23:**440–448.

Wolters, H., Kuipers, F. Slooff, M. J. H., and Vonk, R. J., 1992, Adenosine triphosphate-dependent taurocholate transport in human liver plasma membranes, *J. Clin. Invest.* **90:**2321–2326.

Yamada, T., Niinuma, K., Lemaire, M., Terasaki, T., and Sugiyama, Y., 1997a, Mechanism of the tissue distribution and biliary excretion of the cyclic peptide octreotide, *J. Pharmacol. Exp. Ther.* **279:**1357–1364.

Yamada, T., Niinuma, K., Lemaire, M., Terasaki, T., and Sugiyama, Y., 1997b, Carrier-mediated hepatic uptake of the cationic cyclopeptide, octreotide, in rats: Comparison between *in vivo* and *in vitro*, *Drug Metab. Dispos.* **25:**536–543.

Yamazaki, M., Suzuki, H., Sugiyama, Y., Iga, T., and Hanano, M., 1992a, Uptake of organic anions by isolated rat hepatocytes: A classification in terms of ATP-dependency, *J. Hepatol.* **14:**41–47.

Yamazaki, M., Suzuki, H., Sugiyama, Y., Iga, T., and Hanano, M., 1992b, Utilization of ATP-depleted cells in the analysis of taurocholate uptake by isolated rat hepatocytes, *J. Hepatol.* **14:**54–63.

Yamazaki, M., Suzuki, H., Hanano, M., and Sugiyama, Y., 1993a, Different relationships between cellular ATP and hepatic uptake among taurocholate, cholate, and organic anions, *Am. J. Physiol.* **264:**G693–G701.

Yamazaki, M., Suzuki, H., Hanano, M., Tokui, T., Komai, T., and Sugiyama, Y., 1993b, Na^+-independent multispecific anion transporter mediates active transport of pravastatin into rat liver, *Am. J. Physiol.* **264:**G36–G44.

Yamazaki, M., Akiyama, S., Nishigaki, R., and Sugiyama, Y., 1996a, Uptake is the rate-limiting step in the overall hepatic elimination of pravastatin at steady-state in rats, *Pharmaceut. Res.* **13:**1559–1564.

Yamazaki, M., Kobayashi, K., and Sugiyama, Y., 1996b, Primary active transport of pravastatin across the liver canalicular membrane in normal and mutant Eisai hyperbilirubinemic rats, *Biopharmaceut. Drug Dispos.* **17:**607–621.

Yamazaki, M., Suzuki, H., and Sugiyama, Y., 1996c, Recent advances in carrier-mediated hepatic uptake and biliary excretion of xenobiotics, *Pharmaceut. Res.* **13:**497–513.

Yamazaki, M., Tokui, T., Ishigami, M., and Sugiyama, Y., 1996d, Tissue-selective uptake of pravastatin in rats: Contribution of a specific carrier-mediated uptake system, *Biopharmaceut. Drug Dispos.* **17:**775–789.

Yamazaki, M., Akiyama, S., Niinuma, K., Nishigaki, R., and Sugiyama, Y., 1997, Biliary excretion of pravastatin in rats: Contribution of the excretion pathway mediated by canalicular multispecific organic anion transporter (cMOAT), *Drug Metab. Disposit.* **25:**1123–1129.

Zhang, L., Dresser, M. J., Gray, A. T., Yost, S. C., Terashita, S., and Giacomini, K. M., 1997, Cloning and functional expression of a human liver organic cation transporter, *Mol. Pharmacol.* **51:**913–921.

Zhang, L., Brett, C. M., and Giacomini, K. M., 1998, Role of organic cation transporters in drug absorption and elimination, *Annu. Rev. Pharmacol. Toxicol.* **38:**431–460.

Zimmerli, B., Valantinas, J., and Meier, P. J., 1989, Multiplicity of Na^+-dependent taurocholate uptake in basolateral (sinusoidal) rat liver plasma membrane vesicles, *J. Pharmacol. Exp. Ther.* **250:**301–308.

15

Molecular and Functional Characteristics of Cloned Human Organic Cation Transporters

Mark J. Dresser, Lei Zhang, and Kathleen M. Giacomini

1. INTRODUCTION

Detoxification mechanisms including enzymes involved in metabolism (e.g., cytochrome P450) and secretory transporters in the kidney, liver, and intestine are critical in the body's defense against deleterious xenobiotics. Organic cation transporters, sometimes termed polyspecific organic cation transporters because of their broad substrate selectivity, play a key role in the elimination of many endogenous amines and xenobiotics including a variety of clinically used drugs. For years, our knowledge of the mechanisms of organic cation transport has been based primarily on studies in isolated tissue preparations including cell culture and purified brush border and basolateral membrane vesicles. From these studies, multiple mechanisms for organic cation transport have been proposed (Zhang *et al.*, 1998a).

Recently, several organic cation transporters were cloned, paving the way to an enhanced understanding of the multiple mechanisms involved in the transport of organic cations. With the availability of the cloned transporters, it is now possible to delineate the functional roles of the various transporters. Furthermore,

---

Mark J. Dresser, Lei Zhang, and Kathleen M. Giacomini • Department of Biopharmaceutical Sciences, University of California San Francisco, San Francisco, California 94143.
Membrane Transporters as Drug Targets, edited by Amidon and Sadée. Kluwer Academic/Plenum Publishers, New York, 1999.

structure–function relationships can be established and antibodies can be developed to localize the transporters to various tissues in the body. From these studies a more complete understanding of the mechanisms of organic cation transport in the body can be established.

A knowledge of the structure, function, tissue distribution, and regulation of organic cation transporters is important in rational drug development and administration. Information about the kinetics of interaction of a candidate drug with organic cation transporters located in the intestine, liver, and kidney will aid in the prediction of the *in vivo* pharmacokinetics of the drug. This information may also be useful in the design and development of drugs targeted to specific tissues. However, for a transporter to represent a target for site-specific delivery of drugs, the tissue distribution of the transporter needs to be determined. If the transporter is widely expressed, e.g., a "housekeeper" transporter, it may not represent a useful drug target. Ideally, the transporter should be localized exclusively to the target tissue. The localization of redundant transporters for the candidate drug needs to be ascertained along with the interaction kinetics of the candidate compounds with the redundant transporters. If a compound interacts with a high selectivity to the target transporter, then it is likely that the transporter in the target tissue will facilitate the site-specific delivery of the candidate compound.

Clearly, our understanding of organic cation transporters relevant to drug dispositon and targeting is rapidly increasing. In this review, we focus largely on the organic cation transporters that have been recently cloned. The methods used to study the molecular characteristics of these transporters will be reviewed. Specific studies of the cloned human organic cation transporters will be reviewed. Finally, future directions in the organic cation transporter field will be discussed with a special emphasis on database searching methods as a tool to identify new members of the organic cation transporter (OCT) gene family, and high-throughput screening as a method to screen for drug–transporter(s) interactions.

2. BACKGROUND

Organic cations are a structurally and pharmacologically diverse group of molecules which contain at least one positively charged amine moiety at physiological pH. As a group, organic cations are estimated to account for approximately 50% of clinically used drugs including antiarrhythmics (e.g., quinidine, procainamide, disopyramide, lidocaine), antihistamines (e.g., promethazine, cimetidine), opioid analgesics (e.g., morphine, codeine, methadone), β-adrenergic blocking agents (e.g, propranolol, timolol, pindolol, acebutolol), and skeletal muscle relaxants (e.g., tubocurarine, pancuronium, vecuronium). In addition, there are many organic cations which are currently at various stages of drug development (e.g., ropinirole, pramipexole, cabergoline) (Gottwald et al., 1997). Many toxic substances (e.g., paraquat and 1-methyl-4-phenylpyridinium, MPP$^+$) as well as endogenous substances (e.g., dopamine, serotonin, noradrenaline, histamine, acetylcholine, choline, and N^1-methylnicotinamide, NMN) are also organic cations. Due to their positive charge, organic cations diffuse poorly across biological membranes and instead rely on membrane transporters for their movement across membranes.

Several model compounds have been used to study the mechanisms of organic cation transport (Fig. 1). These include tetraethylammonium (TEA), NMN, MPP$^+$, and guanidine. The mechanisms of organic cation transport have been studied extensively in different epithelial and nonepithelial tissues including kidney, liver, intestine, placenta, and brain (Iseki et al., 1993; Muller and Jansen, 1997; Prasad et al., 1992; Pritchard and Miller, 1993, 1996; Streich et al., 1996; Ullrich, 1997). Various experimental techniques have been used to study organic cation transport including perfused and nonperfused tissue, vesicles prepared from distinct biological membranes (e.g. apical and basolateral), isolated tissue slices, primary cell cultures and continuous cell lines. Detailed reviews of the various mechaisms of organic cation transport are available and will not be repeated here (Pritchard and Miller, 1993; Pritchard and Miller, 1996; Zhang et al., 1998a). Although the transport mechanisms of organic cations differ from tissue to tissue to

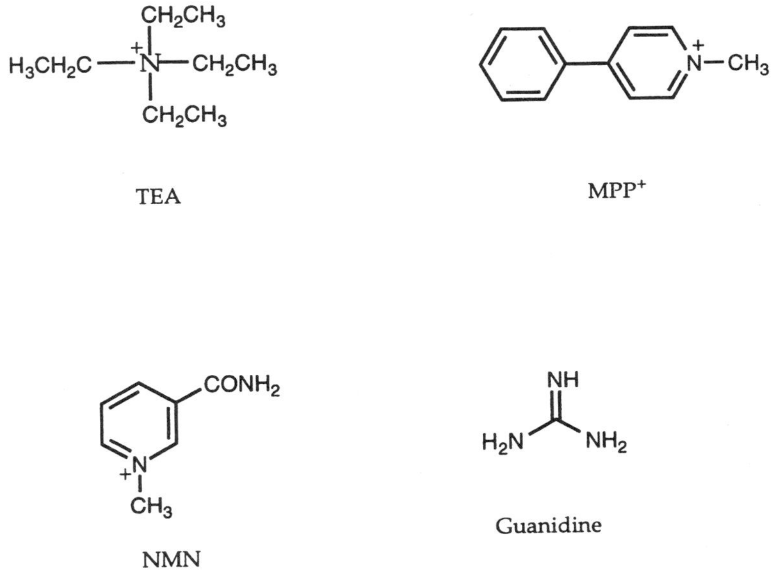

Figure 1. Chemical structures of model organic cations.

varying degrees, one common underlying theme has emerged from these mechanistic studies; namely, that multiple redundant transport mechanisms exist. To characterize fully these overlapping and multiple transport mechanisms it is critical to clone and study individual transporters in isolation. In this way, the individual components may be characterized and then related back to the whole tissue.

In recent years there has been considerable progress in the cloning of organic cation transporters from a number of species and tissues (Table I) (Gorboulev *et al.*, 1997; Grundemann *et al.*, 1994, 1997; Okuda *et al.*, 1996; Tamai *et al.*, 1997; Terashita *et al.*, 1998; Walsh *et al.*, 1996; Zhang *et al.*, 1997a; Zhang *et al.*, 1997b). The first organic cation transporter, termed rOCT1, was cloned from a rat kidney cDNA library in 1994 using the *Xenopus laevis* expression cloning technique (Grundemann *et al.*, 1994). Since then six other mammalian organic cation transporters have been cloned by homology cloning techniques. Two of these are rabbit (rbOCT1) (Terashita *et al.*, 1998) and human (hOCT1) (Gorboulev *et al.*, 1997; Zhang *et al.*, 1997b) homologues of rOCT1; one (rOCT1A) (Zhang *et al.*, 1997a) is a splice variant of rOCT1; and three are OCT2 transporters, which are related to, yet distinct from, OCT1, from the rat (rOCT2) (Okuda *et al.*, 1996; Walsh *et al.*, 1996), pig (pOCT2) (Grundemann *et al.*, 1997), and human (hOCT2) (Gorboulev *et al.*, 1997). Very recently a new transporter was isolated from a human fetal liver library (OCTN1) and may represent a new member of the OCT gene family (Tamai *et al.*, 1997). The use of computational tools (e.g., DNA and protein database searching and sequence alignment analysis) to assist in identifying and cloning new family members is starting to emerge and will likely become critical as more and more sequence information becomes available (Gillen *et al.*, 1996). Computational methods are discussed in Section 6 of this chapter. In this section we will briefly discuss the *Xenopus laevis* expression cloning and homology cloning techniques.

The *Xenopus laevis* expression cloning technique for transporters was developed by Hediger and co-workers and has greatly impacted the transport field (Hediger *et al.*, 1987). Due to the difficulties in isolating membrane proteins such as transporters, primary peptide sequence for DNA probe design has not been available for conventional cloning strategies. The *Xenopus laevis* expression system has provided a novel approach for screening a cDNA library by a transport activity assay. It has been used extensively and successfully as a functional assay to isolate the cDNAs encoding transporters and remains the cloning method of choice for transporter cDNAs when little or no sequence or homology information is available (Fei *et al.*, 1994; Grundemann *et al.*, 1994; Sweet *et al.*, 1997). In this technique (see Fig. 2) poly $(A)^+$-RNA (mRNA) is isolated from a desired cloning source (tissues or cells) by affinity chromatography, then can be size fractionated to concentrate mRNA by size. The different sizes of mRNA are injected into oocytes followed by functional assay to prescreen the mRNA population. Once the desired mRNA fractions are found, they are reverse-transcribed into cDNA for construction of a cDNA library. Alternatively, a cDNA library can be constructed

Table I.
Molecular Characteristics of Cloned Organic Cation Transporters

	OCT1				OCT2			Other:
	rOCT1	rOCT1A	rbOCT1	hOCT1	rOCT2	pOCT2	hOCT2	OCTN1
Species	Rat	Rat	Rabbit	Human	Rat	Pig	Human	Human
Protein (number of amino acids)	556	430	554	554	593	554	554	551
Identity to rOCT1 (%)	100	92	81	78	67	67	68	32
Identity to rOCT2 (%)	67	57	65	64	100	81	81	33

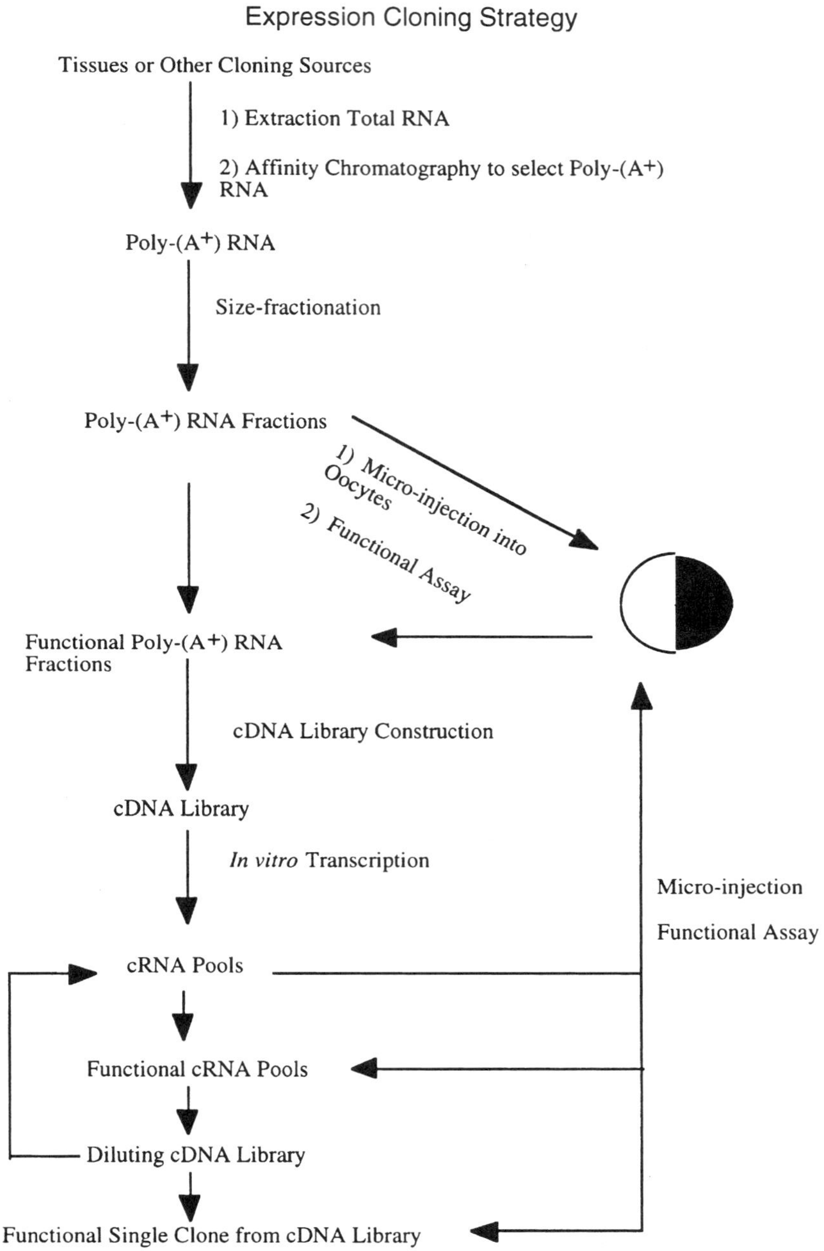

Expression Cloning Strategy

Tissues or Other Cloning Sources

1) Extraction Total RNA
2) Affinity Chromatography to select Poly-(A^+) RNA

Poly-(A^+) RNA

Size-fractionation

Poly-(A^+) RNA Fractions

1) Micro-injection into Oocytes
2) Functional Assay

Functional Poly-(A^+) RNA Fractions

cDNA Library Construction

cDNA Library

In vitro Transcription

cRNA Pools

Functional cRNA Pools

Diluting cDNA Library

Functional Single Clone from cDNA Library

Micro-injection
Functional Assay

Figure 2. Flow chart of the steps involved in expression cloning. Further details are given in the text.

from total mRNA especially when mRNA is from a scarce source. Pools of cDNA from a cDNA library are then transcribed *in vitro* to obtain cRNA pools. The pools of cRNA can then be injected into oocytes for screening by a transport functional assay. Positive cDNA pools are divided and transcribed into cRNA followed by functional assays in oocytes. These steps are repeated until a single cDNA clone is found to be functional. The advantage of this technique is that it confirms the desired clone while screening. However, the technique is very laborious and would only be used as a cloning strategy if no other suitable method were feasible. In addition, if the desired clone does not have high activity, the function of the clone expressed in oocytes could be missed when a large population of other genes is present at the beginning of screening.

Homology cloning refers to techniques which utilize known sequence information to clone isoforms and/or related genes. The two primary homology cloning techniques are library screening and reverse transcriptase-polymerase chain reaction (RT-PCR). Library screening involves screening a cDNA library with a nucleic acid probe to identify related sequences in the library. This technique was used to clone rOCT2, hOCT1, hOCT2, and OCTN1 (Gorboulev *et al.*, 1997; Okuda *et al.*, 1996; Tamai *et al.*, 1997). Homology-based RT-PCR is carried out in three basic steps. First PCR primers are designed from a known sequence(s). These may be either degenerate primers or primers from conserved sequence regions. Second, the PCR primers are used to amplify a DNA fragment. Next 5' and 3' rapid amplification for cDNA ends (RACE) is used to clone the corresponding 5' and 3' regions of the cDNA (Schaefer, 1995). Homology-based RT-PCR was used to clone rOCT1A, rbOCT1, and hOCT1 (Terashita *et al.*, 1998; Zhang *et al.*, 1997a, b). Homology cloning techniques are relatively straightforward and have become an efficient cloning method due to improvements in molecular biology protocols.

3. MOLECULAR CHARACTERISTICS

The cloning of organic cation transporters allows the investigation of their molecular characteristics using sequence analysis and molecular biology techniques. At the molecular level organic cation transporters are homologous to one another and share significant sequence similarity with other xenobiotic transporters including the multispecific organic anion transporters. Sequence analysis of the organic cation transporter polypeptides has provided insight into conserved regions which may be important in the structure, function, and regulation of these proteins. In addition, database searches and further sequence analysis suggest that these xenobiotic transporters evolved from the same gene family. Secondary structures have been proposed for the organic cation transporters using hydropathy analysis. Here we will focus on the human organic cation transporters.

hOCT1 was cloned by homology-based RT-PCR coupled with 5'- and 3'-RACE in our laboratory (Zhang et al., 1997b) and by homology screening of a human liver cDNA library in another laboratory (Gorboulev et al., 1997). The full-length hOCT1 cDNA is 1870 bp and consists of a 53-bp 5'-untranslated region (UTR), a 1665-bp open reading frame (ORF), and a 152-bp 3'UTR. The 1665-bp ORF is predicted to encode a 554-amino acid protein with a calculated molecular mass of 61 kDa (Fig. 3). The secondary structure of hOCT1 is predicted to have

```
                1                         TMD1                    50
hOCT1.pep    MPTxVDDILE QVGESGWFQK QAFLILCLLS AAFAPICVGI VFLGFTPDHH
hOCT2.pep    MPTTVDDVLE HGGEFHFFQK QMFFLLALLS ATFAPIYVGI VFLGFTPDHR

                51                        .                     .100
hOCT1.pep    CQSPGVAELS QRCGWSPAEE LNYTVPGLGP AGEAFLGQCR RYEVDWNQSA
hOCT2.pep    CRSPGVAELS LRCGWSPAEE LNYTVPGPGP AGEASPRQCR RYEVDWNQST

                101                       .                      150
hOCT1.pep    LSCVDPLASL ATNRSHLPLG PCQDGWVYDT PGSSIVTEFN LVCADSWKLD
hOCT2.pep    FDCVDPLASL DTNRSRLPLG PCRDGWVYET PGSSIVTEFN LVCANSWMLD

                151         TMD2                    TMD3          200
hOCT1.pep    LFQSCLNAGF LFGSLGVGYF ADRFGRKLCL LGTVLVNAVS GVLMAFSPNY
hOCT2.pep    LFQSSVNVGF FIGSMSIGYI ADRFGRKLCL LTTVLINAAA GVLMAISPTY

                201              TMD4                             250
hOCT1.pep    MSMLFFRLLQ GLVSKGNWMA GYTLITEFVG SGSRRTVAIM YQMAFTVGLV
hOCT2.pep    TWMLIFRLIQ GLVSKAGWLI GYILITEFVG RRYRRTVGIF YQVAYTVGLL

                251                   TMD6       *                300
hOCT1.pep    ALTGLAYALP HWRWLQLAVS LPTFLFLLYY WCVPESPRRL LSQKRNTEAI
hOCT2.pep    VLAGVAYALP HWRWLQFTVA LPNFFFLLYY WCIPESPRWL ISQNKNAEAM
                                                  *

                301                       *                *     350
hOCT1.pep    KIMDHIAQKN GKLPPADLKM LSLEEDVTEK LSPSFADLFR TPRLRKRTFI
hOCT2.pep    RIIKHIAKKN GKSLPASLQR LRLEEETGKK LNPSFLDLVR TPQIRKHTMI
                                       *

                351  TMD7                        TMD8             400
hOCT1.pep    LMYLWFTDSV LYQGLILHMG ATSGNLYLDF LYSALVEIPG AFIALITIDR
hOCT2.pep    LMYNWFTSSV LYQGLIMHMG LAGDNIYLDF FYSALVEFPA AFMIILTIDR

                401           TMD9                     TMD10      450
hOCT1.pep    VGRIYPMAMS NLLAGAACLV MIFISPDLHW LNIIIMCVGR MGITIAIQMI
hOCT2.pep    IGRRYPWAAS NMVAGAACLA SVFIPGDLQW LKIISCLGR MGITMAYEIV

                451                    TMD11                      500
hOCT1.pep    CLVNAELYPT FVRNLRMMVC SSLCDIGGII TPFIVFRLRE VWQALPLILF
hOCT2.pep    CLVNAELYPT FIRNLGVHIC SSMCDIGGII TPFLVYRLTN IWLELPLMVF

                501  TMD12                *                       550
hOCT1.pep    AVLGLLAAGV TLLLPETKGV ALPETMKDAE NLGRKAKPKE NTIYLKVQTS
hOCT2.pep    GVLGLVAGGL VLLLPETKGK ALPETIEEAE NMQRPRKNKE KMIYLQVQKL

                551
hOCT1.pep    EPSGT
hOCT2.pep    DIPLN
```

Figure 3. Sequence alignment of hOCT1 and hOCT2. Putative transmembrane domains are indicted. PKC sites are indicated by an asterisk and N-linked glycosylation sites are indicated by a dot.

12 transmembrane domains with the N- and C-termini facing the cytosol (Figs. 4 and 5). It contains five potential protein kinase C phosphorylation (PKC) sites at residues 285, 291, 327, 340, and 524 which may play a role in the posttranslational regulation of this transporter. In addition, three N-linked glycosylation sites were identified at residues 71, 96, and 112 (Fig. 3).

hOCT2 was isolated from a human kidney cDNA library by homology screening (Gorboulev et al., 1997). The 2257-bp cDNA contains a 144-bp 5'UTR, a 1668-bp ORF, and a 445-bp 3'UTR. The protein encoded by the ORF is predicted to be 555 amino acids in length with a molecular mass of 62 kDa (Fig. 3). hOCT2 is 70% identical to hOCT1 at the amino acid level. Like hOCT1, it is predicted to have 12 transmembrane domains in its secondary structure (Figs. 4 and 5). hOCT2 has two putative PKC sites at residues 286 and 327. As with the hOCT1 peptide, hOCT2 is predicted to have three N-linked glycosylation sites in the large extracellular loop between the first and second transmembrane domains at residues 72, 97, and 113 (Fig. 3).

hOCT1 and hOCT2 are overall strikingly similar at the molecular level (Fig. 3, Table I). They both share a common secondary structure of 12 transmembrane domains as predicted by hydropathy analysis (Figs. 4 and 5). They both have a large extracellular loop between their first two transmembrane domains. Three conserved N-linked glycosylation sites are located in this loop which may be important in protein sorting or in the stability or function of the transporters. This loop is the most conserved secondary structural element in the protein. However, it should be mentioned that an organic cation splice variant, rOCT1A, does not contain the first two transmembrane domains (and hence this extracellular loop)

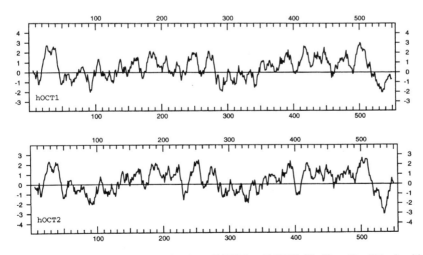

Figure 4. Comparison of the hydropathy plots of hOCT1 and hOCT2. The Kyte–Doolittle algorithm was used to generate the plots using a window of 11.

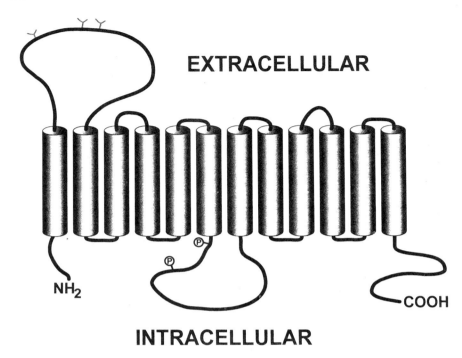

Figure 5. Proposed secondary structure of hOCT1 and hOCT2. The proposed model is based on the hydropathy plots (Fig. 4). The conserved PKC and N-linked glycosylation sites are indicated.

and appears to be functional (Zhang *et al.*, 1997a). Therefore, it is not known why this extracellular loop is so highly conserved in not only the human organic cation transporters, but also the homologous transporters from other species (Taylor *et al.*, 1997). hOCT1 and hOCT2 both contain a large intracellular loop between their sixth and seventh transmembrane domains which has two conserved PKC sites which may play a role in the regulation of these transporters. There has been one report in the literature which suggests that PKC plays a role in the posttranslational regulation of organic cation transporters (Hohage *et al.*, 1994). With the availability of the cloned transporters, we can begin to address this question.

Recently, a cDNA was cloned from a human fetal liver cDNA library which may encode an organic cation transporter (OCTN1) (Table I) (Tamai *et al.*, 1997). Interestingly, OCTN1 is quite different from the other members of the organic cation transporter family; OCTN1 is only 31% identical to hOCT1 and 33% identical to hOCT2. In addition, unlike hOCT1 or hOCT2, OCTN1 has a nucleotide-binding-site sequence motif, which suggests that it might operate as an ATP-dependent transporter. There has been one report in the literature of an ATP-dependent organic cation transport mechanism for tetraethylammonium (TEA) in the

Cloned Human Organic Cation Transporters

renal brush border membrane (McKinney and Hosford, 1993). It is unclear whether OCTN1 has 11 or 12 transmembrane domains (TMDs) in its secondary structure. OCTN1 was proposed to have only 11 TMDs, but this is problematic for two reasons. First, it would force the nucleotide-binding-site sequence motif to be extracellular. Second, it would predict that the conserved PKC sites reside in an extracellular loop. A biochemical analysis of the secondary structure of OCTN1 would help resolve these issues, as it has for other transporters (Chang and Bush, 1997; Ferreira *et al.*, 1990).

4. FUNCTIONAL CHARACTERISTICS OF CLONED TRANSPORTERS

In this section we will focus on the functional characteristics of these transporters. Elucidating the functional characteristics is a key step in understanding the substrate and inhibitor selectivity of the cloned transporters. The functional characteristics of the organic cation transporters have been investigated using several expression systems, including *Xenopus laevis* oocytes and cell lines (Busch *et al.*, 1996a, b; Gorboulev *et al.*, 1997; Grundemann *et al.*, 1994, 1997; Martel *et al.*, 1996; Okuda *et al.*, 1996; Tamai *et al.*, 1997; Terashita *et al.*, 1998; Zhang *et al.*, 1997a, b, 1998b). These systems will be described in detail below. Electrical measurements and radiolabel tracer studies have been used to determine the substrate selectivity of the cloned transporters. The advantages and disadvantages of these techniques will be discussed. Finally, the use of fluorescent compounds to further investigate the functional characteristics of organic cation transporters will be addressed.

4.1. Expression Systems

Several expression systems have been used to study the function of organic cation transporters. There are several desirable characteristics that a suitable expression system must possess. First, the endogenous organic cation transport activity should be negligible. Second, the transporter should be processed by the cellular machinery to produce a functional transporter at the plasma membrane. Third, the transporter must be expressed at sufficiently high levels in the system to provide enough acitivity for the assay.

4.1.1. *XENOPUS LAEVIS* OOCYTES

The *Xenopus laevis* oocyte expression system has been used extensively to study the function of membrane proteins such as transporters, channels, and pumps (Marino, 1996). It has been used to characterize the cloned organic cation transporters including hOCT1 and hOCT2 (Gorboulev *et al.*, 1997; Zhang *et al.*, 1997b). Detailed reviews regarding the *Xenopus laevis* expression system are available, here we will describe a general protocol (Wang *et al.*, 1991). The oocytes are surgically removed from the frog, and then the oocytes are treated with collagenase in a physiologic buffer to remove the tissue surrounding the oocyte. The treated oocytes are washed extensively to remove any remaining collagenase. The oocytes are then size-selected; in general the larger stage V and VI oocytes are most suitable for transporter expression. They are allowed to recover from the collagenase treatment by incubation in Barth solution overnight at 18°C. The next day, the oocytes are injected with the cRNA of a transporter, which is transcribed *in vitro* from an expression vector containing the cDNA of the transporter (Fig. 6). After injection, the oocytes are maintained in the Barth solution for several days (usually at least 2 or 3 days) to allow enough time for translation and processing of the protein.

Functional studies are then carried out by incubating oocytes with radiolabeled model compounds (Fig. 6). Each oocyte serves as an individual expression system. Transport studies can be designed to elucidate the kinetics and characteristics of the cloned transporter. For example, inhibitors can be added to the reaction solution to determine the inhibition profile of the transporter. In addition, the concentrations of substrates and inhibitors can be varied to determine kinetic parameters (e.g., K_m, V_{max}, and K_i). It should be pointed out that protein expression levels vary from batch to batch of oocytes; therefore, it is not possible to compare

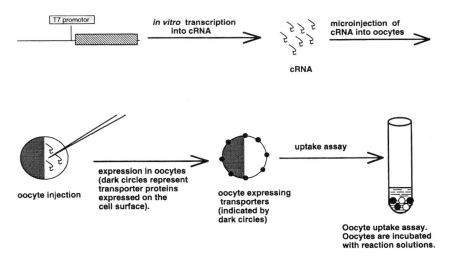

Figure 6. Schematic representation of the steps involved in the *Xenopus laevis* oocyte expression system.

V_{max} values from different experiments. Radiolabeled model compounds are assayed in individual, solubilized oocytes by scintillation counting.

The *Xenopus laevis* expression system provides an excellent way to study the functional characteristics of transport proteins. There are, however, two major drawbacks to the use of this system. The first is seasonal variability; we and other groups have found that during the summer months it is extremely difficult to maintain viable oocytes for more than a day, which is not long enough to allow for expression of the protein. This limits the use of the *Xenopus laevis* expression system to approximately September to May. Second, the assay is time-consuming, labor-intensive, and not suitable for large-scale studies. For these reasons, our laboratory and others have developed mammalian expression systems to study the function fo cloned organic cation transporters (Busch *et al.*, 1996a, b; Grundemann *et al.*, 1997; Martel *et al.*, 1996; Zhang *et al.*, 1998b).

4.1.2. CELL LINES

Although the *X. laevis* oocyte expression system is widely used for functional studies of cloned membrane proteins, more recently mammalian cell expression systems have been developed for the functional characterization of cloned transporters (Grundemann *et al.*, 1997; Martel *et al.*, 1996; Schaner *et al.*, 1997; Zhang *et al.*, 1998b). In comparison to the oocyte expressoin system, mammalian expression systems do not require the *in vitro* handling of mRNA or specialized microinjection equipment and techniques, and thus are more readily adapted to routine use in drug screening processes especially in the development of high-throughput screening. In addition, as noted above, oocytes are subject to seasonal variability in their viability as well as protein translation function, whereas continuous mammalian cell lines are not.

Similar to the oocyte expression system, mammalian cell lines are capable of expressing bioactive and correctly modified exogenous proteins after transfection, i.e., the delivery of foreign molecules such as plasmid DNA containing a gene of interest. In general, a cell line with low endogenous organic cation transport activity is suitable as an expression system. There are two approaches to gene expression in mammalian cells: transient gene expression and stable transformation. Transient gene expression, like expression of protein in the oocytes, is a simple and efficient technology with the advantage of speed. Following transient transfection the DNA is introduced into the nucleus of the cell, but does not integrate into the chromosome. The protein product is transiently produced from the non-integrating, nonreplicating plasmid DNA. The duration of protein expression is relatively short (between 24 and 96 hr). The major advantage of this procedure is that reproducible results can be obtained within days of plasmid construction once the experimental conditions have been optimized. Selection markers are not needed. Furthermore, the transient mode of expression is ideal when the overexpression proteins are toxic to the cells because cell growth is not involved or altered. The major disadvantage is that transfection must be repeated each time.

In contrast to transient transfection, stable transfection will allow DNA either to integrate into the chromosomal DNA or to be maintained as an episome, and the cell will retain the ability to produce recombinant product with each successive generation. However, the generation of stable transformed cell lines is time-consuming, and frequently requires more than 4 weeks. After transfection with plasmid DNA containing the gene of interest as well as a selection marker, cells are grown under selective pressure. Under selective conditions, only cells which have successfully integrated the DNA of interest or have maintained episomal plasmid DNA will survive. Since DNA will incorporate into different positions of the chromosome, functional assays or immunoassays with antibodies to the transporter are needed to select the single clone which has the desired function. Once the stable transformed cell line is obtained, the cell can consistently produce the desired protein and the production level will not decrease as the cells begin to divide. The stable cell lines can therefore be used to characterize the function of the overexpressed protein especially in high-throughput screening processes. In addition, the cell line can be stored for later study. However, the overexpression level can fade away after several passages; it is therefore necessary to deposit stocks of earlier passages for later use.

DNA can be introduced into mammalian cells by a variety of methods including calcium phosphate precipitation or the use of diethylaminoethyl (DEAE)-dextran, electroporation, liposomes, and retrovirus. Methods such as the calcium phosphate precipitation method and the liposomal method can be used for both transient and stable gene expression, while DEAE-dextran is useful only for transient expression.

Functional studies can then be carried out in tissue culture plates or in suspension following either transient or stable transfection of the transporter gene. Since cells have a smaller volume than oocytes, equilibrium is more rapidly achieved. For kinetic studies, earlier time points are needed to obtain accurate initial transport rates.

Organic cation transporters, e.g., rOCT1 and OCT2p, have been transiently expressed in the human cell line HEK-293 (Busch *et al.*, 1996a; Grundemann *et al.*, 1997; Martel *et al.*, 1996). More recently, our laboratory has developed a mammlian cell expression system in HeLa cells for the functional characterization of hOCT1 (Zhang *et al.*, 1998b). In addition, rOCT1 can be transiently expressed in this cell line (K. M. Giacomini, unpublished data). Lipofection, a liposomal method, has been utilized in the transfection of both cell lines.

In preliminary studies, we compared the function of hOCT1 expressed in *X. laevis* oocytes to that in HeLa cells. Similar K_i values of three different organic cations (TEA, MPP^+, and vecuronium) interacting with hOCT1 were obtained from these two systems suggesting that kinetic data can be reliably compared among the expression systems (Zhang et al., 1998b).

4.2. Functional Methods

The functional characteristics of cloned organic cation transporters can be investigated using a variety of experimental techniques. The substrate and inhibitor selectivity of the transporter can be investigated using tracer flux measurements. These same methods can be used to determine kinetic parameters of transport such as K_m, V_{max}, and K_i. Fluorescent compounds are additional tools to study transporter function and will also be addressed below.

4.2.1. RADIOLABELED LIGAND ASSAYS

The key function of a transporter is to facilitate the movement of its substrates from one side of the membrane to the other side. Three basic steps are involved: binding, translocation, and debinding (release). In order to detect this process, radiolabeled compounds with high specific activities are appealing. Namely, only trace amount of compounds are needed. A liquid scintillation counter can detect as low as femtomolar amounts of high-specific-activity radiolabeled compounds. In addition, by measuring transport of radiolabeled model substrates, interaction studies can be carried out using unlabeled compounds.

In elucidating the function of cloned organic cation transporters, ^{14}C-TEA and ^3H-MPP$^+$ are commonly used organic cation model compounds (Fig. 1). Uptake (influx) studies are conducted in either cRNA-injected oocytes or transfected cells. At the end of the incubation period the medium containing the radiolabeled compound is removed, and the oocytes or cells are rinsed with ice-cold buffer to remove nonspecifically bound radioligand. Then the oocytes or cells are lysed. The amount of radiolabeled compound inside the cell (bound plus unbound) can then be determined by liquid scintillation counting. Similarly, efflux studies can be conducted by injecting or preincubating the radiolabeled compounds followed by rapid washing with ice-cold buffer to remove bound radioligands. The efflux of radioactivity into the cell medium at different time points can be determined. Both influx and efflux studies can determine if a radiolabeled compound is a substrate of the transporter, i.e., it is truly translocated (not just bound) by the transporter.

In order to determine whether a compound is a substrate for the transporter, radiolabeled compound is needed to demonstrate directly its translocation. However, radiolabeled compounds are not always available. As an alternative, transstimulation studies can be carried out for those transporters which can operate in both directions. For example, we recently demonstrated that hOCT1 can operate as an exchanger (Zhang et al., 1998b). Preincubating hOCT1 cDNA-transfected HeLa cells with saturating concentrations of unlabeled TEA trans-stimulated TEA uptake (influx). However, if a compound (e.g., cimetidine) is not translocated by the transporter, such trans-stimulation phenomenon is not observed. If a compound (e.g., decunium-22) binds tightly to the transporter, we observed an apparent trans-"inhibition" effect. Similar trans-stimulation in efflux studies has been reported in rOCT1 cRNA-injected oocytes using electrophysiology studies (see the next section) (Nagel et al., 1997). Namely, MPP^+ efflux was trans-stimulated by extracellular MPP^+ (a known substrate of rOCT1), whereas the efflux was trans-inhibited by quinine (which was not translocated by rOCT1, as indicated in the radiolabeled tracer study).

Radiolabeled tracer studies have general limitations. First, scintillation counting alone cannot distinguish between the parent compound and its metabolites. Methods to seperate the parent compound from its metabolites (e.g., thin-layer chromatography, TLC; or high-performance liquid chromatography, HPLC) are needed to specifically detect the parent compound. Alternatively, a shorter period (before metabolism occurs) or specific inhibitors of metabolism can be used. Model organic cation substrates such as TEA and MPP^+ are not generally metabolized by cells. However, most organic cations will be metabolized. Second, it is not possible to monitor dynamically the transport of a radiolabeled compound in the intact tissue (a more physiological condition) or to know its subcellular distribution. Another method which involves fluorescent compounds has been developed to fulfill these goals and will be discussed below. Third, radiolabeled compounds cause environmental hazards.

4.2.2. ELECTROPHYSIOLOGY

Electrical measurements in voltage-clamped oocytes expressing an organic cation transporter, rOCT1, appeared to be a straightforward and powerful method to investigate the substrate selectivity of a transporter, particularly when radiolabeled compounds were not readily available (Busch et al., 1996a, b). The technique was originally used to investigate the substrate selectivity of rOCT1 (Busch et al., 1996a, b). Surprisingly it was found that a large number of compounds, including small hydrophilic molecules such as TEA and choline as well as large hydrophobic molecules such as quinidine and tubocurarine, induced inward currents in oocytes expressing rOCT1 (Busch et al., 1996b). Therefore it was originally concluded that all of these compounds were in fact substrates of rOCT1. However in a recent paper, it was shown that the quinidine-induced inward currents were in fact generated by inhibition of choline efflux (the oocytes had been preloaded with 10 mM choline). The investigators went on to show in tracer flux studies that ^{3}H-quinidine is in fact not transported by rOCT1 (Nagel et al., 1997).

Electrophysiological techniques are available to study the driving force(s) of cloned organic cation transporters. Several organic cation transporter mechanisms use the inside negative potential of a cell as a driving force for transport (Zhang et al., 1998a). It is therefore desirable to know whether or not a cloned organic cation transporter is sensitive to membrane potential. This question is easily addressed with electrophysiological recordings in oocytes expressing an organic cation transporter. For example, a membrane potential of -32 mV was determined in oocytes expressing hOCT1 in normal physiological buffer (Zhang et al., 1997b). When the oocytes were incubated in a potassium-rich buffer, which depolarized the cells, the potential was -13 mV (Zhang et al., 1997b). The uptake of ^{3}H-MPP$^+$ in hOCT1 cRNA-injected oocytes was significantly lower in the potassium medium compared to that in the physiological medium, which suggests that MPP$^+$ transport by hOCT1 is sensitive to membrane potential (Zhang et al., 1997b). This technique has been used to study the potential dependence of other organic cation transporters as well (Grundemann et al., 1994; Terashita et al., 1998; Zhang et al., 1997a).

4.2.3. FLUORESCENCE METHODS

A number of fluorescent dyes carry one or more positively charged amines and therefore by definition are organic cations. These fluorescent compounds may be transported by organic cation transporters and can be used as model organic cations in the study of organic cation transport in intact tissues or cells. To be a suitable model substance, the fluorescent dye (probe) has to yield a good tissue fluorescent signal and to be sufficiently translocated by the transporter, i.e., bind, be translocated, and released. Fluorescent probes in combination with confocal laser scanning microscopy or other detection equipment can enable us to obtain optical sectioning images of organic cation transport in fixed and living biological samples. The measurement of the increase in cellular fluorescence can be considered to reflect transport of the dye via a transporter.

A fluorescent compound, daunomycin (anthracycline), was demonstrated to be transported by organic cation transporters and P-glycoprotein in killifish renal proximal tubules at pH 7.25, using epifluorescence microscopy and video-image analysis (Miller, 1995). Recently, Ullrich and co-workers identified 4-(4-dimethylaminostyryl)-N-methyl-pyridinium (4-Di-1-ASP$^+$), a quaternary amine fluorescent dye, as a test substance for studying organic cation transport in the kidney (Pietruck and Ullrich, 1995). 4-Di-1-ASP$^+$ has a very low lipophilicity and is virtually impermeable to membranes. It has a protein binding of 47%. It was demonstrated that net reabsorption of 4-Di-1-ASP$^+$ occurred in the rat proximal tubule and that several secreted organic cations enhanced this net reabsorption presumably by inhibiting its secretion (Pietruck and Ullrich, 1995). More recently, 4-

Di-1-ASP$^+$ has been used to monitor dynamically transport in LLCPK$_1$ cells by fluorescent microscopy (Stachon et al., 1996). This fluorescence technique provides several advantages. First, radioactive compounds are not needed. Second, it is possible to monitor continuously the uptake rate of 4-Di-1-ASP$^+$ in living cells. Furthermore, paired experiments can be performed by using the first 5 min of the experiment as a control and comparing it to the next 5 min of experimental manipulations; thus each cell can serve as its own control. Finally, fluorescent probes may be more advantageous than radioactive substrates in the high-throughput drug screening process because cells do not need to be lysed in the uptake studies and radioactive waste disposal is avoided.

Fluorescence methods may become more widely used in the near future for organic cation transport studies. However, the main difficulty when applying the technique of fluorescence microscopy is photobleaching of the dye caused by the excitation light itself. The excitation light needs to be reduced to a minimum, and the emitted light needs to be large enough to be detected above the background signal. Using a rotation filter wheel which provides a pulsation excitation and reducing the light of the xenon lamp to 10% of control with a gray filter has been shown to overcome the photobleaching problem (Stachon et al., 1996).

It is not clear whether these fluorescent probes are substrates for the cloned organic cation transporters and whether this technique can also be applied to studies in oocytes and transfected cells.

4.3. Functional Characteristics of the Cloned Human Organic Cation Transporters

The functional characteristics of the cloned human organic cation transporters have been investigated using the *Xenopus laevis* oocyte and mammalian cell line expression systems (Gorboulev et al., 1997; Tamai et al., 1997; Zhang et al., 1997b, 1998b). As described below, these initial studies provide a foundation for further studies in areas such as drug–transporter interactions and high-throughput screening technologies. This discussion focuses primarily on the human transporters hOCT1, hOCT2, and OCTN1 because of their relevance to drug disposition and therapy in humans.

4.3.1. hOCT1

The initial functional characterization of hOCT1 was carried out using the *X. laevis* oocyte expression system (Zhang et al., 1997b). An eightfold enhanced uptake of ^{3}H-MPP$^+$ was observed in hOCT1 cRNA-injected oocytes in comparison to that in water-injected oocytes. ^{3}H-MPP$^+$ uptake was saturable (K_m 14.6 μM) and was sensitive to membrane potential. MPP$^+$ uptake via hOCT1 was inhibited by structurally diverse organic cations. Both small and bulkier organic cations such as TEA, choline, procainamide, verapamil, decynium-22, and quinine potently inhibited ^{3}H-MPP$^+$ uptake mediated by hOCT1 in oocytes. TEA was also a permeant of hOCT1 with K_m value of 163 μM. Besides organic cations, other compounds such as taurocholate (bile acid), inosine, and thymidine (nucleosides) at 1 mM also significantly inhibited ^{3}H-MPP$^+$ uptake, whereas D-glucose and *para*-aminohippurate (PAH) did not (at 5 mM). These functional studies are limited, since only one concentration was used. Concentration dependence studies are needed in order to elucidate the potency of these compounds in interacting with hOCT1. Furthermore, studies of translocation are needed to discern which compounds are inhibitors and which are translocated by the transporter.

For reasons stated above, a mammalian expression system, HeLa cells transiently transfected with the cDNA of hOCT1, was developed in this laboratory to study the interactions of various compounds with hOCT1 (Zhang *et al.,* 1998b). In this system, transient expression of hOCT1 activity was observed between 24 and 72 hr posttransfection, with maximal expression at approximately 40 hr. TEA transport was temperature-dependent and saturable with a K_m of 229 μM. Organic cations including clonidine (K_i 0.55 μM), acebutolol (K_i 96 μM), quinine (K_i 23 μM), quinidine (K_i 18 μM), and verapamil (K_i 2.9 μM) potently inhibited ^{14}C-TEA uptake. In addition, the neutral compounds corticosterone (K_i 7.0 μM) and midazolam (K_i 3.7 μM) also potently inhibited ^{14}C-TEA uptake. Collectively, these data indicate that TEA transport via hOCT1 can be inhibited not only by organic cations, but also by various other compounds. However, as noted above, inhibitors may not be substrates of the transporter. For example, cimetidine, an antihistamine, has been used as a model compound for renal organic cation transport. Although it inhibits ^{14}C-TEA uptake in hOCT1 cDNA-transfected HeLa cells (K_i 166 μM), no significant ^{3}H-cimetidine uptake was observed in the cDNA-transfected cells versus cells transfected with "empty" vector, i.e., vector not containing the cDNA of hOCT1. These data suggest that cimetidine is not translocated by hOCT1 under these experimental conditions. Further studies are needed to determine whether the compounds found to inhibit the transport of the model organic cations are also substrates of hOCT1 and the role that hOCT1 plays in the elimination and disposition of these compounds.

4.3.2. hOCT2

Functional studies of hOCT2 were carried out in *X. laevis* oocytes injected with the cRNA of hOCT2 (Gorboulev *et al.,* 1997). In radiolabel tracer flux experiments it was determined that TEA (K_m 76 μM), MPP$^+$ (K_m 19 μM), NMN (K_m 300 μM), and choline (K_m 210 μM) are permeants of hOCT2. Therefore,

Table II.
Interactions of Various Compounds with hOCT1 and hOCT2

	K_i (μM)	
	hOCT1[a]	hOCT2[b]
TEA	161	76
MPP+	12.3	2.4
NMN	7715	266
Desipramine	5.3	616
Procainamide	73.9	50
Tetrapentylammonium	7.4	61.5
Quinine	22.9	3.4
Decynium-22	2.73	0.1

[a]Data cited from Zhang et al. (1998b).
[b]Data cited from Gorboulev et al. (1997).

hOCT2 is a polyspecific organic cation transporter. Similar to TEA transport in oocytes expressing hOCT1, transport of ^{14}C-TEA in oocytes expressing hOCT2 was also inhibited by various organic cations. Concentration-dependent inhibition of TEA uptake by different organic cations was measured and K_i values were estimated (Table II). The K_i values of several compounds for hOCT2 differ substantially from those for hOCT1. In general, most compounds have a higher affinity for hOCT2 than for hOCT1 (Table II). These data together with information about the tissue distribution of hOCT1 and hOCT2 may aid in the prediction of the whole-body disposition of organic cations.

4.3.3. OCTN1

Very recently, a third subtype of organic cation transporters, termed OCTN1, has been cloned from human fetal liver (Tamai et al., 1997). Initial functional studies have been carried out using transiently transfected HEK-293 cells. ^{14}C-TEA uptake in OCTN1 cDNA-transfected cells was saturable with a K_m of 0.436 mM. A pH dependence of TEA transport was observed in acidic to neutral medium (pH 6.0–7.5), whereas no pH dependence was observed in neutral to alkaline pH (pH 7.5–8.5). Based on these data, the authors proposed that this transporter may represent the pH-sensitive organic cation transporter located on the apical side of the cell and function in the secretion of organic cations from cell into the lumen. However, further studies are clearly needed to support this proposal. Analysis alone of driving force is not conclusive, since several of the cloned OCT1 and OCT2 transporters are also sensitive to pH (Grundemann et al., 1997; Zhang et al., 1997). Immunological studies as well as detailed funtional studies are needed to determine the functional and physiological roles of OCTN1.

In summary, functional studies of the cloned organic cation transporters have begun to provide information about the mechanisms of organic cation transport. With the availability of cloned human organic cation transporters along with various *in vitro* functional assays, it is now possible to develop *in vitro* systems (especially in high-throughput screening) to understand drug interactions with these transporters individually. Ultimately, *in vitro* systems can be used for the prediction of drug disposition and targeting in the organism.

5. TISSUE DISTRIBUTION

A key step in understanding the role of a transporter *in vivo* is the determination of its tissue distribution and level of expression. A transporter which is widely distributed would suggest a "housekeeping" role, whereas a transporter expressed in a single tissue may suggest a more selective function and may serve as a better drug target. Organic cation transport systems have been characterized in many tissues including kidney, liver, intestine, brain, and placenta (Gisclon *et al.,* 1987; Iseki *et al.,* 1993; Muller and Jansen, 1997; Prasad *et al.,* 1992; Pritchard and Miller, 1993, 1996; Streich *et al.,* 1996; Ullrich, 1997; Zhang *et al.,* 1998a). In addition, multiple transport mechanisms, presumably mediated by multiple transporters, have been characterized in several tissues (Miyamoto *et al.,* 1989; Muller and Jansen, 1997). With the availability of the cloned transporters we are able to begin to understand the tissue-specific distribution of each transporter and ultimately its role in drug absorption, disposition, and site-specific drug targeting.

The tissue distribution of mRNA transcripts is determined by three primary methods. These are Northern blot analysis, RT-PCR, and RNase protection assays (RPAs). Norther blot analysis is the most commonly used method to detect an mRNA transcript. There are several commercial kits available to perform Northern blot analysis and several companies now sell ready-to-use multiple human tissue RNA blots for Northern analysis (e.g., Clontech and Invitrogen). Northern blot analysis is an effective way to determine the tissue distribution, approximate transcript size, and relative abundance of an mRNA species. The main drawback of this method is that high levels of expression are necessary in order to detect a positive signal, so if the mRNA of interest is expressed in low abundance, it could be missed by Northern blot analysis. For this reason, it is now very common to carry out RT-PCR analysis along with Northern blot analysis when reporting the tissue distribution of the mRNA transcript of a transporter. RT-PCR analysis is a very sensitive method to detect mRNA transcripts that are expressed in low abundance. The major limitation of this technique is contamination which would result in false positives. Therefore it is necessary to conduct requisite controls and ideally to repeat the RT-PCR analysis on different batches of cDNA. RNase protection assays are a more specialized method to determine tissue distribution. These methods are most often used to determine the existence and abundance level of splice variants. For example, this method was used to determine the expression of rOCT1A, an alternative splice variant of rOCT1 (Zhang et al., 1997a). rOCT1A mRNA transcripts are 104 nt smaller than rOCT1 transcripts and would be indistinguishable by Northern blot analysis. RNase protection assays allowed for the clear resolution of these two transcripts.

The tissue distribution of hOCT1 was determined by Northern blot analysis and RT-PCR (Gorboulev et al., 1997; Zhang et al., 1997b). The results of Northern blot analysis suggest that hOCT1 is expressed predominantly in the human liver, that is, a strong hOCT1 signal was only detectable in liver. The RT-PCR data indicate that hOCT1 is expressed at lower abundance in other tissues including kidney, skeletal muscle, stomach, spleen, placenta, small intestine, colon, and brain. This broad tissue distribution suggests a possible housekeeping function for hOCT1, or it may function principally in the liver, where it is found in greatest abundance.

The expression of the mRNA transcripts of hOCT2 was determined by Northern blot analysis and RT-PCR (Gorboulev et al., 1997). In sharp contrast to hOCT1, hOCT2 mRNA transcripts are primarily expressed in abundance in the human kidney. That is, Northern blot analysis detected two hOCT2 transcripts (2.5 and 4.0 kb) only in the kidney. The more sensitive RT-PCR assay showed that hOCT2 is also transcribed in the spleen, placenta, small intestine, and brain.

In summary, hOCT1 and hOCT2 appear to have very different tissue distributions (Table III). Namely, hOCT1 is expressed primarily in the liver, whereas

Table III.
Tissue Distribution of hOCT1 and hOCT2[a]

	hOCT1	hOCT2	hOCTN1[b]
Stomach	++	—	—
Small intestine	++	+	+
Colon	++	—	—
Kidney	++	++++	++++
Liver	++++	—	—
Brain	++	+++	—
Placenta	++	++	++
Spleen	++	+	++
Skeletal muscle	++	—	+++
Granulocytes	++	—	N.D.
Lymphocytes	++	—	N.D.
Activated lymphocytes	++	—	N.D.

[a] ++++, High abundance; +++, moderate abundance; ++, low abundance; +, very low abundance; —, no expression; N.D., not determined.
[b] Determined from Northern blot only.

hOCT2 is expressed primarily in the kidney. It could be that organ-specific transporters are at least in part responsible for organ-specific transport mechanisms of organic cations. In addition to the organ-specific expression of organic cation transporters in humans, notable species differences in the expression of OCT1 and OCT2 have been observed. For example in the rat, OCT1 is expressed abundantly in the kidney, intestine, and liver, but not in the brain (Grundemann *et al.*, 1994, 1997). However, in the rabbit, OCT1 is expressed primarily in the liver, as is the case in humans (Terashita *et al.*, 1998). Thus, the rabbit may prove to be a closer *in vivo* model of organic cation transport than the rat. Unlike OCT1, OCT2 appears to have similar tissue distributions in the human and the rat; OCT2 is expressed most abundantly in the kidney, but is also expressed at lower levels in the brain (Okuda *et al.*, 1996; Grundemann *et al.*, 1997).

The tissue distribution of OCTN1 was determined by Northern blot analysis (Tamai *et al.*, 1997). It appears to have a broad tissue distribution (Table III). It was reported to be strongly expressed in the fetal kidney, liver, and lung. In adult tissue, it is expressed most abundantly in the kidney, trachea, bone marrow, and pancreas. Weaker signals were observed in skeletal muscle, lung, placenta, prostrate, heart, uterus, spleen, and spinal cord. In addition, OCTN1 transcripts were detected in several human cancer cell lines.

6. FUTURE STUDIES

Membrane transporters provide a new and exciting area of pharmaceutical research, especially in the areas of pharmacokinetics/pharmacodynamics, drug development, and drug targeting. With the large number of clinically used drugs as well as bioactive amines falling within the organic cation class, organic cation transporters are likely to emerge as a highly relevant group of transporter proteins for pharmaceutical research. Developments in molecular biology ushered in the new era of organic cation transporter research. As of this writing eight mammalian organic cation transporters have been cloned, and expression systems to study their functional characteristics have been developed and continue to improve (Grundemann *et al.*, 1994, 1997; Gorboulev *et al.*, 1997; Okuda *et al.*, 1996; Tamai *et al.*, 1997; Terashita *et al.*, 1998; Zhang *et al.*, 1997b). Yet we have only just begun to recognize these transporters as drug research tools. In order to realize organic cation transporters as drug targets several key areas of research must be developed: efficient cloning strategies must be developed to clone the remaining members of the OCT gene family, and high-throughput assay methods need to be developed to allow for the rapid screening of compounds that interact with the transporters. Each of these areas will now be discussed in turn.

Collectively, earlier results from tissue and vesicle studies suggest that multiple organic cation transporters exist in a number of tissues and cell membranes (Zhang et al., 1998a). The eight mammalian organic cation transporters that have been cloned currently fall into one of three organic cation transporter (OCT) groups based on their sequence homology (i.e., OCT1, OCT2, OCTN1) (Table I). Functional and mechanistic data in the literature suggest that there are many more organic cation transporters that remain to be cloned (McKinney, 1982; McKinney and Hosford, 1993; Miyamoto et al., 1989; Pritchard and Miller, 1993, 1996; Zhang et al., 1998a). The traditional cloning methods described earlier were the best and perhaps the only way to clone the first organic cation transporters. However, these methods are very labor- and time-intensive; therefore, new cloning strategies are desirable. One growing area of research which may significantly assist in cloning new transporters is bioinformatics. As more and more sequence information becomes available, the importance of bioinformatics in biological research will continue to grow. For our purposes we will define bioinformatics as computational methods and tools to search and study gene and protein databases and sequences. Bioinformatics offers potential in informing and designing cloning strategies in numerous ways. A recent monograph contains numerous articles on database searching and sequence analysis; here we will only briefly discuss the most relevant databases and methodologies (Madden et al., 1996).

Bioinformatics emerged as a research field only recently in response to the explosive growth of DNA and protein sequence information. Indeed, the human genome project is scheduled to be completed by the year 2005, the entire genomes of several other organisms have already been sequenced, and several expressed sequence tag (EST) and sequenced tagged site (STS) projects are now underway. How can one identify new putative members of the human organic cation transporter family in such a large set(s) of information? There are at least three primary ways to address this problem. First, using a known sequence, a researcher can search a database for new, related sequences. This may seem straightforward, but one must carefully decided which database(s) to search and which search algorithm to use. Second, multiple sequence alignments may indicate conserved sequence(s) motifs that may be specific to that gene family. For example, we have identified several conserved sequence motifs in organic cation transporters. Finally, sequences can be analyzed for splice sites, which if found could allow the isolation of splice variants of organic cation transporters. In this review we will briefly summarize a few methods to search databases.

The most straightforward and obvious way that bioinformatics will assist in cloning new organic cation transporters is database searching. This method has been used recently to clone several human transporter cDNAs. For example, a human potassium–chloride cotransporter was identified by searching the human EST database (Gillen et al., 1996). The information from the database search was then used to assist in RT-PCR to clone the cotransporter.

When conducting a database search to idenfity new gene family members the first step is to decide which database(s) to use. Currently there are a number of databases available to the public on the world wide web. An excellent starting point for a database search is the *Entrez* web site (http://www.ncbi.nlm.nih.gov/Entrez) (Schuler *et al.*, 1996). *Entrez* is a sequence and biological information database developed by the National Center for Biotechnology Information (NCBI). The sequence information in *Entrez* is pooled from a number of sources including GenBank, SWISS-PROT, PIR-International, Brookhaven Protein Data Bank, the Protein Research Foundation database, and the Genomic Sequence Database. In addition to searching the standard *Entrez* database, there are two additional databases that deserve special attention. These are the EST and STS databases, which are accessible through *Entrez* as the dbEST and dbSTS databases. These databases contain cDNA sequence fragments from a variety of tissues and species, including a large number derived from human tissues.

The next step in database searching, after choosing a database to search, is deciding on which search algorithm to use. A popular, easy to use, and publicly available search algorithm is the basic local alignment search tool (BLAST), which is available through *Entrez* or the National Library of Medicine/National Institutes of Health (http://www.ncbi.nlm.nih.gov). BLAST is a powerful and fast search algorithm which can be used to compare an input sequence against an entire database (Madden *et al.*, 1996). After searching a database, BLAST provides the user with an output list of statistically significant related sequences in the database. There are five BLAST programs available which allow one to search either DNA or protein databases with a protein or DNA sequence as the input. BLAST searches range from simple one-sequence, one-database searches to multiple-sequence, multiple-database iterative searches. There are several reviews on BLAST and databases which go into greater detail (Madden *et al.*, 1996).

Database searching has been used to clone several human transporter cDNAs (Gillen *et al.*, 1996). By employing database searching methods, new organic cation transporters will likely be identified and this information may be used to assist in cloning new transporters. Computational methods offer several advantages over traditional cloning methods. Namely, computational methods are rapid and efficient. There are, however, several disadvantages to computational methods. Currently the databases contain redundant sequence information which can complicate the search results. In some cases, the quality of the sequence information is poor and may not be very reliable, which would require a researcher to order and resequence the clone.

High-throughput screening (HTS)-based pharmacological testing has recently been developed as a useful *in vitro* assay to study drug interactions with cloned human metabolizing enzyme and drug receptor targets in a fast and large-scale way. By this method, clinically relevant data can be obtained in early drug discovery. It is a safe and time- and cost-saving method. Similarly, development of HTS for transporters, important determinants for drug absorption, elimination, and disposition, will be in demand. Major factors for such systems are cloned transporters, sensitive analytical methods, and automation systems for handling larger numbers of compounds. Several organic cation transporters have been cloned and transfected into mammalian cell lines. The development of sensitive analytical methods such as liquid chromatography-mass spectrometry (LC/MS) and LC-NMR are still needed for detection of organic cation transport. As noted previously, some organic cations, such as 4-Di-1-ASP$^+$, are fluorescent, and may be used to develop fluorescent detection in microplates.

7. CONCLUSIONS

Within the last 5 years the organic cation transport field has moved from tissue and cellular levels to a molecular level of investigation. As we gain a better understanding of the genes and proteins responsible for the multiple mechanisms of organic cation transport, we can begin to apply this knowledge to pharmaceutical and drug discovery research. For example, in the future, we will know all of the transporters that are expressed in each tissue and have a substrate/inhibitor profile for each transporter. With this information, we could begin to predict, for example, organ-specific clearances or organ-specific distributions of drugs. Perhaps we will discover a brain-specific organic cation transporter which could serve as a drug target. It will be years and perhaps several more decades before all of these transporters are cloned and characterized. We have made a good start and there will be years of exciting work and discovery in the organic cation transport field in the future.

ACKNOWLEDGMENTS

This work was supported by grants from the National Institutes of Health (GM36780 and GM57656). We thank Patricia C. Babbitt for helpful discussions on bioinformatics and Karin M. Gerstin for critically reading the manuscript.

REFERENCES

Busch, A. E., Quester, S., Ulzheimer, J. C., Gorboulev, V., Akhoundova, A., Waldegger, S., Lang, F., and Koepsell, H., 1996a, Monoamine neurotransmitter transport mediated by the polyspecific cation transporter rOCT1, *FEBS Lett.* **395**:153–156.
Busch, A. E., Quester, S., Ulzheimer, J. C., Waldegger, S., Gorboulev, V., Arndt, P., Lang, F., and Koepsell, H., 1996b, Electrogenic properties and substrate specificity of the polyspecific rat cation transporter rOCT1, *J. Biol. Chem.* **271**:32599–32604.

Chang, H. C., and Bush, D. R., 1997, Topology of NAT2, a prototypical example of a new family of amino acid transporters, *J. Biol. Chem.* **272**:30552–30557.

Fei, Y. J., Kanai, Y., Nussberger, S., Ganapathy, V., Leibach, F. H., Romero, M. F., Singh, S. K., Boron, W. F., and Hediger, M. A., 1994, Expression cloning of a mammalian proton-coupled oligopeptide transporter, *Nature* **368**:563–566.

Ferreira, G. C., Pratt, R. D., and Pedersen, P. L., 1990, Mitochondrial proton/phosphate transporter. An antibody directed against the COOH terminus and proteolytic cleavage experiments provides new insights about its membrane topology, *J. Biol. Chem.* **265**:21202–21206.

Gillen, C. M., Brill, S., Payne, J. A., and Forbush III, B., 1996, Molecular cloning and functional expression of the K–Cl cotransporter from rabbitt, rat, and human, *J. Biol. Chem.* **271**:16237–16244.

Gisclon, L., Wong, F. M., and Giacomini, K. M., 1987, Cimetidine transport in isolated luminal membrane vesicles from rabbit kidney, *Am. J. Physiol. ;B2253:*F141–F150.

Gorboulev, V., Ulzheimer, J. C., Akhoundova, A., Ulzheimer-Teuber, I., Karbach, U., Quester, S., Baumann, C., Lang, F., Busch, A. E., and Koepsell, H., 1997, Cloning and characterization of two human polyspecific organic cation transporters, *DNA Cell Biol.* **16**:871–881.

Gottwald, M. D., Bainbridge, J. L., Dowling, G. A., Aminoff, M. J., and Alldredge, B. K., 1997, New pharmacotherapy for Parkinson's disease, *Ann. Pharmacother.* **31**:1205–1217.

Grundemann, D., Gorboulev, V., Gambaryan, S., Veyhl, M., and Koepsell, H., 1994, Drug excretion mediated by a new prototype of polyspecific transporter, *Nature* **372**:549–552.

Grundemann, D., Babin-Ebell, J., Martel, F., Ording, N., Schmidt, A., and Schomig, E., 1997, Primary structure and functional expression of the apical organic cation transporter from kidney epithelial LLC-PKC1 cells, *J. Biol. Chem.* **272**:10408–10413.

Hediger, M. A., Coady, M., Ikeda, T. S., and Wright, E. M., 1987, Expression cloning and cDNA sequencing of the Na$^+$/glucose co-transporter, *Nature* **330**:379–381.

Hohage, H., Morth, D. M., Quierl, I. U., and Greven, J., 1994, Reglation by protein kinase C of the contraluminal transport system for organic cations in rabbit kidney S2 proximal tubules, *J. Pharmacol. Exp. Ther.* **268**:897–901.

Iseki, K., Sugawara, M., Saitoh, N., and Miyazaki, K., 1993, The transport mechanisms of organic cations and their zwitterionic derivatives across rat intestinal brush-border membrane. II. Comparison of the membrane potential effect on the uptake by membrane vesicles, *Biochim. Biophys. Acta* **1152**:9–14.

Madden, T. L., Tatusov, R., and Zhang, J., 1996, Applications of network BLAST server, *Meth. Enzymol.* **266**:131–140.

Marino, M. H., 1996, Protein expression in *Xenopus* oocytes, in: *Protein Engineering* (J. L. Cleland and C. S. Craik, eds.), Wiley, New York, pp. 219–248.

Martel, F., Vetter, T., Russ, H., Grundemann, D., Azevedo, I., Koepsell, H., and Schomig, E., 1996, Transport of small organic cations in the rat liver, *Naunyn-Schmiedeberg's Arch. Pharmacol.* **354**:320–326.

McKinney, T. D., 1982, Heterogeneity of organic base secretion by proximal tubules, *Am. J. Physiol.* **243**:F404–F407.

McKinney, T. D., and Hosford, M. A., 1993, ATP-stimulated tetraethylammonium transport by rabbit renal brush border membrane vesicles, *J. Biol. Chem.* **268**:6886–6895.

Miller, D. S., 1995, Daunomycin secretion by killfish renal proximal tubules, *Am. J. Physiol.* **269**:R370–R379.

Miyamoto, Y., Tiruppathi, C., Ganapathy, V., and Leibach, F. H., 1989, Multiple transport systems for organic cations in renal brush-border membrane vesicles, *Am. J. Physiol.* **256**:F540–F548.

Muller, M., and Jansen, P. L., 1997, Molecular aspects of hepatobiliary transport, *Am. J. Physiol.* **272**:G1285–G1303.

Nagel, G., Volk, C., Friedrich, T., Ulzheimer, J. C., Bamberg, E., and Koepsell, H., 1997, A reevaluation of substrate specificity of the rat cation transporter rOCT1, *J. Biol. Chem.* **272**:31953–31956.

Okuda, M., Saito, H., Urakami, Y., Takano, M., and Inui, K., 1996, cDNA cloning and functional expression of a novel rat kidney organic cation transporter, OCT2, *Biochem. Biophys. Res. Commun.* **224**:500–507.

Pietruck, F., and Ullrich, K. J., 1995, Transport interactions of different organic cations during their excretion by the intact rat kidney, *Kidney Int.* **47**:1647–1657.

Prasad, P. D., Leibach, F. H., Mahesh, V. B., and Ganapathy, V., 1992, Specific interaction of 5-(N-methyl-N-isobutyl)amiloride with the organic cation–proton antiporter in human placental brush-border membrane vesicles. Transport and binding, *J. Biol. Chem.* **267**:23632–23639.

Pritchard, J. B., and Miller, D. S., 1993, Mechanism mediating renal secretion of organic anions and cations, *Physiol. Rev.* **73**:765–796.

Pritchard, J. B., and Miller, D. S., 1996, Renal secretion of organic anions and cations, *Kidney Int.* **49**:1649–1654.

Schaefer, B. C., 1995, Revolutions in rapid amplification of cDNA ends: New strategies for polymerase chain reaction cloning full-length cDNA ends, *Anal. Biochem.* **227**:255–273.

Schaner, M. E., Wang, J., Zevin, S., Gerstin, K. M., and Giacomini, K. M., 1997, Transient expression of a purine-selective nucleoside transporter (SPNTint) in a human cell line (HeLa), *Pharmaceut. Res.* **14**:1316–1321.

Schuler, G. D., Epstein, J. A., Ohkawa, H., and Kans, J. A., 1996, Entrez: Molecular biology database and retrieval system, *Meth. Enzymol.* **266**:141–161.

Stachon, A., Schlatter, E., and Hohage, H., 1996, Dynamic monitoring of organic cation transport processes by fluorescence measurements in LLC-PK1 cells, *Cell Physiol. Biochem.* **6**:72–81.

Streich, S., Bruss, M., and Bonisch, H., 1996, Expression of the extraneuronal transporter (uptake2) in human glioma cells, *Naunyn Schmiedeberg's Arch. Pharmacol.* **353**:328–333.

Sweet, D. H., Wolff, N. A., and Pritchard, J. B., 1997, Expression cloning and characterization of ROAT1. The basolateral organic anion transporter in rat kidney, *J. Biol. Chem.* **272**:30088–30095.

Tamai, I., Yabuuchi, H., Nezu, J., Sai, Y., Oku, A., Shimane, M., and Tsuji, A., 1997, Cloning and characterization of a novel human pH-dependent organic cation transporter, OCTN1, *FEBS Lett.* **419**:107–111.

Taylor, C. A. M., Stanley, K. N., and Shirras, A. D., 1997, The Orct1 gene of *Drosophila melanogaster* codes for a putative organic cation transporter with six or twelve transmembrane domains, *Gene* **201**:69–74.

Terashita, S., Dresser, M. J., Zhang, L., Gray, A. T., Yost, S. C., and Giacomini, K. M., 1998, Molecular cloning and functional expression of a rabbit renal organic cation transporter, *Biochim. Biophy. Acta* **1369**:1–6.

Ullrich, K. J., 1997, Renal transporters for organic anions and organic cations. Structural requirements for substrates, *J. Membr. Biol.* **158**:95–107.

Walsh, R. C., Sweet, D. H., Hall, L. A., and Pritchard, J. B., 1996, Expression cloning and characterization of a novel organic cation transporter from rat kidney, *Exp. Biol.* **10**:A127.

Wang, H.-C., Beer, B., Sassano, D., Blume, A. J., and Ziai, M. R., 1991, Gene expression in *Xenopus* oocytes, *Int. J. Biochem.* **23**:271–276.

Zhang, L., Dresser, M. J., Chun, J. K., Babbitt, P. C., and Giacomini, K. M., 1997a, Cloning and functional characterization of a rat renal organic cation transporter isoform (rOCT1A), *J. Biol. Chem.* **272**:16548–16554.

Zhang, L., Dresser, M. J., Gray, A. T., Yost, S. C., Terashita, S., and Giacomini, K. M., 1997b, Cloning and functional expression of a human liver organic cation transporter, *Mol. Pharmacol.* **51**:913–921.

Zhang, L., Brett, C. M., and Giacomini, K. M., 1998a, Role of organic cation transporters in drug absorption and elimination, *Annu. Rev. Pharmacol. Toxicol.* **38**:431–460.

Zhang, L., Schaner, M. E., and Giacomini, K. M., 1998b, Functional characterization of an organic cation transporter (hOCT1) in a transiently transfected human cell line (HeLa), *J. Pharmacol. Exp. Ther.* **286**:354–361.

16

Organic Anion Transporters

Akira Tsuji and Ikumi Tamai

1. INTRODUCTION

The characteristics and mechanisms of carrier-mediated transport of organic anions have been studied in various tissues by use of the perfused organ, isolated cells, and purified membrane vesicles. The transport mechanism across intestinal epithelial cells is of great interest because the transport process determines the rate and/or extent of bioavailability of orally administered drugs which are highly dissociated into hydrophilic ionized forms at the gastrointestinal pH. Although the intestinal transport of weakly acidic drugs by simple diffusion definitely occurs, recent investigations have provided direct and indirect evidence for participation of carrier-mediated intestinal epithelial transport (Tsuji and Tamai, 1996; Tamai and Tsuji, 1996a). The intestinal secretion of several lipophilic drugs, amphipathic drugs, and oligopeptides has also been proved to occur via specialized efflux transporters (Hunter and Hirst, 1997). Excretion of organic anions from the circulation after absorption from the gastrointestinal tract is an essential function of the liver and kidney. Bile acids and other organic anions are taken up by hepatocytes from portal blood via specific transport systems localized in the sinusoidal plasma membranes, in Na^+-dependent and/or Na^+-independent processes (Oude Elferink *et al.*, 1995; Yamazaki *et al.*, 1996; Muller and Jansen, 1997). Renal tubular cells play an important role in secretion and reabsorption and they have an extremely wide substrate selectivity covering not only endogenous anionic substances, but also a

---

Akira Tsuji and Ikumi Tamai • Department of Pharmacobio-Dynamics, Faculty of Pharmaceutical Sciences, Kanazawa University, 13-1 Takara-machi, Kanazawa 920-0934, Japan.
Membrane Transporters as Drug Targets, edited by Amidon and Sadée. Kluwer Academic/Plenum Publishers, New York, 1999.

number of clinically important drugs (Pritchard and Miller, 1993; Dantzler and Wright, 1997). Kinetic evidence shows that the distribution of organic anions, including endogenous substrates such as lactate and long chain fatty acids, into various tissues such as lung, muscle, brain, kidney, and liver certainly occurs via specialized transport system(s), as is the case in the intestine [for transport across the blood–brain barrier, see Tamai ad Tsuji (1996b) and Tsuji and Tamai (1998, 1999)].

A number of groups have attempted to isolate and characterize cDNAs encoding organic anion transporters participating in intestinal absorption/secretion, hepatic uptake/biliary secretion, renal secretion/reabsorption and uptake/efflux in other tissues, and several transporters have been successfully cloned and proved to be involved in the organic anion influx and/or efflux in these tissues.

This review focuses on the transport characteristics of cloned organic anion transporters for drugs rather than endogenous substrates. The transporters mentioned are summarized schematically in Fig. 1. Bile acid transporters, ATP-dependent multispecific organic anion transporters, and sulfate and phosphate transporters are discussed in other chapters in this book.

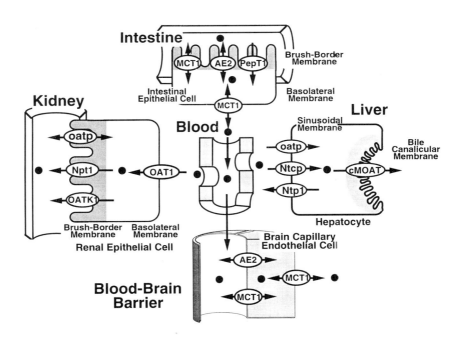

Figure 1. Summary of anion transporters in intestine, kidney, liver, and other tissues.

2. MONOCARBOXYLATE TRANSPORTERS IN INTESTINE AND BRAIN

2.1. Molecular Characterization of MCT1 and MCT2

Mammalian cells take up and excrete lactate, pyruvate, and other monocarboxylate anions by means of proton-coupled monocarboxylate transporters (MCTs). Previous kinetic studies have shown that lactate transport is stereoselective and pH-sensitive, resembling proton–lactate symport, with a Michaelis constant K_m of 10–20 mM in rat skeletal muscle (McDermott and Bonen, 1994), 13 mM in rabbit intestine (Tiruppathi et al., 1988), 4 mM in human placenta (Balkovetz et al., 1988), and 2 mM in rat blood–brain barrier (Pardridge, 1983), though the values change with the age of the animals (Cremer et al., 1979).

Recently, MCTs have been characterized at the molecular level. Complementary DNAs that encode two MCTs (MCT1 and MCT2) were cloned from Chinese hamster ovary (CHO) cells (Garcia et al., 1994) and from Syrian hamster liver (Garcia et al., 1995), respectively. MCT1 and MCT2 encode two similar proteins having ca. 494 amino acids and 12 putative transmembrane segments. Subsequently, rat, mouse, and human homologues were cloned and sequenced (Carpenter et al., 1996; Garcia et al., 1994; Jackson et al., 1995; Takanaga et al., 1995). MCT1 is widely distributed and abundant in various tissues and cell types, including erythrocytes, lung, eye, testis, skeletal muscle, heart, intestine, kidney, and brain. MCT2, which is 60% homologous to MCT1, is not detectable in erythrocytes or intestinal epithelial cells, but is expressed much more in liver than is MCT1. MCT2 is also expressed in kidney, stomach, and skin (Garcia et al., 1995). Both transporters are expressed in mitochondria-rich (oxidative) skeletal muscle fibers and cardiac myocytes. Most recently, four new MCTs (MCT3, MCT4, MCT5, and MCT6) have been cloned for human tissues (Price et al., 1998).

2.2. Intestinal Transport of Lactate and Short-Chain Fatty Acids via MCT1

By Northern blot analysis using CHO MCT1 as the probe, we identified MCT1 in rat and rabbit intestines and in Caco-2 cells (Tamai et al., 1995b). We screened a rat intestinal cDNA library by using CHO MCT1 as the probe and finally obtained rat MCT1 cDNA, which encodes a 494-amino acid protein with 12 putative membrane-spanning domains. The amino acid sequence showed 93.1% and 84.6% identity to the hamster and human MCT1, respectively. When cRNA encoding rat MCT1 was injected into *Xenopus laevis* oocytes, stereospecific and pH-dependent lactate transport activity appeared (Takanaga et al., 1995), suggesting that MCT1 expressed in the small intestine contributes to proton-coupled transport of lactate.

The MCT1 may also play a role in the proton cotransport of short-chain fatty acids (SCFAs) such as acetate, propionate, and butyrate because *Xenopus laevis* oocytes expressing rat MCT1 cDNA showed transport activity for acetic acid as well as for pyruvic acid and lactic acid (Takanaga et al., 1995).

2.3. Transport of Weak Organic Acids by MCT1

Intestinal absorption of benzoic acid derivatives has been studied in detail, though the participation of a carrier-mediated transport system is controversial. There are several lines of evidence for carrier-mediated intestinal absorption. When the permeation of salicylic acid was measured across rat intestine, unidirectional mucosal-to-serosal flux was observed at neutral pH, where salicylic acid predominantly exists as an ionized form (Turner et al., 1970). The accumulation against a concentration gradient and the significant effect of metabolic inhibitor on the salicylic acid uptake by intestinal tissues also suggest active transport (Fisher, 1981). In a similar experiment using rat intestinal tissues, transport of *p*-aminobenzoic acid, but not *m*-aminobenzoic acid, showed unidirectional mucosal-to-serosal flux, concentration dependence, and metabolic inhibitor sensitivity (Yamamoto et al., 1991). Although the difference of lipophilicity between *p*- and *m*-isomers may influence the passive transport of these two acidic compounds, participation of a carrier-mediated transport mechanism is suggested.

Several monocarboxylic acids such as benzoic acid, salicylic acid, 3-hydroxy-3-methylglutaryl-coenzyme A (HMG-CoA) reductase inhibitors, and valproic acid as well as SCFAs such as acetic acid and propionic acid significantly reduced the transcellular transport of both benzoic acid and salicylic acid across human small intestinal epithelial cell-like Caco-2 cells, whereas dicarboxylic acids were not inhibitory (Takanaga et al., 1994; Tsuji et al., 1994). It was also demonstrated that the uptake of acetic acid by intestinal brush-border membrane vesicles is inhibited by salicylic acid and benzoic acid as well as other SCFAs (Tsuji et al., 1990; Simanjuntak et al., 1991). Furthermore, pravastatin transport in brush-border membrane vesicles was specifically inhibited by several monocarboxylates (Tamai et al., 1995a). Accordingly, it seems probable that these monocarboxylates share a common transporter(s) existing in the small intestine. Recently, we have confirmed that MCT1 exhibits distinct transport activities for benzoic acid, salicylic acid, and nicotinic acid as well as for lactic acid and SCFAs by using MDA-MB231 cells stably transfected with rat MCT1 (Tamai et al., 1999), whereas the intestinal proton-coupled oligopeptide transporter PepT1, which has recently been cloned from human, rabbit, and rat intestines (Liang et al., 1995; Fei et al., 1994;

Boll et al., 1994; Miyamoto et al., 1996) and rat kidney (Saito et al., 1996), cannot transport these monocarboxylates (Miyamoto et al., 1996).

If MCT1 is expressed at the apical membrane of the intestinal epithelial cells, these monocarboxylates may be absorbed via MCT1. The direction of net transport of monocarboxylates via the proton cotransporter MCT1 is strongly dependent on the pH gradient across the epithelial membrane. The microclimate pH is in the range of 5.5–6.6 at the apical surface of the intestinal epithelium (Lucus, 1983), which provides a driving force for transport by MCT1 from the apical-to-basolateral direction, i.e., absorption. Although a previous study indicated that in the gastrointestine, MCT1 is expressed at the basolateral epithelium of stomach and cecum (Garcia et al., 1994), the precise cellular location of this transporter in small intestine remains to be examined. We have confirmed by Western blot analysis that the brush-border membrane vesicles from rat small intestine express MCT1 protein, whereas the MCT1 protein was rather abundant on the basal or lateral plasma membrane of the intestinal epithelial cells (Tamai et al., 1999). Under conditions where the intestinal flora produce large amounts of fermentation metabolites, including acetic, propionic, butyric, and lactic acids, MCT1 may play a role in the absorption of such compounds.

Thus, in the presence of a sufficient proton gradient, monocarboxylates are expected to be transported across the intestinal brush-border membrane via MCT1 into the intraepithelial space at pH around 7.0 and then be transported via the basolateral MCT1, again utilizing a proton gradient into portal blood at pH 7.4.

2.4. Monocarboxylate Transport via MCT1 at the Blood–Brain Barrier

It is well known that the brain capillary endothelial cells, which are connected together through tight junctions, form a barrier to the exchange of solute between the blood and brain interstitial fluid. Specific transport systems mediate the passage of nutrients and other solutes across this blood–brain barrier (Pardridge, 1986; Tamai and Tsuji, 1996b).

Previous studies have demonstrated that the monocarboxylate transport system plays an important role in the transport across the blood–brain barrier of SCFAs, lactate, and ketone bodies, which are essential for brain metabolism in both normal and pathological situations such as ischemia and hypoglycemia (Pardridge,1986). Moreover, we have reported that the monocarboxylate transport system functions at the blood–brain barrier as a transporter of drugs having a monocarboxylate moiety within the molecule and confirmed the expression of MCT at the blood–brain barrier (Takanaga et al., 1995). Recently, MCT1 was found on the luminal and abluminal membranes of brain capillary endothelial cells (Gerhart et al., 1997). Under normal physiological conditions, this MCT1 presumably facilitates efflux of lactate, which is produced from glucose in the brain, from the brain into blood. The abundance of MCT1 in cerebral microvessels of suckling rats suggests an important role of the transporter in delivery of energy substrates to the neonatal brain (Gerhart et al., 1997). Recently, salicylate, benzoate, and probenecid have been suggested to be transferred via the monocarboxylate transport system from the brain interstitial fluid to plasma across the blood–brain barrier (Deguchi et al., 1997). The restricted distribution of probenecid in the brain may be ascribed to efficient efflux from the brain via MCT1.

This bidirectional transport characteristic of MCT1 at the blood–brain barrier makes MCT1 a promising target for attempts to modify the brain entry of monocarboxylate drugs.

2.5. Anion Exchange Transport of Monocarboxylates

Anion exchange transport with HCO_3^- for acetic acid in rabbit and for propionic acid in human intestinal brush-border membrane vesicles has been reported (Simanjuntak et al., 1991; Harig et al., 1991). Interestingly, in these studies, such exchange transport of SCFAs with HCO_3^- was apparently enhanced at acidic extravesicular pH (Tamai et al., 1997). That the SCFAs influx is coupled with an efflux of HCO_3^- was suggested in rat and human colon by the observation of enhanced alkalinization in intestinal luminal fluid upon bathing the serosal side with HCO_3^- when the luminal surface of the colon was bathed with an SCFA such as acetic acid, propionic acid, or butyric acid (Dohgen et al., 1994). We also observed similar enhancement of luminal alkalinization upon addition of acetic acid to the luminal bathing solution using rabbit small intestinal segments (Tamai et al., 1997). These observations support the functional presence of an SCFA/HCO_3^- exchange mechanism in the small intestine as well as the colon.

Figure 2 shows the initial uptake of several monocarboxylates by rabbit intestinal brush-border membrane vesicles when either an inwardly directed proton or an outwardly directed HCO_3^- gradient was imposed. The uptakes of endogenous monocarboxylates such as acetic acid, L-lactic acid, and nicotinic acid and xenobiotic monocarboxylates such as benzoic acid and valproic acid were greatly enhanced by imposition of either a proton or an HCO_3^- gradient compared to those in the absence of any ion gradient. On the other hand, stimulation of the uptake of mevalonic acid and pravastatin was observed upon imposition of a proton gradient, but not an HCO_3^- gradient (Tamai et al., 1997). The proton cotransporter- or HCO_3^- antiporter-mediated uptake of nicotinic acid was confirmed by the observation of enhanced uptake in *Xenopus laevis* oocytes injected with mRNA from rabbit intestinal epithelial cells (Takanaga et al., 1996).

As mentioned above, MCT, but not PepT1 is likely to participate in the trans-

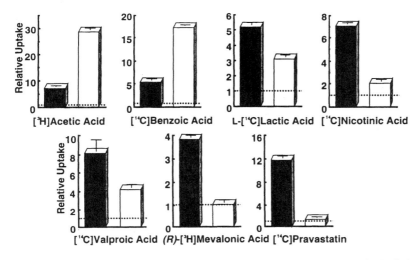

Figure 2. Proton and bicarbonate dependences of uptake of monocarboxylic acids by isolated intestinal brush-border membrane vesicles. Uptake of each compound in the presence of a proton gradient (closed column) or bicarbonate (open column) was measured at 10 sec and is represented as relative uptake normalized by the uptake in the absence of any ion gradient.

port of several monocarboxylates stimulated by an inwardly directed proton gradient. What kind of antiporter mediates the transport stimulated by an outwardly directed HCO_3^- gradient? The most likely candidates are the proteins encoded by the anion exchanger (AE) gene family, which consists of three members, AE1, AE2, and AE3. These genes encode plasma membrane Cl^-/HCO_3^- exchange proteins, which have homologous bipartite structures with a hydrophilic, cytoplasmic amino-terminal domain and a hydrophobic, membrane-spanning carboxyl-terminal domain. The membrane-spanning domains share approximately 65% amino acid identity among family members, whereas the cytoplasmic domains are less closely related. Each isoform has a different tissue distribution (Kudrycki et al., 1990), but characterization of their specific functions is still in its early stages.

Among the AE family, AE2 was originally isolated from the human leukemic cell line K562 (Demuth et al., 1986) and subsequently from several other tissues, including mouse and human kidney (Alper et al., 1988; Gehrig et al., 1992), rat gastric mucosa (Kudrycki et al., 1990), mouse choroid plexus (Lindsey et al., 1990) and rabbit small intestine (Chow et al., 1992). Interestingly, AE2 protein was identified by immunoblot analysis in brush-border membrane of rabbit small intestine (Chow et al., 1992). AE2 and monocarboxylate/anion exchanger in the intestinal brush-border membrane vesicles are very likely to be functionally related because both of them are susceptible to inhibition by stilbene disulfonates. As shown in Fig.

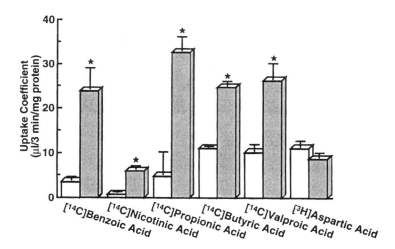

Figure 3. Uptake of monocarboxylic acids in AE2-transfected HEK293 cells. Uptake of each monocarboxylic acid was measured by incubating AE2- (closed column) and mock-transfected HEK293 cells (open column) for 3 min. Asterisk indicates significantly different from the uptake by mock-transfected cells ($p < 0.05$).

3, we have confirmed that AE2-transfected HEK293 cells exhibit significantly enhanced transport activities for benzoate, nicotinate, propionate, butyrate and valproate compared with mock-transfected cells (Yabuuchi et al., 1998). Similar anion exchange transport was also observed in Caco-2 cells (Ogihara et al., in press).

These results indicate that AE2 functions as an anion antiporter for various organic monocarboxylate drugs as well as inorganic anions. AE2 as well as MCT1 are supposed to play an important role in the carrier-mediated intestinal absorption of organic weak acids.

3. ORGANIC ANION TRANSPORTERS IN LIVER AND KIDNEY

Two multispecific transporter families for organic anions have been identified. One is the ATP binding cassette (ABC) superfamily, which includes P-glycoprotein, a multidrug-resistance protein (Gottesmann and Pastan, 1993), and a canalicular multispecific organic anion transporter (cMOAT) (Paulusma et al., 1996; Ito et al., 1997). The other is the oatp (organic anion-transporting polypeptide) family, which includes Na^+-independent oatp isolated from rat liver (Jacquemin et al., 1994), OATP from human liver (Hagenbuch et al., 1995), prostaglandin transporter (PGT) (Kanai et al., 1995), kidney-specific oatp (OAT-K1) (Saito et al., 1996), OAT1 (Sekine et al., 1997), and oatp2 (Noe et al., 1997).

3.1. Organic Anion Transporters in Liver

The liver plays an essential role in the extraction of numerous organic anions such as bilirubin and bromosulfophthalein (BSP) from the circulation. Although such anions are tightly bound to plasma albumin, they are rapidly taken up by hepatocytes. It was found that this sinusoidal (basolateral) uptake process involved both Na^+-dependent and Na^+-independent transport systems.

Bile acids conjugated with taurine or glycine are predominantly taken up into hepatocytes by Na^+-dependent taurocholate-cotransporting polypeptide, which has been cloned from rat liver (Ntcp) and human liver (NTCP) (Hagenbuch et al., 1991; Hagenbuch and Meier, 1994).

Based primarily on kinetic cis-inhibition studies using hepatic perfusion, isolated hepatocytes, and sinusoidal membrane vesicles, one or more Na^+-independent uptake mechanisms are suggested to be responsible for the uptake of unconjugated and conjugated bile salts, other amphipathic organic anions such as BSP, indocyanine green (ICG), cardiac glycosides such as ouabain, neutral steroids, certain organic cations, and numerous other drugs including the HMG-CoA reductase inhibitor pravastatin. Data on the uptake of individual organic anions have been extensively reviewed elsewhere (Oude Elferink et al., 1995; Yamazaki et al., 1996; Muller and Jansen, 1997). Several transporter proteins have been proposed to be involved in sinusoidal Na^+-independent transport of organic anions, including a 55-kDa BSP/bilirubin-binding protein (BBBP), a 55-kDa organic anion-binding protein (OABP), and a 37-kDa bilitranslocase (BTL). BBBP and OABP have never been reconstituted in liposomes to demonstrate a transport function. BLT has been reconstituted and mediates electrogenic transport of BSP (Torres et al., 1993), though its molecular identification has not been completed.

An Na^+-independent basolateral organic anion-transporting polypeptide (oatp1) cDNA which encodes 670 amino acids (80-kDa glycoprotein) with 10 putative transmembrane domains has been cloned from rat liver (Jacquemin et al., 1994). Oatp is unrelated to several cloned Na^+-dependent bile acid transporters, ntcp in rats (Hagenbuch et al., 1991) and NTCP in humans (Hagenbuch et al., 1994). The oatp1 mediated Na^+-independent and Cl^--dependent uptake of BSP in the presence of serum albumin, and Cl^--dependent uptake was completely abolished in the absence of serum albumin or a high concentration of BSP, behavior similar to that described previously in hepatocytes (Min et al., 1991). The oatp1 seems to exhibit a broad substrate specificity since it can also mediate Na^+-independent uptake of bile acids such as cholate and taurocholate as well as a wide range of amphipathic substrates, including BSP, estrogen conjugates, neutral steroids, certain organic cations (e.g., N-propylajmalinium), peptidomimetic drugs, and the mycotoxin ochratoxin A (Bossuyt et al., 1996; Eckhardt et al., 1996). The oatp1-mediated transport of taurocholate in oatp1 cDNA-transfected

HeLa cells is Na^+ independent, saturable with K_m of 19 μM, temperature dependent, and observed only when organic anion/HCO_3^- exchanger is coexpressed (Saltin et al., 1997). Northern blot analysis suggests that oatp1 is expressed in liver, kidney, brain, lung, skeletal muscle, and proximal colon. Interestingly, immunomorphological examination of kidney revealed apical plasma membrane localization in the S3 segment of the proximal tubule of the outer medulla, while this oatp1 is localized at the basolateral membrane of hepatocytes (Bergwerk et al., 1996).

Human OATP exhibits a 67% amino acid sequence identity and 82% similarity with the rat oatp1 (Kullak-Ublick et al., 1995). Hence, the anion transporters most probably belong to the same transporter gene family. However, in comparative functional expression studies the affinities and transport activities for selected substrates differed in some respects between OATP and oatp1 (Bossuyt et al., 1996). For example, the affinity of human OATP for BSP is 13-fold lower than that of rat oatp, suggesting that high BSP affinity of rat oatp is probably required for efficient extraction of BSP from albumin in rat liver. The most distinct feature of OATP as compared with oatp1 was the markedly higher mRNA level as a proportion of total brain mRNA, suggesting high-level expression of OATP or closely related polypeptides in human brain as well as in lung and kidney (Kullak-Ublick et al., 1995). These results indicate that human OATP plays an important role in the cellular transport of endogenous organic anions.

To evaluate the extent to which oatp1 and ntcp account for overall hepatic bile acid and BSP uptake, an antisense oligonucleotide approach has been used (Hagenbuch et al., 1994). When total rat liver mRNA prehybridized with ntcp-specific and oatp1-specific antisense oligonucleotides was injected into *Xenopus laevis* oocytes, Na^+-independent taurocholate uptake was inhibited by 80% compared to that in the absence of the oligonucleotides, indicating that no additional Na^+-independent carrier would be required for hepatocellular uptake of this major conjugated natural bile acid. In contrast, the same antisense oligonucleotides inhibited BSP uptake only up to 50%, thus supporting the presence of additional Na^+-independent BSP uptake system(s) other than oatp1 at the basolateral plasma membrane of rat hepatocytes. When oatp1 was transfected into HeLa cells, unconjugated bilirubin, bilirubin monoglucuronide, and bilirubin diglucuronide appeared not to be transported by oatp1, either with or without carrier albumin (Kanai et al., 1996). Alternative sinusoidal transporters, including BBBP, OABT, BLT, or other oatp isoforms, are required for the hepatocellular uptake of not only BSP, bilirubin, and its conjugates, but also more water-soluble organic anions such as *p*-aminohippurate and various mono- and dicarboxylates.

Similar but distinct oatp isoforms appear to be involved in the elimination pathways for amphipathic compounds in different organs. OATP/oatp isoforms are located in the basolateral membrane of liver parenchymal cells and in the apical membrane of kidney epithelial cells. Due to their ability to facilitate bidirectional transport, OATP/oatp isoforms may work as "uptake" systems or as "export" systems in the kidney. In the liver, OATP/oatp isomers might play an important role in uptake of amphipathic compounds into hepatocytes from the circulation. In recent studies, oatp isoforms from liver and kidney were compared in more detail (Bergwerk et al., 1996; Wolkoff, 1996).

Intrahepatocellular compounds can be secreted back into the sinusoidal space. Sinusoidal efflux has been demonstrated for bilirubin (Wolkoff et al., 1987), dibromosulfophthalein (DBSP) (Nijssen et al., 1991), and harmol-sulfate (de Vries et al., 1985), and may be mediated by the same carrier that is responsible for their uptake. However, indirect evidence exists that the efflux may involve separate transporters (Nijssen et al., 1991) in addition to the mechanism by which plasma albumin strongly binds a large amount of organic anions and thereby stimulates sinusoidal efflux (Proost et al., 1993). Npt1 (NaPi-1), which has been cloned as the renal inorganic phosphate transporter, is present both in kidney and liver (Werner et al., 1991; Chong et al., 1995) and rabbit Npt1 was shown to transport organic anions such as benzylpenicillin, probenecid, and phenol red (Busch et al., 1996). We have previously demonstrated that β-lactam antibiotics are taken up by hepatocytes and excreted into bile via multiple organic anion transporters (Tsuji et al., 1986; Tamai and Tsuji, 1987; Tamai et al., 1990). The tissue distribution and organic anion-transporting function of Npt1 suggest that it may participate in the hepatic transport of β-lactam antibiotics. Npt1 was demonstrated to be localized in the sinusoidal membrane of mouse liver, and the transport of benzylpenicillin by mouse Npt1 examined in *Xenopus laevis* oocytes was Na^+ independent and was inhibited by several organic anions (Yabuuchi et al., 1998b). However, since its transport activity was reduced by a high concentration of chloride ion, it was supposed that Npt1 may function for the efflux of organic anions such as β-lactam antibiotics from hepatocytes to blood across the sinusoidal membrane (Yabuuchi et al., 1998b).

Sinusoidal efflux of glutathione (GSH) is a physiologically relevant process because over 90% of GSH in the circulation originates from the liver and it plays an important role in detoxification and protection against oxidative stress (DeLeve and Kaplowitz, 1991). The rat sinusoidal membrane GSH transporter sGshT has been cloned and confirmed to facilitate Na^+-independent bidirectional transport of GSH (Yi et al., 1995).

Unlike the sinusoidal organic anion transport systems, most canalicular transport systems belong to the ABC transporter superfamily. Five ATP-dependent organic substrate transporters have been identified at the bile-canalicular front of rat hepatocytes (Muller and Jansen, 1997): (1) a copper-transporting P-type ATPase that is probably defective in Wilson's disease; (2) mdr2 (for human MDR3), which mediates ATP-dependent canalicular secretion of phosphatidylcholine; (3) mdr1 (for human MDR1), which mediates canalicular excretion of hydrophobic organic cations such as daunomycin, vinblastine, and verapamil; (4) cMOAT (for human MRP2), a multispecific organic anion transporter, which primarily transports divalent organic anions such as oxidized glutathione (GSSG), leukotrienes, bilirubin diglucuronide, dinitrophenylglutathione, BSP-glutathione conjugate, and p-nitrophenylglucuronide, and many glucuronide and/or sulfate conjugates of drugs; (5) cBST, a bile salt transporter which mediates transport of monovalent bile acids. For more detailed information about canalicular organic anion transporters, see other chapters of this book.

3.2. Organic Anion Transporters in Kidney

Renal tubular cells play an important role in basolateral extraction from the circulation and secretion/reabsorption in the apical membrane of anionic endogenous substances and xenobiotics. The basolateral uptake of organic anions has been extensively investigated using p-aminohippuric acid (PAH) as a test substrate. The most characteristic feature of the organic transport system responsible for PAH is its extremely wide substrate selectivity, covering not only endogenous anionic substrates, but also a number of clinically important drugs (Pritchard and Miller, 1993). It has been recognized, as shown schematically in Fig. 1, that the basolateral uptake of PAH is mediated by an organic anion/dicarboxylate exchanger (Shimada et al., 1987). An outwardly directed dicarboxylate gradient is essential to energize the transport activity of this exchanger. Recently, a multispecific organic anion transporter OAT1 cDNA which encodes a 551-amino acid protein with 12 putative membrane-spanning domains was successfully cloned from rat kidney (Sekine et al., 1997). This was achieved by using a functional expression method together with coexpression of rat sodium dicarboxylate transporter, rNaDC-1, in *Xenopus laevis* oocytes to energize organic anion transport in the oocytes. There is no significant sequence similarity between OAT1 and other members of the multispecific transporter superfamily. OAT1 mediated sodium-independent and saturable PAH uptake with a K_m value of 14 μM. The uptake rate of PAH was increased by an outwardly directed dicarboxylate gradient (Sekine et al., 1997), in accordance with the proposed mechanism. OAT1 exhibits a remarkably wide substrate selectivity, covering endogenous substrates such as cyclic nucleotides, a prostaglandin, and uric acid, and a variety of drugs with different structures such as β-lactam antibiotics (cephaloridine, benzylpenicillin), a quinolone antibacterial agent (nalidixic acid), a nonsteroidal antiinflammatory drug (indomethacin), diuretics (furosemide, ethacrynic aci), a uricosuric drug (probenecid), an antiepileptic drug (valproic acid), and an anticancer drug (methotrexate). Northern blot analysis revealed that OAT1 is exclusively expressed in the kidney. The overall pattern of in situ hybridization suggests that OAT is most strongly expressed in the middle portion of the proximal tubule (S2 segment). These data suggest that OAT1 is a multispecific organic anion transporter at the basolateral membrane of the proximal tubule.

Interestingly, another organic anion transporter, OAT-K1, was cloned from a rat kidney cDNA library as a homologue of rat liver oatp (Saito *et al.,* 1996). When the transport activity was measured in stably transfected renal cells expressing OAT-K1, this transporter recognized endogenous folate and an anticancer drug, methotrexate, but not PAH. The rat OAT-K1 cDNA encodes a 669-amino acid protein with 12 putative transmembrane domains which shows 72% amino acid identity with the rat oatp and 35% identity with the rat prostaglandin transporter. OAT-K1 was confirmed to be expressed especially at the apical membrane of renal proximal tubules, but not in liver (Masuda *et al.,* 1997). Recently cloned rat OAT-K2 cDNA encoding a 498-amino acid protein shows 91% identity with rat OAT-K1. OAT-K2 mediates the uptake of hydrophobic organic anions, such as taurocholate, methotrexate, folate, and prostaglandin E2, although its homologue OAT-K1 transports methotrexate and folate, but not taurocholate and prostaglandin E2 (Matsuda *et al.,* 1999). It is likely that OAT-K2 participates in epithelial transport of hydrophobic anionic compounds in the kidney and that both OAT-K1 and OAT-K2 work in secretion into lumen of intracellular organic anions taken up by the basolateral OAT1.

Npt1, found as a renal inorganic phosphate transporter, as mentioned above localized at the apical membrane of kidney epithelial cells (Biber *et al.,* 1993) and transports several organic anions (Busch *et al.,* 1996). Therefore, this transporter presumably also contributes to secretion of organic anions from renal tubular epithelial cells into urine (Yabuuchi *et al.,* 1998b, Uchins *et al.,* in press).

3.3. Organic Anion Transporters in Other Tissues

Recently, a novel multispecific organic anion transporting polypeptide, oatp2, has been cloned from rat brain (Noe *et al.,* 1997). The oatp2 encodes a 661-amino acid protein with 12 membrane-spanning domains and exhibits an amino acid sequence identity of 77% to oatp1 and OAT-K1 and of 73% to human OATP. The corresponding identity of oatp2 to the prostaglandin transporter is 35% (Noe *et al.,* 1997). As judged from Northern blot analysis, oatp2 is highly expressed in brain, liver, and kidney, but not in heart, spleen, lung, skeletal muscle, or testes. By Western blot and immunohistochemical analyses, oatp2 was confirmed to be expressed in rat brain and basolateral membrane of hepatocytes (Kakyo *et al.,* 1999). Oatp2 appears to be involved in the multispecificity of the uptaking substrates in the liver and brain.

In functional expression studies in *Xenopus laevis* oocytes, oatp2 mediated uptake of bile acids such as taurocholate (K_m = 35 μM) and cholate (K_m = 46 μM), estrogen conjugates such as estradiol-17β-glucuronide (K_m = 0.24 μM), and

HMG-CoA reductase inhibitors such as pravastatin (K_m = 37.5 μM) and cardiac glycosides such as ouabain (K_m = 470 μM) and digoxin (K_m = 0.24 μM) (Noe et al., 1997; Tokui et al., 1999). Although most of the compounds indicated above are common substrates of several oatp-related transporters, high-affinity uptake of digoxin is a unique feature of oatp2 (Noe et al., 1997).

Digoxin elimination from the body after administration occurs to the extent of ca. 25% by hepatobiliary excretion into bile, 15% via gut mucosa, and about 60% via renal excretion into urine (Mayer et al., 1996). Digoxin overdose induces several types of side effects in the central nervous system (Kelly and Smith, 1992). These results indicate that oatp2, which is highly expressed in brain, liver, and kidney, may play an especially important role in the brain accumulation and toxicity of digoxin and in the hepatobiliary and renal excretion of cardiac glycosides from the body. The detoxification of digoxin in brain and facilitated excretion into bile and urine may proceed at low doses of digoxin owing to active efflux by P-glycoprotein expressed at the blood–brain barrier (Schinkel et al., 1995) and in the bile canalicular membrane of hepatocytes and the brush-border membrane of renal tubular epithelial cells, in combination with the transport function of oatp2 in these tissues. Both of oatp2 and recently cloned oatp3 (Abe, et al., 1999) are multifunctional transporters involved in the transport of thyroid hormones in the brain, retina, liver, and kidney.

Although prostaglandins (PGs) are often assumed to traverse membranes by simple diffusion, they are charged anions at physiological pH and diffuse poorly across model biological membranes such as rabbit erythrocytes. This limited simple diffusion appears to be augmented, at least in some circumstances, by specialized carrier-mediated transport systems. Selective PG clearance by the pulmonary vascular bed is suggested to be attributed to differential transport across the plasma membrane of pulmonary cells. Additional evidence supports the presence of specific active transporters in lung and in many epithelial cells including liver, kidney, choroid plexus, and uterus. A new organic anion transporter, PGT, has been cloned as a member of the oatp family (Kanai et al., 1995). Hydropathy analysis of PGT revealed strong similarities to oatp, suggesting that the two are closely related members of the transporter superfamily characterized by 12 membrane-spanning domains. Although PGs were not transported by oatp, PGT transfected into HeLa cells or expressed in *Xenopus laevis* oocytes caused a rapid and marked increase in the transport of PGE1, PGE2, and PGE3. In contrast, the stable prostacyclin analogue iloprost and the prostacyclin metabolite 6-ketoPGF1 were not transported to any significant degree by PGT. Transport of prostanoids was inhibited by several organic anions such as bromocresol green, bromosulfophthalein, indocyanine green, and furosemide with K_i values in the range of 4–50 μM, but was hardly inhibited by other organic anions such as glutathione, *p*-aminohippurate, taurocholate, urate, unconjugated and conjugated bilirubin, and estradiol glucuronide at concentrations of less than 400 μM (Kanai et al., 1995).

PGE2 transport in HeLa cells transfected with PGT cDNA was not affected either by extracellular Na^+ or by an inwardly directed proton gradient, indicating that PGP does not function either as a PG/Na^+ cotransporter or as a PG/H^+ cotransporter. Extracellular Cl^- had no effect on the transport, which suggests that PGT is not a PG/Cl^- exchanger. PGT mRNA is most abundant in the lung, liver, kidney, and testes, moderately abundant in the brain, stomach, ileum, and jejunum, and hardly present in heart and skeletal muscle. PGT may mediate the clearance of prostanoids from the circulation.

4. CONCLUSION

Organic anion transporters were first identified in intestine, liver, and kidney based on extensive studies of membrane transport phenomena mainly of xenobiotics. Recent advances in membrane transport studies with the use of molecular biological approaches have confirmed the participation of such carrier-mediated transport in these tissues and the pharmacological and/or pharmacokinetic relevance of these transporters to the disposition of xenobiotics. Furthermore, the tissue distributions of these transporters suggest the physiological significance of carrier-mediated transport of nutrients, endogenous substrates, and xenobiotics in many tissues in addition to intestine, kidney, and liver. Since at least some of the organic anion transporters probably have physiologically crucial roles, manipulation of drug disposition by utilization of these transporters is expected to be useful for regulating the pharmacokinetics of anionic drugs not only in intestinal absorption and renal and hepatic elimination, but also in distribution into pharmacological target tissues.

REFERENCES

Abe, T., Kakyo, M., Sakagami, H., Tokui, T., Nishio, T., Tanemoto, M., Nomura, H., Hebert, S. C., Matsuno, S., Kondo, S. and Yawo, H., 1998, Molecular characterization and tissue distribution of a new organic anion transporter subtype (oatp3) that transport thyroid hormones and taurocholate and comparison with oatp2, *J. Biol. Chem.* **273:** 22395–22401.

Alper, S. L., Kopito, R. R., Libresco, S. M., and Lodish, H. F., 1988, Cloning and characterization of a murine band 3-related cDNA from kidney and from a lymphoid cell line, *J. Biol. Chem.* **263:**17092–17099.

Balkovetz, D. F., Leibach, F. H., Mahesh, V. B., and Ganapathy, V., 1988, A proton gradient is the driving force for uphill transport of lactate in human placental brush-border membrane vesicles, *J. Biol. Chem.* **263:**13823–13830.

Bergwerk, A. J., Shi, X. Y., Ford, A. C., Kanai, N., Jacquemin, E., Burk, R. D., Bai, S.,

Novikoff, P. M., Stieger, B., Meier, P. J., Schuster, V. L., and Wolkoff, A. W., 1996, Immunological distribution of an organic anion transporter protein in rat liver and kidney, *Am. J. Physiol.* **271**:G231–G238.

Biber, J., Custer, M., Werner, A., Kaissling, B., and Murer, H., 1993, Localization of NaPi-1, a Na/Pi cotransporter, in rabbit kidney proximal tubules, *Pflugers Arch.* **424**:210–215.

Boll, M., Markovich, D., Weber, W.-M., Korter, H., Daniel, H., and Murer, H., 1994, Expression cloning of a cDNA from rabbit small intestine related to proton-coupled transport of peptides, β-lactam antibiotics and ACE-inhibitors, *Pflugers Arch.* **429**:146–149.

Bossuyt, X., Muller, M., Hagenbuch, B. and Meier, P. J., 1996, Polyspecific steroid and drug clearance by an organic anion transporter of mammalian liver, *J. Pharmacol. Exp. Ther.* **276**:891–896.

Busch, A. E., Schuster, A., Waldegger, S., Wagner, C. A., Zempel, G., Broer, S., Biber, J., Murer, H., and Lang, F., 1996, Expression of a renal type I sodium/phosphate transporter (NaPi-1) induces a conductance in Xenopus oocytes permeable for organic and inorganic anions, *Proc. Natl. Acad. Sci. USA* **93**:5347–5351.

Carpenter, L., Poole, R. C., and Halestrap, A. P., 1996, Cloning and sequencing of the monocarboxylate transporter from mouse Ehrilich Lettre tumor cell confirms its identity as MCT1 and demonstrates that glycosylation is not required for MCT1 function, *Biochim. Biophys. Acta* **1279**:157–163.

Chong, S. S., Kozak, C. A., Liu, L., Kristjansson, K., Dunn, S. T., Bourdeau, J. E., and Highes, M. R., 1995, Cloning, genetic mapping, and expression analysis of a mouse renal sodium-dependent phosphate cotransporter, *Am. J. Physiol.* **268**:F1038–F1045.

Chow, A., Dobbins, J. W., Aronson, P. S., and Igarashi, P., 1992, cDNA cloning and localization of band 3-related protein from ileum, *Am. J. Physiol.* **263**:G345–G352.

Cremer, J. E., Cunningham, J. V., Pardridge, W. M., Braun, L. D., and Oldendorf, W. H., 1979, Kinetics of blood–brain barrier transport of pyruvate, lactate and glucose in sucking, weaning and adult rats, *J. Neurochem.* **33**:439–445.

Dantzler, W. H., and Wright, S. H., 1997, Renal tubular secretion of organic anions, *Adv. Drug Deliv. Rev.* **25**:217–230.

Deguchi, Y., Nozawa, K., Yamada, S., Yokoyama, Y., and Kimura, R., 1997, Quantitative evaluation of brain distribution and blood–brain barrier efflux transport of probenecid in rats by microdialysis: Possible involvement of the monocarboxylic acid transport system, *J. Pharmacol. Exp. Ther.* **280**:551–560.

DeLeve, L. D., and Kaplowitz, N., 1991, Glutathione metabolism and its role in hepatotoxicity, *Pharmacol. Ther.* **52**:287–305.

Demuth, D. R., Showe, L. C., Ballantine, M., Palumbo, A., Fraser, P. J., Cioe., L., Rovera, G., and Curtis, P. J., 1986, Cloning and structural characterization of human non-erythroid band 3-like protein, *EMBO J.* **5**:1205–1214.

de Vries, M. H., Groothuis, G. M. M., Mulder, G. J., Nguyen, H., and Meijer, D. K. F., 1985, Secretion of the organic anion harmol sulfate from liver into blood, *Biochem. Pharmacol.* **34**:2129–2135.

Dohgen, M., Hayashi, H., Yajima, T., and Suzuki, T., 1994, Stimulation of bicarbonate secretion by luminal short-chain fatty acids in the rat and human colon *in vivo*, *Japn. J. Physiol.* **44**:519–531.

Eckhardt, U., Horz, J. A., Petzinger, E., Stuber, W., Reers, M., Dickneite, G., Daniel, H., Wagener, M., Hagenbuch, B., Stieger, B., and Meier, P. J., 1996, The peptide-based thrombin inhibitor CRC220 is a new substrate of the basolateral rat brain organic anion-transporting polypeptide, *Hepatology* **24**:380–384.

Fei, Y.-J., Kanai, Y., Nussberger, S., Ganapathy, V., Leibach, F. H., Romero, M. F., Singh, S. K., Boron, W. F., and Hediger, M. A., 1994, Expression cloning of a mammalian proton-coupled oligopeptide transporter, *Nature* **368**:563–566.

Fisher, R. B., 1981, Active transport of salicylate by rat jejunum, *Q. J. Exp. Physiol.* **66**:91–98.

Garcia, C. K., Li, X., Luna, J., and Francke, U., 1994, cDNA cloning of the human monocarboxylate transporter 1 and chromosomal localization of the SLC16A1 locus to 1p13.2–p12, *Geomics* **23**:500–503.

Garcia, C. K., Brown, M. S., Pathak, R. K., and Goldstein, J. L., 1995, cDNA cloning of MCT2, a second monocarboxylate transporter expressed in different cells than MCT1, *J. Biol. Chem.* **270**:1843–1849.

Gehrig, H., Muller, W., and Appelhans, H., 1992, Complete nucleotide sequence of band 3 related anion transport protein AE2 from human kidney, *Biochim. Biophys. Acta* **1130**:326–328.

Gerhart, D. Z., Enerson, B. E., Zhdankina, O. Y., Leino, R. L., Drewes, L. R., 1997, Expression of monocarboxylate transporter MCT1 by brain endothelium and glia in adult and suckling rats, *Am. J. Physiol.* **273**:E207–E213.

Gottesmann, M. M., and Pastan, I., 1993, Biochemistry of multidrug transporter, *Annu. Rev. Biochem.* **62**:385–427.

Hagenbuch, B., and Meier, P. J., 1994, Molecular cloning. Chromosomal localization, and functional characterization of a human liver Na^+/bile acid cotransporter, *J. Clin. Invest.* **93**:1326–1331.

Hagenbuch, B., Stieger, B., Forguet, H., Lubbert, H., and Meier, P. J., 1991, Functional expression cloning and characterization of the hepatic Na^+/bile acid cotransport system, *Proc. Natl. Acad. Sci. USA* **88**:10629–10633.

Hagenbuch, B., Scharschmidt, B. F., and Meier, P. J., 1994, Use of antisense oligonucleotides to assess the role of the cloned transporters Ntcp and oatp in a Na^+-dependent and Na^+-independent hepatic uptake of taurocholate and non-bile acid organic anions, *Hepatology* **20**:201A.

Hagenbuch, B., Stieger, B., Schteingart, C. D., Hofmann, A. F., Wolkoff, A. W., and Meier, P. J., 1995, Molecular and functional characterization of an organic anion transporting polypeptide cloned from human liver, *Gastroenterology* **109**:1274–1282.

Harig, J. M., Soergel, K. H., Barry, J. A., and Ramaswamy, K., 1991, Transport of propinate by ileal brush-border membrane vesicles, *Am. J. Physiol.* **260**:G776–782.

Hunter, J., and Hirst, B. H., 1997, Intestinal secretion of drugs. The role of P-glycoprotein and related efflux systems in limiting oral drug absorption, *Adv. Drug Deliv. Rev.* **25**:129–157.

Ito, K., Suzuki, H., Hirohashi, T., Kume, K., Shimizu, T. and Sugiyama, Y., 1997, Molecular cloning of canalicular multispecific organic anion transporter defective in EHBR, *Am. J. Physiol.* **272**:G16–G22.

Jackson, V. N., Price, N. T., and Halestrap, A. P., 1995, cDNA cloning of MCT1, a monocarboxylate transporter from rat skeletal muscle, *Biochim. Biophy.: Acta* **1238**:193–196.

Jacquemin, E., Hagenbuch, B., Stieger, B., Wolkoff, A. W., and Meier, P. J., 1994, Expression cloning of rat liver Na^+-independent organic anion transporter, *Proc. Natl. Acad. Sci. USA* **91**:133–137.

Kakyo, M., Sakagami, H., Nishio, T., Takai, D., Nakagomi, R., Tokui, T., Naitoh, T., Matsuno, S., Abe, T. and Yawo, H., 1999, Immunohistochemical distribution and functional characterization of an organic anion transporting polypeptide 2 (oatp2), *FEBS-Lett.* **26**:343–346.

Kanai, N., Lu, R., Satriano, J. A., Bao, Y., Wolkoff, A. W., and Schuster, V. L., 1995, Identification and characterization of a prostaglandin transporter, *Science* **268**:866–869.

Kanai, N., Lu, R., Bao, Y., Wolkoff, A. W., and Schuster, V. L., 1996, Transient expression of oatp organic anion transporter in mammalian cells: Identification of candidate substrates, *Am. J. Physiol.* **270**:F319–F325.

Kelly, R. A., and Smith, T. W., 1992, Recognition and management of digitalis toxicity, *Am. J. Cardiol.* **69**:108G–119G.

Kudrycki, K. E., Newman, P. R., and Shull, G. E., 1990, cDNA cloning and tissue distribution of mRNAs for two proteins that are related to the band 3 Cl^-/HCO_3^- exchanger, *J. Biol. Chem.* **265**:462–471.

Kullak-Ublick, G. A., Hagenbuch, B., Stieger, B., Schteingart, C. D., Hosmann, A. F., Wolkoff, A. W., and Meier, P. J., 1995, Molecular and functional characterization of an organic anion transporting polypeptide cloned from human liver, *Gastroenterology* **109**:1274–1282.

Liang, R., Fei, Y.-J., Presad, P. D., Ramamoorthy, S., Han, H., Yang-Feng, T. L., Hediger, M. A., Ganapathy, V., and Leibach, F. H., 1995, Human intestinal H^+/peptide cotransporter: Cloning, functional expression, and chromosomal localization, *J. Biol. Chem.* **270**:6456–6463.

Lindsey, A. E., Schneider, K., Simmons, D. M., Baron, R., Lee, B. S., and Kopito, R. R., 1990, Functional expression and subcellular localization of an anion exchanger cloned from choroid plexus, *Proc. Natl. Acad. Sci. USA* **87**:5278–5282.

Lucus, M., 1983, Determination of acid surface pH *in vivo* in rat proximal jejunum, *Gut* **24**:734–739.

Masuda, S., Saito, H., Nonoguchi, H., Tomita, K., and Inui, K., 1997, mRNA distribution and membrane localization of the OAT-K1 organic anion transporter in rat renal tubules, *FEBS Lett.* **407**:127–131.

Masuda, S., Ibaramoto, K., Takeuchi, A., Saito, H., Hashimoto, Y. and Inui, K. 1999, Cloning and functional characterization of a new multispecific organic anion transporter, OAT-K2 in rat kidney, *Mol. Pharmacol.* **55**:743–752.

Mayer, U., Wageneaar, E., Beijnen, J. H., Smit, J. W., Meijer, D. K. F., van Asperen, J., Borst, P., and Schinkel, A. H., 1996, Substantial excretion of digoxin via the intestinal mucosa and prevention of long-term digoxin accumulation in the brain by the mdr1a P-glycoprotein, *Br. J. Pharmacol.* **119**:1038–1044.

McDermott, J. C., and Bonen, A., 1994, Lactate transport in rat sarcolemmal vesicles and intact skeletal muscle, and after muscle contraction, *Acta Physiol. Scand.* **151**:17–28.

Min, A. D., Johansen, K. L., Campell, C. G., and Wolkoff, A. W., 1991, Role of chloride and intracellular pH on the activity of the rat hepatocytes organic anion transporter, *J. Clin. Invest.* **87**:1496–1502.

Miyamoto, K., Shiraga, T., Yamamoto, H., Haga, H., Taketani, Y., Morita, K., Tamai, I., Sai,

Y., and Tsuji, A., 1996, Sequence, tissue distribution and developmental changes in rat intestinal oligopeptide transporter, *Biochim. Biophys. Acta* **1305**:34–38.

Muller, M., and Jansen, P. L., 1997, Molecular aspects of hepatobiliary transport, *Am. J. Physiol.* **276**:G1285–G1303.

Nijssen, H. M. J., Pijning, T., Meijer, D. K. F., and Groothuis, G. M. M., 1991, Mechanistic aspects of uptake and sinusoidal efflux of dibromosulfophothalein in the isolated perfused liver, *Biochem. Pharmacol.* **42**:1997–2002.

Noe, B., Hagenbuch, B., Stieger, B., and Meier, P. J., 1997, Isolation of multispecific organic anion and cardiac glycoside transporter from rat brain, *Proc. Natl. Acad. Sci. USA* **94**:10346–10350.

Ogihara, T., Tamai, I., and Tsuji, A., 1999, Structural characteriztion of substrates for the anion exchange transporter in Caco-2 cells, *Pharm. Res.*, in press.

Oude Elferink, R. P. J., Meier, D. K. F., Kuipers, F., Jansen, P. L. M., Groen, A. K., and Groothuis, G. M. M., 1995, Hepatobiliary secretion of organic compounds; molecular mechanisms of membrane transport, *Biochim. Biophys. Acta* **1241**:215–268.

Pardridge, W. M., 1983, Brain metabolism: A perspective from the blood–brain barrier, *Physiol. Rev.* **63**:1481–1535.

Pardridge, W. M., 1986, Blood–brain barrier transport of nutrients, *Nutr. Rev.* **44**:15–25.

Paulusma, C. C., Bosma, P. J., Zaman, G. J., Bakker, C. T., Otter, M., Scheffer, G. L., Scheper, R. J., Borst, P., and Oude Elferink, R. P., 1996, Congenital jaundice in rats with a mutation in a multidrug resistance-associated protein gene, *Science* **271**:1126–1128.

Price, N. T., Jackson, V. N. and Halestrap, A. P., 1998, Cloning and sequencing of four new mammalian monocarboxylate transporter (MCT) homologues confirms the existence of a transporter family with an ancient past, *Biochem. J.* **329**:321–328.

Pritchard, J. B., and Miller, D. S., 1993, Mechanisms mediating renal secretion of organic anions and cations, *Physiol. Rev.* **73**:765–796.

Proost, J. H., Nijssen, H. M. J., Strating, C. B., Meijer, D. K. F., and Groothuis, G. M. M., 1993, Pharmacokinetic modeling of the sinusoidal efflux of anionic ligands from the isolated perfused rat liver. The influence of albumin, *J. Pharmacokinet. Biopharm.* **21**:375–349.

Saito, H., Masuda, S., and Inui, K., 1996, Cloning and functional characterization of a novel rat organic anion transporter mediating basolateral uptake of methotrexate in the kidney, *J. Biol. Chem.* **271**:26719–20725.

Saltin, L. M., Amin, V., and Wolkoff, A. W., 1997, Organic anion transporting polypeptide mediate organic anion/HCO_3^- exchange, *J. Biol. Chem.* **272**:26340–26345.

Schinkel, A. H., Wagenaar, E., van Deemter, L., Mol, C. A. A. M., and Borst, P., 1995, Absence of the mdr1a P-glycoprotein in mice affects tissue distribution andpharmacokinetics of dexamethasone, digoxin, and cyclosporin A, *J. Clin. Invest.* **96**:1698–1705.

Sekine, T., Watanabe, N., Hosoyamada, M., Kanai, Y., and Endou, H., 1997, Expression cloning and characterization of a novel multispecific organic anion transporter, *J. Biol. Chem.* **272**:18526–18529.

Shimada, H., Moewes, B., and Bucrckhardt, G., 1987, Iindirect coupling to Na^+ of p-aminohippuric acid uptake into rat renal basolateral membrane vesicles, *Am. J. Physiol.* **253**:F795–F801.

Simanjuntak, M. T., Terasaki, T., Tamai, I., and Tsuji, A., 1991, Participation of monocarboxylic anion and bicarbonate exchange system for the transport of acetic acid and monocarboxylic acid drugs in the small intestinal brush-border membrane vesicles, *J. Pharmacobio-Dyn.* **14**:501–508.

Takanaga, H., Tamai, I., and Tsuji, A., 1994, pH-dependent and carrier-mediated transport of salicylic acid across Caco-2 cells, *J. Pharm. Pharmacol.* **46**:567–570.

Takanaga, H., Tamai, I., Inaba, S., Sai, Y., Higashida, H., Yamamoto, H., and Tsuji, A., 1995, cDNA cloning and functional characterization of rat intestinal monocarboxylate transporter, *Biochem. Biophys. Res. Commun.* **217**:370–377.

Takanaga, H., Maeda, H., Yabuuchi, H. Tamai, I., Higashida, H. and Tsuji, A., 1996, Nicotinic acid transport mediated by pH-dependent anion antiporter and proton cotransporter in rabbit intestinal brush-border membrane, *J. Pharm. Pharmacol.* **48**:1073–1077.

Tamai, I., and Tsuji, A., 1987, Transport mechanism of cephalexin in isolated hepatocytes, *J. Pharmacobio-Dyn.* **10**:632–638.

Tamai, I., and Tsuji, A., 1996a, Carrier-mediated approaches for oral drug delivery, *Adv. Drug Deliv. Rev.* **20**:5–32.

Tamai, I., and Tsuji, A., 1996b, Drug delivery through the blood–brain barrier, *Adv. Drug Deliv. Rev.* **19**:401–424.

Tamai, I., Maekawa, T., and Tsuji, A., 1990, Membrane potential-dependent and carrier-mediated transport of cefpiramide, a cephalosporin antibiotic, in canalicular rat liver plasma membrane vesicles, *J. Pharmacol. Exp. Ther.* **253**:537–544.

Tamai, I., Takanaga, H., Maeda, H., Ogihara, T., Yoneda, M., and Tsuji, A., 1995a, Proton-cotransport of pravastatin across intestinal brush-border membrane, *Pharmaceut. Res.* **12**:1727–1732.

Tamai, I., Takanaga, H., Maeda, H., Sai, Y., Ogihara, T., Higashida, H., and Tsuji, A., 1995b, Participation of a proton-cotransporter, MCT1, in the intestinal transport mechanism for monocarboxylic acids, *Biochem. Biophys. Res. Commun.* **214**:482–489.

Tamai, I., Takanaga, H., Maeda, H., Yabuuchi, H., Sai, Y., Suzuki, Y., and Tsuji, A. 1997, Intestinal brush-border membrane transport of monocarboxylic acids mediated by proton-coupled transport and anion antiport mechanism, *J. Pharm. Pharmacol.* **49**:108–112.

Tamai, I., Sai, Y., Ono, A., Kido, Y., Yabuichi, H., T. H., Satoh, E., Ogihara, T., Amano, O., Iseki, S. and Tsuji, A., 1999, Immunohistochemical and functional characterization of pH-dependent intestinal absorption of organic weak acids by monocarboxylic acid transporter MCT1, *J. Pharm. Pharmacol.* In press.

Tiruppathi, C., Balkovetz, D. F., Ganapathy, V., Miyamoto, Y., and Leibach, F. H., 1988, A proton gradient, not a sodium gradient, is the driving force for active transport of lactate in rabbit intestinal brush-border membrane vesicles, *Biochem. J.* **256**:219–223.

Tokui, T., Nakai, D., Nakagomi, R., Yawo, H., Suzuki, H., Abe, T. and Sugiyama, Y., 1999, Pravastatin, an HMG-Co A reductase inhibitor, is transported by rat organic anion transporting polypeptide, oatp2. *Pharm. Res.* **16**:904–908.

Torres, A. M., Lunazzi, G. C., Stremmel, W., and Tiribelli, C., 1993, Bilitranslocase and sulsobromophthalein-binding protein are both involved in the hepatic uptake of organic anions, *Proc. Natl. Acad. Sci. USA* **90**:8136–8139.

Tsuji, A., and Tamai, I., 1996, Carrier-mediated intestinal transport of drugs, *Pharmaceut. Res.* **13**:963–977.

Tsuji, A., and Tamai, I., 1998, Blood–brain barrier transport of drugs, in: *An Introduction to the Blood–Brain Barrier* (W. M. Pardridge, ed.), Plenum Press, New York, pp. 238–247.

Tsuji, A. and Tamai, I., 1999, Carrier-mediated or specialized transport of drugs across the blood–brain barrier, *Adv. Drug Deliv. Rev.* **36**:277–299.
Tsuji, A., Terasaki, T., Takanosu, K., Tamai, I., and Nakashima, E., 1986, Uptake of benzylpenicillin, cefpiramide, and cefazolin by freshly prepared rat hepatocytes, *Biochem. Pharmacol.* **35**:151–158.
Tsuji, A., Simanjuntak, M. T., Tamai, I., and Terasaki, T., 1990, pH-dependent intestinal transport of monocarboxylic acids: Carrier-mediated and H^+-cotransport mechanism versus pH-partition hypothesis, *J. Pharmaceut. Sci.* **79**:1123–1124.
Tsuji, A., Takanaga, H., Tamai, I., and Terasaki, T., 1994, Transcellular transport of benzoic acid across Caco-2 cells by a pH-dependent and carrier-mediated transport mechanism, *Pharmaceut. Res.* **11**:30–37.
Turner, R. H., Mehta, C. S., and Benet, L. Z., 1970, Apparent directional permeability coefficients for drugs ions: *In vitro* intestinal perfusion studies, *J. Pharm. Sci.* **59**:590–595.
Uchino, H., Tamai, I., Yabuuchi, H., China, K., Miyamato, K., Takeda, E., and Tsuji, A., 1999, Faropenem transport across the renal epithelial luminal membrane via inorganic phosphate transporter Npt1. *Antimicrob. Agents Chemother.* in press.
Werner, A., Moore, M. L., Mantei, N., Biber, J., Semenza, G., and Murer, H., 1991, Cloning and expression of cDNA for a Na/Pi cotransport system of kidney cortex, *Proc. Natl. Acad. Sci. USA* **88**:9608–9612.
Wolkoff, A. W., 1996, Hepatocellular sinusoidal membrane organic transport and transporters, *Semin. Liver Dis.* **16**:121–127.
Wolkoff, A. W., Samuelson, A. C., Johansen, K. L., Nakata, R., Withers, D. M., and Sociak, A., 1987, Influence of Cl^- on organic anion transport in short-term cultured rat hepatocytes and isolated perfused rat liver, *J. Clin. Invest.* **79**:1259–1268.
Yabuuchi, H., Tamai, I., Sai, Y., and Tsuji, A., 1998a, Possible role of anion exchanger AE2 as the intestinal monocarboxylic acid/anion antiporter, *Pharmaceut. Res.* **15**:409–414.
Yabuuchi, H. Tamai, I., Morita, K., Kouda, T., Miyamoto, K., Takeda, E. and Tsuji, A. 1998b, Hepatic sinusoidal membrane transport of anionic drugs mediated by anion transporter Npt1, *J. Pharmacol. Exp. Ther.* **286**:1391–1396.
Yamazaki, M., Suzuki, H., and Sugiyama, Y., 1996, Recent advances in carrier-mediated hepatic uptake and biliary excretion of xenobiotics, *Pharmaceut. Res.* **13**:497–513.
Yamamoto, A., Sakane, T., Shibukawa, M., Hashida, M., and Sezaki, H., 1991, Absorption and metabolic characteristics of *p*-aminobenzoic acid and its isomer, *m*-aminobenzoic acid, from the rat small intestine, *J. Pharmaceut. Sci.* **80**:1067–1071.
Yi, J. R., Lu, S. L., Fernandez-Checa, J. C., and Kaplowitz, N., 1995, Expression cloning of the cDNA for a polypeptide associated with rat hepatic sinusoidal reduced glutathione transport: Characteristics and comparison with the canalicular transporter, *Proc. Natl. Acad. Sci. USA* **92**:7690–7694.

17

Vitamin B$_{12}$ Transporters

Gregory J. Russell-Jones and David H. Alpers

1. INTRODUCTION

Oral administration of peptide and protein pharmaceuticals so that they are absorbed from the intestinal lumen and can elicit their pharmacological effect systemically is a problem that has vexed the pharmaceutical industry for many years. Apart from the obvious problems of protecting the pharmaceutical from proteolysis in the intestine, perhaps the biggest barrier to oral delivery is that due to the intestinal wall itself. Thus, while the intestinal epithelial cells possess carriers and transporters that are highly efficient in the uptake of the small products of digestion, such as vitamins, minerals, and amino acids, the epithelial cell layer presents itself as an almost impenetrable barrier to peptides larger than five or six amino acids in size.

Normally the absorption of most nutrients occurs via a variety of different mechanisms. Fat-soluble materials such as vitamins A, D, E and K are generally solubilized in mixed micelles (containing bile salts, phospholipids, monoglycerides, and fatty acids) from which they are subsequently absorbed. Water-soluble vitamins such as vitamin C, thiamine, nicotinic acid, riboflavin, pyridoxine, biotin, pantothenic acid, and folic acid are absorbed via facilitated diffusion. Amino acids, on the other hand are absorbed by active transport with a separate transport system for basic, neutral, and imino acids. Many low-molecular-weight antibiotics "piggy-back" onto the amino acid transport system and are taken up due to simi-

---

Gregory J. Russell-Jones • C/- Biotech Australia Pty Ltd, Roseville, NSW, 2069 Australia. *David H. Alpers* • Gastroenterology Division, Washington University School of Medicine, St. Louis, Missouri 63110.

Membrane Transporters as Drug Targets, edited by Amidon and Sadée. Kluwer Academic/Plenum Publishers, New York, 1999.

larities in structure to amino acids (Russell-Jones, 1999; Walter *et al.,* 1996: Swaan and Tukker, 1997).

In contrast to the vitamins mentioned above, vitamin B_{12} (VB_{12}, cobalamin, Cbl) is a large, water-soluble molecule (molecular weight 1356) (for structure see Fig. 1) which cannot be absorbed directly from the intestine, as it is too big to diffuse across the intestinal wall. Its absorption occurs via a markedly different mechanism from that of the other vitamins and involves several transport proteins each having different functions in uptake. It is these vitamin B_{12} transporters that are the subject of this review. As will be shown in the review, these transporters have particular relevance in the field of oral peptide and protein delivery, as recently it has been shown that it is possible to utilize the uptake mechanism for VB_{12} to enhance the oral uptake of various peptide and protein pharmaceuticals. Thus, Russell-Jones and co-workers have shown that this uptake system can potentially increase the oral uptake of molecules such as leutinizing hormone-releasing hormone (LHRH) analogues, (α-interferon, erythropoietin (EPO), granulocyte colony-stimulating factor

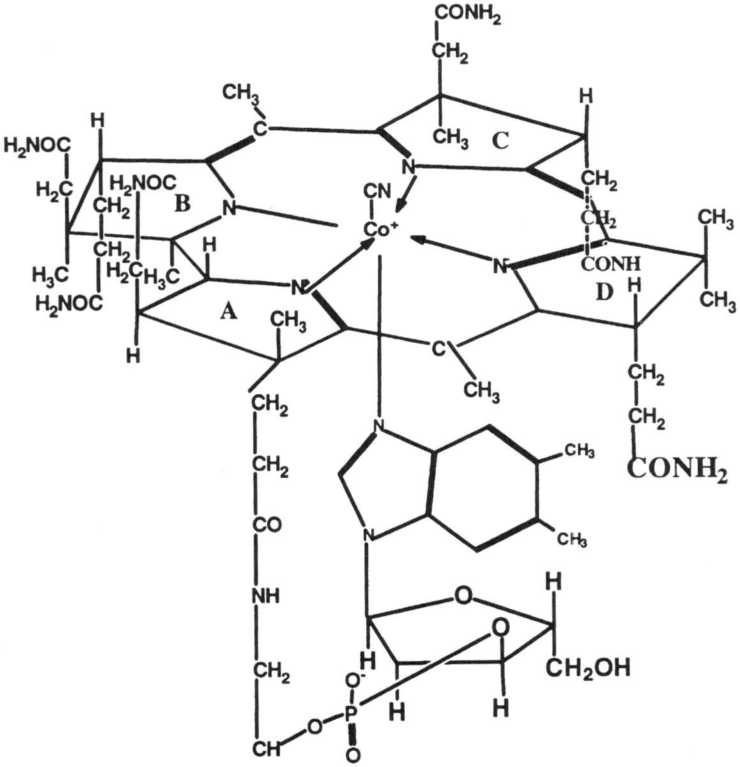

Figure 1. Structure of vitamin B_{12}.

(G-CSF), consensus interferon, and nanoparticles which have been covalently linked to the VB_{12} molecule (Russell-Jones, 1996a,b). It is the purpose of this review to describe the structural characteristics of the various transport proteins involved in the uptake of VB_{12} and to define the relative importance of each of the proteins in the absorption of VB_{12} and how this ultimately relates to the potential use of the uptake system for the oral delivery of pharmaceuticals.

2. GENERAL MECHANISM OF DIETARY ABSORPTION OF VITAMIN B_{12}

Before VB_{12} is absorbed it must first be released from food by the action of pepsin in the stomach. The VB_{12} is then bound by a corrin-specific transport protein, haptocorrin (Hc, non-intrinic factor, IF), that is secreted into saliva and is able to bind to VB_{12} in the acid environment of the stomach. The functional role of Hc appears to be to protect the vitamin from acid hydrolysis and possibly to protect the vitamin from being "scavenged" by intestinal fauna. The [Hc:VB_{12}] complex passes out of the stomach, and pancreatic proteases (possibly trypsin and chymotrypsin) present in the duodenum cleave the Hc molecule and release the VB_{12}. A second carrier protein, intrinsic factor (IF), then binds the VB_{12} and remains complexed to the vitamin until it reaches the ileum where the [IF:VB_{12}] complex is recognized by an IF-specific receptor (IF receptor, IFR) located on the luminal surface of the intestinal cells lining the gut wall. Binding of the [IF:VB_{12}] complex to the IFR triggers the internalization of the complex, presumably due to receptor-mediated endocytosis. Once inside the cell the IF is cleaved by an intracellular protease, cathepsin L, and the VB_{12} is released. The free VB_{12} is then bound by another transport protein, transcobalamin II (TCII), and the [TCII:VB_{12}] complex is subsequently transported across the cell and released into the circulation (Allen and Majerus, 1972; Booth *et al.*, 1957; Fyfe *et al.*, 1991, 1992; Guéant *et al.*, 1992; Linnell, 1975; Okuda, 1960; Ramanujam *et al.*, 1991a).

3. STRUCTURE AND BIOLOGY OF THE VB_{12} TRANSPORT PROTEINS

3.1. Haptocorrin (Hc Cobalophilin, R-Binders, R Protein, Nonintrinsic Factor)

Hc is produced by many cells including those in tissues deriving from foregut anlage, white blood cells, leucocytes (Simons and Wever, 1966), amnion lining cells (Stenman, 1974), mammary glands (Stenman, 1975), and accessory digestive glands (Lee et al., 1989b). In the salivary glands and the pancreas it is largely the ductal cells that contain the protein, but in the rat Hc is found in acini as well (Allen and Mehlman, 1973). Hc is found in the secretions from these organs, and can also be detected in tissues (such as intestine), where it is thought to be taken up by endocytosis (Hu et al., 1991).

3.1.1. SYNTHESIS AND SECRETION

Hc (M_r 66,000) is synthesized in the salivary glands and is secreted into saliva and can be found bound to Cbl in the stomach, where it is postulated to function to protect Cbl from acid degradation in the milieu of the stomach. At the lower pH encountered in the stomach the higher binding affinity of Hc for VB_{12} means that VB_{12} present in the stomach would be preferentially bound by Hc rather than IF (Sourial and Waters, 1991). Hc found in the stomach and small intestine is derived from salivary glands or foregut tissue. Growth hormone (GH) stimulates growth of submandibular glands in dwarf rats, and triggers upregulation of Hc, synergic with the GH-stimulated increase in IF (Lobie and Waters, 1997). This protein is also found to be secreted from parietal cells of all species examined, apart from hog (Lee et al., 1989). Hc is also found localized within granules in white cells, from which it can be released by lithium (Scott et al., 1974) and by the calcium ionophore A23187 (Simon et al., 1976). It is also localized to granules of pancreatic acinar cells in the rat, but is not in granules in other cell types from foregut tissues (Lee et al., 1989a). The Hc derived from foregut tissues is largely unsialylated (transcobalamin I, TCI), different from the situation in white cells which produce TCIII (Simon et al., 1976; Stenman, 1974).

After passing through the stomach, Hc undergoes degradation by pancreatic proteases (Allen et al., 1978; Marcoullis et al., 1980). In the newborn animal (and in the mouse and rat prior to weaning) synthesis and secretion of pancreatic enzymes is quite limited, and much of the Hc derived from milk, salivary, gastric, or pancreatic sources survives intact in the intestinal lumen. Although Hc is degraded by pancreatic proteases, Cbl bound to Hc is relatively resistant to removal by human pancreatic juice (Carmel et al., 1983). In humans when pancreatic proteases are limited a significant percentage of dietary Cbl is found complexed with Hc in the upper jejunum (Marcoullis et al., 1980).

Recently it has been found that this protein is also secreted from liver cells and has been found to be secreted *in vitro* by cell lines such as the Opossum Kidney cell line (OK). These cells synthesize and secrete two Cbl-binding proteins, TCII and Hc (M_r 66,000 and 43,000, respectively). The addition of colchicine, a microtubule-disruptive drug, increases the apical, but not the basolateral secretion of both TCII and Hc. In contrast, Caco-2 cells appear to only synthesize TCII and not HC (Ramanujam et al., 1991a).

3.1.2. STRUCTURE AND FUNCTION

Haptocorrins of quite similar amino acid and sugar content, molecular mass, and amino-terminal sequence have been isolated from hog gastric mucosa (Allen and Mehlman, 1973), plasma (Burger *et al.,* 1975a), and amniotic fluid (Stenman, 1974). Human TCI (Johnson *et al.,* 1989) and hog gastric Hc (Hewitt *et al.,* 1990) have been sequenced by isolation of cDNA clones. The primary structure of these two Hc's is only 45% identical. However, there are at least six areas of extensive homology (not identity) distributed evenly along the sequences of IF, Hc, and transcobalamin II (TCII) (Fig. 2) (Li *et al.,* 1993). Thus, it is likely that a folded structure will be essential for the Cbl (and probably receptor) binding to Hc, as well as to other Cbl-binding proteins. The purpose of the extensive glycosylation (30–40% w/w) of Hc is not known, although apo-Hc is lost from the lumen of the suckling pig intestine to a much greater extent than for holo-Hc (with bound Cbl). There is a single Cbl-binding site per molecule of Hc (B. Seetharam and Alpers, 1994).

Although Hc accounts for about 80% of Cbl bound in human serum (Hall, 1977), the function of Hc is not clear, although it is commonly elevated in conditions of hyperleukocytic syndromes (Johnson *et al.,* 1989). Methylcobalamin accounts for most of the Cbl bound to serum Hc (TCI and TCIII), whereas 5'-deoxyadenosylcobalamin is the major derivative found on TCII, and the majority of the binding capacity of Hc is saturated (Nexø and Anderson, 1977). It has also been suggested that Hc may act to scavenge cobalamin analogues produced by bacteria and prevent their uptake from the intestine (Lee *et al.,* 1989b). This protein differs from other VB$_{12}$-binders in that it also binds corrins other than cobalamins (Nexø

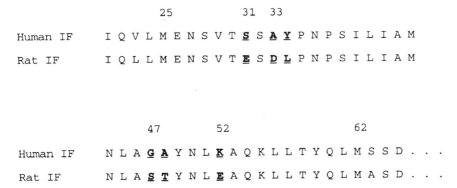

Figure 2. Sequence of the N-terminus of human and rat IF showing differences in sequence in the putative IF receptor binding site. The N-terminal sequence of rat and human IF is as described by Tang *et al.* 1992.

and Olesen, 1981). Other suggested functions for Hc include a role in regulating the microflora of the upper gastrointestinal tract and clearance from the body of Cbl derivatives that do not support the two essential Cbl-requiring reactions (B. Seetharam and Alpers, 1994).

3.1.3. UPTAKE AND TRANSPORT OF CBL BOUND TO HC

While Hc has not been implicated in intestinal uptake of VB_{12}, there is evidence that neonatal enterocytes may possess the asialoglycoprotein receptor that may be able to take up Hc nonspecifically (Hu *et al.*, 1991). Cbl recovered from intracellular extracts of pig (Marcoullis *et al.*, 1977) and dog (Marcoullis *et al.*, 1981) mucosa was found entirely protein-bound, equally divided between IF and Hc. This suggests that when IF is rate-limiting, as occurs in the newborn or in patients with IF deficiency, the intact Hc–Cbl complex that escapes proteolysis in the lumen may be taken up either by nonspecific pinocytosis (Boass and Wilson, 1963) or by a receptor for Hc–Cbl that is expressed in suckling (Hu *et al.*, 1991; Trugo *et al.*, 1985) and adult (Hu *et al.*, 1991; Mu *et al.*, 1993) intestine, as well as in many other tissues (Mu *et al.*, 1993). This receptor may be the asialoglycoprotein receptor (ASGP-R), which is expressed in intestine as well as liver (Hu *et al.*, 1991). Moreover, the minor component (47–50 kDa) of the receptor is expressed apically in the intact rat intestine (Hu *et al.*, 1991) as well as apically in one colon carcinoma cell line, HT-29 (Mu *et al.*, 1997). This situation is different from the hepatocyte (Geffen and Spiess, 1992; Schwartz, 1990) and another colon carcinoma cell line, Caco-2 (Mu *et al.*, 1994), in which the receptor is basolaterally located and the major subunit (43–46 kDa) is the predominant subunit. The apical location of the minor subunit in rat intestine and in HT-29 cells does suggest, however, that natural ligands may bind to this receptor. The binding maximum of Hc–Cbl complexes to rat intestinal brush border membranes is 4.6 fmole/mg protein, which is about 10% that of the IF–Cbl complex (Hu *et al.*, 1991). Cbl absorption in humans occurs in the absence of IF, but is not clearly dose dependent (Berlin *et al.*, 1968).

The presence of the ASGP-R on the apical membrane adds a second possible system for lumenal uptake to that of IF–Cbl (see below), although it functions at only 10% capacity of the latter system. A proportion of TCII receptors has also been found to be located apically in renal tubular cells, although 90% of the receptor activity remained basolateral (Bose *et al.*, 1995). Although these systems appear to be limited in capacity, Hc is involved in the recruitment of Cbl from the body stores to the tissues by a high-capacity uptake system in the liver, possibly via the ASGP-R. The Hc–cbl complexes are rapidly internalized and degraded in the lysosomes with subsequent release of Cbl (Burger *et al.*, 1975a). This system can clear 3–9 µg of Cbl into human bile each day, which is much less than the production of Hc by granulocytes, which has been estimated at 150 μg/day (B. Seetharam and Alpers, 1994).

The function of serum transcobalamin I is not known, although it is commonly elevated in conditions of hyperleukocytic syndromes (Johnston *et al.*, 1989).

3.2. Intrinsic Factor

3.2.1. OCCURRENCE AND DISTRIBUTION

IF, like Hc, is produced by tissues derived from the foregut, although the gastric mucosa is the predominant tissue (Glass, 1974). In the rat, IF is produced primarily in chief cells (Schepp *et al.*, 1983a), but also significantly in parietal cells (Lee *et al.*, 1989a). In the mouse, IF is found only in chief cells (Hoedemaeker *et al.*, 1966; Lorenz and Gordon, 1993). In the dog the pancreas is the principal source of IF (Batt *et al.*, 1989), which is produced by the ductular cells in response to appropriate hormonal stimuli (e.g., cholecystokinin) (Simpson *et al.*, 1993). In the dog (Valliant *et al.*, 1990) and human stomach (Howard *et al.*, 1996) IF is found in a subset of enteroendocrine cells.

3.2.2. SYNTHESIS AND SECRETION

All substances that increase acid secretion also increase IF production in those species in which the parietal cell is the principal source of IF. Substances that have been shown effective in humans include histamine (Jeffries and Sleisenger, 1965), gastrin (Irvine, 1965), and methacholine (Vatn *et al.*, 1975). The regulated step in IF secretion appears to be secretion from the cell, as the rate of IF synthesis is constant, independent of the addition of stimulators of secretion (Serfilippi and Donaldson, 1986; Donaldson *et al.*, 1967), and as IF mRNA content does not change during intense secretory stimulation (Dieckgraefe *et al.*, 1989). Both prostaglandin E2 (Zucker *et al.*, 1988) and somatostatin (Oddsdotter *et al.*, 1987) inhibit histamine-stimulated IF secretion. Drugs that directly affect acid secretion in human can markedly inhibit IF secretion, including histamine2 receptor antagonists (Binder and Donaldson, 1978) and proton pump inhibitors (Kitting *et al.*, 1985).

The regulation of IF secretion differs in rodents because the chief cell is the major source of IF, although about 3% of parietal cells produce IF under normal circumstances (Shao *et al.*, 1998). In the rat, carbachol and dibutyryl cAMP, not histamine or gastrin, stimulate IF secretion (Schepp *et al.*, 1983b, 1984).

3.2.3. STRUCTURE AND FUNCTION

Gastric intrinsic factor is a 399-amino acid glycoprotein essential for the majority of uptake of Cbl from the intestine of most vertebrates. IF purified by affinity chromatography using a Sepharose-coupled monocarboxylic acid derivative of Cbl has a molecular mass of 45–47 kDa, but is retarded on gel filtration (59–66 kDa), perhaps due to its carbohydrate content of 15% (Allen and Mehlman, 1973). The molecular mass predicted from recombinant rat IF produced in CHO cells is 46 kDa (Dieckgraefe *et al.*, 1988), while rat IF produced in the baculovirus system was observed to be 47 kDa (Gordon *et al.*, 1992). Thus, the 45- to 47-kDa size appears to be correct, although the native IF has been reported to be 59 kDa (Guéant *et al.*, 1985). The sequence of human IF deduced from its cDNA clone is 80% identical with that of rat IF (see Table III). Nonglycosylated IF produced in a cell-free system (Dieckgraefe *et al.*, 1988) or in COS-1 cells by transfection and tunicamycin treatment bound cbl and the IF receptor equally to the glycosylated form (K_a in the range of 0.26–0.35 nM) (Gordon *et al.*, 1991).

The entire Cbl molecule appears to be important for IF binding. Binding of IF to cobalamin is dependent upon interactions with both the 5,6-dimethylbenzimidazole and the corrin ring (Fig. 1). Following binding to Cbl, IF has been shown to shrink as the molecule closes around the Cbl molecule (Lien *et al.*, 1974). This change in structure presumably accounts for the higher affinity of the [Cbl:IF] complex for the IF receptor than for the IF molecule alone. The ribazole fragment in the cobalamin side chain is needed to promote a conformational change allowing corrinoid binding (Andrews *et al.*, 1991). Cobamide binding to IF is strongly affected by the Co—N coordination bonds of the lower cobalt nucleotide ligands (Stupperich and Nexø, 1991).

Although there are six regions of high sequence homology between IF, Hc, and TCII, the predicted secondary structure of these proteins is not well conserved (Li *et al.*, 1993). These differences may account for the different binding affinity of these proteins for cobamides (Table I) or for their individual receptors. It is well established that there are two domains on IF, one for binding Cbl and the other for IF–Cbl receptor binding (Ardeman and Chanarin, 1963). The existence of these domains has been confirmed using monoclonal antibodies that inhibit Cbl, but not receptor binding (Smolka and Donaldson, 1990). Similarly, Tang and co-workers (1992) have shown that Cbl binding by IF is dependent upon the C-terminal fragment of IF, with the IF receptor-binding region at the N-terminus of the molecule. Data from Alpers and Russell-Jones (1999) also show that the Cbl-binding site is located in the C-terminal portion of the IF molecule. Thus, these workers isolated two fragments of IF (produced in *Pichia*) (M_r 28 and 32 kDa) which were found to bind to Cbl, but not to the IFR (Wen *et al.*, 1999). N-terminal sequence analysis of these fragments showed them to be C-terminal fragments of IF.

The N-terminal domain of IF appears to be more important for binding to the IF R, as expression of IF constructs in COS-1 cells which were lacking up to 25%

Table I.
Affinity of Different VB_{12}-Binders for VB_{12} and Various VB_{12} Analogues

	Affinity of given VB_{12}-binder (M^{-1})			
	Hc	IF	TcI	TcII
Cbl	1×10^{11a}	6×10^{9b}	3×10^{11b}	1.8×10^{11b}
bVB_{12} analogue	0.5^c	0.004^c	—	0.02^c
dVB_{12} analogue	0.1^c	0.0006^c	—	0.006^c
eVB_{12} analogue	0.6^c	0.3^c	—	0.3^c

[a]Schneider and Stroinski (1987, p. 320)
[b]Hippe and Olesen (1971).
[c]Binding of analogue relative to binding to Cbl.

of the C-terminal domain of IF produced a truncated IF with no demonstrable Cbl-binding activity (Gordon *et al.*, 1991) (Fig. 2). Loss of 12% of the C-terminal of IF produced in a cell-free system also abolished Cbl binding (Tang *et al.*, 1992). Although the IFs from different species are 80% identical, the conformation of these proteins may be quite different. In fact only rat IF can bind to the rat IF–Cbl receptor, while IF from dog and humans cannot. The amino terminus of IF (residues 25–62 from the mature end) is important for receptor binding, whereas a fragment containing the carboxy-terminal 232–338 residues (out of 399) does not support receptor binding (Tang *et al.*, 1992). Furthermore, altering the charge of the amino-terminal segment of IF by mutating E31 to R and D33 to R produces a protein with a 50- to 100-fold lower affinity for the receptor than the intact recombinant protein (Wen *et al.*, 1997). On the other hand, linear peptides identical to the sequence from amino acids 25–44 do not inhibit IF binding to its receptor, and creating a recombinant human IF mutant with residues identical to rat IF between amino acids 25 and 44 did not permit binding to the rat receptor (Wen *et al.*, 1997). Thus, it seems likely that the amino terminus of IF is very important in the configurational changes required for receptor binding, but that it itself is not sufficient to determine the binding.

3.2.4. CELLULAR UPTAKE AND UTILIZATION OF IF–CBL

During the process of uptake, IF binds to VB_{12} in the upper duodenum of the small intestine. The IF–VB_{12} complex survives the proteolytic environment within the duodenum and ileum due to the resistance of IF to acidic and neutral proteases (Ramanujam *et al.*, 1992). The IF–VB_{12} complex passes down the small intestine and is bound by an IF receptor located on the ileal epithelium. Binding appears to take place in the villous and not crypt cells (Kapadia and Essandoh, 1988). Apart from the ileal receptor for IF, significant IF–VB_{12} binding activity also has been found in brush border membranes isolated from rat placenta and kidney (Lee *et al.,* 1989b). The work of Fyfe *et al.* (1991) suggests that the intestinal and kidney IF receptors are the product of a single gene, although the kidney receptor is found in much greater abundance than the intestinal receptor (Lee *et al.,* 1989b; Stenman, 1974, 1975). After internalization by endocytosis, the intracellular fate of the IF–Cbl complex involves proteolysis and release of Cbl from IF. Cultured cell lines derived from either colon cancer cells such as Caco-2 (Ramanujam *et al.,* 1991a; Dix *et al.,* 1990) or HT-29 (Guéant *et al.,* 1992), or derived from proximal renal tubules of opossum (OK cells) (Ramanujam *et al.,* 1992) or pig (LLC-PK1 cells) (Gordon *et al.,* 1995) have been used to delineate the process. These studies have shown that rapid internalization of IF–Cbl occurs, and that lysosomotropic agents can inhibit transcytosis of Cbl. Some evidence suggests that liberation of Cbl and degradation of IF can occur independently of each other (B. Seetharam *et al.,* 1985) at pH 5.0. Prelysosomal vesicles have been implicated in this process. Radioautographic studies using [^{125}I]-IF demonstrated localization in endosomes of the apical cytoplasm, consistent with this interpretation (Guéant *et al.,* 1988). Moreover, IF is recovered within enterocytes (Marcoullis and Rothenberg, 1981), suggesting that IF degradation is a slow process. A half-life of 4 hr for degradation of internalized IF has been estimated from studies in Caco-2 cells (Dan and Cutler, 1994), and this time frame is consistent with the 2–4 hr needed to transfer Cbl from the luminal IF–Cbl complex to the serum TCII–Cbl complex (B. Seetharam and Alpers, 1994). The initial mechanism for intracellular IF degradation appears to be via cathepsin L activity (Gordon *et al.,* 1995). Following intracellular release of Cbl, it exits the cell bound to TCII (B. Seetharam and Alpers, 1994; Nicolas and Guéant, 1995).

All the Cbl bound to IF from the apical side of polarized epithelial cells is eventually transcytosed (Bose *et al.,* 1997). Evidence suggests that in the adult human the only uptake of VB_{12} from the intestine occurs via the IFR pathway of Cbl transport, as TCII has never been found in the gastrointestinal lumen and because patients with inherited disorders of IF or IFR develop Cbl deficiency.

Therapeutic ingestion of heterologous IF has produced antibodies to IF in the host (Nicolas and Guéant, 1995) that have inhibited uptake of IF and thence VB_{12}. These issues must be considerd before heterologous IF is used as a vehicle for drug delivery, or if modified Cbl is found to alter the intracellular processing of the IF–Cbl complex.

3.3. Intrinsic Factor Receptor

3.3.1. LOCATION

The IF receptor (IFR) is expressed on the apical surface of ileal mucosal cells of all higher mammals and mediates the endocytosis and subsequent transport of

IF-bound Cbl from the luminal to the basal surface of intestinal epithelial cells. In humans the IFR can be found along the terminal three-fifths of the small intestine (Hagedorn and Alpers, 1977).

3.3.2. SYNTHESIS AND SECRETION

The synthesis of the receptor in the endoplasmic reticulum begins in crypt enterocytes and continues to occur in cells throughout the villus. The receptor directly inserts into the microvillus pits, where it remains. The receptor (230–240 kDa) (B. Seetharam et al., 1988, Ramanujam et al., 1992) appears to be predominantly located on the outside of the cell, with a small, 36-kDa fragment in the membrane (Ramanujam et al., 1992; Levine et al., 1984). The IFR appears to be identical with gp280, a protein closely associated with the early endocytic system and necessary for preventing fetal malformations (B. Seetharam et al., 1997).

3.3.3. BINDING TO IFR

Binding of the IF–Cbl complex to the IF receptor is energy independent and requires Ca^{++} and a neutral pH (Levine et al., 1984; Beesley and Bacheller, 1980). Following binding of IF to the IFR, the IFR undergoes an irreversible conformational change (Marcoullis et al., 1981). In all species except the rat, the IFR binds IF complexed to Cbl and binds free IF poorly, if at all (Mathan et al., 1974). In the rat, both free rat IF and rat IF–Cbl complexes bind to the rat IFR (B. Seetharam et al., 1983); however, the Cbl-complexed IF binds with a higher affinity than free IF. In contrast, free IF from species such as the hog, dog, and human is not able to bind to the rat IFR (B. Seetharam et al., 1983). The IFR has also been found to be present in kidney of humans, rats, and dogs (B. Seetharam et al., 1988, Ramanujam et al., 1990; 1992), where it is expressed at much higher levels than in the intestine. In the renal tissue the receptor is located on the apical surface membrane of proximal tubular cells. Within species the intestinal and renal receptors are immunologically identical, however the receptors vary from species to species (B. Seetharam et al.,. 1988). Similarly the receptor has been found in polarized Opossum Kidney cells (OK), where it is expressed at very high levels (Ramanujam et al., 1993), and human intestinal epithelial Caco-2 cells, where it is expressed at lower and somewhat more variable levels. Surface binding of Cbl–IF to the OK cell IFR is inhibited by growth of cells in tunicamycin, while Cbl transcytosis is inhibited by both tunicamycin and cerulenin (Ramanujam et al., 1994).

3.3.4. TRANSCYTOSIS

The process of transcytosis of IF-bound VB_{12} in OK cells is very slow (48 hr half-life—Ramanujam et al., 1994) in comparison to cells which have been treated with cerulenin or tunicamycin (48 hr) (Ramanujam et al., 1994), in which transcytosis occurs in 24 hr. Transcytosis of Cbl across OK cells is also inhibited by leupeptin and ammonium chloride (Ramanujam et al., 1992) due to inhibition of intracellular breakdown of IF. The IFR from different species has a similar molecular weight (230 kDa), but has distinct interspecies differences in binding affinity for different IFs, as well as different sensitivities to endoglycosidases. Thus, OK IFR exists as a single protein of $M_r \sim 230$ on sodium dodecyl sulfate–polyacrylamide gel electrophoresis (SDS–PAGE) (Ramanujam et al., 1993), which has a 10-fold higher affinity for the IF–Cbl complex formed with OK IF than that from rat IF.

3.4. Transcobalamin II

3.4.1. SYNTHESIS AND SECRETION

TCII is synthesized by both the intestinal epithelial cells as well as the vascular endothelium (Quadros et al., 1996b; Chanarin et al., 1978), liver (Green et al., 1976), fibroblasts (Green et al., 1976), and kidney (Li et al., 1994). The level achieved in human kidney reaches 14 times that found in liver (Li et al., 1994). Secretion (at least from fibroblasts) is blocked by cycloheximide and puromycin (Green et al., 1976). A TCII-like molecule is synthesized and secreted from HT 29 cells and Caco-2 cells (Schohn et al., 1991; Dix et al., 1990). The protein has a molecular weight of 44 kDa with an isoelectric point of 6.4, and is immunologically cross-reactive with human TCII (Tables I and II) (Schohn et al., 1991).

3.4.2. FUNCTION

Cbl in serum which is destined for uptake into all tissues and cells is bound to a plasma transporter, transcobalamin II (TCII), of molecular weight 43–45 kDa

Table II.
Characteristics of Various Cbl-Binding Proteins

Binding protein	M_r (kDa)	Isoelectric point	Affinity (CN–Cbl) (M^{-1})
Hc	58 (human)	4.5–5.2(saliva)[a] 3.0–5.0(stomach)[a]	1.5×10^{10}
IF	45 (human) 55 (hog)	4.12–5.6[b]	$1.5–5 \times 10^{10}$
TCI	60[c]	3.2–3.9[a]	3.0×10^{11}
TCII	44[d]	6.4	$1.0–3.0 \times 10^{11}$

[a]Schneider and Stroinski (1987, pp. 313–333).
[b]Schneider and Stroinski (1987, pp. 267–297).
[c]Rabbit serum Hc (Nexø and Olesen, 1981).
[d]Schohn et al, (1991).

(Quadros et al., 1996b; Li et al., 1994; Lindemans et al., 1989) (Table III). Uptake of the Cbl into tissues occurs via receptor-mediated endocytosis (RME) in which Cbl bound to TCII is bound by a surface receptor for TCII (TCIIR) (En-Nya et al., 1993; Regec et al., 1995). TcII is the only serum VB_{12}-binding protein known to promote uptake of VB_{12} by tissues and cells *in vivo* and *in vitro* (Green et al., 1976). TCII internalized by RME is degraded in the acidic environment of the lysosomes and free Cbl is transported out of the lysosomes and into the cytoplasm of the cell. Once in the cytoplasm the Cbl is converted into methyl-Cbl and 5'-deoxyadenyosyl-Cbl (Li et al., 1994). TCII has a high affinity for VB_{12} (10^{11} M^{-1}; Pathare et al., 1996).

Apart from its role in uptake of Cbl into the cells of the body, TCII also has a critical role in the uptake of VB_{12} from the intestine. Following uptake of IF–VB_{12} bound to the IFR by the enterocyte, the IF is cleaved by intracellular cathepsins and VB_{12} is released. While it is not known for sure, it is likely that the free VB_{12} is then bound to membrane-bound TCII within the cell and is then transported across the cell bound to TCII (Chanarin et al., 1978). While the possibility also exists that VB_{12} is bound by free TCII present on the basolateral surface of the cell, this seems unlikely, as a Cbl-binding protein would be required to direct the Cbl released from the IF to the basal surface of the cell.

3.5. Transcobalamin II Receptor

This receptor occurs as a noncovalent homodimer (molecular mass of 124 kDa) in the tissue plasma membranes of humans, rats, and rabbits (Bose et al.,

Table III.
Sequence of VB_{12}-Binders from Various Species[a]

	10	20	30	40
Human IF	STSQTQSSCSVPSAQEPLVNGIQVLMENSVTSSAYPNPSILIAMN			
Rat IF	STR AQRSCSVPPDQQPWVNGLQLLMENSVTESDLPNPSILIAMN			
Human TCI	EICEVSEENYIRLKPLLNTMIQSNYNRGTSAVNVVLSLK			
Human TCII	EMCEIPEMDSHLVEKLGQHLLPWMDRLSLEHLNPSIYVG			

	50	60	70	80	90
Human IF	LAGAYNLKAQKLLTYQL MSSDNNDLTIGHLGLTIMALTSSCR				
Rat IF	LASTYNLEAQKLLTYQL MASDSADLTNGQLALTIMALTSSCR				
Human TCI	LVGIQIQTLMQKMIQQIKYNVKSRLSDVSSGELALIILALGVCRN				
Human TCII	LRLSSLQAGTKEDLYLHSLKLGYQQCLLGSAFSEDDGDQGKPSMG				

	100	110	120	130
Human IF	DPGDKV	SILQRQMENWAPSSPNAEASAFYGPSLAI		
Rat IF	DPGSKV	SILQKNMESWTPSNLGAESSSFYGPALAI		
Human TCI	AEENLIYDYHLTDKLENKFQAEIENMEAHN GTPLTNYYQLSLDV			

(continued)

Table III. (*Continued*)
Sequence of VB$_{12}$-Binders from Various Species[a]

	100	110	120	130
Human TCII	QLALYLLALRANCEFVRGHKGDRLVSQLKW FLEDEDRAIGHHEH			

	140	150	160	170	180
Human IF	LALCQKNSEATLPIAVRFAKTLLAN SSPFNVDTGAMATLALT				
Rat IF	LALCQKNSEATLPIAVRFAKTLMME SSPFSVDTGAVATLALT				
Human TCI	LALCLFNGNYSTAEVVNHFTPENKNYYFGSQFSVDTGAMAVLALT				
Human TCII	EHKGHHTSTTQTGLGILALCLHQKRVRDSVVDKLLYAVEPFHQ				

	190	200	210	220
Human IF	CMYNKI PVGSEEGYRSLFGQVLKDIVEKISMKIKDNGIIGDI			
Rat If	CMYNRI PVGSQENYRDLFGQALKVIVDNISLRIDADGIIGDI			
Human TCI	CVKKSLINGQIKADEGSLKNISIYTKSLVEDILSEKKENGLIGNT			
Human TCII	GHHSVDTAAMAGLAFTCLKRSNFNPGRRQRITMAIRTVREEILKA			

	230	240	250	260	270
Human IF	YSTGLAMQALSVTPE PSKKEWNCKKTTDMILNEIKQGKFHNPMS				
Rat IF	YSTGLAMQALSVTPE QPTKEWDCEKTMYTILKEIKQGKFQNPMS				
Human TCI	FSTGEAMQALFVSSDYYNENDWNCQQTLNTVLTEISQGAFSNPNA				
Human TCII	QTPEGHFGNVYSTPLALQFLMTSPMPGAELGTACLKARVALLASL				

	280	290	300	310
Human IF	IAQILPSLKGKTYLDVPQVTCSPDHEVQPTLPSNPGPGPTSASNI			
Rat IF	IAQILPSLKGKTYLDVPQVTCGPDHEVPPTLTDYPTPVPTSISNI			
Human TCI	AAQVLPALMGKTFLDINKDSSCVSASGNFNISA DEPITVTPPDS			
Human TCII	QDGTQNALMISQLLPVLNHKTYIDLIFP DCLAPRVMLEPAAET			

	320	330	340	350	360
Human IF	TVIYTINNQLRGVELLFNETINVSVKSGSVLLVVLEEAQPKNP M				
Rat IF	TVIYTINNQLRGVDLLFNVTIEVSVKSGSVLLAVLEEAQRRNH M				
Human TCI	QSYISVNQSVRINETYF TNVTVLNGSVFLSVMEKAQKMNDTI				
Human TCII	IPQTQEIISVTLQVLSLLPPYRWSISVLAGSTVEDVLKKAHELGG				

	370	380	390	400
Human IF	FKFETTMTSWGLVVSSINNIAENVNHKTYWQFLSGV TPLNEGVAD			
Rat IF	FKFETTMTSWGLIVSSINNIAENVKHKTYWEFLSGK TPLGEQVAY			
Human TCI	FGFTMEERSWGPYITCIQGLCANNNDRTYWELLSGG EPLSQQAGS			
Human TCII	FTYETQASSSGPYLTSVMGKA AGEREFWQLLRDPNTPLLQGIAD			

	410
Human IF	YIPFNHEHITANFTQY
Rat IF	TIPFNYEHITANFTQY
Human TCI	TVVRNGENLEVRWSKY
Human TCII	YRPKDGETIELRLVSW

[a]Dissimilar sequences between rat and human IF are identified by underlining. Human IF, Hewitt *et al.* (1991); rat IF, Dieckgraefe *et al.* (1988) and Hewitt *et al.* (1991); human TCI, Johnston *et al.* (1989); human TCII Regec *et al.* (1995).

1997; Quadros et al., 1994, 1996b). The TCIIR is expressed in its dimeric form on both the apical and basal surfaces of epithelial cells such as those in the kidney, small intestine, placenta, and, at very low levels, the liver (Bose and Seetharam, 1994). There is an eightfold enrichment of the receptor on the basal surface in comparison with the apical surface (B. Seetharam et al., 1995). Formation and maintenance of intramolecular —S—S— bonds is important in ligand binding and trafficking to basolateral membranes (Bose and Seetharam, 1997). Studies by Bose and co-workers (1997) have shown that in the polarized human intestinal epithelial Caco-2 cell, the apically located receptor can initiate RME and transcytosis of Cbl–TCII via a nonlysosomal pathway such that the Cbl–TCII complex is secreted undigested from the basal surface. In contrast, the basolaterally located receptor initiates RME via a lysosomal pathway such that the TCII is degraded and the released Cbl is transferred to other cellular proteins and is then retained by the cell. Addition of chloroquine or leupeptin to cell cultures has no effect on apical-to-basal transport of TCII or Cbl; however, material endocytosed from the basal surface of the cell is not degraded in the presence of these two lysosomal inhibitors.

3.6. Cell Models of Vitamin B_{12} Transport

The study of the intracellular events involved in the transcytosis of VB_{12} has been greatly aided by the availability of a number of cell lines which have been shown to bind, internalize, and transcytose VB_{12} in an IF-dependent fashion. IFR expression and functional activity has been found on the polarized cells of intestinal origin such as the human colon carcinoma cell lines Caco-2 (Dix et al., 1990; Ramanujam et al., 1991a; 1992, Wilson et al., 1990; Hassan and Mackay, 1992; S. Seetharam et al., 1991) and HT29 (Guéant et al., 1992; Schohn et al., 1991). Expression has also been observed in two kidney lines, Opossum Kidney and the LLCPK1 line from porcine kidney (Ramanujam et al., 1991a, 1994). When these cells are cultured in vitro on permeable membranes they form a confluent monolayer with tight junctions between neighboring cells, which restricts nonspecific transport of small macromolecules. The cell cultures eventually polarize to form apical and basal surfaces and exhibit unidirectional, IF-dependent transport of VB_{12} from the apical to the basolateral chamber.

Studies using OK cells have shown that they secret two Cbl-binding proteins, Hc and TCII. These cells secrete both proteins into the apical medium, but only Hc into the basal medium (Ramanujam et al., 1994). Similarly the TCII receptor is expressed both basally and apically in Caco-2 cells (Bose et al., 1997) with a six- to sevenfold enrichment of the receptor in the basal side of the cells (Bose et al., 1997). Ramanujam and co-workers (1991a) measured 250 fmole of IF-dependent $^{57}CoVB_{12}$ transport in 4 hr in CaCo-2 cell cultures. This level of transport appears to be dependent upon the source of the cells and the culture method, since other groups have only been able to demonstrate transport of 40 fmoles of $^{57}CoVB_{12}$ in 18–21 hr (Dix et al., 1979; Wilson et al., 1990).

IF-dependent VB_{12} transport is much more stable and uniform in the OK and LL-CPK1 cell lines. These cell lines have a vastly different cell surface than both the CaCo-2 and HT29 cell lines as well as normal intestinal epithelial cells in that they do not possess microvilli. The OK and LL-CPK1 cell lines do, however, show a high degree of uniformity, as well as specific, IF-dependent unidirectional transport of VB_{12} from the apical to the basal surface of the cells (Ramanujam et al., 1991b).

OK cells have been shown to synthesize two VB_{12}-binding proteins, haptocorrin (Hc; 66 kDa) and TcII (43 kDa). Ramanujam and co-workers (1991b) have shown that these cells secrete both proteins from the apical surface of the cells, while only TcII is secreted from the basal surface (Ramanujam et al., 1991a). Roughly equal levels of VB_{12}-binders are secreted into the apical and basal media, with a twofold higher level of TcII than Hc. In contrast, Ramanujam and co-workers observed that Caco-2 cells only secrete TcII into the basal medium of transwell cultures with no apparent secretion of VB_{12}-binders into the apical medium (Ramanujam et al., 1991b). In contrast, Hc was found to be secreted into the cell medium of HT-29 cell cultures (Schohn et al., 1991).

4. THE USE OF VITAMIN B_{12} FOR TRANSPORT OF PHARMACEUTICALS

Recently it has been shown that it is possible to utilize the natural, receptor-mediated uptake mechanism for VB_{12} to deliver peptide and protein pharmaceuticals from the intestine into the circulation. This mechanism has the ability to overcome the natural barrier presented by the intestinal epithelial cell layer. The formation of an endocytotic pit during uptake of VB_{12} enables molecules of up to several hundred nanometers to be cotransported across the intestinal epithelium.

4.1. Conjugation of Vitamin B_{12} to Peptides and Proteins

In order for VB_{12}-mediated uptake of peptide and protein pharmaceuticals to occur these molecules must first be covalently linked to VB_{12}. Native VB_{12} does not contain any easily modifiable chemical groups for conjugation and so it must first be chemically modified to provide suitable functional groups for conjugation to these molecules. Conjugation to the VB_{12} molecule may be to groups located either below (α-ligands) or above (β-ligands) the corrin ring. Substitution of the α-ligands has a dramatic effect on the affinity of the modified Cbl molecule for IF. In contrast, the β-ligands can vary substantially without affecting the affinity of the molecule for IF. Molecules may be conjugated to the α-position of VB_{12} by modification of the native molecule by acid hydrolysis to yield a monocarboxylic acid derivatives. The VB_{12} can then be conjugated to amino groups of the peptide or protein directly using a suitable carbodiimide. This type of linkage was originally used by Allen and others to conjugate VB_{12} to Sepharose resins for the purification of VB_{12}-binding proteins (Allen and Majerus, 1972; Francis et al., 1977).

For conjugation of peptides to VB_{12} via axial substitution of functional groups into the Co atom of the corrin ring (conjugation to β-ligands), the axial CN ligand of VB_{12} can be replaced with a functionalized alkyl chain. This group can then be used for conjugation to a peptide or protein using chemistry described above (Russell-Jones et al., 1995). Conjugates formed in the β-position are relatively unstable as they contain a light-sensitive Co~C bond that can be cleaved upon exposure to visible light!

4.2. *In Vitro* Transport of Peptides and Proteins Linked to VB_{12}

As mentioned above, it has been found that the colon carcinoma cell line CaCo-2, the HT29 human colon carcinoma cell lines, and the Opossum Kidney cell line OK can all bind, internalize, and transcytose VB_{12} in an IF-dependent fashion. Habberfield and co-workers were able to demonstrate VB_{12}-mediated transport of 180–300 fmole of G-CSF (20 kDa) linked to VB_{12} ([VB_{12}–G-CSF]) from the apical to the basolateral chamber of Caco-2 cells grown in transwell cultures (Habberfield et al., 1996). Control wells in which the unconjugated protein was placed alone showed less than 14 fmole of transport (Fig. 3). Similarly, P. C. de Smidt and co-workers (personal communication) demonstrated VB_{12}-mediated transport of an LHRH analogue, interferon, and a protease-resistant octapeptide across Caco-2 cell cultures (Fig. 4).

4.3. *In Vivo* Transport of Peptides and Proteins Linked to VB_{12}

Data obtained in *in vitro* culture experiments have been mirrored in *in vivo* uptake experiments. Thus, Russell-Jones et al. (1995) found that when they linked VB_{12} to a D-Lys_6 analogue of LHRH, the resultant conjugate was active at stimulating ovulation in mice at an oral or intravenous dose of 10 ng of the D-Lys_6 LHRH analogue, which was significantly higher ($p < 0.05$, Student's t test) than in control mice. Oral administration of the 10-ng dose of the analogue alone or the VB_{12}–

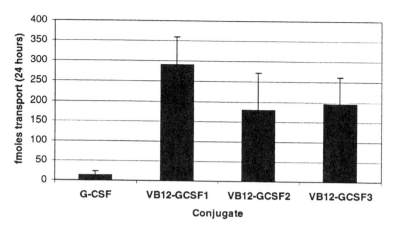

Figure 3. Vitamin B_{12}-mediated transport of G-CSF across Caco-2 monolayer cultures. Reprinted from Habberfield *et al.* (1996), copyright 1996, with permission from Elsevier Science.

D-Lys$_6$ LHRH conjugate administered in the presence of excess VB_{12} did not stimulate ovulation significantly above that of control (pregnant mare's serum gonadotropin-primed) mice.

Oral administration to rats of VB_{12} conjugated to a D-Lys$_6$-ethylamide analogue of LHRH either directly or by the use of spacers resulted in a systemic

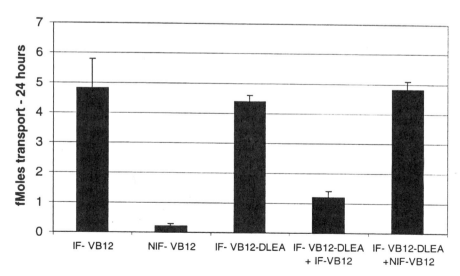

Figure 4. Vitamin B_{12}-mediated transport of an LHRH analogue across Caco-2 cell monolayer cultures.

bioavailability of 20–45%, depending upon the spacer used (de Smidt et al., n.d.; J. Sandow, personal communication, 1993).

Habberfield and co-workers (1996; Russell-Jones et al., 1996) have shown that it is possible to link VB_{12} to EPO and G-CSF in such a manner as to maintain the bioactivity of these substances in vivo. These workers administered the VB_{12}–EPO and VB_{12}–G-CSF complexes intraduodenally to rats using Alzet™ miniosmotic pumps which were surgically implanted into the duodenum of anaesthetized rats. Control animals received the same quantity of nonconjugated EPO or G-CSF under the same delivery conditions. Under the conditions described above, serum levels of G-CSF reached 300 ng/ml at 24 hr in the animals that received the VB_{12}–G-CSF conjugate. Levels were reduced to 554 pg/ml after 2 days. In contrast, serum levels of G-CSF only reached 1.94 pg/ml after 24 hr in animals receiving G-CSF alone and declined to undetectable levels after 2 days (Table IV). Similarly, intraduodenal administration of VB_{12} conjugated to EPO resulted in serum levels which were two to three orders of magnitude higher than for administration of EPO alone (Fig. 5).

4.4. VB_{12}-Mediated Oral Delivery of Nanoparticles

Although the VB_{12}-mediated uptake mechanism offers the opportunity of delivering small peptides and proteins to the circulation following oral administration, the drugs to be delivered still must survive the hostile environment of the gut, including the low pH of the stomach, attack by bile acids, and digestion by intestinal enzymes. It would be preferable if the material to be delivered could be shielded from the intestinal milieu during its passage down the small intestine. Perhaps the most attractive solution would be to incorporate the peptide or protein pharmaceutical to be delivered within a biodegradable microsphere or nanosphere. Uptake could then be stimulated by linking VB_{12} to the surface of the particle to be delivered. Encapsulation within nanoparticles would also be of benefit in cases where it is not possible to link the material to be delivered to the targeting agent without losing functional activity. One disadvantage of the encapsulation system, however, is that once the nanoparticles enter the circulation, they would have to degrade rapidly in order to release their internal pharmaceutical load.

We have recently shown that it is indeed feasible to deliver nanoparticles to the circulation by linking them to VB_{12}. VB_{12}-coated fluorescent nanoparticles have been administered to rats either orally or directly into intestinal loops. Upon histological examination, fluorescent particles were initially found bound to the surface of the intestinal villous cells. Some time later the nanoparticles were found to have crossed the villous epithelial cells and were found congregating in the central lacteal gland.

Experiments in Caco-2 cells have shown that these cells are able to take up

Table IV.
Serum Levels of G-CSF in Rats Receiving G-CSF or
VB_{12}–G-CSF Conjugate Intraduodenally

	Day 1	Day 2
G-CSF	1.94±7 pg/ml	Undetectable[a]
VB_{12}–G-CSF3	299±101 ng/ml	554±440 pg/ml

[a]Data from Russell-Jones (1995).

and transport nanoparticles across the cells in both a VB_{12}-dependent, IF-independent manner as well as a VB_{12}/IF-dependent manner. Cells are able to take up 50-, 100-, and 200-nm particles in a fashion that is dependent upon surface VB_{12} density, particle numbers, and particle size (Russell-Jones et al., 1998).

5. SUMMARY

The uptake of vitamin B_{12} from the intestine into the circulation is perhaps the most complex uptake mechanism of all the vitamins, involving no less than five separate VB_{12}-binding molecules, receptors and transporters. Each molecule involved in uptake has a separate affinity and specificity for VB_{12} as well as a separate cell receptor. Thus VB_{12} is initially bound by haptocorrin in the stomach, then by IF in the small intestine. An IF receptor is then involved in uptake of the

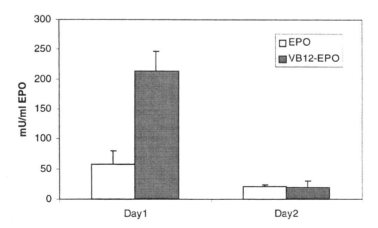

Figure 5. Plasma levels of EPO in rats administered EPO intraduodenally. Reprinted from Habberfield et al. (1996), copyright 1996, with permission from Elsevier Science.

IF–VB$_{12}$ complex by the intestinal epithelial cell, with the subsequent proteolytic release of VB$_{12}$ and subsequent binding to TcII. The TcII receptor then transports the TcII–VB$_{12}$ complex across the cell, whence it is released into the circulation. It is surprising, then, that despite its complexity, it has been possible to harness the vitamin VB$_{12}$ uptake mechanism to enhance the oral uptake of peptides, proteins, and nanoparticles.

REFERENCES

Allen, R. H., and Majerus, P. W., 1972, Isolation of vitamin B$_{12}$-binding proteins using affinity chromatography. III. Purification and properties of human plasma transcobalamin II, *J. Biol. Chem.* **247**:7709–7717.

Allen, R. H., and Mehlman, C. S., 1973, Isolations of gastric vitamin B12 binding proteins using affinity chromatography. I. Purification and properties of human intrinsic factor, *J. Biol. Chem.* **248**:3660–3669.

Allen, R. H., Seetharam, B., Podell, E., and Alpers, D. H., 1978, Effect of proteolytic enzymes in the binding of cobalamin to R protein and intrinsic factor, *J. Clin. Invest.* **61**:47–54.

Alpers, D. H., and Russell-Jones, G. J., 1999, Intrinsic factor and haptocorrin and their receptors. in: *Chemistry and Biochemistry of B12: Part II: Biochemistry of B12* (R. Banerjee, ed.) Wiley, New York, in press.

Andrews, E. R., Pratt, J. M., and Brown, K. L., 1991, Molecular recognition in the binding of vitamin B12 by the cobalamin-specific intrinsic factor, *FEBS Lett.* **281**: 90–92.

Ardeman, S., and Chanarin, I., 1963, A method for the assay of human gastric intrinsic factor for the detection and titration of antibodies against intrinsic factor, *Lancet* **ii**:1350–1354.

Batt, R. M., Horadagoda, N. U., McLean, L., Morton, D. B., and Simpson, K. W., 1989, Identification and characterization of a pancreatic intrinsic factor in the dog, *Am. J. Physiol.* **256**:G517–523.

Beesley, R. C., and Bacheller, C. D., 1980, Divalent cations and vitamin VB$_{12}$ uptake by intestinal brush-border membrane vesicles, *Am. J. Physiol.* **239**:G452–G456.

Berlin, H., Berlin, R., and Brante, G., 1968, Oral treatment of pernicious anemia with high doses of vitamin B12 without intrinsic factor, *Acta Med. Scand.* **184**:247–258.

Binder, H. J., and Donaldson, R. M., 1978, Effect of cimetidine on intrinsic factor and pepsin secretion in man, *Gastroenterology* **74**:371–375.

Boass, A., and Wilson, T. H., 1963, Development of transport mechanisms for intestinal absorption of vitamin B12 in growing rats, *Am. J. Physiol.* **204**:101–104.

Booth, C. C., Chanarin, I., Anderson, B. B., and Mollin, D. L., 1957, The site of absorption and tissue distribution of orally administered ^{57}Co-labelled vitamin B$_{12}$ in the rat, *Br. J. Haematol.* **3**:253–261.

Bose, S., and Seetharam, B., 1994, Transcobalamin II receptor (TC II-R) expression in human tissues, *Gastroenterology* **106**:A801.

Bose, S., and Seetharam, B., 1997, Effect of disulfide bonds of transcobalamin II receptor on its activity and basolateral targeting in human intestinal epithelial Caco-2 cells, *J. Biol. Chem.* **272**:20920–20928.

Bose, S., Seetharam, S., Hammond, T. G., and Seetharam, B., 1995, Regulation of expression of transcobalamin II receptor in the rat, *Biochem. J.* **310**:923–929.

Bose, S., Seetharam, S., Dahms, N. M., and Seetharam, B., 1997, Bipolar functional expression of transcobalamin II receptor in human intestinal epithelial Caco-2 cells, *J. Biol. Chem.* **272**:3538–3543.

Burger, R. L., Mehlman, C. S., and Allen, R. H., 1975a, Human plasma R-type vitamin B_{12} binding proteins. I. Isolation and characterization of transcobalamin I, transcobalamin III, and the normal granulocyte vitamin B_{12} binding protein, *J. Biol. Chem.* **250**:7700–7706.

Burger, R. L., Schneider, R. J., Mehlman, C. S., and Allen, R. H., 1975b, Human plasma R-type vitamin B12 binding proteins. II. Role of TCI and TCII and normal granulocyte vitamin B12 binding proteins in plasma transport of vitamin B12, *J. Biol. Chem.*, **250**:7707–7713.

Carmel, R., Abramson, S. B., and Renner, I. G., 1983, Characterization of pure human pancreatic juice: cobalamin content, Cobalamin-binding proteins and activity against human R binders of various secretions, *Clin. Sci.* **64**:193–205.

Chanarin, I., Muir, M., Hughes, A., and Hoffbrand, A. V., 1978, Evidence for intestinal origin of transcobalamin II during vitamin B_{12} absorption, *Br. Med. J.* **1**:1453–1455.

Dan, N., and Cutler, D. F., 1994, Transcytosis and processing of intrinsic factor-cobalamin in Caco-2 cells, *J. Biol. Chem.*, **269**:18849–18855.

de Smidt, P. C., Russell-Jones, G. J., Steffen, H., and Alsenz, J., n.d., Recognition and transport of a cobalamin–LHRH conjugate by the cobalamin uptake pathway *in vitro* (in preparation).

Dieckgraefe, B. K., Seetharam, B. Banaszak, L., Leykam, J. F., and Alpers, D. H., 1988, Isolation and structural characterization of a cDNA clone encoding rat gastric intrinsic factor, *Proc. Natl. Acad. Sci. USA* **85**:46–50.

Dieckgraefe, B. K., Seetharam, B., and Alpers, D. H., 1989, Developmental regulation of rat intrinsic factor mRNA, *Am. J. Physiol.* **254**:913–919.

Dix, C. J., Hassan, I. F., Obray, H. Y., Shah, R., and Wilson, G., 1990, The transport of vitamin VB_{12} through polarized monolayers of Caco-2 cells, *Gastroenterology* **88**:1272–1279.

Donaldson, R. M. Jr., Mackenzie, I. L., and Trier, J. S., 1967, Intrinsic factor-mediated attachment of vitamin VB_{12} to brush borders and microvillous membranes of hamster intestine, *J. Clin. Invest.* **46**:1215–1228.

En-Nya, A., Guéant, J. L., Nexø, E., Gerard, A., Boukhzer, E. B., Nicolas, J. P., and Gerard, H., 1993, Endocytosis of transcobalamin in male rabbit germ cells: Electron microscope radioautography study, *Int. J. Dev. Biol.* **37**:353–357.

Francis, G. L., Smith, G. W., Toskes, P. P., and Sanders, E. G., 1977, Purification of hog gastric intrinsic factor by a simple two-step procedure based on affinity chromatography and a selective guanidine hydrochloride gradient, *Gastroenterology* **72**:1304–1307.

Fyfe, J. C., Ramanujam, K. S., Ramasamy, K., Patterson, D. F., and Seetharam, B., 1991,

Defective brush-border expression of intrinsic factor-cobalamin receptor in canine inherited intestinal cobalamin malabsorption, *J. Biol. Chem.* **266**:4489–4494.

Fyfe, J. C., Giger, U., Hall, C. A., Jezk, P. F., Klumpp, S. A., Levine, J. S., and Patterson, D. F., 1992, Inherited selective intestinal cobalamin malabsorption and cobalamin deficiency in dogs, *Paediatric Res.* **29**:24–31.

Geffen, I., and Spiess, M., 1992, Asialoglycoprotein receptor, *Int. Rev. Cytol.* **137B**:181–219.

Glass, G. B. J., 1974, *Gastric Intrinsic Factor and Other Vitamin B12 Binders*, Georg Thieme, Stuttgart.

Gordon, M., Hu, C., Chokshi, H., Hewitt, J. E., and Alpers, D. H., 1991, Glycosylation is not required for ligand or receptor binding by expressed rat intrinsic factor, *Am. J. Physiol.* **260**(*Gastrointest. Liver Physiol.* **23**):G736–G742.

Gordon M., Chokshi H., and Alpers D. H., 1992, In vitro expression and secretion of functional mammalian intrinsic factor using recombinant baculovirus, *Biochim. Biophys. Acta* **1132**:276–283.

Gordon, M. M., Howard, T., Becich, M. J., and Alpers, D. H., 1995, Cathepsin L mediates intracellular ileal digestion of gastric intrinsic factor, *Am. J. Physiol.* **268** (*Gastrointest. Liver Physiol.* **31**):G33–G40.

Green, P. D., Savage, R., Jr., and Hall, C. A., 1976, Mouse transcobalamin II: Biosynthesis and uptake by L-929 cells, *Arch. Biochim. Biophys. Acta* **176**:683–689.

Guéant, J. L., Kouvonen, I., Michalski, J. C., *et al.*, 1985, Purification of human intrinsic factor using high performance ion exchange chromatography as the final step, *FEBS Lett.* **184**:14–19.

Guéant, J. L., Gerard, A., Monin, B., *et al.*, 1988, Autoradiographic localization of iodinated human intrinsic factor (IF), *Gut* **29**:1370–1378.

Guéant, J.-L., Masson, D., Schohn, J., Girr, M., Saunier, M., and Nicolas, J. P., 1992, Receptor-mediated endocytosis of the intrinsic factor-cobalamin complex in HT 29, a human colon carcinoma cell line, *FEBS J.* **297**:229–232.

Habberfield, A., Jensen-Pippo, K., Ralph, L., Westwood, S. J., and Russell-Jones, G. J., 1996, Vitamin B12-mediated uptake of erythropoietin and granulocyte colony stimulating factor *in vitro* and *in vivo Int. J. Pharmacol.* **145**:1–8.

Hagedorn, C. H., and Alpers, D. H., 1977, Distribution of intrinsic factor-vitamin B_{12} receptors in human intestine, *Gastroenterology,* **73**:1019–1022.

Hall, C. A., 1977, The carriers of native vitamin B12 in normal human serum, *Clin. Sci. Mol. Med.* **53**:453–457.

Hassan, I. F., and Mackay, M., 1992, The transport of vitamin VB_{12} across monolayers of Caco-2 cells grown on permeable supports, *In vitro Cell. Dev. Biol.* **28**:T-12.

Hewitt, J. E., Seetharam, B., Leykam, J., and Alpers, D. H., 1990, Isolation and characterization of a cDNA encoding porcine gastric haptocorrin, *Eur. J. Biochem.* **189**:125–130.

Hewitt, J. E., Gordon, M. M., Taggart, T., Mohandas, T. K, and Alpers, D. H., 1991, Human gastric intrinsic factor: Characterization of cDNA and genomic clones and localization to human chromosome 11, *Genomics* **10**:432–440.

Hippe, E., and Olesen, H., 1971, Nature of vitamin VB_{12} binding. III. Thermodynamics of binding to human intrinsic factor and transcobalamins, *Biochem. Biophys. Acta* **243**:38–39.

Hoedemaeker, P. J., Abels, J., Wachters, J. J., Arends, A., and Nieweg, H. O., 1966, Further investigations about the site of production of Castle's gastric intrinsic factor, *Lab. Invest.* **15:**1163–1173.

Howard, T. A., Misra, D. N., Grove, M., Becich, M. J., Shao, J.-S., Gordon, M., and Alpers, D. H., 1996, Human gastric intrinsic factor expression is not restricted to parietal cells, *J. Anat.* **189:**303–313.

Hu, C., Lee, E. Y., Hewitt, J. E., Baenziger, J. U., Mu, J.-Z., DeSchryver-Kecskemeti, K., and Alpers, D. H., 1991, The minor components of the rat asialoglycoprotein receptor are apically located in neonatal enterocytes, *Gastroenterology* **101:**1477–1487.

Irvine, W. J., 1965, Effect of gastrin I and II on secretion of intrinsic factor, *Lancet* **i:**736–737.

Jeffries, G. H., and Sleisenger, M. H., 1965, The pharmacology of intrinsic factor secretion in man, *Gastroenterology* **48:**444–448.

Johnson, J., Bollekens, J., Allen, R. H., and Berliner, N., 1989, Structure of the cDNA encoding transcobalamin I, a neutrophil granule protein, *J. Biol. Chem.* **264:**15754–15757.

Kapadia, C. R., and Essandoh, L. K., 1988, Active absorption of vitamin B12 and conjugated bile salts by guinea pig ileum occurs in villous and not crypt cells, *Dig. Dis. Sci.* **33:**1377–1382.

Kitting, E., Aadland, E., and Schjonsby, H., 1985, Effect of omeprazole on the secretion of intrinsic factor gastric acid and pepsin in man, *Gut* **26:**594–598.

Lee, E. Y., Seetharam, B., Alpers, D. H., and DeSchryver-Kecskemeti, K., 1989a, Cobalamin binding proteins (IF and R): An immunohistochemical study, *Gastroenterology* **97:**1171–1180.

Lee, E. Y., Seetharam, B., Alpers, D. H., and DeSchryver-Kecskemeti, K., 1989b, Immunochemical survey of cobalamin-binding proteins, *Gastroenterology* **97:**1171–1180.

Levine, J. S., Allen, R. H., Alpers, D. H., and Seetharam, B., 1984, Immunocytochemical localization of the intrinsic factor-cobalamin receptor in dog-ileum: Distribution of intracellular receptor during cell maturation, *J. Cell Biol.* **98:**1111–1118.

Li, N., Seetharam, S., Lindemans, J., Alpers, D. H., Arwert, F., and Seetharam, B., 1993, Isolation and sequence analysis of variant forms of human transcobalamin II, *Biochim. Biophys. Acta* **1172:**21–30.

Li, N., Seetharam, S., Rosenblatt, D. S., and Seetharam, B., 1994, Expression of transcobalamin II mRNA in human tissues and cultured fibroblasts from normal and transcobalamin II-deficient patients, *Biochem. J.* **301:**585–590.

Lien, E. L., Ellenbogen, L., Law, P. Y., and Wood, H. M., 1974, Studies on the mechanism of cobalamin binding to hog intrinsic factor, *J. Biol. Chem.* **249:**890–894.

Lindemans, J., Kroes, A. C. M., V. Geel, J., van Kapel, J., Schoester, M., and Abels, J., 1989, Uptake of transcobalamin II-bound cobalamin by HL-60 cells: Effects of differentiation induction, *Exp. Cell Res.* **184:**449–460.

Linnell, J. C., 1975, The fate of cobalamins *in vivo*, in: *Cobalamin: Biochemistry and Pathophysiology* (B. M. Babior, ed.), Wiley, New York, pp. 287–333.

Lobie, P. E., and Waters, M. J., 1997, Growth hormone (GH) regulation of submandibular gland structure and function in the GH-deficient rat—Upregulation of haptocorrin, *J. Endocrinol.* **154:**459–466.

Lorenz, R. G., and Gordon, J. I., 1993, Use of transgenic mice to study regulation of gene expression in the parietal cell lineage of gastric units, *J. Biol. Chem.* **268**:26559–26570.

Marcoullis, G., and Rothenberg, S. P., 1981, Intrinsic factor-mediated intestinal absorption of cobalamin in the dog, *Am. J. Physiol.* **241**(*Gastrointest. Liver Physiol.* **4**):G294–G299.

Marcoullis, G., Grasbeck, R., and Salonen, E. M., 1977, Identification and characterization of intrinsic factor and cobaphilin from pig ileal and pyloric mucosa, *Biochim. Biophys. Acta* **497**:663–672.

Marcoullis, G., Parmentier, Y., Nicolas, J. P., Jimenez, M., and Gerard, P., 1980, Cobalamin malabsorption due to nondegradation of R proteins in the human intestine. Inhibited cobalamin absorption in exocrine pancreatic dysfunction, *J. Clin. Invest.* **66**:430–440.

Marcoullis, G., Rothenberg, S. P., and Jimenez, M., 1981, Conformational changes of the receptor molecule in its interaction with intrinsic factor, *Fed. Proc.* **40**:489.

Mathan, V. I., Babior, B. M., and Donaldson, R. M., Jr., 1974, Kinetics of attachment of intrinsic factor-bound cobamides to ileal receptors, *J. Clin. Invest.* **54**:598–608.

Mu, J.-Z., Tang, L.-H., and Alpers, D. H., 1993, Asialoglycoprotein mRNAs are expressed in most extrahepatic rat tissues during development, *Am. J. Physiol.* **264**(*Gastrointest. Liver Physiol.* **27**):G752–G762.

Mu, J.-Z., Fallon, R. J., Swanson, P. E., Carroll, S. B., Danaher, M., and Alpers, D. H., 1994, Expression of an endogenous asialoglycoprotein receptor in a human intestinal epithelial cell line, Caco-2, *Biochim. Biophys. Acta* **1222**:483–491.

Mu, J.-Z., Gordon, M., Shao, J.-S., and Alpers, D. H., 1997, Apical expression of functional asialoglycoprotein receptor in the human intestinal cell line HT-29, *Gastroenterology* **113**:1501–1509.

Nexø, E., and Andersen, J., 1977, Unsaturated and cobalamin saturated transcobalamin I and II in normal human plasma, *Scand. J. Clin. Lab. Invest.* **37**:723–728.

Nexø, E., and Olesen, H., 1981, Purification and characterization of rabbit haptocorrin, *Biochim. Biophys. Acta* **667**:370–376.

Nicolas, J. P., and Guéant, J. L., 1995, Gastric intrinsic factor and its receptor, *Bailliere's Clin Haematol.* **8**:515–531.

Oddsdotter, M., Ballantyne, G. H., Adrian, T. E., Zdor, M. J., Zucker, K. A., and Modlin, I. M., 1987, Somatostatin inhibition of intrinsic factor secretion from isolated guinea pig gastric glands, *Scand. J. Gastroenterol.* **22**:233–238.

Okuda, K., 1960, Vitamin B_{12} absorption in rats, studied by a 'loop' technique, *Am. J. Physiol.* **199**:84–90.

Pathare, P. M., Wilbur, D. S., Heusser, S., Quadros, E. V., McLoughlin, P., and Morgan, A. C., 1996, Synthesis of cobalamin-biotin conjugates that vary in the position of cobalamin coupling. Evaluation of cobalamin derivative binding to transcobalamin II, *Biconjugate Chem.* **7**:217–232.

Quadros, E. V., Sai, P., and Rothenberg, S. P., 1994a, Characterization of the human placental membrane receptor for transcobalamin II-cobalamin, *Arch. Biochem. Biophys.* **308**:192–199.

Quadros, E. V., Regec, A., Khan, F., and Rothenberg, S. P., 1994b, Intrinsic factor (IF) mediated transepithelial transport of cobalamin (Cbl) in the distal ileum also requires transcobalamin II (TC II) that is synthesized *in situ*, *Blood,* **84**:503A.

Quadros, E. V., Rothenberg, S. P., and McLoughlin, P., 1996b, Characterization of monoclonal antibodies to epitopes of human transcobalamin II, *Biochem. Biophys. Res. Commun.* **222**:149–154.

Ramanujam, K. S., Seetharam, S., Ramasamy, M., and Seetharam, B., 1990, Renal brush border membrane bound intrinsic factor, *Biochim. Biophys. Acta* **1030**:157–164.

Ramanujam, K. S., Seetharam, S., Samasamy, M., and Seetharam B., 1991a, Expression of cobalamin transport proteins and cobalamin transcytosis by colon adenocarcinoma cells, *Am. J. Physiol.* **260**(*Gastrointest. Liver Physiol.* **23**):G416–G422.

Ramanujam, K. S., Seetharam, S., and Seetharam, B., 1991b, Synthesis and secretion of cobalamin binding proteins by opossum kidney cells, *Biochem. Biophys. Res. Commun.* **179**:543–550.

Ramanujam, K. S., Seetharam, S., and Seetharam, B., 1992, Leupeptin and ammonium chloride inhibit intrinsic factor mediated transcytosis of [57Co]cobalamin across polarized renal epithelial cells, *Biochem. Biophys. Res. Commun.* **182**:439–446.

Ramanujam, K. S., Seetharam, S., and Seetharam, B., 1993, Intrinsic factor-cobalamin receptor activity in a marsupial, the American opossum (*Didelphis virginiana*), *Comp. Biochem. Physiol.* **104A**:771–775.

Ramanujam, K. S., Seetharam, S., Dahms, and Seetharam, B., 1994, Effect of processing inhibitors on cobalamin (Vitamin VB_{12}) transcytosis in polarized opossum kidney cells, *Arch. Biochem. Biophys.* **315**:8–15.

Regec, A., Quadros, E. V., Platica, O., and Rothenberg, S. P., 1995, The cloning and characterization of the human transcobalamin II gene, *Blood* **10**:2711–2719.

Russell-Jones, G. J., 1996a, Utilisation of the natural mechanism for vitamin B12 uptake for the oral delivery of therapeutics, *Eur. J. Pharm. Biopharm.* **42**:241–249.

Russell-Jones, G. J., 1996b, The potential use of receptor-mediated endocytosis for oral drug delivery, in: *Carrier Mediated Approaches for Oral Drug Delivery* (S. Øie, F. C. Szoka, P. W. and Swaan, eds.), *Advanced Drug Delivery Reviews,* **20**:83–86.

Russell-Jones, G. J., 1996c, Utilisation of the Natural Mechanism for Vitamin B_{12} uptake for the oral delivery of therapeutics. *Eur. J. Pharm. Biopharm.* **42**:241–249.

Russell-Jones, G. J., 1998, Use of Vitamin B_{12} conjugates to deliver protein drugs by the oral route. *Critical Reviews in Therapeutic Drug Carrier Systems* **16**:557–558.

Russell-Jones, G. J., 1999, Carrier-mediated transport for oral drug delivery, in: *Encyclopedia of Controlled Drug Delivery* (E. Mathiowitz, ed.), in press.

Russell-Jones, G. J., Westwood, S. W., and Habberfield, A. D., 1995, Vitamin B_{12} mediated oral delivery systems for granulocyte-colony stimulating factor and erythropoietin, *Bioconjugate Chem.* **6**:459–465.

Russell-Jones, G. J., Arthur, L., and Walker, H., 1999, Vitamin B_{12}-mediated transport of nanoparticles across Caco-2 cells. *Int. J. Pharm.* **179**:247–255.

Schepp, W., Heim, K. H., and Ruoff, H.-J., 1983a, Comparison of the effect of PGE2 and somatostatin on histamine stimulated 14C-aminopyrine uptake and cyclic AMP formation in isolated rat gastric mucosal cells, *Agents Actions* **13**:200–206.

Schepp, W., Rouff, H. J., and Miederer, S. E., 1983b, Cellular origin and release of intrinsic factor from isolated rat gastric mucosal cells, *Biochim. Biophys. Acta* **763**:426–433.

Schepp, W., Miederer, S. E., and Ruoff, H.-J., 1984, Intrinsic factor secretion from isolated gastric mucosal cells of rat and man—Two different patterns of secretagogue control, *Agents Actions* **14**:522–528.

Schneider, Z., and Stroinski, A., 1987, *Comprehensive B_{12}*, de Gruyter Berlin, pp. 297–330.

Schohn, J., Guéant, J.-L., Girr, M., Nexø, E., Baricault, L., Zweibaum, A., and Nicolas, J.-P., 1991, Synthesis and secretion of a cobalamin-binding protein by HT 29 cell line, *Biochem. J.* **280**:427–430.

Schwartz, A. L., 1990, Cell biology of intracellular protein trafficking, *Annu. Rev Immunol.* **8**:195–229.

Scott, J. M., Bloomfield, F. J., Stebbins, R., and Herbert, V., 1974, Studies on derivation of transcobalamin III from granulocytes: Enhancement of lithium and elimination by fluoride of *in vitro* increments in vitamin B_{12}-binding capacity, *J. Clin. Invest.* **53**:228–239.

Seetharam, B., and Alpers, D. H., 1994, Cobalamin binding proteins and their receptors, in: *Vitamin Receptors: Vitamins as Ligands in Cell Communication* (K. Dakshinamurti, ed.), Cambridge University Press, Cambridge, pp. 78–105.

Seetharam, B., Bagur, S. S., and Alpers, D. H., 1982, Isolation and characterization of proteolytically derived ileal receptor for intrinsic factor-cobalamin, *J. Biol. Chem.* **257**:183–189.

Seetharam, B., Bakke, J. E., and Alpers, D. H., 1983, Binding of intrinsic factor to ileal brush border membrane in the rat, *Biochem. Biophys. Res. Commun.* **115**:283–241.

Seetharam, B., Presti, M., Frank, B., Tiruppathi, C., and Alpers, D. H., 1985, Intestinal uptake and release of cobalamin complexed with rat intrinsic factor, *Am. J. Physiol.* **248**(*Gastrointest. Liver Physiol.* **11**):G306–G313.

Seetharam, B., Levine, J. S., Ramasamy, M., and Alpers, D. H., 1988, Purification, properties, and immunochemical localization of a receptor for intrinsic factor-cobalamin complex in the rat kidney, *J. Biol. Chem.* **263**:4443–4449.

Seetharam, B., Bose, S., Hammond, T. G., and Seetharam, S., 1995, Surface and intracellular membrane distribution of transcobalamin II receptor (TC II-R) in the rat intestine and kidney, *FASEB J.* **9**:1416.

Seetharam, B., Christensen, E. I., Moestrup, S. K., Hammond, T. G., and Verroust, P. J., 1997, Identification of rat yolk sac target protein of teratogenic antibodies, gp 280, as intrinsic-factor receptor, *J. Clin. Invest.* **99**:2317–2322.

Seetharam, S., Ramanujam, K. S., and Seetharam, B., 1991, Synthesis and brush border expression of intrinsic factor-cobalamin receptor from rat renal cortex, *J. Biol. Chem.* **267**:7421–7427.

Serfilippi, D., and Donaldson, R. M., 1986, Production and secretion of intrinsic factor by isolated rabbit gastric mucosa, *Am. J. Physiol.* **14**:287–292.

Shao, J.-S., Schepp, W., and Alpers, D. H., 1998, Expression of intrinsic factor and pepsinogen in the rat glandular stomach demonstrates the presence of a subset of parietal cells, *Am. J. Physiol. (Gastrointest. Liver Physiol)* **274**:962–970.

Simon, J. D., Houck, W. E., and Albala, M. M., 1976, Release of unsaturated vitamin B_{12} binding capacity from human granulocytes by the calcium ionophore A23187, *Biochem. Biophys. Res. Commun.* **73**:444–450.

Simons, K., and Wever, T., 1966, The vitamin B12 binding protein in human leucocytes, *Biochim. Biophys. Acta* **117**:201–208.

Simpson, K. W., Alpers, D. H., DeWile, J., Swanson, P., Farmer, S., and Sherding, R. G., 1993, Cellular localization and hormonal regulation of pancreatic intrinsic factor secretion in the dog, *Am. J. Physiol.* **265**(*Gastrointest. Liver Physiol.* **28**):G178–G188.

Smolka, A., and Donaldson, R. M., 1990, Monoclonal antibodies to human intrinsic factor, *Gastroenterology* **98**:607–614.

Sourial, N. A., and Waters, A. H., 1991, Effect of pH on binding of B_{12} to IF and R-protein, *British J. Haematol.,* **81**:136–138.

Stenman, U. H., 1974, Amniotic fluid vitamin B_{12} binding protein. Purification and characterization with isoelectric focusing and other techniques, *Biochim. Biophys. Acta* **342**:173–184.

Stenman, U. H., 1975, Vitamin B_{12}-binding proteins of R-type, cobalophilin: Characterization and comparison of cobalophilin from different sources, *Scand. J. Haematol.,* **14**:91–107.

Stupperich, E., and Nexø, E., 1991, Effect of the cobalt-N coordination on the cobamide recognition by the human vitamin B_{12} binding proteins intrinsic factor, transcobalamin and haptocorrin, *Eur. J. Biochem.* **199**:299–303.

Swaan, P. W., and Tukker, J. J., 1995, Carrier-mediated transport mechanism of foscarnet (trisodium phophonoformate hexahydrate) in rat intestinal tissue, *J. Pharmacol. Exp. Ther.* **27**:242–247.

Tang, L., Chokshi, H., Hu, C.-B., Gordon, M. M., and Alpers, D. H., 1992, The intrinsic factor (IF)-cobalamin receptor binding site is located in the amino-terminal portion of IF, *J. Biol. Chem.* **267**:22982–22986.

Trugo, N. M., Ford, J. E., and Salter, D. N., 1985, Vitamin B_{12} absorption in the neonatal piglet. 3. Influences of vitamin B12-binding protein in sows' milk on uptake of vitamin B12 by microvillus membrane vesicles prepared from small intestine of the piglet, *Br. J. Nutr.* **54**:269–283.

Vaillant, C., Horadagoda, N. U., and Batt, R. M., 1990, Cellular localization of intrinsic factor in pancreas and stomach of the dog, *Cell Tiss. Res.* **260**:117–122.

Vatn, M. H., Schrumpf, E., and Myren, J., 1975, The effect of carbachol and pentagastrin on the gastric secreton of acid, pepsin, and intrinsic factor in man, *Scand. J. Gastroenterol.* **10**:55–58.

Walter, E., Kissel, T., and Amidon, G. L., 1996, The intestinal peptide carrier: A potential transport system for small peptide derived drugs. *Adv. Drug Delivery Rev.* **20**:33–58.

Wen, J., Gordon, M. M., and Alpers, D. H., 1997, A receptor binding site on intrinsic factor is located between amino acids 25–44 and interacts with other parts of the protein, *Biochem. Biophys. Res. Commun.* **231**:348–351.

Wen, J., Kinnear, M. B., Richardson, M. A., Willetts, N. S., Russell-Jones, G. J., Gordon, M. J., and Alpers, D. H., 1999, Functional expression of human and rat intrinsic factor in *Pichia pasatoris. J. Biol. Med.* (Submitted).

Wilson, G., Hassan, I. F., Dix, C. J., Williamson, I., Shah, R., Mackay, M., and Artusson, P., 1990, Transport and permeability of human Caco-2 cells: An *in vitro* model of the intestinal epithelial cell barrier, *J. Controlled Release* **11**:25–40.

Zucker, K. A., Adrian, T. E., Ballantyne, G. H., and Modlin, I. M., 1988, Prostaglandin analogue inhibition of intrinsic factor release, *Scand. J. Gastroenterol.* **23**:650–654.

Index

Acetylcholine receptors, 7–9
Active transport, 10–12, 31
Active transporters, 10–12, 91
  primary vs. secondary, 10, 12
Acyclovir drugs, intestinal permeability of, 74, 75
Adenosine transporters, 65–66
Adenosine/P1 purinoceptor subtypes, 336–337
ALP1–YEAST, 48
Alternating confirmation model, 206–208
Amino acid carriers, 186
  neutral, 185–186
Amino acid ester prodrugs, 74
Amino acid transporters, 48–49, 67, 76, 77
Angiotensin-converting enzyme (ACE) inhibitors, 73, 418
Antibiotics
  cephalosporin, BBM transport of, 274, 275
  β-lactam, 278, 279
Anticancer drugs, 338–343
Antigenic peptide transporter (TAP), 289, 305–306
  as drug target, 301–305
  and MHC class I antigen processing pathway, 289–292
  structure and function, 293–297
  viral inhibition, 297–301
Antigenic peptide transporter (TAP) activity, allosteric modulation of, 305
Antigenic peptide transporter (TAP) inhibitors, 302–303
Antiporters, 31
Arginine, jejunal absorption of, 77
Arphamenine A, 65
Asialogylcoprotein (ASGP-R), 498
ATPase activity, 358
ATPases, ion-motive, 10–11

ATP-binding cassette (ABC), 50
ATP-binding cassette (ABC) transport proteins
  in canaliculus, 103–104
  and cLPM, 116–121
ATP-binding cassette (ABC) transporters, 3, 91, 93, 103–104, 293, 355; *see also under* Cationic drugs
ATP-binding transport protein family, 50–52
ATP-dependent transport systems, 120, 417, 418
ATP/P2 purinoceptors, 337
Azidoprocainamide methoiodide (APM), 99–101
Azopentyldeoxyajmalinium (APDA), 95, 99

Bestatin, basolateral transport of, 273, 274
Bicarbonate transporter, 69
Bile, 107–111; *see also* Canaliculus
Bile acid transporter, 66–67, 76–77, 394–395, 424
  as target for delivery, 395–396
Bile canalicular membrane, 403–404; *see also* Canalicular liver plasma membrane
  ATP-dependent efflux of TC across: *see* cBAT/cBST
  organic anion transport across, 407–408, 414, 416–417, 419–424
  transport of bile acids across, 404, 406–407
  transport of organic anions across, 407–408, 414, 416–417, 419–424, 423
Bile canalicular transport of cationic drugs, 97–104
Biliary excretion
  efficient, mediated by cMOAT, 416–417
  of temocaprilat, 408, 414–416
BLAST (Basic Local Alignment Search Tool) analysis, 38, 39, 41–43, 45, 46

521

Blood-brain barrier (BBB), 181–183, 193
　endothelium, 181–182
　permeability, 193
　　carrier-mediated transport, 183–190
　　transcytosis, 190–192
　P-gp and, 368
Blood-brain barrier (BBB) macromolecule and transcytotic systems, 191
Blood-brain barrier (BBB) transport systems, 184, 189
BQ-123, 390, 391
Brush border membrane (BBM) transport, 270–271, 274, 275
Brush border membrane vesicles (BBMVs), 251, 252
Bulk transport, 12
Buthathione sulfoximine (BSO), 372

Caco-2 cell monolayers, 272–273
Calcium-selective voltage-gated channel, 7, 8
Canalicular bile salt transporter (cBST/sPgp), 120–121
Canalicular extrusion of bile acids and non-bile acid organic anions, 405
Canalicular liver plasma membrane (cLPM), 97; see also Bile canalicular membrane
　ABC transport proteins and, 116–121
Canalicular membrane vesicles (CMVs), 404, 406, 407, 414, 415, 417, 419
Canalicular multispecific organic anion transporter (cMOAT), 103, 104, 372, 404
　molecular cloning, 419–422
　role in drug disposition, 408, 414, 416–417
　substrate specificity, 408–413, 409
Canalicular multispecific organic anion transporter (cMOAT) function, 417
Canalicular organic cation carriers, 122–140
　substrate specificity, driving forces, and multiplicity, 105–107
Canaliculus, ABC transport proteins in, 103–104
Cancer
　anticancer drugs, 338–343
　anticancer nucleosides, 339
　P-gp expression in, 360–361
Carboxylic (COOH) groups, 89, 90
Carrier-mediated facilitated diffusion, models for, 13, 14
Cationic amino acid transporter (CAT) proteins
　expression in NO-producing cells, 239–241
　transport properties, 233–237

Cationic amino acid transporters (CATs), 231, 243–244
Cationic amino acids, 229–230
　carrier proteins for, 231–233
　CATs and transport systems for, 237–239
Cationic drugs
　candidate proteins for carrier-mediated hepatic uptake, 93–97
　candidate proteins involved in bile canalicular transport, 97–104
Cation/proton antiport secretory/reabsorptive system, 97–99
Cations, organic: see Organic cations
CBAT/cBST, 404, 406–407
Cbl (cobalamin): see Vitamin B_{12}
Cbl-binding proteins, 504
CCRF-CEM, 340
Cell culture, 22–23
Cell lines, mammalian, 453–455
Cephalosporin antibiotics, BBM transport of, 274, 275
Channel proteins, 91
Channels, 1–2, 91
Chlorambucil, 395
Cholic acid (CA), 388–389
Choline chloride, 234
Cloning, 444; see also Human organic cation transporters
　homology, 447
　molecular, 391–394, 399–403, 419–422, 423–424
　peptide transporter, 274–281
Cobalamin: see Cbl; Vitamin B_{12}
Concentration, substrate
　transport and, 6–9
Concentrative nucleoside transporter (CNTs), 319–321, 327–328
　CNT2 (cif) subfamily, 319, 330–332
　CNT1 (cit) subfamily, 328–330
Confocal microscopy, 21
Contraluminal NMeN$^+$/TEA$^+$ transporter, interaction of substrates with, 171–172
Contraluminal organic cation (PAH) transporter, interaction of substrates with, 170–172
Contraluminal sulfate transporters, interaction with substrates, 171
Cotransporters, 31
CPT-11, 416, 418
Cultured cells, 22–23
Cyclic dipeptide, chemical structure of, 283

Index

Cystic fibrosis conductance regulator (CFTR), 30, 52

D-amino acids, incorporation into TAP structures, 304–305
Defects, 3, 4
D-glucose transporters: see Facilitative glucose transporters (GLUT1-5)
Dicarboxylate transporters, 48, 49
  interaction with substrates, 171
Diethylpyrocarbone (DEPC), 278, 281
Diffusion, passive/simple, 5–6
Dipeptide symporters, 40–43
Dipeptide transporters, 269; see also PEPT1; PEPT2
Dipeptides, 76, 77
  cyclic, 283
DNP-SG, 407, 421
Drug efflux systems, 187–188
Drug excretion/elimination, 107–112, 159, 173
  renal, 159, 173
Drug resistance, nucleoside transporters and, 339–341
Drugs
  absorption, 60
  crossing BBB, 187–188
  that interact with organic anion and cation transporters, 172
DTPT–LCLA, 40–43

Electrolyte transporters, 75
Energy needed for transport, 12–13, 31
Epithelial transport
  electrophysiological study of, 23–24
  in intestines, 64
Equilibrative nucleoside transporter (ENTs), 319–321, 340–343
  ENT2 *(ei)* subfamily, 324–327
  ENT1 *(es)* subfamily, 321–323
  molecular and functional characteristics, 321–327
Erythrocyte glucose transport: see Glucose transport
ESTs (expressed sequence tags), 33, 39
Excitatory amino acid transporters (EATs), 48–49

Facilitated diffusion, 9–10, 12
Facilitated-diffusion adenosine carriers, 65–66

Facilitative glucose transporters (GLUT1-5), 31, 44–45, 66, 183, 185, 201–203; see also Glucose transport
  recent work and future directions regarding, 220
  structure, 203–205
  tissue-specific distribution, 211–220
F-ATPase, 11
Fatty acid transporter, 69
Fatty acids, short-chain, 474–475
FCCP, 271–273
Ferryboat transport, 13, 14
Fick's law, 5–6
Fluorescence digital imaging microscopy, 20
Folate metabolism, 77
Foscarnet (phosphonoformic acid), 264

Gastrointestinal transport: see Intestinal transport
Gated pore transport, 13, 14
Gene expression, transporter, 36
Genomic sequences, 32–36
Genomics of transporters, 32–36
Glucose transport, erythrocyte, 9–10; see also Sugar transport
  dynamics, 205–209
Glucose transporter photoaffinity labeling, 209–211
Glucose transporters (GTRs), 44–47; see also Sodium/glucose cotransporters
GLUT, see also Facilitative glucose transporters (GLUT1-5)
  biophysical investigation of, 204–205
  GLUT1, 183, 185, 205, 210
  kinetic scheme, 208
  oligomerization, 208–209
  secondary structure, 203–204
  tissue-specific distribution, 211–220
Glutathione conjugate leukotriene C_4 (LTC_4), 371
Glutathione (GSH), 481
Glutathione-S-conjugate (GS-X) pump, 371, 372

Haptocorrin (Hc), 495–499
HCO_3^-, 476–477
H^+/dipeptide symporters, 40–43
Hepatic drug transporters, 110, 112
Hepatic transporters, 387–388, 424; see also Liver; Sinusoidal membrane
Hepatic uptake of organic anions, 112–116, 424

Hepatobiliary secretion of amphiphilic drugs, 110, 112
Hepatocytes, MRP/mrp isoforms in, 118–119
Herpes simplex virus (HSV) ICP47 protein, 298–300
Hexose, schematic model of, 217
Histidine residues, 278, 280, 281
HMG-CoA, 396, 398
HNP36, 324, 325, 327
Homology cloning, 447
Human CAT-protein (hCAT)-mediated transport, 236–237
Human CAT-proteins (hCATs), 233
Human concentrative nucleoside transporter (hCNT), 330
Human cytomegalovirus (HCMV), 298
Human cytomegalovirus (HCMV) US6 protein, 300–301
Human equilibrative nucleoside transporter (hENT), 321–327
Human organic cation transporters (hOCTs), cloned, 441–442, 444, 466
  expression cloning strategy, 444, 446
  expression systems, 451–455
  functional characteristics, 451, 458–461
    electrophysiology, 456–457
  functional methods of investigating, 455–461
  future studies, 463–466
  interactions with various compounds, 460
  molecular characteristics, 444, 445, 447–451
  tissue distribution, 461–463

IF (intrinsic factor), 495–498, 512–513
  occurrence and distribution, 499
  structure and function, 500–501
  synthesis and secretion, 499
IF-Cbl, 498, 501–502
IFR (intrinsic factor receptor), 495, 502–504
IF-VB$_{12}$, cellular uptake and utilization of, 501–502
Ileal-localized NTCP (INTCP), 114
Impedance analysis, 23
Insulin, 191, 192
Intestinal drug absorption, prodrugs to increase, 70–75, 78
Intestinal permeability, 74, 75
Intestinal transport, 59–60
  epithelial, 64
  of lactate and short-chain fatty acids via MCT1, 473–474

Intestinal transporters, 60–64, 77–78, 215, 217; *see also specific transporters*
  diseases related to, 75–77
  targeting with prodrugs, 70–75, 78
Intestines, 107–111
  P-gp and, 365–367
  transporters in, 60–63
Intrinsic factor: *see* IF
Ion channels, 6–9
  gating and selectivity, 7–8
  purified and reconstituted, 22
Ion concentration, intra- *vs.* extracellular, 6, 7
Ion-selective channels, 7–8
Isoforms of transporters, 60, 64

Kidney, *see also* Renal drug elimination
  P-gp and, 368
Kinetic equation, 13

Lactate, intestinal transport of, 473–474
L-a-methyldopa, 73
  and its dipeptidyl derivatives, 73
Large neutral amino acids (LNAA), 185–186
L-arginine, 229, 235, 239
  transport, 235
  and NO synthesis, 241–243
L-dopa, 186
Leucine, jejunal absorption of, 77
Leukemia viruses, murine ecotropic, 231
LHRH, 509–510
Ligand-gated (neurotransmitter-gated) channels, 7, 8, 22
Liver, *see also* Bile; Hepatic transporters; Sinusoidal membrane
  carrier-mediated transport of cationic compounds into, 113
  cholestatic, 120, 121
  OATs in, 478–482
  P-gp and, 367–368
LLC-PK$_1$ cells, 111, 357
Lovastatin, 396, 398
Low-density lipoproteins (LDL), 2, 12, 192
Lung resistance-related protein (LRP), 373–374

Macromolecular transport, 12
Major histocompatibility complex (MHC) class I antigen processing, 291–292
Major histocompatibility complex (MHC) class I antigen processing pathway, 289–292

Index

Maximum rate (V_{max}) of carrier transport, 14, 15, 17
MDR gene family, 354–355
MDR transporters, 30, 39, 51–52, 113, 130–135, 354–355, 374
MDR-reversal agents, 364
MDR1 P-gp, 113; *see also* P-glycoprotein
  expression in cancer, 360–361
  as pump of cytotoxic drugs, 356–357
MDR1 tissue distribution, 359–360
mdr1a, 365–369
mdr1a gene disruption and organic cation elimination, 101
mdr1-type P-gp, 113
mdr1-type P-gp deficiency, 112
mdr1-type P-gp secretory system, 99–103
mdr2, 116–119
MDR2, tissue distribution of, 361
MDR2 P-gp, as phospholipid transporter, 359
MDR3 P-gp, 116–118
Membrane permeability, determination of, 18
Membrane transport: *see* Transport
1-methyl-4-phenylpyridinium (MPP), 93, 95, 172
Michaelis-Menten constant (K_t), 14–17
Microscopy, 20–21
Mitochondria, 2
Molecular defects, 3, 4
Monocarboxylate transporters (MCTs)
  anion exchange transport, 476–478
  MCT1
    intestinal transport of lactate and short-chain fatty acids via, 473–474
    molecular characterization, 473
    monocarboxylate transport via, at BBB, 475–476
    transport of weak organic acids, 474–475
  MCT2, molecular characterization of, 473
Monocarboxylic acids at BBB, 185
Mouse transporter protein (MTP), 336
MRP (multidrug resistance-related protein), 103, 110, 117, 135–140, 370
  as ATP-dependent drug transporter, 371–372
  homologues, 118–120
MRP expression, tissue distribution of, 370–371
MRP family, molecular cloning of, 423–424
MRP homologues, 373
MRP-mediated resistance, modulation of, 372–373
Multispecific bile acid transporters, 388–389

Multispecific organic anion transporter, 396–397; *see also* Canalicular multispecific organic anion transporter
  role in drug disposition, 398–399
Murine CAT-protein (mCAT)-mediated transport, 235–236
Murine CAT-proteins (mCATs), 231–233
  substrate specificity, 234–235
Murine ecotropic leukemia viruses (MuLVs), 231
Mutations, 3

Nanoparticles, VB_{12}-mediated delivery of, 511–512
Natrium chloride, 234
NIH 3T3 cells, 371
Nitrobenzylmercaptopurine ribonucleoside (NBMPR), 317–318, 333–335, 340
NO synthase (NOS), 239, 241–243
NO-producing cells, CAT protein expression in, 239–241
Nucleoside transport process, 343
  heterogeneity, 316–319
Nucleoside transporter (NT) proteins, 319
Nucleoside transporter(s) (NT), 45–47, 65–66, 314, 343
  chemical structures, 315
  CNT family, 319–321, 327–335
  as drug targets, 336–343
  ENT family, 319–327, 340–343
  "orphan," 332–336
  recent advances in molecular biology of, 318–321
Nucleosides, anticancer, 339

Octreotide, 390
Opioid peptides, 190
Organellar transporters, 335–336
Organelles, 2
Organic anion transport
  across bile canalicular membrane, 407–408, 414, 416–417, 419–424
  molecular aspects, 111–112, 122–140
Organic anion transporter peptide (oatp), 95–96, 108, 109, 111, 113, 125–126
Organic anion transporters (OATs), 67–68, 403, 483–485
  contraluminal, 171–172
  in kidney, 482–483
  in liver, 478–482

Organic anion-transporting polypeptide (oatp), 124–125, 481–482
  human homologue, 403
  molecular cloning, 399–403
Organic anion-transporting polypeptides (OATPs), 113, 115–116, 123–124
Organic cation transport, 104–105, 442–443
  molecular aspects, 122–140
Organic cation transporters, 69, 108–110, 113, 126–130, 471–472; *see also* Human organic cation transporters; Monocarboxylate transporters
Organic cations, 89–90, 95–97
  carrier-mediated transport, 96
  chemical structures, 89–90, 443
  excretory patterns, 107–110

Para-aminohippurate (PAH), 160, 171
Passive transporters, 31, 91; *see also* Facilitated diffusion
Patch clamping, 19–20
P-ATPase, 11
PEPT1, 35, 36, 40–43, 275–278, 280, 283
  characteristics, 281, 282
PEPT2, 275, 276, 278, 280, 283
  characteristics, 281, 282
Peptide bonds, reduced
  incorporation into TAP structures, 303–304
Peptide transporters, 64–65; *see also* Antigenic peptide transporter; Dipeptide transporters
  in apical membranes, 273
  application to drug delivery, 281–282
  in basolateral membranes, 273–274
  cloning, 274–281
  structural features, 278, 280, 281
  structure, 274–275
  substrate specificity and recognition, 276, 278
  tissue distribution, 276, 277
Peptide-like compounds, chemical structures of, 283
Peptide-like drugs, 271–272
  transcellular transport, 272–274
Peptides, *see also* Dipeptides; Organic anion transporter peptide
  at BBB, carriers for, 188–190
  binding to TAP, 303
  opioid, 190
  small, 270–271
  VB_{12} and, 508–511
Peptidomimetics, 301–302

Peptyl prodrugs, 72–73
p-glu-l-dopa, 74, 76
P-glycoprotein (P-gp), 69–70, 104–105, 353–354, 374
  as ABC family transporter, 355
  as ATP-dependent pump, 358
  and BBB, 368
  compounds that interact with, 356–357
  in drug absorption and disposition, 364–365
  expression in cancer, 360–361
  intestines and, 365–367
  kidney and liver and, 367–368
  modulation, 361–363
  role in normal tissues and drug disposition, 365
  structure, 356
PGT, 484–485
Phosphate transporter, 68–69
Phosphatidylcholine (PC) flippases, 116, 117
Phosphonoformic acid (PFA), 264
Photoaffinity labeling, 209–211
Plasma membranes, 2
Polypeptides: *see* Organic anion-transporting polypeptides; Sodium/taurocholate contransporting polypeptide
Polyspecific organic anion transporters (OATPs), 115–116, 123–125
Pore-forming proteins, 91
Potassium-selective channel, 7, 8
Pravastatin, 396, 397, 399
  hepatic uptake, 398–399
Prodrugs to increase intestinal drug absorption, 70–75, 78
  schematic representation, 72
Progressive familial intrahepatic cholestasis (PFIC-3), 118
Prostaglandins (PGs), 484–485
Protein kinase C (PKC) modulation, 111
Protein kinase C (PKC)-mediated phosphorylation, 104–105
Proximal tubule cell, 160, 161
PSI-BLAST (positron-specific iterated BLAST), 43
Purigenic mechanisms, therapeutic implications of, 337–338
Purigenic receptors, 336–338

Rat concentrative nucleoside transporter (rCNT), 328–332
Rat equilibrative nucleoside transporter (rENT), 320, 321–327, 322, 323

Index

Rat organic cation transporter 1 (roct1), 93–95
Reconstitution, 21
Renal drug elimination, 159, 173
Renal Na$^+$-coupled divalent anion transporters, 251–252; *see also* Type II sodium/phosphate cotransporter
Renal phosphate (P$_i$) apical reabsorption, 252
Renal tubule, proximal
 crossing membrane barriers during net reabsorption and secretion, 162, 169
 drugs that interact with organic anion and cation transporters in, 162–168
 location of transport processes and transporters in, 160–162
Roatp, 95, 97

Sequence analysis, 36–39
Short-chain fatty acids (SCFAs), 474–475
Simvastatin, 396, 398
Sinusoidal membrane (liver), transport across, 388
 Na$^+$-dependent, 388–396
 Na$^+$-independent, 396–403
Sinusoidal uptake of bile acids and non-bile acid organic anions, 391, 392
Small intestine transport, 63–64
SN-38 glucuronide, 416–418
SNST1 protein, 334–335
Sodium purine nucleoside transporter (SPNT): *see* Concentrative nucleoside transporter, CNT2
Sodium slippage, 256
Sodium-coupled P$_i$ cotransport, kinetic scheme for, 262–264
Sodium-dependent transporters, nucleoside, 333–336
Sodium/dicarboxylate supporter family, 48
Sodium/glucose cotransporters (SGLTs), 45–47, 60, 66
 SGLT1, 256–257, 260, 262, 334–335
Sodium/neurotransmitter symporters, 49–51
Sodium/nucleoside cotransporters, 45–47
Sodium/phosphate cotransport: *see* Type II sodium/phosphate cotransport
Sodium-taurocholate contransporting polypeptide (Ntcp), 113–115, 222–223, 391
 human homologue of, 394
 molecular cloning and functional analysis, 391–394
Sodium/taurocholate cotransporter (NTCP), 113–115, 122, 394–395

Substrates, 31
Sugar transport, *see also* Glucose transport
 in erythrocytes, 202
 kinetic parameters, 15–16, 206
Sugar transporters, 66, 76
Sulfate transporters, contraluminal, 171
Sulfonic acid groups, 89–90
Sulfonylurea receptors (SURs), 51–52
Symporters, 31

TAP: *see* Antigenic peptide transporter
Taurocholate-chlorambucil complex, 395
Taurocholic acid (TC), 388–391, 393–395, 400, 402
 ATP-dependent uptake, 404, 406–407
Temocaprilat, 398, 408, 414–416
Tetraethylammonium (TEA), 93, 94, 160, 161, 162
Thyrotropin-releasing hormone derivative with lauric acid (Lau-TRH), 74
Toxin-extruding antiporters (TEXANS), 98–99
Transcobalamin I (TCI), 496–497
Transcobalamin II (TCII), 496–498, 504–505
Transcobalamin II receptor (TCIIR), 505, 507
Transferrin, 191, 192
Transmembrane domains (TMDs), 30, 36, 37, 43
Transport, membrane, *see also specific topics*
 analysis, 18
 disorders, 3, 4
 energetics, 12–13
 kinetics, 13–17
 mechanisms, 5
 modes, 3, 5–12
Transport defects, 3, 4
Transport experiments, 16–17
Transport processes, 1
 linear *vs.* saturable, 6
Transport research, methods in, 18–24
Transporter associated with antigen processing (TAP): *see* Antigenic peptide transporter
Transporter classification, 52–53
 by function and substrate specificity, 30–32
 by genomics, 32–36
 by sequence analysis, 36–39
Transporters, membrane, 1; *see also specific topics*
 evolution, 36–39
 mechanisms used to move solutes across membranes, 30–31
 pharmacogenomics, 35–36

Tributylmethylamonium (TBuMA), 95, 97–98, 100, 104, 106, 111
Type II sodium/phosphate (Na^+/P_i) cotransport
  sequential, alternating access model, 262–263
Type II sodium/phosphate (Na^+/P_i) cotransporters
  expressed in *Xenopus* oocytes, 253–254
  pre-steady-state electrophysiological characteristics, 256–262
  site of interaction of protons and foscarnet with, 263–264
  steady-state electrophysiological characteristics, 253–256

Urine, 107–111

V-ATPase, 11
Vesicular monoamine transporters (VMATs), 98
Vinblastine, 106
Viral inhibition, TAP, 297–301
Viruses, 231, 298–301

Vitamin B_{12} (VB_{12})
  chemical structure, 494
  as drug transporter, 508–512
  mechanism of dietary absorption, 495
Vitamin B_{12} (VB_{12}) binders, 497, 500, 501, 505–506
Vitamin B_{12} (VB_{12}) transport, 512–513
  cell models, 507–508
Vitamin B_{12} (VB_{12}) transport proteins, 495–508
Vitamin B_{12} (VB_{12}) transporters, 68, 494–495
Vitamin transporters, 68, 493–495
Voltage-gated channels, 7, 8, 22

Water transporters, 75

Xenopus laevis expression cloning technique, 444, 447
Xenopus laevis oocyte expression system, 21–22, 252, 451–453
  steps involved in, 452
Xenopus laevis oocyte(s), 21–22, 400, 451–453; *see also* Type II sodium/phosphate cotransporter
  recording P_i-induced currents from, 253–254

Pharmaceutical Biotechnology
Chronological Listing of Volumes

Volume 1   PROTEIN PHARMACOKINETICS AND METABOLISM
Edited by Bobbe L. Gerraiolo, Marjorie A. Mohler, and Carol A. Gloff

Volume 2   STABILITY OF PROTEIN PHARMACEUTICALS, Part A: Chemical and Physical Pathways of Protein Degradation
Edited by Tim J. Ahern and Mark C. Manning

Volume 3   STABILITY OF PROTEIN PHARMACEUTICALS, Part B: *In Vivo* Pathways of Degradation and Strategies for Protein Stabilization
Edited by Tim J. Ahern and Mark C. Manning

Volume 4   BIOLOGICAL BARRIERS TO PROTEIN DELIVERY
Edited by Kenneth L. Audus and Thomas J. Raub

Volume 5   STABILITY AND CHARACTERIZATION OF PROTEIN AND PEPTIDE DRUGS: Case Histories
Edited by Y. John Wang and Rodney Pearlman

Volume 6   VACCINE DESIGN: The Subunit and Adjuvant Approach
Edited by Michael F. Powell and Mark J. Newman

Volume 7   PHYSICAL METHODS TO CHARACTERIZE PHARMACEUTICAL PROTEINS
Edited by James N. Herron, Wim Jiskoot, and Daan J. A. Crommelin

Volume 8   MODELS FOR ASSESSING DRUG ABSORPTION AND METABOLISM
Edited by Ronald T. Borchardt, Philip L. Smith, and Glynn Wilsion

Volume 9   FORMULATION, CHARACTERIZATION, AND STABILITY OF PROTEIN DRUGS: Case Histories
Edited by Rodney Pearlman and Y. John Wang

Volume 10   PROTEIN DELIVERY: Physical Systems
Edited by Lynda M. Sanders and R. Wayne Hendren

Volume 11   INTEGRATION OF PHARMACEUTICAL DISCOVERY
            AND DEVELOPMENT: Case Histories
            Edited by Ronald T. Borchardt, Roger M. Freidinger,
            Tomi K. Sawyer, and Philip L. Smith

Volume 12   MEMBRANE TRANSPORTERS AS DRUG TARGETS
            Edited by Gordon L. Amidon and Wolfgang Sadée